Improving World Energy Production and Productivity

Volume Two in the
INTERNATIONAL ENERGY SYMPOSIA SERIES
A 1982 WORLD'S FAIR EVENT

Improving Worl

Energy Production and Productivity

Proceedings of the
International Energy Symposium II
November 3-6, 1981

Edited by

LILLIAN A. CLINARD • MARY R. ENGLISH • ROBERT A. BOHM

Energy, Environment, and Resources Center
The University of Tennessee

Ballinger Publishing Company • *Cambridge, Massachusetts*
A SUBSIDIARY OF HARPER & ROW, PUBLISHERS, INC.

International Standard Book Number: 0-88410-877-5

Library of Congress Catalog Card Card Number: 82-6869

Printed in the United States of America

Library of Congress Cataloging in Publication Data

International Energy Symposium (2nd : 1981 : Knoxville, Tenn.)
  Improving world energy production and productivity.

  (International energy symposia series ; v. 2)
  1. Power resources—Congresses. 2. Energy policy—Congresses.
I. Clinard, Lillian A. II. English, Mary R. III. Bohm, Robert A.
IV. Title. V. Series.
TJ163.15.I5635 1981   333.79   82-6869
ISBN 0-88410-877-5   AACR2

This Symposia Series and the people who participate in it are the soul and substance of The 1982 World's Fair. As President of The 1982 World's Fair and as a private citizen of the world, I hope that this Series serves as a means by which solutions to the world's energy problems can be found and new energy opportunities can be sought. Further, it is my hope that the entire World's Fair experience will benefit mankind.

S. H. Roberts, Jr.
President
The 1982 World's Fair

Editors' Preface

Symposium II is the second in the three-part International Energy Symposia Series. To clarify not only the rationale and framework of this Symposium but that of the Series as a whole, much of what was contained in the preface to Symposium I's *Proceedings* is given here in basically unchanged form, together with some new material on Symposium II's structure. It is hoped thereby to familiarize those new to the Series with its intent and to bring others up to date as it progresses. In addition, this preface to the second volume attempts to synopsize Symposium II's message, an attempt which stresses brevity over definitiveness and which therefore can convey only a sense of some central aspects of the Symposium but not the full panoply of views and expertise present there. The preface continues with a review of the Series' future directions: Symposium III and its resultant communiqué. Finally, the preface includes notes about editorial considerations that affected this volume—some of them common to the prior volume as well; some unique to this one. These considerations and constraints should be recognized in seeing these *Proceedings* for what they are: not a verbatim record of all the formal and informal discussions that took place at Symposium II—for these discussions, although the yeast of the conference, were far too extensive for inclusion here—but a compilation of the papers and formal addresses given during the Symposium and of the integrating reports (with adjunct comments) presented at its final plenary session.

Purpose and Structure of the International Energy Symposia Series

The International Energy Symposia Series (IESS) is intended to provide a broad-based forum for addressing basic concerns about mankind's need for energy. Carried out under the auspices of an international energy exposition entitled the 1982 World's Fair, the IESS began with the convening of Symposium I on October 14–17, 1980, as a prelude to the Fair. Symposium I was followed by Symposium II held November 3–6, 1981, and the Series will be concluded with a third Symposium held May 24–27, 1982, after the Fair has officially opened.

The IESS is designed to enable the deliberation and debate of alternative energy policies. Its broad theme, "Increasing World Energy Production and Productivity," is global in scope and provides room for analysis of both energy supply and energy demand. Set within the context of the World's Fair, a symbol of worldwide cooperation and sharing of knowledge, the IESS is intended to help rectify world energy problems by bringing many different viewpoints together in order to assess areas of conflict and reach areas of agreement. The IESS thus undertakes an evaluation of energy issues that includes but goes beyond solely technical considerations. It seeks to understand and resolve energy issues in light of the human condition as a whole. In doing so, it hopes to have a three-pronged impact—one which affects the work of scientific and technical personnel on energy issues, the public's understanding of the international energy situation, and policymakers' decisions on both a national and a global basis.

As structured by the Program Committee, the Organizing Committee, and the Symposia staff (see Appendix III), each Symposium in the Series is intended to complement the others, ultimately leading to a comprehensive set of energy proposals. The first Symposium focused on defining the nature of the energy problem. Building on the results of Symposium I, Symposium II has served as the analytical component of the Series from which general proposals and specific recommendations will be carried forward to Symposium III. Deliberations at Symposium III will be directed toward arriving at resolutions that will help rectify the world energy crisis.

This sequenced approach is a distinguishing aspect of the IESS, for the Series not only addresses issues but also provides a means of moving toward their resolution. At the end of Symposium III, a communiqué will be issued to the world's nations highlighting areas of agreement and conflict. The relationship of the three Symposia and the anticipated flow of ideas can therefore be shown as:

SYMPOSIUM I   →   SYMPOSIUM II   →
issue identification   issue analysis

SYMPOSIUM III   →   COMMUNIQUÉ
issue resolution

Symposium I's Message

As evidenced in the first volume of the Series' *Proceedings*, Symposium I achieved two primary things in fulfillment of its function of issue identification: it defined the energy crisis—on both a broad, conceptual level and a more specific, immediate plane—and it suggested areas in which possible solutions should be sought. In doing so, it laid the groundwork for Symposia II and III to follow.

With respect to the first achievement, that of defining the problem, Symposium I found that for the near and mid term the energy crisis is primarily a political and economic problem and only secondarily a technological one: in general, the needed technologies exist but the infrastructures to use them do not. In the long term, exotic technologies might offer alternative solutions, but they cannot be expected to make a substantial contribution for several decades. At the conceptual level, it was also agreed that the energy problem is unavoidably tied to other problems—in particular, problems attending the overuse or misuse of other natural resources and problems arising from disparities in standards of living both within and between countries.

These general points of agreement about the nature of the problem were, during the course of the Symposium, specified into a number of issues that need to be addressed. For example, the following questions were raised and discussed: How should the high demand for fossil fuel and its economic effects on oil-exporting and oil-importing nations be handled? What are the best methods for redressing the lack of technological sophistication in some nations? To what extent can energy conservation help offer a solution; to what extent might it impede that solution?

Although there was dissent on the precise elements of an overall answer to the energy problem, Symposium I did reach an informal but strong consensus about how that answer should be shaped. It was generally agreed that there was no *one* solution, hard path or soft path, centralized or decentralized; the answer instead lies in a *mix* of solutions appropriate to each nation and dependent upon the kind of economy, level of development, and array of energy resources the nation has.

It was, however, acknowledged that there are some issues which transcend national boundaries and which must be considered from a global standpoint if the appropriate mix is to be found. These supranational considerations include (a) energy resources such as nuclear power and biomass whose potential effects are of worldwide concern, (b) environmental conflicts and constraints affecting more than one nation, (c) the correlation between energy factors—both supply and consumption—and equitable economic distribution, and (d) appropriate global or multinational institutional arrangements to address these and other considerations.

Symposium I's message was not conclusive, but it did provide a tangible sense of mission. As shown below, it thus helped to structure the analysis phase taking place in Symposium II.

Symposium II Organization

The Symposium II program was built around a format of work and plenary sessions.* The focus of the Symposium was a series of seven concurrent work sessions. Four of the sessions were organized around world divisions according to national perspective: industrialized nations with market economies, industrialized nations with nonmarket economies, energy-surplus industrializing nations, and energy-deficit industrializing nations. The remaining three sessions were focused on issues of global importance: the role of nuclear power, biomass and its energy/nonenergy conflict, and energy for rural development. The major areas addressed by all work sessions included:

- an identification of the nature of the energy problem for the topic at hand,
- a preliminary evaluation of the options available and their technical and economic considerations,
- an analysis of institutional and societal factors affecting the problem,
- a review of pertinent education and information needs, and
- a final evaluation of the options available.

*See Appendix I for a fuller review of the Symposium II program.

The point of departure for each work session was a comprehensive position paper covering these topics. Each work session also featured a separately authored case study or special paper. Participants were assigned to work sessions based on their individual perspective, expertise, and interest. Although they were to consider their assigned work sessions as their centers of activity, they were free to present ideas and join in discussions of other work sessions. These work sessions offered opportunity for intense debate and analysis of options by participants with the chairman/integrator for each work session charged with summarizing and reporting its essence in the final plenary session. Again, the issues were open for debate and discussion as well as minority reports from the work session panel (i.e., its assigned participants). For the sake of candor in the sessions and for expediency to provide a document for Symposium III participants, these *Proceedings* do not include the work session discussions. However, the chairmen/integrators' summaries and the final plenary session discussion are included in Chapters 6, 9, 12, 15, 18, 21, and 24.

The remaining plenary sessions at Symposium II concentrated on broad-ranging topics: alternative energy futures, including the case for electricity and appropriate energy strategies for industrializing countries, and the development of international institutional arrangements for increased energy production and productivity.

Symposium II's Message

There is a central paradox to the world energy problem which became evident during Symposium I and was a governing factor in Symposium II: all countries are embroiled in this problem, but each country is different. Consequently, it is difficult to generalize about either factors contributing to the problem or how these factors might be handled. However, a number of findings emerged from Symposium II's papers and discussions—findings which help to crystallize both intractable areas of conflict and areas of promise in the search for adequate and sustainable energy supplies. A few of these key findings are synopsized below by work session, followed with a brief review of these areas of conflict and of promise. Finally, mention is made of the demand side of the world energy equation—sometimes overlooked but critically important.

Synopsis of findings.

The role of nuclear power. Globally, nuclear power is in

economic and political trouble. Its survival will depend on several factors: on an increased demand for electricity (forecasted in the special plenary paper entitled "The Case for Electricity," but not without debate); on the maintenance of a nuclear safety record sufficiently impeccable to assure the public that the technologies involved—including those dealing with spent fuel reprocessing or storage—are adequate and well under control; and on choices to be made on a country-by-country basis about the economic and environmental appropriateness of nuclear power. With regard to the last factor, it was found that in the past nuclear power has not seemed appropriate for most less developed countries (LDCs), but, with the advent of the LDCs' search for energy self-sufficiency and of the possibility of smaller nuclear plants more in proportion with these countries' electricity grids, this situation may change. With that change, the pros and cons of technology transfer and nuclear nonproliferation become major issues.

The biomass energy/nonenergy conflict. This conflict is potentially a major one, involving not only the deflection of biomass from its needed use as food for a growing global population but also the environmental degradation—deforestation, desertification, and loss of soil nutrients—that can occur if biomass is improperly managed. On the other hand, for those countries with rich biomass resources (and especially for LDCs that are otherwise resource poor), biomass offers the possibility of profitably utilizing lands that are unsuitable or only marginally productive agriculturally and of attaining a measure of energy independence. To reap the full benefits of biomass for energy while minimizing its drawbacks, decisions about employing biomass for such purposes should be made on a country-by-country basis and should be carried out at a decentralized (i.e., regional or community) level, employing multipurpose systems that integrate such end uses as food, fuel, feed, and fertilizer. While such approaches could greatly defuse the potential conflict, the issue of global food needs versus national interests remains a haunting one.

Energy for rural development. The problem of rural development in Third World countries is multifaceted, entailing not only energy inputs but inputs from many other sources (human, animal, environmental, financial, etc.). And above all, it entails the need to remove barriers that hinder socioeconomic development—barriers involving inadequate education and capital, customs and institutional infrastructures that are inconducive or antagonistic to development, and the tendency on the part of governments to favor large, prestigious projects over smaller but potentially more productive ones. Furthermore, a critical problem for many Third World rural people—one specific to energy but permeating their lives—is that of a fuelwood shortage. Approximately one-half of the world's population depends on wood, dung, and vegetative waste for its heating and cooking needs; as this population grows, these resources become increasingly scarce, to the severe detriment of both the people and their environments. To ameliorate this situation and to increase energy's contribution to rural development, two main possibilities (not necessarily mutually exclusive) are foreseen: (1) the utilization of alternative energy sources such as the sun, the wind, and biogas; and (2) the establishment of decentralized electricity systems based on either small-scale hydro power or mini diesel-fired generators. In establishing these, the barriers to change noted above must be taken into account: proposed rural energy projects must be appropriate to the recipient group and its locality, and that group must be not just recipients but participants in the project. However, with even the simplest technologies, a lack of capital may be a major obstacle, particularly if the intended beneficiaries cannot afford even minimal new household or agricultural equipment.

Industrialized market-economy nations. The primary problem for this group of nations is the transition from cheap to expensive energy—a transition which began with the 1973–74 oil crisis and which is likely to continue inexorably, despite respites from temporary oil gluts due to market imbalances. Such market-induced imbalances can work both ways, at times provoking false panic, at other times a false sense of security, and either way, they point out two key concerns for market-economy nations: (1) how to temper the effects of energy supply and demand imbalances on related economic factors, and, as an important part of this, (2) how, despite temporary supply gluts, to promote continued energy efficiency, conservation, and conversion from scarce to less scarce fuels. In both of these concerns, government plays a central but delicate role, attempting to correct market imperfections without dictating resource allocations or prices. Another equally important but equally delicate role of government is that of arbitrating among social goals involving economic growth, employment, equitable income distribution, national security, and environmental quality—goals that are all affected by energy needs and resources and that may all come into conflict. Finally, and most elusive to deal with, are the international equity questions, particularly that of the appropriate allocation between the developed and the developing nations of the world's dwindling petroleum resources.

Industrialized nonmarket-economy nations. This category may be a misnomer: its work session emphasized that the nations included here—primarily East European centrally planned economies—do not operate on a uniformly nonmarket basis, and they are in various stages of development rather than being fully industrialized. While this indicates their lack of homogeneity, the latter point, especially, reveals a shared characteristic of these nations: many started industrializing as recently as World War II, and with the 1973–74 oil crisis and its aftermath, many were caught in mid stride. Their common problem, then, is not one of making adjustments in order to maintain the economic status quo (as is the case with many of the nations in the industrialized market-economy group), but one of pumping growth into their economic systems in order to bring their development processes to maturation and raise their standards of living. However, the root problem—expensive energy—is the same for this group, as are many of the related issues: a past overreliance on oil and, as a corollary, energy infrastructures that are sufficiently well established to be difficult to change; a need for a mixture of solutions to attain resilient energy systems; and a need for attention to environmental problems that, if ignored, bode ill for the future. The roles of government obviously differ between the two groups, but this could be an asset: it could offer enlightenment about alternative political/institutional possibilities and their ramifications.

Energy-surplus industrializing nations. This is another broad category which includes diverse nations: those with high per capita incomes, those with low; those with vast oil reserves, those with scant. But the commonalities they share are (1) they currently are able to derive export income from their oil resources and (2) those resources are eventually depletable. To the extent that the oil-exporting nations are otherwise resource poor, they thus must translate their ephemeral oil riches into more lasting forms of investment or development if they are to attain secure economic bases. Their needs in doing so are to some extent shared by other developing nations—in addition to needing opportunities for foreign investments, they need technological assistance and foreign markets for their domestically produced goods. However, their international relationships are subject to particular pressures: in terms of energy, they are seemingly the haves, the rest of the world's nations the have nots. They thus have an especial need for cooperation with regional and international institutions, in order to help resolve oil supply/price issues and to obtain the support necessary to attain stable socioeconomic systems. A central contradiction remains, however: it is generally in the best economic interest of these nations to maximize their oil returns by getting the highest prices possible over the longest period possible, but for the other nations of the world—especially for the LDCs, who need large and cheap energy supplies to initiate their development processes—this interest is in direct conflict. In particular, it may cause the latter's development to founder unless ways (such as investment of petrodollars in those countries) can be found to relieve the problem.

Energy-deficit industrializing countries. As suggested above, these nations are caught in a twofold dilemma: as noted in the special plenary session paper entitled "Appropriate Energy Strategies for Industrializing Countries," the financial burden of increasingly expensive oil imports is particularly great for the LDCs since it absorbs such a great proportion of their GNPs, but in order to increase their GNPs, they need energy to fuel their nascent development. This category of nations includes both middle- and low-income developing countries; thus their remedies to this dilemma differ. Nevertheless, their prerequisites are similar: the need to obtain technological assistance; the need to have foreign markets for their goods; the need to attract foreign investments; the need for energy planning that incorporates multiple approaches, both of energy supply modes and of their centralized or decentralized application. In the poorest of these countries, however, their problems' severity makes those problems particularly difficult. For example, the fuelwood crisis is especially urgent in many such countries but has no ready solution, and while other forms of energy—for example, electrification—are considered imperative to raise living standards above subsistence levels, these energy systems are in many instances very difficult to adopt because of their expense. For the extremely poor nations, then, direct aid and the transfer of basic technology will continue to be critical.

Areas of conflict. Within the overall world energy problem, there are areas of existing or potential conflict that are particularly intractable but that must be addressed if this problem is to be rectified. These conflict areas frequently occur in the interplay of national and supranational interests; they also occur over time, when short-term benefits are weighed against prospective long-term liabilities (or vice versa). Symposium II's findings illuminated a number of such conflict areas; a sampling of them might include the following:

- The desirability of national energy self-sufficiency by all means available, including nuclear power, vs. the desirability of nuclear nonproliferation.
- The need of oil-exporting countries to maximize the economic value of their resources vs. the need of oil importers, especially those that are LDCs, to have adequate and affordable supplies of conventional fuels during the coming energy transition period.
- As a corollary, the appropriate allocation of dwindling fossil fuel resources—in particular, who has the greatest ability to pay for such resources vs. who has the most urgent need for them.
- Trade protectionism out of national self-interest vs. the need of developing countries—energy-deficit or -surplus—to have open foreign markets for their domestically produced goods.
- National decisions about using arable land to produce biomass for fuel vs. global food needs.
- The increasingly limited ability of many nations to provide foreign aid in the face of their own rising costs vs. the continued need of the poorest, subsistence-level nations for direct technological and fiscal assistance, to help secure both energy supplies and other essentials.
- The fact that today's energy solutions—even if adopted out of necessity—may provoke severe environmental problems tomorrow (as exemplified, most immediately and critically, by the present fuelwood shortage, but also by the environmental risks thought to be attendant with either coal or nuclear power).

Areas of promise. The areas of conflict are not without redress; during Symposium II, other more promising trends were identified—trends which contain the seeds of remedies to the above and other aspects of the global energy problem. At present, these trends generally take the form of a deepening understanding of the global energy problem's complex nature and of the need for integrated approaches to its resolution at national and international levels. It is also increasingly recognized that whereas the former level usually can be handled within existing governmental structures, the latter requires special multilateral relationships. Such relationships need further development but are not completely lacking: as noted in the special plenary session paper entitled "International Cooperation and the Development of Institutional Arrangements on Energy," some steps have already been taken to facilitate international cooperation and attain a sense of global interdependency. Especially when combined with these institutional arrangements, the growing areas of greater understanding, or of perceived need, hold the promise of new and more lasting solutions to the world's energy problem. Several such areas were highlighted during Symposium II in both specific and general contexts; a few of them include the following:

- The need to incorporate into national energy decisionmaking an understanding of the complex relationship of energy supply and demand and the corresponding role of government.
- The need to understand the environmental effects of proposed energy supply actions and to modify supply decisions accordingly.
- The need for energy efficiency and conservation, for fuel switching, for the development of new and renewable energy sources, and for emphasis on indigenous rather than imported energy resources.
- The need for flexible, multifaceted approaches to energy supply—approaches which incorporate a mix of energy production modes, which integrate several purposes (e.g., combined heat and power systems or food/fuel/feed/fertilizer biomass systems), and in which decisions about centralization or decentralization are made pragmatically rather than arbitrarily.
- The need to analyze the interdependent relationships among nations on energy and related issues, and to foster cooperation and multilateral arrangements in order to resolve international problems and facilitate mutual aid.
- As part of the above, the need for technology transfer that is appropriate in type and scale to the locality where it is to be applied, and for appropriate infrastructures to assure the technology's continued workability.
- The need for energy and energy-related information and education at all levels, from national and international decisionmakers to remote rural poor, with the information selected to pertain specifically to the group it is to inform.

The importance of demand. During Symposium II, as in Symposium I, the focus was largely on energy supply issues, with demand treated as a given—modifiable in some respects, through more rational and efficient energy use, but otherwise an inexorable force: 4 billion people now, with subsistence levels of living in much of the world; the possibility of a doubled global population within the next fifty years, with even worse and more widespread poverty inevitable unless major remedies are found. Thus, the focus on energy supply is appropriate. However, absolute demand was not disregarded; it was also recognized that underlying the world's energy problem is the problem of population growth, and that this larger problem can only be rectified if well-considered and effective population planning measures are adopted. While the issue of population planning is not within the direct scope of this Symposia Series, it nonetheless cannot be avoided. As Ioan Ursu noted in his summary, "There cannot be peace in a hungry world."

Future Directions

Symposium III, a World Energy Congress, will be held May 24–27, 1982, and is designed to bring to a conclusion the work performed in Symposia I and II. A group of distinguished energy experts from around the world will meet a final time to ratify the findings of the previous Symposia. Ministerial level individuals and technical experts will seek a platform from which an international body for the continuation of an exchange of technology and resources may be created. Through a set of speeches and case studies by energy ministers and energy experts, recommendations will be presented to the Symposium III participants. Participants will then vote on these recommendations. The final outcome of the Series will be an international consensus on global energy commonalities and an agenda for continuation of the deliberations. A worldwide communiqué announcing the conclusions will be issued at the close of Symposium III.

Editorial Considerations for Symposium II Proceedings

Although editorial changes have been held to a minimum in these *Proceedings*, some alterations have been necessary to insure consistency among the papers and the prepared or spontaneous remarks. Throughout, however, the overriding aim has been to retain the original tone and intended meaning. Insofar as this has been achieved, the editors extend thanks to the authors and other participants for their cooperation. Insofar as this aim has not been fully met, the editors request the understanding of those involved, who undoubtedly realize the complexity of the task and the rapidity with which it had to be executed.

One aspect of the *Proceedings*—the "selected comments" sections of the summary chapters—requires special explanation. These sections consist of (1) oral comments presented by participants during the formal discussion periods of the Symposium's final, summarizing plenary session; (2) written comments submitted by participants at or following the close of the Symposium; and (3) abridged versions of two short reports volunteered during the Symposium. In all instances, the material included in these sections has been organized by topic, grouped with the work sessions to which they most closely pertain. Thus, some of the comments' points of reference may not be immediately apparent—partly because the summarizing plenary session's sequence has not been fully retained; partly because the comments may contain references to informal or work session discussions not included in the *Proceedings*. Furthermore, it should be noted that all comments, oral and written, had as their context the work session chairmen/integrators' oral summaries presented during the summarizing session but not the written versions submitted later and appearing in these *Proceedings*. And finally, it should be mentioned that a few oral comments were omitted from the *Proceedings* either at their authors' request or because quotation permission was not received in time for their inclusion here. Nevertheless, it is hoped that despite these impediments the selected comments can, when taken together with the papers and summaries, bring to the *Proceedings* reader an understanding of the wealth of opinion present at the Symposium.

Acknowledgments

Our sponsoring organization, the 1982 World's Fair, is to be commended for assuming the demanding task of hosting the International Energy Symposia Series. Through the Fair's unfailing support, this Series promises to make a much more substantial contribution to the body of knowledge about energy policy and prospects than would otherwise be possible. In addition, the editors would like to acknowledge the important support of the Symposium's cosponsors: the International Energy Agency and its parent organization, the Organisation for Economic Co-operation and Development; the United States Department of Energy; the Tennessee Valley Authority; and The University of Tennessee. The International Energy Symposia Series has been facilitated by a grant from the United States Department of Energy and by the support of its offices and officials. The International Energy Symposia Series is also cosponsored by a diverse group of foundations, organizations, and corporations concerned with the global energy problems that are associated with high levels of demand coupled with uncertain future supplies.

The *Proceedings* could not have been completed without the enthusiastic and professional efforts of the staffs of the 1982 World's Fair and The University of Tennessee's Energy, Environment, and Resources Center. Our editorial team included Jim Billingsley, who developed the book and cover design and layout specifications; Joyce Rupp Troxler, who prepared graphics; Carolyn Srite, who provided editorial assistance; Dan Hoglund and Sylvia Hoglund, who proofed several drafts of this volume; and Rica Swisher, Nancy Gibson, and Peggy Taylor, who prepared manuscripts.

A special thanks is offered to Earl Williams and to Tana Lawson and the rest of the staff of the Williams Company, who patiently worked with us to typeset and compose this volume on a very tight schedule. Finally, to those at the Ballinger Company with whom we worked—Carol Franco, Editor; Gerry Galvin, Production Manager; Robert Entwistle, Marketing Director; and Leslie Zheutlin, Marketing—we would like to express our appreciation for their professional support.

LAC, MRE, and RAD
Knoxville, TN
May 1, 1982

Contents

SECTION X   CLOSING SESSION

Figures

Tables

Abbreviations

bbl	barrel
bpd	barrel per day
Btu	British thermal unit
BWR	boiling water reactor
CANDU	Canadian deuterium-uranium reactor
cm	centimeter
CPE	centrally planned economy
ECM	European Common Market
EEC	European Economic Community
ESCAP	Economic and Social Commission for Asia and the Pacific
FAO	UN Food and Agricultural Organization
FRG	Federal Republic of Germany
FY	fiscal year
GDP	Gross Domestic Product
GJ	gigajoule
GNP	Gross National Product
GW	gigawatt
GWe	gigawatt-electric

ha	hectare
IAEA	International Atomic Energy Agency
IEA	International Energy Agency [an agency of the Organisation for Economic Co-operation and Development]
IIASA	International Institute for Applied Systems Analysis
IMF	International Monetary Fund
INFCE	International Nuclear Fuel Cycle Evaluation
ISFM	International Spent Fuel Management
kcal	kilocalorie
kg	kilogram
km	kilometer
kw	kilowatt
kwh	kilowatt-hour
LDC	less developed country
LNG	liquefied natural gas
LPG	liquefied petroleum gas
LWR	light-water reactor
m^3	cubic meter
mbd	million barrels per day
mbdoe	million barrels per day of oil equivalent
mtce	million tons of coal equivalent
mtoe	million tons of oil equivalent
mwe	megawatt-electric
NPT	nuclear nonproliferation treaty
OECD	Organisation for Economic Co-operation and Development

OLADE	Organisaçion Latinoamericana de Energia [the Latin America energy organization]
OPEC	Organization of Petroleum Exporting Countries
PWR	pressurized water reactor
quad	quadrillion (10^{15}) British thermal units
rad	radiation absorbed dose
R&D	research and development
RD&D	research, development, and demonstration
rem	roentgen equivalent man
SNG	substitute natural gas
SWU	separative work unit [a measure of the work required to separate uranium isotopes in the enrichment process]
t	metric ton [1000 kilograms, or 2,205 pounds avoirdupois]
tce	ton of coal equivalent
therm	100,000 British thermal units
toe	ton of oil equivalent
TPE	total primary energy
TW	terawatt
UNCTAD	United Nations Conference on Trade and Development
UNEP	United Nations Environment Programme
UK	United Kingdom
US	United States
USSR	Union of Soviet Socialist Republics
W	watt

WAES	Workshop on Alternative Energy Strategies
WCC	World Council of Churches
WOCOL	World Coal Study

The following terms and prefixes have been used to express quantities given in powers of ten:

Quantity	Term	Prefix
10^{15}	one quadrillion	peta
10^{12}	one trillion	tera
10^{9}	one billion	giga
10^{6}	one million	mega
10^{3}	one thousand	kilo

Unless otherwise specified, all tons refer to metric tons and all dollars [$] refer to US dollars.

SECTION I

Opening Session

Opening Statement

Charles E. Fraser
United States Commissioner General
The 1982 World's Fair

On behalf of President Reagan, Secretary of Commerce Baldrige, Secretary of Energy Edwards, and all of the federal energy agencies, I would like to extend our warmest welcome to this notable gathering of energy experts. We hope that the insights and knowledge that you bring here will be a meaningful contribution toward the world's energy solutions and toward the world's collective human progress over the next several years.

Since your last gathering in October of 1980, you will notice much progress in attaining public notice of and support for the 1982 World's Fair and in proceeding with construction here in Knoxville. A tremendous number of dedicated, skillful individuals in this city, in this state, and across the entire nation have been working to make this great event possible. Professionals and experts like those of you participating in this Symposium represent the intellectual core, the seeds of ideas, and the germ of progress for International Energy Symposium II and for the 1982 World's Fair.

I am truly proud of the organization of the Knoxville International Energy Exposition (KIEE) for its accomplishments, and I feel honored to be a part of this worldwide event that the Knoxville community is hosting. Just as in past great world's fairs, where nations have joined together to share with the world their unique cultural heritages and technological developments, so it will be in Knoxville, where the international community will explore and exchange the inventions of science and the splendid creativity of human energy. Together we will make the 1982 World's Fair a remarkable and memorable experience.

Opening Remarks

Lewis E. Striegel
Deputy Administrator
Economic Regulatory Commission
United States Department of Energy

Secretary Edwards has asked me to express his sincere regret that he cannot be here to welcome you to what I know he considers to be one of the most important events of this administration. You thus can understand how privileged I felt when he asked me to stand in for him tonight, and you should also know with what trepidation I now try to convey his message to you. On behalf of Secretary Edwards and the Department of Energy, then, I welcome you to the second International Energy Symposium.

I understand that many of you participated last year in the first of this three-part Symposia Series, and I welcome you back. To those of you who are here for the first time, I welcome the points of view on energy that you bring with you. You've committed your talents and time to participate in these discussions, underscoring an important agreement of last month's meeting in Cancún, Mexico—that we must find ways to improve the exchange of valid, valuable energy information. The views each of you hold are important, and they take on added significance when they are exchanged in a forum such as this. It is through debate and thoughtful examination that the real sense of your views can be revealed; if they stand in part or in whole under this rigorous review, then they become valuable. I challenge each of you to hear your peers with the same openness that will mark the expression of your own thoughts on energy. If you accept this challenge, there will be a true exchange of information; if you do not, it's just another symposium.

It is well recognized that energy constitutes a problem of global proportions, primarily because of the heavy reliance on oil as an energy source. Even though the problem transcends national boundaries, the solutions, in most cases, are peculiarly national. I believe that a strong national commitment to develop indigenous energy resources is the most important first step towards easing international oil dependency.

In the United States, greater reliance on the private sector's capital and expertise will undoubtedly lead to rapid, orderly development of sufficient energy supplies. To insure this, this administration is striving to provide a free and unregulated environment to stimulate industrial development. Technical and business creativity—which are themselves priceless—and indigenous resources cannot function effectively when they are inhibited by cumbersome government regulations. We in this administration are committed to removing these encumbrances and to providing a healthy climate in which private enterprise can function and lead the way out of the energy morass.

We've made great strides towards the above goal in recent months by removing price controls from oil, and we are considering plans to accelerate the decontrol of natural gas prices. Our actions to decontrol oil prices have paid big dividends already. In June more than 6,900 wells were drilled in the United States, a 40 percent increase over last year. More than 575 new rigs began operations during the first five months after decontrol. The number of seismic crews at work also has increased sharply: in August we had 735 crews at work, 30 percent more than one year earlier.

It is unreasonable to think that the United States, or any nation for that matter, can solve all its energy problems in isolation. However, each country must establish a national environment that is conducive to energy development before reasonable solutions can take hold. Once this happens, countries will work together as necessary to enhance their own energy productivity. National governments can serve as catalysts in this cooperative process, but real progress can only come from applying expertise to meet precise energy needs. In the United States, that expertise lies not in the government but in private industry. I'm convinced that US industry stands ready to apply its talents where requested if there is a reasonable assurance that a successful venture will result. As noted above, we are beginning to see how proper incentives to industry are helping to increase US energy productivity. This approach should also have broad international applications.

As you debate and consider your ideas for the next three days, you will share some points of view and you undoubtedly will disagree on others. This is a healthy process, one that will certainly lead to a better understanding of the goals of each country represented here. It is my hope that some common goals will emerge from your deliberations which can serve as a basis for further discussions. You may even find areas of sufficient agreement which can be applied toward meeting each nation's respective national energy needs.

In looking at the *Proceedings* of Symposium I, I noted that John Sawhill, that Symposium's chairman and the former US Deputy Secretary of Energy, quoted the great American patriot Patrick Henry, who said, "I like dreams of the future better than the history of the past." By your continued participation in this Symposia Series, you are in fact committing yourself to be a force in turning dreams into reality.

CHAPTER ONE

Opening Charge

David J. Rose
Professor of Nuclear Engineering
Massachusetts Institute of Technology

We are moving from the topics and attitudes of Symposium I last year, whose main purpose was identification of issues, into the hard work of Symposium II, called analysis of those issues, whose end product is to be the statement, as best we can manage, of available options, and of findings upon which will follow recommendations.

This is no simple linear progression. Exploring the option space, by which I mean the range of technical, economic, political, environmental, and social opportunities, conditions, and restrictions within which reasonable actions are to be found, involves trial positions, stepping forward, sometimes stepping back, sometimes sideways, comparing options, comparing consequences.

The difference between Symposium I and Symposium II might be oversimply but not wrongly stated as the difference between goals and policy, a difference that is sometimes overlooked. An amusing and instructive example of the difference occurs in Ben Jonson's "The Alchemist," when the alchemist says,

> . . .this night I'll change
> All, that is metal, in my house, to gold.

That's a goal, not a policy. What was his policy? His policy was to lure people in and take all their money; and that is policy.

People, cultures, aspirations being diverse, the goals themselves conflict: short time perspectives vis-à-vis long ones; conservation vis-à-vis exploitation; other topics familiar to you all. Goals contain paradoxes, absolutely essential to recognize. Were it not for the paradoxes inherent in issues surrounding energy, there would be no need for this Symposium. We do not have great symposia now to calculate the first twenty-seven figures of pi.

So Symposium I illuminated those paradoxes and tradeoffs in what to me were rich and agreeable terms. Many solutions and options were called for, sometimes reminiscent of Shakespeare's *Henry IV, Part I*, where Glendower says, "I can call spirits from the vasty deep," and Hotspur replies, "Why, so can I, and so can any man, but will they come when you do call for them?" More seriously, the published account of Symposium I captured its excellent spirit in many places. Here are two brief extracts. The very first sentence in the Editors' Preface says, "The availability of energy affects the attainment or deferment of many human goals: economic development, environmental quality, and, most critically, world peace." That written by my friends, Robert Bohm, Lil Clinard, and Mary English. Also, part way through the book Hans Landsberg says eloquently, "In reality, we are discussing the state of mankind locally, nationally, regionally, globally: who we are; who we wish to be; what is right and what is wrong." These words capture the spirit of that time.

I rejoice in seeing much holistic thought here in the prepared papers; hence in the discussions to come. Not long ago and still in some places, energy issues were looked upon as just one or at most a few things: mostly technology, mostly economics, optimization of particular single options, or whatever. A thousand plans flourished, too often suggestive of a thousand candidates volunteering to run a centrally planned economy, but often narrow, rigid, supposedly complete, claiming to be revealed truth—and all of them, of course, different.

In Symposium II, I see much less of that. Economic and other analyses have become integrated into a more complex and more true whole. Let us try to preserve that better balance. Now, let me turn to some topics that, while most of them are joined, admit of separate titles.

Time Perspectives

Activities designed for only one time in the future generally clash with activities designed for another time: a simplistic statement, but true and often ignored. We need to live with a multiplicity of them.

The goals of Symposium I tended often to be long range. The more specific topics of Symposium II tend on the average to be of shorter range, I hope not to the detriment of the longer range issues which, if ignored now, catch us later. A problem that will only appear in full bloom fifty years from now, but also took fifty years to get that way, needs attention now, not forty years from now. We suffer from the lack of earlier thinking ahead. We live in the future of twenty-five years ago, when we didn't pay enough attention to how to use coal well, and now we try to play catch-up technology. Coal versus nuclear power is one such issue. But some other examples are more trenchant.

The global carbon dioxide problem was, unfortunately, almost unmentioned in these papers. Responses to the acid rain phenomenon should alert us to how long it takes to get action on these things, even when the technological facts are pretty clear and the repair is not all that difficult. Carbon dioxide in every respect will be much worse.

Biomass is a good thing in many ways, but long-term effects on fragile tropical soils, for example, must be guarded against.

Consider the characteristic time to develop new and different societies compared with the time to deplete principal resources, a central problem for the OPEC countries at present. Extending time horizons of availability and reducing conflict can be beneficial. Respecting that, our colleague, Mary English, yesterday recalled these appropriate lines from Wordsworth's poem, "The Rainbow:"

> The Child is father of the Man;
> And I could wish my days to be
> Bound each to each by natural piety.

Responsibilities

Where do they fall? In different places, but the topics are similar everywhere. For example:

- How are they to be handled, and by whom? By the government? By the private sector? The answer differs in different places but still needs to be obtained.
- Outreach. No good idea will flourish without being actively propagated.
- Research and development, especially if it falls too far in the future to be attractive to the private sector.
- Education—For what? By whom?
- Handling the commons: deforestation and carbon dioxide, for example.
- Resolving disparate time perspectives, already mentioned.
- Capital formation—more on that a little later.
- Seeking consensus.
- Speeding action on new agreed-upon issues.

Action on many of these things is local. The effects are both local and extensive and extend in time as well as space. Our neighbors exist all over the world and in future generations. Two thousand years ago, Titus Lucretius Carus, in his book *de Rerum Natura*, said, "Resources will be needed to sustain all future generations. We do not own the earth in freehold but only in usufruct." Use the fruits, but not the tree that yields them.

Rational Utilization

The real purpose of energy is to be used, and not just supplied. Rational use appears in these Symposium papers, incorporated in natural ways. Many cost-attractive and socially attractive opportunities exist, ranging from improvements in US industry to better cookstoves in rural lands. But adoption is, alas, slow, as we all know. What impedes the adoption, and what can be done about it? This is a question I leave for your earnest considerations.

In this trend to lower energy use, how much of it arises from each of these: (1) better technology, (2) better housekeeping, (3) changing product mix, (4) simple curtailment? There's confusion on those issues, and some enlightenment would be very helpful. The apparent deemphasis on conservation in some governments these days is regrettable. The work is just fairly begun and will need continued attention. Even if there is an oil glut at the moment for some, it surely isn't there for many others, and it may not last very long.

Electric Power

Many of the new energy options to be considered, both for provision and for utilization, tend to be more electric than before. They also tend to be more capital intensive. This raises again the whole question of rational and effective use of energy—especially of what the thermodynamicists call "availability," a quality that incorporates the idea of what is appropriate for what purpose. Where is electricity more appropriate and where is it less? Also, where does the capital come from?

Nuclear power, now and for some time in the future a part of the electric sector, merits a separate workshop. Is a new international grouping of participants likely to arise? It might be a rearrangement of the present participants, but with changing emphasis and changing importance of the various members.

Capital Formation

Not only electric power but most of the other new energy options tend to be much more capital intensive than before. Except for some housekeeping tasks—albeit important ones—to avoid waste, energy conservation costs money, too; the cost to save a barrel of oil may be considerably less than the cost to provide a new one, but still expensive by standards of a decade ago.

The roles of the World Bank, the International Monetary Fund, the United Nations and other international organizations will surely need modifying, as was argued recently by many heads of state at Cancún, and at many other meetings on energy and related topics. We have an opportunity to say something here.

But there is much more to it than high international finance. In the less industrialized countries, "do it yourself" is an essential ingredient. Please do not mistake this advice for an excuse for inaction or unconcern by the industrialized sector. What I mean is that people must be full partners in the preparation of their own future, so we seek mutually beneficial collaboration and cooperative interdependence. Unfortunately, inequity exists in both the industrialized and less industrialized countries and needs rooting out.

People talk about enlightened self-interest and humanitarianism, often in passing. They are central to these issues, and convergent if the time perspectives are appropriate.

Information

Here is a multiplicity of topics which I will try to summarize briefly under three subheadings.

(1) Clearly good. Here we find mutually beneficial education and training, for example. New options demand new skills for both provision and use. Even new biomass, whatever its exact form, will not be a matter of gathering up sticks. A village using local hydro power needs skilled people. The collaboration of industrialized and less industrialized countries in defining, analyzing, and resolving such matters concerns my own institute, the Massachusetts Institute of Technology, a great deal, and your suggestions would be appreciated. Many Symposium II papers stress the importance of these activities.

But how will they be accomplished?

(2) Necessary and good, but confusing. A prime example here is the whole range of activities called "access to technology," "technology transfer," "technology adaptation," and so on. There's something important here, but what is it? It stands not for mere access to a mountain of reports but for a complex interactive task of sharing options in ways not hitherto possible, and it is not accomplished by any simplistic phrases. It requires collaboration and good will.

People, even experts in the fields, hypnotize themselves with comfortable phrases that seem to describe what is to be done but in fact do not. Here is an example from a topic not on the agenda of this Symposium: controlled nuclear fusion. Some years ago, in a paroxysm of planning, the US Energy Research and Development Administration—the forerunner to the present Department of Energy—laid out a route to controlled fusion almost as precise as Plato's plan in his *Republic* for how the world would forever run henceforth. It envisaged a sequence of ever larger fusion reactors, one of which was called TNS. That sounded very in-group. Detailed planning meetings took place about this, some lore built up about it, but all those initials TNS stood for was The Next Step. The more one analyzed the ideas, the less real they appeared.

Even perfectly good and useful words sometimes need de-mythologizing, even depoliticizing. One such is "appropriate," a word captured for a time by particular groups. I hope it has now become liberated again, as seems to be the case here in Symposium II.

(3) Bad. Consider proliferation, not only of nuclear weapons —certainly a bad thing, even an ultimate idiocy—but also of some other things germane to how we deal with problems of energy, food, and so forth. We find proliferation of uncertainty, of non-sense, of information overload. It has been a principal cause for stagnation of many activities and for confusion about others. Uncertainty and risk-averseness do much more than introduce data-spread; they bias decisions, especially long-term ones, and tend to bias them toward inaction. James Tobin of Yale University received the Nobel Prize in Economics a month ago for developing these and related ideas in more formal ways.

Proliferation of uncertainty contributed to stopping nuclear power in the United States dead in the water. Proliferation of nonsense has led to many energy proposals that were fashionable for a time but withered after exposure to the candor of objective scrutiny. Respect for the intellectually dead says I shouldn't give you any examples right there. But even after failing, those things left their marks and a legacy of confusion, distrust, and much wasted intellectual effort.

Consider, also, information overload: it shows not always vigor or intellectual industry but often intellectual sloth. A thousand pages of computer printout is easier to produce than one incisive assessment. A week ago I received a 4-pound, 2-inch-thick report from the US government entitled *Volume Reduction of Low Level Waste Materials.* To be sure, it had an executive summary, but it gave no guidance whatsoever of whether any useful or adoptable ideas were to be found inside.

Toward a More Just, Participatory, and Sustainable Society

It is my favorite topic. Many papers emphasized the necessity of working in that direction, both last year and this.

We are alleviating the root causes of war; we are undermining them, and what could be a more honorable occupation?

Talks and arguments on arms limitations is one route. Another, essential to follow, is to reduce the tensions that lead people and nations to feel insecure, exploited, marginalized—tensions that lead people and nations to want to build bombs in the first place. Out of internal unrest comes also external unrest. And both together, via alliances, client states one to another, and so forth, could lead to Armageddon. Here, enlightened self-interest and humanitarianism converge a fortiori.

Some say: let the economic costs decide. Well and good for those in the economic sector, but such decisions leave out the billion or more people who are not benefited by any economic system whatever but may be exploited nevertheless. The origins of that word—economics—is a mini-sermon. It comes from the same Greek root as the word ecumenical: *oikoumenikos,* of the whole world, from *oikoumenē,* the inhabited world, from *oikos,* house. Economics should be the management of our global house and all that is therein. In managing that household, we must manage our affairs as best and most expeditiously as we can. That task is one for the committee of the whole, which involves more than money, which like energy is also an intermediate good and not a final thing in itself.

The debate often called north/south becomes better described as also rich/poor, urban/rural, and so forth, as it is in many of these papers of Symposium II. That lets us see more clearly the real plight of the rural and urban poor and the need for global collaboration. There is something amiss in the global allocations. Here is just one example. From data in several of the Symposium II papers, it seems that an investment of, say, 5 to 10 billion dollars plus local educational infrastructure applied to cookstoves with two or three times the efficiency of traditional ones would substantially reduce a global problem, much discussed, of deforestation of many major populous tropical and subtropical lands around the world. Such an investment in its own way would go far to make a better world, a world with less tension. That would make much better use of the money than guns for us, for them, or for anyone.

Some will say, "Don't make such comparisions; if we didn't spend the money on arms, we wouldn't have spent it on the stoves anyway."

Indeed, there is something amiss. We are here to help make the world run better, and there will be plenty of markets for our good products.

SECTION II

Special Plenary Sessions

CHAPTER TWO

Part A: The Case for Electricity

Umberto Colombo
Chairman
Italian Atomic Energy Authority

Part B: Appropriate Energy Strategies for Industrializing Countries

Enrique V. Iglesias
Secretary-General
United Nations Conference on
New and Renewable Sources of Energy

INTRODUCTION

These two papers were prepared for the plenary session entitled *Alternative Energy Futures.* The papers were prepared independently; therefore, each paper's views should be attributed only to its author.

PART A: THE CASE FOR ELECTRICITY

INTRODUCTION: ENERGY FUTURES

Energy Scenarios, Hard and Soft

Regarding the size of energy systems required to carry the world's population through the next century, current thinking spans at least an order of magnitude. High energy futures explored by Häfele at the International Institute for Applied Systems Analysis (IIASA) imply a three to fourfold increase in global primary energy consumption over the coming fifty years.[1,2] Low energy futures recently developed by the Lovinses[3,4] conceive that fifty years hence, it will be possible to survive comfortably and with an improved standard of living on even as little as one-half of the energy that is consumed in the world today, notwithstanding a likely doubling of the population during the same period. At the heart of the divergence between these extreme visions of the future is a deep-rooted question of ethics and values: a question that is evident from the existence of two irreducible philosophies, the hard and the soft, disagreeing with one another on practically everything related to energy technology and its interactions with the economy and society.

While the hard philosophy holds absolute faith in the ability of technology to meet problems of any size—the bigger the more welcome—the soft philosophy emphasizes that energy systems must be within the grasp of the ordinary citizen and therefore must be small. Although in appearance the opposition between the two schools centers on questions of basically a technological and economic nature, and centers particularly on the dichotomy between big and small, in reality the question at stake behind the dispute is related to the social control of energy systems.

In the hard-path, high-forecast energy futures explored by the Häfele group, extreme centralization of energy systems leads to ever higher energy conversion, transport, and distribution losses which fifty years from now may equal all the energy used in the world today. In the Lovinses' soft-path scenarios, most of the energy saving is achieved by decentralizing the energy system to provide an optimum match between supply and demand (with minimum conversion), through "small, benign, and convivial" technologies close to citizen control. The hard and soft scenarios assume opposite views not only on the nature of the energy system but on society, lifestyles, and values. The hard scenario demands a strong top-down approach leading to world coordination at the political level. It attributes a great importance to transport (in fact, in its proposed approaches most of the energy is produced in segregated peripheral areas and conveyed to consumer countries far away). The soft scenario, on the other hand, assumes that energy systems will be small, decentralized, and locally managed. This faith in decentralization is extended to cover the economy and society at large, with voluntary simplicity as a dominant feature. In this sense the struggle as we know it up to now is basically ideological.

Both these visions of the future are possible in a "thermodynamic" sense. The long-term horizons they represent may in fact be considered internally consistent in all their major static equilibrium features, and it would certainly be possible to identify coherent paths from the present state to these futures.

In real life, however, social and economic systems are not equilibrated and are highly dynamic in nature. In view of social, economic, and technological change this is certainly the case in the industrialized countries, but it is even more so in developing countries, where very rapid population growth is accompanied with extreme levels of urbanization and sustained industrial development.

The world system is, in fact, a strongly interlinked assemblage of independent decisionmaking elements, each aiming at different objectives frequently unrelated to energy as such. Social, economic, and political factors rather than enlightened choice eventually dominate the process of selecting the actual courses taken by nations in their quests for fulfillment of their basic needs.

Both Häfele and the Lovinses have addressed the problem of "how to get from here to there." The Häfele group openly recognizes that its scenarios imply a number of very optimistic assumptions relating to international cooperation and to social, political, economic, and other nontechnical features, each by their present situations heavily influencing the potential evolution of energy systems. However, when taken together with prevailing human nature, the Lovinses' assumption that all constraints can be abolished through common sense, cultural adaptation, and a sophisticated degree of organization is in the end tantamount to admitting the existence of a formidable transition problem.

Whatever the practical limitations of these visions of the future, both contain elements of truth, and both highlight two essential aspects of energy systems management. Each reminds us that what is required is not energy itself but the services rendered by energy. Lovins makes it quite clear that a lot of energy is wasted today while technologies are, or will soon be, available to substantially improve energy efficiency in end uses, decreasing potentially severe impacts on the environment and climate. The hard approach emphasizes the importance of the functionality of the energy vectors to be favored in the future. The hard and soft approaches, taken together, leave us with the important message that energy is not necessarily required in large quantities but needs to be available under forms of increasingly high quality.

Perhaps surprisingly, a common feature of both energy futures is the relatively unobtrusive position of electricity as an energy vector. In the high energy scenarios of Häfele, a major energy vector is represented by synfuels from coal, including methanol and hydrogen, while electricity increases its share in overall primary energy consumption only slightly above its present level in the world as a whole, although, of course, it increases considerably in absolute terms. In the Lovinses' constant energy demand scenario the share of electricity gains in importance, but only because most energy savings occur in nonelectric uses; in their decreasing energy demand scenario, on the other hand, electricity requirements stay roughly constant or even decline relative to today.

Häfele appears to have overemphasized the importance of substituting oil as an energy vector with coal-derived liquids and gases and to have concentrated on the overall thermodynamic efficiency of combined coal and nuclear systems, losing sight of the enormous potential and versatility of the electric vector. The Lovinses, on the other hand, are enthusiastic about the "thermodynamic elegance" of employing electricity only in such end uses as can be satisfied exclusively by this form of energy, ignoring a whole host of social, economic, and strategic factors that favor this energy vector above others.*

'In a written comment submitted to the editors following Umberto Colombo's presentation at Symposium II, Amory Lovins noted that several statements made by Professor Colombo about the study cited in his fourth reference—a study to be published as *Least-Cost Energy: Solving the CO_2 Problem* (Andover, MA: Brick House, 1982)—required clarification. In particular, Mr. Lovins indicated the following: (1) The study's global "efficiency scenario" showed that it was possible to do far more than "survive comfortably" on less than half the energy used today —instead, it assumed increases of 2.4-fold in the per capita GDPs of the developed countries and 4.3-fold in those of the developing countries— and the criterion used in the analysis was not "voluntary simplicity," smallness, "conviviality," nor value or lifestyle changes, but *least direct* *economic cost.* (2) The study did not assume decentralized settlement patterns; it only assumed that further centralization would not take place and that the fraction of people living in cities would thus remain constant. (3) Based on Colombo and Bernardini's original analysis, the energy savings achieved by such population patterns is only one-third as big as that achieved by technical improvements in energy productivity, not more than the latter. (4) The study's choice between electricity and other energy vectors rested solely on economics (with the condition that energy services be provided no less conveniently or reliably than they are today)—thermodynamic elegance was therefore a consequence of cost minimization; not the reverse. [The foregoing is an abridged version of Mr. Lovins's written comment. The abridgment is the editors'.]

A Pragmatic Approach to Energy Futures

It may be taken for granted that the most likely future course of the world energy system will lie somewhere between the extremes of Häfele and the Lovinses, which is the reason why it may be useful to refer to these scenarios as outlying boundary cases. The author [Colombo], in collaborating with Oliviero Bernardini, has suggested a different, more pragmatic approach to the study of energy futures—one which leads, fifty years from now, to a world scenario with double the current population and energy consumption and which assumes a substantial redistribution in the per capita energy consumption among the industrialized and developing countries.[5,6]

Our scenario is based on a careful study of the past energy trends and on an analysis of the likely evolution of events over the next fifty years, taking into consideration technological progress and its energy implications and the availability of investment capital. In addition, our scenario discards unrealistic hypotheses that concern the structure of political decisionmaking power and that demand a tight coordination at the world level.

Our scenario does not assume that urbanization is a fatal process. In the past, urbanization was favored by population pressure, the availability of centralized energy systems—systems which were far cheaper in urban surroundings than the decentralized energy systems they displaced—and the development of large-scale technologies which have dominated the world scene. The prospects for the future appear quite different. In industrialized countries, population growth is practically nil and potentially declining; in developing countries, population pressure still exists but the context of energy and technology is quite different. The diseconomies of the very large scale due to its excessive rigidity and to environmental impacts and other factors need not be discussed in detail—the new waves of fundamental innovations address areas (such as microelectronics and biotechnologies) that make possible and sometimes even favor decentralization, and the nature of energy systems (in particular, the important role of electricity) make it possible to conceive a pattern of economic and social growth for the developing countries characterized by less centralization than might otherwise be assumed. These factors, incidentally, will also favor the exploitation of renewable energy sources.

In studying the historical relation between energy and GDP in industrial countries, we have found a pattern which is common to all major countries. During the initial stage of a country's economic and industrial take-off, when base industries and infrastructures are l uilt and societal demand is concentrated on material goods, the energy content of GDP increases dramatically. This process continues until the country reaches industrial maturity. At this point, basic needs become saturated and demand shifts towards less material, more sophisticated goods and services. It is thus intuitive that the proportional energy content of GDP cannot increase indefinitely but must at a certain stage stabilize and start decreasing. To this one should add the consideration that the technological efficiency of energy production and utilization keeps increasing, thus contributing to a further drop in the energy intensity of GDP. Figure 2A-1 allows an appreciation of this trend, as illustrated by the past and projected experience in different industrial countries. This trend towards a proportionally lower content of energy in GDP is likely to continue in the industrialized world over the next decades and will mitigate the developing countries' general growth in energy demand required by their continuing population growth and by their process of industrialization.

When this scenario was first developed, the author and Dr. Bernardini had assumed the total energy consumption in 2030 to be held to a level of 16 TW, compared with 8 TW in 1975, thus leaving the world's per capita energy consumption unchanged. This constraint appeared to be a severe one in comparison with the IIASA forecasts. In particular, it was feared that the developing countries might not be assured of having a fair share of the global energy supply. However, the results obtained by our model indicated a substantial redistribution of per capita energy. In fact, the amount of energy used by the developing countries in 2030 is projected by the model to equal that consumed by the industrial world. Taking demographic data into account, this means that in

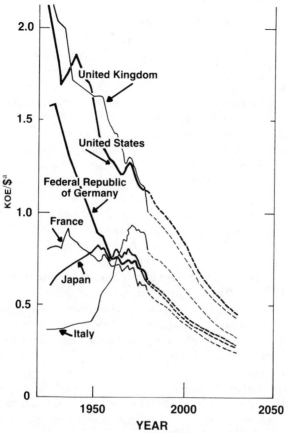

Source: U. Colombo and O. Bernardini, *A Low Energy Growth 2030 Scenario and the Perspectives for Western Europe,* Report prepared for the Commission of the European Communities, Panel on Low Energy Growth, July 1979.
[a]Kilograms of oil equivalent per dollar (US $ at constant 1973 value).

Figure 2A-1. **Energy/GDP Ratios in Major OECD Countries**

the year 2030, the average inhabitants of the industrialized world will use four times more energy than do their counterparts in the developing world. At present, the comparable ratio of energy consumption is about 11:1.

THE CASE FOR ELECTRICITY

The Position of Electricity in the Energy System

The electricity intensity of GDP has invariably increased in all countries since the beginning of electrification a century ago, despite the attainment of relatively advanced degrees of industrial development. Even during the early stages of growth, represented today by the less developed countries and yesterday by the now industrialized countries, growth in electricity consumption has been—with a very few exceptions due to particular local circumstances—considerably greater than energy consumption in general. This holds true even during a country's developing phases when overall energy consumption tends to increase considerably faster than GDP. Fifty years ago, in the industrialized countries, electricity consumption represented only 4 percent of total primary energy consumption; this figure has now risen to 27 percent. In 1980 in the Third World countries, 25 percent of all commercial primary energy resources were transformed into electricity, a portion that is expected to rise to 31 percent by 1990.[7]

The unique position of electricity relative to energy as a whole is further emphasized by developments subsequent to the energy crisis of the 1970s. Despite the fact that all energy prices rose substantially, electricity consumption continued to grow faster than GDP in almost all countries, while overall energy consumption remained at a standstill or even declined. Moreover, although the price of hydrocarbon fuels is likely to keep on increasing, the price of electricity will tend to decline from today's level, as nuclear power and to some extent coal increase their contributions to supplies. Thus, independent of arguments related to thermodynamic convenience, this situation should further enhance the penetration of the electric vector in the energy systems of the coming two decades. Figures 2A-2 through 2A-7 show that in several industrialized countries (the United States, Japan, the Federal Republic of Germany, France, the United Kingdom, and Italy), there has been a substantial decoupling since 1973 of the growth of GDP and the increased consumption of energy, but with a continued close relationship between the trends of GDP and electricity consumption. It is worth noting that the per capita index of electricity consumption is particularly high in France (Figure 2A-5), a country which is actively implementing a policy of strong electro-nuclear development.

This is further evidenced by the data of Figure 2A-8 on the average values of per capita income elasticities of total energy, electric energy, and nonelectric energy, as seen in the periods before and after the oil crisis (1960–73 and 1973–80). The United Kingdom is the only country where electric power elasticity to GDP is lower than 1 in the latter period. However, the point to be noted

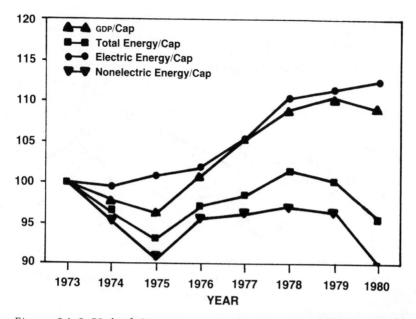

Figure 2A-2. United States—per Capita Growth Indices of GDP, Total Energy, Electric Energy, and Nonelectric Energy

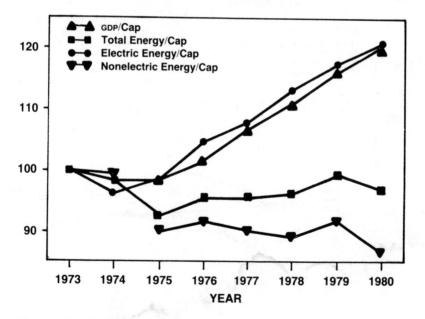

Figure 2A-3. Japan—per Capita Growth Indices of GDP, Total Energy, Electric Energy, and Nonelectric Energy

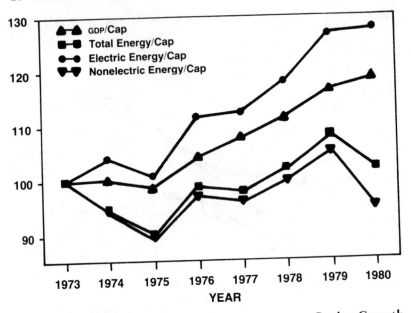

Figure 2A-4. **Federal Republic of Germany—per Capita Growth Indices of** GDP, **Total Energy, Electric Energy, and Nonelectric Energy**

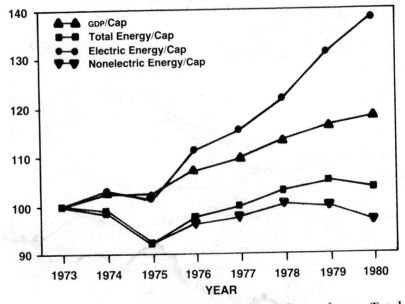

Figure 2A-5. **France—per Capita Growth Indices of** GDP, **Total Energy, Electric Energy, and Nonelectric Energy**

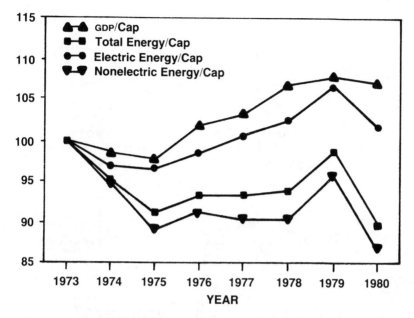

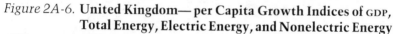

Figure 2A-6. United Kingdom— per Capita Growth Indices of GDP, Total Energy, Electric Energy, and Nonelectric Energy

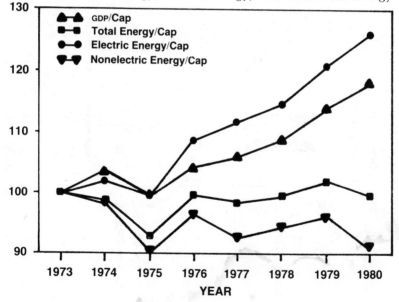

Figure 2A-7. Italy—per Capita Growth Indices of GDP, Total Energy, Electric Energy, and Nonelectric Energy

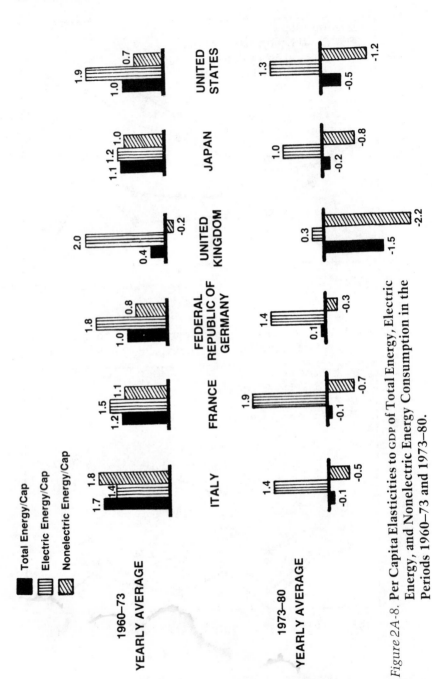

Figure 2A-8. Per Capita Elasticities to GDP of Total Energy, Electric Energy, and Nonelectric Energy Consumption in the Periods 1960–73 and 1973–80.

is the large difference between the total energy elasticities and the electric energy elasticities.

On the demand side, among the most important features of the electric vector are its versatility of use, its easily controllable nature, its relative safety in end uses, and its uniqueness in satisfying certain essential requirements. Electricity growth and industrial productivity increases have been closely correlated in all countries and are likely to continue being so in the future, as societies develop from material-intensive to information-intensive stages of growth (see Figure 2A-9).

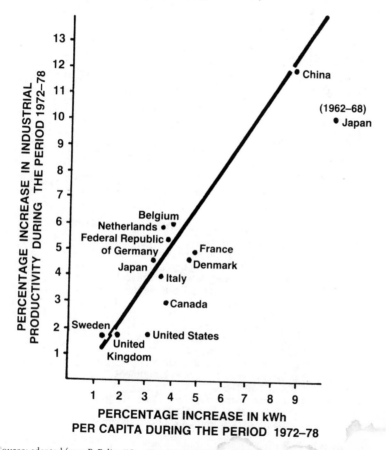

Source: adapted from F. Felix, "Our Top Priority: Expanded Electrification Will Substantially Reduce Oil Use, While Propelling Economic Recovery," paper presented at The Economy, Energy, and Electricity Conference 1980, Toronto, Canada, October 14–15, 1980.

Figure 2A-9. **Industrial Productivity versus Electricity Consumption**

Moreover, there appears to be a distinct relationship between the increase in a product's degree of technological sophistication, the decrease in its total energy content, and the penetration of electricity within it. This is shown by the graph in Figure 2A-10, which relates the total energy content of various kinds of products with the share of electricity needed to manufacture them. Figure 2A-11 shows that the wider the use of electricity in a country, the smaller on the average is the energy content of its GDP.

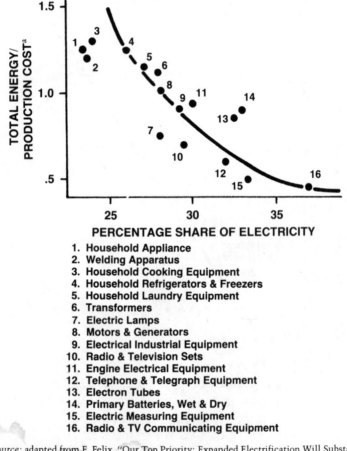

1. Household Appliance
2. Welding Apparatus
3. Household Cooking Equipment
4. Household Refrigerators & Freezers
5. Household Laundry Equipment
6. Transformers
7. Electric Lamps
8. Motors & Generators
9. Electrical Industrial Equipment
10. Radio & Television Sets
11. Engine Electrical Equipment
12. Telephone & Telegraph Equipment
13. Electron Tubes
14. Primary Batteries, Wet & Dry
15. Electric Measuring Equipment
16. Radio & TV Communicating Equipment

Source: adapted from F. Felix, "Our Top Priority: Expanded Electrification Will Substantially Reduce Oil Use, While Propelling Economic Recovery," paper presented at The Economy, Energy, and Electricity Conference 1980, Toronto, Canada, October 14–15, 1980.
[a]US $ at constant 1975 value.

Figure 2A-10. **Energy and Electric Content of Some Electrical Equipment**

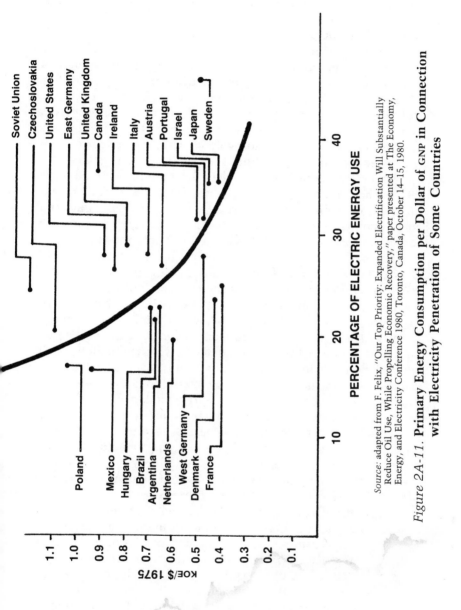

Source: adapted from F. Felix, "Our Top Priority: Expanded Electrification Will Substantially Reduce Oil Use, While Propelling Economic Recovery," paper presented at The Economy, Energy, and Electricity Conference 1980, Toronto, Canada, October 14–15, 1980.

Figure 2A-11. **Primary Energy Consumption per Dollar of GNP in Connection with Electricity Penetration of Some Countries**

On the supply side, electricity is attractive because it can be produced from many different sources: hydro power, geothermal steam, fossil fuels, nuclear power. This is a strategically significant factor since it increases the possibilities of geopolitical diversification of energy supplies. The recent development of new technologies, stimulated by the energy crisis, has added other sources to the potential suppliers of electricity: urban and agricultural wastes, biomass, wind, thermodynamic and photovoltaic solar, ocean thermal gradients, and so on. In general, these new technologies are compatible with small decentralized conversion systems, although several lend themselves also to large-scale production (e.g., consider photovoltaics and the solar power satellite project proposed by Arthur D. Little, Inc., and P. E. Glaser).[8]

Different technologies for electricity production and transmission have many elements in common, which is another factor making the electric vector extremely palatable to the industrial and economic system as a whole and leading to greater simplicity through the greater interchangeability of supply infrastructures.

But perhaps the most critical feature in determining the electric vector's level of penetration lies in its convenience of transport and distribution, from the producers all the way to the end consumers. In making their energy decisions, consumers give high consideration to the bulk quality and use ease of the energy forms with which they come into direct contact—a consideration which may frequently extend beyond that which price alone might counsel. Solids, which must be moved in trucks, carried in buckets, and fed with shovels, are clearly inferior to liquids, which still must be delivered by trucks within city boundaries but otherwise can be piped. Gases, in turn, present an even greater quality jump, since their delivery throughout the urban fabric can be based entirely on a pipe network. Once the piping structures are laid down, they can last for many decades with only minor repairs or layout changes, and they clearly represent a fundamental asset to urban societies.

If gas distribution systems are favored over liquid and solid systems, electric systems represent the ultimate in distribution, at least from the consumer's point of view. Not only does the distribution and use of electricity involve a minimum of physical effort (plugging in cords and turning on switches), but this energy vector is also more ecologically acceptable to the end user (cleaner, less noisy) as well as more versatile in its potential for meeting all end-use requirements.

Prospects for Electricity in Industrial Countries

Despite these very favorable circumstances, it is not likely that electricity consumption in industrial countries will achieve in the next fifty years the long-term growth rates experienced in the past. The growth of industrial electricity use has, in fact, traditionally been related to the growth of heavy industries such as the manufacturing of steel and other basic materials. Although there are still a number of opportunities to substitute electricity for other fuels in many energy-intensive sectors of industry, this trend may be moderated if the price of competing fuels becomes low enough to cover the economic bonus connected with the convenience of using electricity. Furthermore, major growth sectors of the future—for example, electronics, fine chemicals, synthetic materials, and biotechnologies—although characterized by proportionally high electricity requirements, generally have a relatively lower level of overall energy intensity per value added, so that the increased electricity requirements of these sectors are likely to have a greater effect on electricity penetration per se than on the level of overall energy consumption.

Whatever the future of electricity in industrial countries may be, there appears to be little doubt that during the next decades it will be considerably better than that for alternative fuels. In particular, as income rises to accommodate easily the price differential between electricity and alternative fuels, the trend towards increased electricity growth will persist, justifying relatively higher levels of investment in this energy sector compared with others. Moreover, in some countries that are not plentifully endowed with alternative energy sources, electricity may represent a strategic element of energy policy, thus amplifying the growth potential of this vector over and above what may be expected for the industrial countries as a whole.

The expansion of electronics and robotization, abetted by the growing demand for safer and more reliable industrial products and processes, favors the evolution towards an information society and is a factor that promotes electricity over less noble forms of energy. Decentralization, which in the past has been hindered by the periphery's lack of opportunities and infrastructures comparable with those available at the center, is made possible by an information society with flexible and widespread automation and by an energy system largely based on electricity.

Prospects for Electricity in Developing Countries

Turning now to the developing countries, forecasts in electricity are difficult to make because of the extreme variability of relevant parameters. Differences in stages of development, in social and political structures, in patterns of industrialization, and in resource endowments are such as to defy any generalization. It is clear, however, that given the availability of adequate supplies, these countries will be pushing for electricity consumption at rates well above those of economic growth and total energy consumption—rates that are probably even higher in absolute terms than those which historically characterized industrial countries at comparable stages of development.

Among the developing countries, many have ample hydro power potential and some have geothermal potential—both resources which, because they are indigenous, represent important development factors that are very likely to lead to enhanced electricity growth rates compared with the rates of countries unendowed with such resources. The basic assumption here is that these well-endowed countries will be able to secure the financial and technical assistance necessary to develop their resources and that, above all, the economic climate will improve in the world as a whole.

In developing countries with large rural sectors, major attention should be given to the question of improving local living conditions through energy resource development. Despite electricity's high development costs in rural areas, it may provide a major strategic element to help create local economic opportunities and satisfy basic needs, allaying urbanization pressures with their implied high energy requirements, and ameliorating, through substitution for traditional energy sources, the impact of rural energy use on land resources and the environment.

In fact, many developing countries face two energy crises. On the one hand, due to increasing population pressure and insufficient land use management, the decreasing availability and increasing costs of fuelwood and other local renewable resources create great hardships for the rural sector. Accompanied with decreasing opportunities for survival on the land, this leads to rapid urbanization rates. On the other hand, this very growth of the urban sector leads to high demands on conventional energy resources—particularly oil, which many countries can no longer afford to import in sufficient quantities. In the absence of alternative coal and nuclear resource development, which is difficult in the short term because of financing and the often small size of developing countries' energy systems, these countries even today face a real energy resource gap, with implications for both economic development and social stability in general.

An important palliative to these problems is afforded by rural electrification. Although at present, the connection of villages and towns to a central electricity grid frequently represents a very costly option, it may in the longer run result in the lowest cost alternative, considering the very high cost implications of expanding the urban infrastructure to accommodate satisfactorily the swelling exodus from the countryside.

However, rural electrification need not be seen, as it frequently is, as a question of connecting villages, towns, and other localities to a central electricity grid. The extreme versatility of current, and even more so of future, electricity supply systems —both technologically and in terms of size—allows electricity to penetrate everyday use from the bottom up as well as from the top down.

In the rural societies of the developing countries, it would seem convenient to develop electric systems using the most suitable local inputs and technologies. Local electricity grids currently are based largely on expensive diesel generation. However, technologies already exist or are being developed that allow local electricity generation from renewable resources, extending from biogas systems and gas from agricultural wastes through wind turbines and microhydro converters to more complex systems based on solar energy conversion such as photovoltaics. Rural electrification based on local resources is a new element of energy strategy in developing countries—an element which is becoming increasingly popular, particularly to the extent that it seems suited to addressing both of the aforementioned energy crises of developing countries at the same time.

Rural electricity development through local renewable energy resources also has strategic implications for the longer term future. In later phases of electric network development, local electric systems based on renewable energy sources naturally become connected to the central electricity grid, resulting in a unified and resilient structure that has the best of both centralization and decentralization: a degree of local self-sufficiency as well as the benefits of centrally coordinated load management.

The particular reliability aspects of such a mixed structure cannot be overemphasized. Present electric systems in many developing countries are highly centralized and vulnerable structures. Partial reliance on local renewable electric systems decreases the danger of generalized malfunctioning and must be considered a strategic element of future resources development. It does, however, require the creation of a local capacity for electric system planning and management—a requirement which may be a difficult aspect of this strategy for developing countries and which urgently needs to be addressed.

The case for increasing linkages between centralized and decentralized technologies can, of course, be made for energy vectors other than electricity. Gas production and distribution, for example, could follow the same principle. On the whole, however, only the electric vector has a sufficiently high quality in local as well as centralized production, together with a comparatively low cost, making it readily amenable to transport and distribution over the distances involved in most countries. Because electricity is more amenable to capillary diffusion of networks and infrastructures, it can even be said that it represents a crucial element of energy strategy.

Considerations on Penetration and Supply

The discussion in the preceding paragraphs has shown the peculiar character of the electric vector and its prospects in industrial as well as developing countries. Our group in Italy is now working on the optimization of the Colombo-Bernardini scenario, with particular attention to the dynamics of the energy system's evolution and to the penetration of electricity up to the year 2000. Our preliminary results indicate electricity penetration values on the order of 40 to 45 percent for the industrial countries and on the order of 30 to 35 percent for the developing countries (excluding noncommercial sources of energy from the base total, as these sources are difficult to account for). This forthcoming increase in electricity demand must be matched by an adequate supply, and the problem to be dealt with is one of the choice of primary sources.

Fuel oil was, until the 1973 crisis, the most widespread and convenient primary energy source to convert into electricity, but it is now too expensive, and only under exceptional conditions can new power plants be based on it. For base-load supply, nuclear power and coal are today the most convenient sources in practically all countries, once the local hydro power potential has been exploited. The choice between coal and nuclear is not simple and depends on several factors, such as the local availability of convenient coal deposits, the perception of environmental and safety problems, and strategic and balance of payment considerations.

In relation to the above factors, each country has defined, more or less explicitly, a strategy for solving its electricity supply problem based on a mix of coal and nuclear power. France, for example, has made a clear-cut choice for nuclear power and has made it the crux of a wider policy of technological development; at the opposite extreme is the example of Denmark, which is rapidly shifting from oil to coal for electricity production. But it should be observed that small countries like Denmark may find it quite expensive to set up whole systems of authorization, licensing, and safety controls on what would amount to a very limited number of nuclear power plants.

The problem of electricity supply in the developing countries has a somewhat different connotation, because the availability of electricity is a precondition of economic and social development. The electrification of rural areas requires precise policy actions and is not brought about by market forces alone.

As far as energy sources for electricity are concerned, besides the obvious convenience of exploiting untapped hydro power, coal is becoming the most convenient and easily implementable fuel for *new* power plants. Whether or not *existing* oil-fired plants should be converted to coal is a more complex matter, since the choice depends on considerations about the available infrastructures as well as on such factors as the shape of the load curve and the number and sizes of existing plants. The erection of nuclear power plants, which are now commercially available only in relatively large capacities (above 600 MW), is convenient only in countries that have power systems of 5,000 MW or more, in consideration of the need for reliable and stable base-load supply.[9]

Connected with the problem of electricity supply is the question of load management. While coal-generated and nuclear-generated electricity represent the most convenient solutions for base loads, peak loads can be tackled with a variety of sources and technologies, some of which are still in a development stage. Nontechnical measures such as differential time-of-day pricing (which might, incidentally, stimulate new uses of off-peak electricity) are also elements of a load management policy aimed at reducing the need for peak power generation by gas or diesel turbines.

Short-Term Problems

As a consequence of economic recession, the recent experience in several industrialized countries has been characterized by reduced rates of growth in electricity demand. In some cases, this has resulted in overcapacity of electricity generation, even in countries where there is little need to replace oil-fired plants. Additions to capacity during the second half of the 1970s in fact reflected the optimistic estimates, during the late 1960s or early 1970s, of anticipated electricity growth.

In other countries, public opinion, pressures from minority groups, and environmental concerns have held back the development of large electric systems based on nuclear power and, to some extent, on coal. Increasing lead times for the development of these large systems have led to an escalation of costs at a time when the size and distribution of electricity loads required increasing contributions from reliable and low-cost base-load power.

From a strategic point of view, both coal and nuclear power are essential elements of the transition to future energy systems, and, in the medium term, their exploitation must rely largely on conversion into electricity. All these features—and others, needless to say—add up to a very uncertain climate for electric power development, both in countries with an excess electricity generation capacity and in those without. Facing these issues, utilities have become increasingly dismayed and disoriented, frequently to the point of immobility, while the gravity of the problem calls for immediate action.

In programming future actions, we must remember that there are two fundamental issues to pursue: (1) maximum energy conservation through a rational use of energy sources and (2) the supply of electricity. In seeking a viable solution for the latter issue, utilities in many countries will have to cope with an enduring situation of high interest rates and of shortages in capital markets. It seems essential that governments and, more generally, policymakers must closely consider the risks that all these circumstances entail. It is their primary role to insure that conditions are once again established in their countries that can facilitate the proper functioning of the market in relation to its actors—specifically the utilities, industries, and financial institutions—as these actors move to satisfy the growing demand for electricity, an essential ingredient for the wealth and progress of the economy and society.

NOTES

1. W. Häfele, "World Energy Productivity and Production: The

Nature of the Problem," in *World Energy Production and Productivity*, Proceedings of the International Energy Symposium I, ed. R. A. Bohm, L. A. Clinard, and M. R. English (Cambridge, MA: Ballinger, 1981).

2. W. Häfele et al., "Energy in a Finite World: A Global Energy Systems Analysis," report by the Energy Systems Program Group of the International Institute for Applied Systems Analysis (IIASA 1980).

3. A. B. and L. H. Lovins, "If I Had a Hammer," in *World Energy Production and Productivity*, ed. R. A. Bohm et al.

4. A. B. and L. H. Lovins, F. Krause, and W. Bach, *Energy Strategy for Low Climatic Risks*, report for the German Federal Environmental Agency (June 1981).

5. U. Colombo and O. Bernardini, *A Low Energy Growth 2030 Scenario and the Perspectives for Western Europe*, report prepared for the Commission of the European Communities, Panel on Low Energy Growth (July 1979).

6. U. Colombo and O. Bernardini, "A Low Energy Growth Scenario for the Year 2030," to be published in the Proceedings of the Study Week on Mankind and Energy: Needs, Resources, Hopes, Pontifical Academy of Sciences, Vatican City, November 10–15, 1980.

7. World Bank, *Energy in the Developing Countries* (Washington, DC: World Bank, 1980).

8. P. E. Glaser, "Progress in Photovoltaics," Arthur D. Little, Inc. impact letter (May 27, 1980).

9. See reference cited in note 7.

PART B: APPROPRIATE ENERGY STRATEGIES
FOR INDUSTRIALIZING COUNTRIES

INTRODUCTION

In recent years the international community has witnessed, in both political and technical circles, an unprecedented surge of concern about energy. The higher prices of hydrocarbons and the prospect of their exhaustion in the more or less immediate future have caused this concern over the energy future of mankind and have led to a review of policies in the short term and a new look at the energy field's long-term prospects. In its initial stages, discussion of this theme was surrounded by a great deal of passion and inexactness. Fortunately, however, many myths have been disappearing recently, and the topic is now treated with much greater objectivity and realism, thus considerably facilitating the discussions.

On the one hand, it is very clear that the energy question is above all a technical and scientific problem. It involves a dual scientific and technical challenge: that of creating a future supply of energy sources which is varied and abundant enough to satisfy the demands of human economic and social development, and that of finding substitute sources to provide an answer to the exhaustion bound to be suffered sooner or later by the hydrocarbons on which mankind has relied too heavily in the last few decades.

On the other hand, with the end of the era of cheap energy sources, the energy question has also become an economic problem. In past decades, the international community enjoyed depressed hydrocarbon prices, but we are now entering a new stage in which those prices are much higher. This has had a strong economic impact which the international community must absorb by making the necessary structural adjustments. Looking ahead, the economic issue is clearly to be seen in the whole investment structure of the developed and developing countries, because of the large amount of resources which will have to be devoted to the solution of the energy problem, thus raising a major challenge in planning economic and social development.

Furthermore, the traditional relationship between "energy" and "development" has changed its focus in recent years and is now in full evolution. Whereas up to now we were basically accustomed to view the relationship between energy and development as something direct and proportional, today we stand on the threshold of much more complex situations. It is no longer so clear that this relationship is always direct and positive: quite the contrary. There have been clear proofs that there is ample room for reducing this dependence of development on energy, in some cases even to the point of making it a negative relationship. This has resulted from a much clearer awareness of the need for policies aimed at greater conservation and efficiency, both of which have given unexpected and sometimes spectacular results. All these elements have changed the previous direct and simplistic approach to the relationship between development and energy.

Whatever the dynamics of the development/energy relationship in the short and medium term, however, it is clear that in the long term this relationship necessarily must be positive: the world's countries will need growing amounts of energy resources in order to meet the demands of their economic and social development. If this is true for the developed countries, it is even more true for the developing countries, which are starting from much lower levels of energy consumption and production. For the latter, then, the adequate supply of energy resources is a challenge of the first magnitude and a serious restriction on their process of economic and social development.

It unfortunately must be recognized that the energy issue has arisen at a time when economic theory is still ill-prepared to provide an answer to the problems connected with the economics of nonrenewable resources. However, new criteria and developments in economic science have been emerging in recent years, and fortunately there are already signs of greater scientific and technical preparedness to treat with scientific objectivity the delicate problems of nonrenewable resources to which traditional economic science has paid so little attention. It is against this background, made evident by the energy crisis of recent years, that the trends in the so-called energy transition become clear.

The energy transition is under way and has been throughout the history of the human race, that species so greedy for energy resources. In the last few years, however, the nature of this transition has changed because of the complexity of the current energy situation, particularly when viewed from medium- and long-term perspectives. Mankind needs to plan ahead for the transition from a supply structure where nonrenewable hydrocarbons are the main source of energy to a new structure where hydrocarbons are relatively less important—one that allows them to be replaced by other, renewable resources and thus allows for a mixed, self-sustaining, and more balanced structure.

It is safe to say that the transition is inevitable and that it will involve significant structural adjustments in both the economies and the societies of developed and developing countries. But it can also be said that this energy transition is viable. The latest IIASA studies show quite properly that there are no legitimate grounds for mankind to have fears about the energy potential of the Earth. Sources of energy exist, and they are available. The fundamental problem is how to adequately exploit them and place them at the disposal of mankind, so as to sustain economic and social progress.

It has also been clearly shown that the energy transition is essentially a process of dynamic interaction between "technologies," "natural resources," and "final uses" of energy. The terms of the current transition will change in the future, as technology progresses, as new energy resources are discovered, or as consumption or production patterns are changed depending on the final use that is made of energy.

The current energy transition is, moreover, much more complex economically than the past one. It is a transition that is going to have to take place within the context of both rising energy costs and huge investments to enable the exploitation of new resources. These features of the current energy transition show how complex it is and clearly justify the need for anticipating and planning the course of events, thus complementing the spontaneous trends in the market forces.

The transition will undoubtedly be much easier in the developed countries than in the developing countries. The developed countries have an abundance of conventional natural resources or the necessary financial means to acquire them on the market. These countries are making great progress in the use and development of nuclear energy and have adequate technologies for increasingly exploiting new and renewable sources of energy. With adequate technical resources and financial means as well as appropriate technologies, the energy transition will be viable and relatively easy for them. The experience of recent years has shown the extent to which the economic systems of the industrial countries have the flexibility to face the technical and economic circumstances resulting from the energy crisis.

The situation is not the same, however, for the developing countries, which in many cases have neither sufficient natural resources, adequate savings and investment capacities, nor suitable technologies for using new methods to create energy resources. In the developing countries, as in the advanced countries, the energy transition must be based on three fundamental pillars:

an effort at conservation and efficiency in the use of energy resources; increased exploitation and utilization of conventional resources, including, in a special way, petroleum and coal; and finally, increased utilization of new and renewable sources of energy.

The terms on which the transition should be approached will, for the reasons expressed above, be much more complex in the developing countries than in the industrialized nations. There is thus an urgent need to design adequate strategies for energy programs that will lead to a suitable allocation of the developing countries' scarce resources and to international cooperation efforts in order to meet the great energy challenge of the developing countries.

THE STARTING POINTS OF THE THIRD WORLD ENERGY STRATEGIES

The consideration of energy strategies in the Third World is certainly not a new topic. The starting point of an energy strategy for any country, and especially for the developing countries, must be based on three premises which, in my opinion, have been made very clear following the intense energy debate of the last few years.

The first premise is that the energy variable should be programmed as an integral part of a country's economic development policy. Energy resources thus should not be programmed outside the context of the economic development process in general, as they are an input and a product of the same process.

The second premise is that different energy sources should not be developed in isolation; rather, an integral strategy should be developed that will cover all the possible energy sources in a country. The recent resurgence of interest in energy sources that involve renewable resources has made it clear that the isolated programming of one or more energy sources can lead to costly errors in the application of relevant economic policies. In that context, therefore, strategies for conventional or nonconventional sources of energy should not be designed separately; instead the approach should be to look at the whole range of opportunities presented by energy and technological resources. The latter is the concept that has been implemented as a result of the so-called hydrocarbons crisis.

With regard to the third premise, it is important to remember that there are significant differences among countries and that it is therefore not possible to speak of a single energy development strategy. Economic policies must always take into account the particular situation of each nation, including—in addition to the availability of natural resources—the economic structure, income levels, and degree of industrialization of the country.

These three premises provide significant starting points that must be borne in mind in any energy strategy, especially in the developing countries.

Considerations Regarding an Energy Strategy

I would like to begin by stressing the five fundamental considerations that justify the developing countries' concern with the energy problem and point to the need to anticipate and envisage solutions. These considerations lie behind the definition of objectives and the identification of tools for an energy plan.

(1) The first consideration is *the explosion of the demand for energy* in the developing countries.

There is widespread recognition of the low energy consumption level of the Third World countries—a level which is to a large extent a consequence, and at the same time a cause, of their low per capita income. This low energy consumption level must be increased considerably if a policy of encouraging the rate of economic and social development is to be implemented. In addition to starting at this low level, the developing countries are experiencing a very high population growth rate, one which will continue over the next decades despite population policies that might be adopted and before the development process itself promotes a drop in the birth rate.

The demand for energy resulting from the high population growth rate will be explosive over the next few decades. At the same time, the continuing urbanization process will bring about increasing energy needs linked with the growth of services in the urban environment and with the lifestyles that go along with that growth. Finally, it must be recognized that the industrialization process is and will be a powerful factor in the rising demand for energy.

Whatever their economic development processes or strategies may be, industrialization—in the broad sense of the word—must be accelerated in the developing countries in a manner consistent with their economic and social progress. All these elements together lead to the conclusion that there will be a very significant explosion in the demand for energy over the next few years. By the end of the century, the 1970s' level of energy consumption in the developing countries will probably have tripled, a projection which shows how tremendous an effort must be made to meet increased energy demands in the developing countries.

(2) A second consideration which must be borne in mind in addressing the developing countries' energy planning is *the economic and social structure* of these countries.

This entails not merely proposing a significant increase in energy resource supplies, but rather identifying the types of energy required to accelerate development in the Third World countries. The latter involves the programming and expansion of energy sources required for particular economic development processes in the developing countries. On the one hand, this means taking into account both the important role played by agriculture and, particularly, the fact that the rural population is scattered and needs certain types of energy, many of which can only be generated by agricultural production itself. On the other hand, other types of energy are required by urban populations (populations which are growing rapidly in some developing countries) or by the industrialization processes on which some of the Third World countries have embarked in recent years. All these elements together affect the quality of energy resources required for the Third World countries' development and should be borne in mind in a dynamic projection of the future.

(3) The third consideration is *the structure of known energy resources.*

It is not possible to make generalizations in this field, as the inventory of known resources in the Third World is very deficient. It is well known that there is a great lack of information in the case both of conventional fuels and of new and renewable sources of energy. However, on the basis of the information that is available, it may be said that there is a shortage of hydrocarbon resources in the developing world (except for the OPEC countries). It is also true that in recent years an increasing number of underdeveloped countries have been found to have potential hydrocarbon resources, but those resources are far from being adequate to satisfy the volume and quality of energy required by the Third World. If to these facts is added the fact that these countries have only a limited market for utilizing nuclear energy at its current technological level, it may be assumed that in the Third World, according to the known data, the so-called conventional resources will not in theory be adequate in themselves to meet the great demand for energy in the next few years.

The same does not appear to be true with regard to new and renewable resources, some of which, such as hydroelectric power and biomass (to mention some of the most conspicuous ones in the area of renewable resources), appear to be large and significant sources which could be included in an effort to program energy supplies in the developing countries. Nevertheless, any increase in the use of these resources is limited by some elements that go hand in hand with their existence. In the case of hydroelectric power, its exploitation requires substantial investments which, as I said before, are not readily available in the developing countries. In the case of traditional forest resources, there are serious problems associated with their premature exhaustion as a result of existing exploitation methods—problems which also make it necessary to be careful in making projections for the future. Despite these restrictions, there is a growing conviction that biomass contains a very significant quantity of potential resources that should be carefully programmed, taking into consideration the ecological conditions under which they are exploited and on the basis of adequate technological research regarding possible uses of these natural resources.

(4) The fourth consideration is of a social nature: it entails *the lifestyles and values* that are behind the developing countries' consumption structure.

This consideration is clearly significant in the case of countries having a large population mass in rural areas where traditions and habits can impede the introduction of certain types of energy that meet with a social resistance grounded in traditional values. Thus, energy strategy can disregard these social factors without running the risk of becoming inoperative, thereby leading to a waste of efforts and resources. The particular importance of this sociocultural question in the utilization of new and renewable resources was clearly stressed in the studies and analyses prepared in conjunction with the United Nations Conference on New and Renewable Sources of Energy.

(5) The fifth and final consideration in any effort to design energy strategies for the Third World must be *the dynamic and changing state of energy technologies and the frequent lack of sufficient experience to demonstrate their uses* in the developing countries.

Indeed, in some cases, work is being done with new nascent technologies that could show a spectacular growth in the future since they are evidently in a developing stage. In other cases, the energy sources and the technologies for using them have been proven, but it is not well known how these approaches can be adapted to the developing countries, and a significant effort must be made to carry out a rational analysis of their applicability to these countries' social and economic environment. Thus, any long-term strategy must address itself to this double limitation stemming from the current state of different technologies: the dynamic growth that some are undergoing and the lack of sufficient experience with regard to others.

These five considerations, which in my opinion are very significant, must be taken into account to different degrees dependent on different circumstances in *any* future energy programming, but particularly in countries that are in the initial or intermediate stages of development.

THE OBJECTIVES AND TOOLS OF AN ENERGY STRATEGY

The central objective of any long-term energy programming in the developing countries can be none other than the adoption of economic policies aimed at increasing the supply of energy resources in sufficient amounts and of adequate quality to sustain economic and social development. However, as described below, other factors must also be taken into account.

In recent times, this objective of expanding the supply of energy as a key element of economic and social development strategies has been accompanied by other objectives of a political nature that carry great weight in some of the Third World countries. In view of the importance of energy, some countries in their national policies emphasized the question of energy supply security and proposed to meet those political objectives by becoming increasingly self-sufficient in the production of energy resources.

Other countries have attempted to complement the traditional objective of increasing energy resource supplies with the objective of orienting the development process itself towards the creation of new resources. This has been the case in the agricultural and industrial development experiences of the developing countries that were particularly oriented towards the creation and increase of energy resources based on biomass.

Finally, there is a growing awareness that, together with the aim of increasing energy supplies, clear environmental conservation objectives must be pursued, in view of the serious environmental implications of a disorderly and predatory expansion of natural resource usage for energy purposes. For example, these implications have become evident in the use of forest resources for energy purposes by large social sectors of the Third World.

This broadening of conventional objectives has increasingly linked energy planning with a review of development styles. Two development/energy options are available for fulfilling the purposes of an energy strategy. One would be to operate within known limits in the production and use of a conventional economic resource, by proceeding on the basis of existing economic structures and by adopting ordinary policies. Another would be to analyze the energy factor by reviewing production and consumption patterns with a view towards turning such a review into a source of new opportunities both for expanding energy resource supplies and for orienting and rationalizing future demand.

The latter option is certainly much more complicated, but is nonetheless challenging. Thus considered, the design of the energy policy objectives admits much more complex factors to which, we must confess with all honesty, we do not yet possess well-developed economic and social approaches. This is one of the main tasks which will have to be accomplished in the next few years in order to carry forward energy planning.

I would now like to propose systematically what I consider could be some of the concrete objectives of an energy strategy. I have identified some broad areas to which the different elements of an energy strategy for developing countries might be directed.

Supply Considerations

1. **Expansion of conventional resources supplies.** The prime objective of an energy strategy should be the expansion of all possible conventional energy sources; in other words, oil and coal, which are basic resources for the economic and social development of any country and whose low accessibility and utilization in the developing countries requires vigorous rectification. I have already pointed out the considerable limitations imposed by the lack of information about the availability of these resources in the underdeveloped countries, and the fact that in recent years the exploration for and exploitation of conventional resources in developing countries has been very limited and imbalanced compared with the same activities in the developed countries. When the amount of oil drilling in recent years in the developed and developing countries is compared, a serious imbalance between the two can be seen, especially in those countries which do not produce oil and which are in the first stages of the search for this

precious resource. Whatever the objectives of the developing countries' energy strategies, it is necessary to make a serious and sustained effort to discover and considerably expand all the opportunities existing in the field of conventional resources.

2. Rationalization and expansion of traditional sources. This objective is especially valid for those countries which for many years will have to keep on depending largely on conventional energy sources for broad segments of their populations. I refer particularly to the use of charcoal and firewood.

In this respect, it is well known that more than 50 percent of the world's population, living mainly in rural areas, basically depend for their energy resources on what they can obtain from the forests by extracting wood and charcoal from them. It is also well known that, whatever the possible alternatives, for several decades the countries with these large rural populations will have to continue to depend largely on these traditional energy sources.

In very extensive areas, inhabited by millions of people, the exploitation of these resources has reached an alarming level of depredation which has resulted in the deforestation of millions of hectares. This phenomenon, known as the "second energy crisis," is one of the urgent and prime concerns in the economic strategies of those countries which are experiencing this serious ecological problem. The main objective should be to increase the ability of forest resources to be used rationally until new, abundant, and economically feasible substitutes become available to these vast masses of humanity.

In addition, it has been very rightly noted that for these developing countries, biomass is a potential resource which could be considerably expanded in the future. The opportunities recently opened up through the production of alcohol in countries such as Brazil are opening new frontiers which could greatly expand during the present critical time, the utility of the so-called traditional energy sources in key areas through such processes as the production of liquid fuels as substitutes for oil.

3. Expansion of new and renewable energy sources. The third front should be the expansion of the so-called new and renewable sources of energy, which were especially stressed at the UN conference held in Nairobi in August of 1981.

Under this heading, hydroelectricity is, of course, particularly important, since it constitutes one of the great and abundant resources of the Third World countries and is currently being used relatively little. In continents such as Latin America, only 8 or 10 percent of the hydroelectric sources' total potential is being used, and the situation may be even worse in Asia and Africa.

Apart from this, however, the use of renewable energy sources through an integrated approach has acquired significant importance. Recent experience in India's rural area in the combined use of renewable resources, such as biomass, wind, and solar energy, mark new frontiers which will have to be explored seriously in the future. Among other promising fields for future energy supply expansion are minihydroelectricity, marine energy resources, and geothermal energy. All of these proposed new alternatives, although in some cases still undeveloped or very specific in their applications, are nevertheless possibilities which should be considered very carefully, especially in the developing countries where these alternative resources are particularly well adapted to the economic and social structure.

4. **Expansion of minisources.** Another objective should be the expansion of minisources of energy. This means not only the so-called conventional sources (oil and gas) but also the nonconventional ones. In the past, exploitation of these sources has been abandoned, and only recently have we begun to give due weight to the enormous global importance which the rational and intense exploitation of dispersed, varied minisources could have, especially for the developing countries. The UN conference on this subject held recently through UNITAR pointed out the significance of this potential resource, to which investments should be devoted and regarding which policies should be prepared in order to achieve maximum exploitation. This message is critically important and should be kept in mind.

These four broad ways of expanding energy supplies should be explored with realism and flexibility, taking into account the previously mentioned dynamic evolution being experienced by technology. It would be wise not to overemphasize any of these resources, especially the new and renewable ones, but neither should the potential relative importance of integrated planning of these resources be underestimated. Basically, what each country needs is its own efficient blend of available and potential resources—a blend that makes it possible to maximize the resources' use according to economic criteria while seeking self-sustaining mechanisms that can expand energy supplies in line with the quantitative and qualitative requirements of economic and social development.

It is important to note that the production structures and the predominantly rural populations of the developing countries give special value in these countries to locally produced energy resources, especially renewable ones. It should further be noted that the options for the developing countries are on very different terms from those for the developed countries.

Planning Demand

Along with planning supply, demand objectives must also be established. As in the developed countries, demand planning is a basic goal in the underdeveloped countries—a goal which includes projections based on current development styles and which should consider both consumption policies and production structures. Proper demand planning should cover, on the one hand, conservation objectives, and on the other, objectives concerning the efficiency with which energy resources are produced or used.

It is well known that the developing countries' low level of energy consumption limits their scope for conservation, but this fact does not exclude the possibility of carrying out vigorous conservationist policies. Some developing countries' recent experience shows that conservation considerations are a necessary part of any energy strategy and cannot be excluded from present and future overall energy policies. The conservation experience gained in areas such as urban and rural transport strategies, the location of economic activities, and the design of industrial habitats and production methods forms a basis on which conservation objectives and criteria should operate, with significant results for the developing countries.

For the same reason, the need for greater energy resource use efficiency in the Third World countries should be borne in mind. This entails greater efficiency in both the production of energy and its final use.

Although there are serious limitations on the opportunities for substituting some energy types for others, changes in the forms of production and the final uses of energy resources have been achieved with undeniable success. Consider, for example, the case of recent Brazilian experiments involving a return to the use of charcoal in steel production—experiments which have achieved great technical efficiency and economy compared with traditional modes of using fossil fuels.

Finally, possible energy strategies must incorporate an analysis of new production and consumption styles appropriate to the developing countries. These styles are today strongly influenced by the industrial countries' patterns involving high energy consumption, especially of liquid fuels (fuels which are scarce and expensive for Third World countries).

Analyses of development strategies that are less energy intensive are now in progress in different countries and centers. In many cases, these strategies run up against attitudes and habits that are not easy to change, but nevertheless these habits can be directed positively towards the adoption of new production and consumption forms which could help solve the developing countries' energy problem.

Similarly, it is important to consider strategies for the developing countries that are based on more intensive uses of their natural environments and on these environments' particular ecological conditions. As part of this, efforts should be undertaken that are oriented towards reorganizing the spatial configuration of production, transport, and human settlements as a function of strategies that do not use energy intensively.

The field for all these approaches is enormous and extremely promising. The increased recognition of the energy crisis has awakened interest in these approaches—approaches which cannot be omitted from alternative energy development strategies while still, of course, acknowledging the notable differences existing among individual countries and situations.

Policy Instruments

How can a set of economic and social policies be implemented that are directed towards achieving the above-mentioned supply and demand objectives? The answer to this question centers on the most important and inevitable but delicate and difficult instruments of any Third World energy strategy: a considerable increase in investments in the energy sector.

All the investment projections involve enormous figures and anticipate a substantial increase in the Third World countries' ratios of energy investments to total investments. Together with this increase, clear technological policies must be followed that are oriented towards testing and developing those technologies best adapted to the particular economic, social, and ecological conditions of the developing countries. Further investments will be required in order to obtain and have access to all available information in all fields of new energy technologies, especially those concerning new and renewable sources of energy.

Along with addressing investments and technologies, considerable emphasis must be placed on relative price policies, especially in order to deal with the structural adjustments imposed by the oil price changes in recent years. The management of relative price policies acquires particular importance in those countries operating under the market economy system, but it also applies to any economy.

As part of the above, it should be noted that the experience of recent decades reveals the implementation of unsatisfactory price and tariff policies regarding energy production and use—policies which, on the one hand, have given rise to an unsuitable allocation of investment resources and, on the other, have encouraged a great deal of waste of precious energy resources.

The use of price mechanisms should not exclude, of course, the possible use of a subsystem of subsidies and incentives within a general system of social costs and benefits. Each economy can carry out this subsystem according to the particular national situation; for example, at given times it may be used to stimulate forms of production or consumption that are considered to have top priority in channeling energy supply and demand. However, let it be said in passing that the conventional criteria of costs and benefits used by neoclassical economists frequently do not appear to take sufficient account of the multiple objectives involved in energy strategies. Thus, in dealing with the social evaluation of these policies of stimuli and incentives, it is necessary to broaden considerably the conventional marketplace criteria.

This leads us to a delicate subject: that of the relative roles which should be played by the government and the marketplace in the design and execution of energy strategies. The problems vary, of course, according to whether the country in question operates as a market economy or as a centrally planned economy; in the former, the problems also vary according to the individual country, its degree of development, and how well the market is operating.

Without denying the importance of the marketplace's role in mixed economies such as those prevailing in many developing countries, it must be recognized that the public sector cannot neglect the promotion of a suitable global strategy for solving the overall energy problem. Furthermore, the key importance of the energy sector, the size of the investments required, the uncertainty of the various technological options, and the marketplace's imperfect ability to achieve long-term objectives all justify the need for a system of economic policies based on the general interest and guided by government policies, although each country must determine the specific form and degree of such policies. This need for government policy becomes particularly obvious in the case of such basic crises as the so-called charcoal and firewood crisis, which market mechanisms alone seem incapable of solving. This combined approach is, in my opinion, the pragmatic and sensible way to approach the delicate and ever controversial subject of the state's and market's relative roles in the solution of the energy problem.

THE OBSTACLES

The implementation of an energy strategy that pursues the preceding objectives will encounter, however, a series of obstacles with which it will be necessary for specific policies to deal.

Without doubt, the first obstacle is the scant infrastructure available for decisionmaking, especially the lack of adequate information. Scarcity of data, deficiencies in resource inventories, and scarcity of skilled manpower for handling energy problems are all common to the developing countries. In this connection, the Nairobi conference brought out the importance of information programs, inventories, and manpower training to improve the infrastructure of available data for making correct energy decisions.

In addition to problems attributable to a lack of information, a lack of knowledge about technologies appropriate to the economic, social, and ecological conditions of the developing countries also presents difficulties. In this connection, the demonstration programs aimed at increasing knowledge of appropriate technologies become particularly significant.

Perhaps the most visible and daunting obstacles are the limitations of a financial and economic nature. The expansion of the energy supply will require, as I have already stated, enormous investments. The World Bank has estimated the amount of resources required to make progress in the developing countries' energy programs. It is worth bearing in mind that the challenge to these countries is especially great because their energy infrastructures must be expanded at costs far higher than those which the present industrialized economies had to bear in the past, when they achieved the extraordinary economic and social growth of the post-World War II period. The latter had cheap energy resources owing to very low petroleum prices—a situation which no longer obtains anywhere in the world. For the latecomers, this phenomenon has been reversed, and their economic efforts to expand their energy supplies will have to be made at costs very much higher than in the past.

In order to attempt to meet this increased demand for energy investments, a vigorous effort to mobilize domestic resources is required, together with the possibility of seeking a suitable form of international cooperation. In this attempt to mobilize resources, it is not enough to think only of resources that are of public origin: it will also be necessary to achieve the maximum degree of mobilization, both internal and external, of the private sector's capacities, for which purpose the promotion of savings and investment policies in the energy sector is of particular urgency.

Institutional problems, especially those due to a lack of suitable governmental focal points to push ahead with energy policies, will present another obstacle. In this regard, it may be said that programming is in a general sense an instrument for reducing uncertainties and for anticipating the phenomena with which national economies will have to deal. It is apparent that the errors committed in the developing countries are relatively much more costly than those of the industrialized countries. For the developing countries, the scarcity of resources makes it much more essential to adopt adequate measures, and hence the need to introduce energy planning is that much greater. In this respect, the developing countries lack appropriate organizations or institutions for decisionmaking, and this is one of the great challenges facing them. It is therefore essential to create suitable institutions in the developing countries for handling their energy problems, whatever the country's economic and social philosophy might be.

THE OPPORTUNITIES

But the energy alternatives of the developing world do not involve only problems. This field also provides very significant opportunities.

Opportunities to Review Styles of Development

The domestic adoption of energy strategies also gives developing countries an opportunity to review their styles of growth on the basis of energy policies that are integrated with their development processes. This in turn makes it possible to use the energy factor in two ways: first by making it a powerful instrument in the economic and social development process, with which it has indissoluble links (for as I said before, appropriate energy strategies make it possible to reorient styles of economic and social development, turning consumption and production patterns into recourses per se); and second, by reorienting the economic growth process so as to make energy resources one of the very consequences of development. In this second sense, energy resources thus would concurrently be inputs and outputs of the development process.

It should be borne in mind, however, that by suitably orienting the production and use of available natural resources, it may be possible to stimulate economic and social development and broaden the energy base, thereby making it possible to address the great demands which developing countries must meet. Such is the case in the use of biomass, for example, which, while increasing the production of liquid fuels, at the same time promotes agricultural, industrial, and commercial development. The same might be said of other resource utilization approaches that train and employ large numbers of people at very low opportunity costs.

In these cases, it is economic development itself which is oriented in such a way as to create a new energy base. Not only are new energy resources generated; at the same time human resources, natural resources, and the production base in general are also developed, thus contributing in a positive way to economic and social development.

In order to make progress with all these policies, appropriate economic criteria must be available. In general, there is a lack of suitable criteria on energy programming, particularly in the developing countries—so much so that it may be concluded that economic theory has been rather weak with regard to the formulation of economic policies on scarce and exhaustible resources. For these purposes, conventional economic theory proves to be rather limited. However, all this is a great challenge which has begun to be taken up on a substantial scale in recent years as one of the positive by-products of the so-called energy crisis.

In conclusion, it might be said that the opportunities presented by the energy question are many and varied. On the one hand, and as a consequence of the recent energy crisis, nations have, in one way or another, had to come up with structural adjustments in their internal economies in order to bring them in line with the new realities of the world energy picture. On the other hand, analyses and studies of possible alternatives are under way in all countries, and these are sure to result in a considerable reduction in the costliness of the mistakes made. Finally, the global energy problem is now being tackled with greater realism and flexibility, with the result that the new strategies are sufficiently broad to accommodate the future technological innovations which are bound to occur in mankind's future energy scenario.

Opportunities for International Cooperation

The energy question also opens up big new opportunities for international cooperation. It unquestionably constitutes one of the main topics of the so-called New International Economic Order. In international fora centered around energy, it has been made abundantly clear that the world's energy resources are badly distributed and that the developing countries have limited and inequitable access to these resources as a means of strengthening their economic and social development processes.

From the above point of view, it is obviously worth making the energy question one of the great items of negotiation in a long-term perspective of international economic relations, but it must also be treated as an element closely tied with the other components of a new world economic order. Furthermore, it is clear that the energy question is a topic where the mutual interests involved are more obvious than in other fields, for it is in the direct interest of the entire international community to have the energy supply of all nations, and of developing nations in particular, show adequate growth. The more that energy resources can be explored and exploited in the developing countries, the fewer the pressures on the international demand for hydrocarbon resources will be— resources which are bound to become relatively more scarce in the decades ahead. It is in all mankind's general interest to increase and diversify energy resource supplies in all countries, especially in the Third World.

It is worth noting that there are numerous opportunities for international cooperation. These opportunities should be considered from both a short-term and a long-term viewpoint.

In the short term, the problem is obvious, and attention has already been drawn to it on many occasions. It is a matter of supporting the developing countries in the trials they are undergoing with regard to their balances of payment as a consequence of the energy price increases. International aid and cooperation must provide an indispensable counterpart to the necessary structural adjustments which these countries will have to make in their domestic economies. Some of the experience gained in the support (in the form of financial assistance and energy supplies) extended by some petroleum-exporting countries to underdeveloped countries has been most encouraging. In the long term, the problem becomes much more complex. It is basically one of supporting the less developed countries in the preinvestment and investment efforts called for by the energy strategies. Regarding preinvestment, large contributions are needed for the efforts being carried out in such fields as education; resource evaluation and inventory preparation; and technological development, adaption, and demonstration—as was clearly stated in the Nairobi Plan of Action in the field of new and renewable sources of energy. Still more important is the cooperation which can be extended in the field of investments. The developing countries require and demand big investment efforts—efforts which must compete in their economic programming with other economic and social investment efforts and which inevitably call for a major increase in the international community's awareness.

The World Bank has very rightly drawn attention to these investment issues, and it is evident that the international community should make every possible effort to support heavier flows of resources to these countries so that they may set about their energy programs. There are many institutional possibilities and other channels for doing this, including the mobilization of both the public and the private sectors. In view of the nature of the energy investments, however, it appears that in this case, more than any other, there is ample justification for a significant increase in public resources, especially those channeled through the World Bank, the regional economic banks, and the bilateral agencies for official cooperation. It is to be hoped that when the United Nations' new round of global negotiations is held in the next few months, pragmatic and urgent agreements will be reached in this field.

Finally, let us say that the developing world's implementation of integrated energy strategies provides excellent opportunities for horizontal cooperation. The similarities in the developing countries of problems and alternatives, of patterns and levels of development, and of resource shortages makes it particularly vital to promote subregional cooperation in such fields as the joint development of technology and the exchange of experience and experts. The example which Latin America has provided in this field through the creation of OLADE, the Latin American energy organization, is extremely promising and represents a model that would be well worth extending to other developing regions.

CHAPTER THREE

International Cooperation and the Development of Institutional Arrangements on Energy

D. W. Campbell
Director General
International Energy Relations Branch
Canadian Department of Energy, Mines and Resources

J. A. D. Holbrook
Chief, North/South Energy Group
International Energy Relations Branch
Canadian Department of Energy, Mines and Resources

"The issue is not whether an energy transition will take place but whether the international community will achieve it in an orderly, peaceful, progressive, fast and integrated manner."[1]

In considering the question of international institutional arrangements, the message of Symposium I—namely, that for the near and medium term the energy crisis is primarily a political and economic problem and only secondarily a technological one—is very relevant. Furthermore, the energy crisis is inextricably linked to other issues, particularly problems pertaining to the misuse or overuse of natural resources and problems arising from disparities in standards of living both within and between countries. This leads to the conclusion that there is no single solution, no hard or soft path, no centralized or decentralized institution to solve the problem. The answers lie in a mix of solutions appropriate to each action and dependent upon the kind of economy, level of development, and array of energy resources that nations possess.[2] With

The authors note that the opinions expressed in this paper are solely their own and do not necessarily reflect the policy of the government of Canada.

these considerations in mind, what role is there for international cooperation and the international framework such cooperation implies?

It is clear that the nature of the problem is different for different countries although the ultimate goals are the same. At the risk of oversimplifying, it is fair to say that there is a greater commonality of issues within each of the major international groups (industrialized countries, the oil-exporting developing countries, and the oil-importing developing countries) than there is between these groups. The institutions that have been developed to date reflect this, and they bear some examination in order to look to the future. There are clearly two sets of institutional arrangements and possibilities—those whose existence is a result of a commonly perceived self-interest, such as the International Energy Agency (IEA) and the Organization of Petroleum Exporting Countries (OPEC), and those who have broader interests and accordingly deal or could deal with the resolution of problems on a greater, worldwide scale. It is, of course, the latter that are integral to the North-South dialogue and upon which any examination of international institutional arrangements for energy cooperation inevitably must concentrate.

International Energy Agency

Initially, however, it might be useful to review the role of the IEA, the second major international institutional arrangement established to deal with these issues. This organization, founded in the wake of the events of 1973, has reached maturity reflecting in no small part the development of the energy policies of its membership, the western industrialized societies. The IEA's need, as a group, to reduce dependence on imported sources of oil through restructuring of the energy economies of its members has been addressed, and an ambitious strategy to effect this has been mapped out. The strategy includes a larger role for coal and nuclear power, higher levels of indigenous production of hydrocarbons, and major efforts in conservation and fuel use efficiency. A second and important role for this institution has been the development of a system to help shield against supply disruptions. This mechanism, which provides for the sharing of oil, fortunately has not had to be tested.

The IEA and the Organisation for Economic Co-operation and Development (OECD) undertook a major study following decisions made at the 1979 Tokyo and 1980 Venice economic summits on the commercialization of energy technologies.[3] A high-level group was established to examine technologies that will be needed for the transition to economies using minimum amounts of oil. At earlier stages of the study, there were expectations that major international cooperative arrangements would be put in place. As the studies developed, and as some governments took new directions, it became clear that more modest goals would be all that could be achieved. No new institutional arrangements emerged for the commercialization of technologies, and the group contented itself with recommendations for stronger national actions complemented by commercial-scale international cooperative projects and more international collaboration on research and development.

On the whole, it is reasonable to conclude that the institutional arrangements of the IEA are, at least in the short to medium term, sufficient to address the specific issues that are of concern to the energy economies of western industrialized countries. The real test will be, however, the period of the 1980s and 1990s, when it will become evident whether or not there has been real, permanent, and structural change in the economies of this group of nations.

Organization of Petroleum Exporting Countries

As with the IEA, OPEC is a special interest group. Its raison d'être has been to provide a forum where a number of oil-exporting countries can agree on pricing and production matters. The considerable success of OPEC in doing just that over the past decade has presented its membership with vast new resources, and these have inevitably changed, and for the most part broadened, the OPEC countries' perceptions of their place among nations.

The organization is made up of a wide spectrum of countries whose per capita GNP in 1979 ranged from $380 per year to over $17,000.[4] Notwithstanding these vastly different standards of living, all OPEC nations have been, within recent memory, developing countries, and many remain so. The organization as a whole identifies strongly with the economic and political viewpoints of all Third World countries, which explains OPEC's steadfast insistence that any discussions with the industrialized world on energy pricing and production must be coupled with a discussion of North/South issues in general. Moreover, OPEC nations practice what they preach: while the OECD countries struggle to raise toward 1 percent the ratios of their Official Development Assistance (ODA) to GNP, the average ODA/GNP ratio for all OPEC nations was 1.44 percent in 1979.[5]

Despite its deep philosophical roots in the Third World, OPEC's bounty over the past few years, and the corresponding responsibility to manage its petrodollars wisely, has inexorably drawn its membership into closer and more substantial financial ties with the industrialized countries. The entry of the OPEC nations into the global financial community has fostered among them a very sophisticated appreciation of and concern for the state of the world's economy. Does any other nation in the world, for instance, take an account of the world economic situation in establishing its own domestic and international energy policies equal to that taken by Saudi Arabia?

Finally, at the same time as OPEC is being drawn closer to the IEA nations, the very success of its own oil-pricing policies has imposed ever increasing burdens on its fellow Third World countries. Thus, OPEC at present is in a quandary, with its various interests pulling it in several directions at once. Despite the strains that this situation imposes on OPEC itself, the OPEC nations are in the key position, with a foot in both camps, in any future development of international energy cooperation and of corresponding institutions.

The North/South Dialogue

In recent years, we have all been involved in, or at least observers of, what has become popularly known as the North/South dialogue. In various forms, this dialogue has addressed itself to activities related to exchanges of goods, people, services, capital, ideas, technology, and power. The basic issue in the North/South relationship is no longer "whether" but "how" these transfers should take place.

Energy questions have been, and will continue to be, a fundamental and integral part of these negotiations. The energy issues pose most of the same questions as do other elements of the dialogue, such as food and agricultural development. Lack of indigenous resources and ever mounting debt, resulting from the need to obtain ever more expensive imported energy to meet basic human needs and to struggle for economic development, has placed energy resource issues at the head of the list.

The Portillo Plan

President Portillo of Mexico placed before the 1979 UN General Assembly a comprehensive proposal for a world energy plan that would become an institutional part of the UN system.[6] Elements included in the plan were:

- guarantees of each nation's sovereignty over its natural resources;
- financial and technical assistance to insure rational exploration, production, distribution, consumption, and conservation of energy resources;
- accelerated and systematic exploitation of energy reserves of all types, traditional and nonconventional;
- the drafting for all nations of national energy plans compatible with a world energy policy;
- measures to promote the formulation in developing countries of auxiliary industries in the energy field;
- a short-term system for oil-importing developing countries which would guarantee oil supply, provide compensation for price increases, and insure "considerate treatment" by oil exporters;
- the establishment of financing and development funds to meet both the long-term objectives and the immediate needs of oil-importing developing countries;
- a system for disseminating and transferring technologies with attendant training programs; and
- an information registry and research and development component.

President Portillo called for the implementation of this plan through the establishment of an international energy institute. As a first step he proposed the establishment of a working group with representation from western industrialized nations, socialist states, the oil-exporting countries, and the oil-importing developing countries. No action has taken place on this comprehensive and imaginative proposal, although Mexico, along with other Latin American countries, has put some of the elements into place in a modest way in regional cooperative arrangements.

Was the proposal an approach whose time had not yet come, or was it too sweeping and too ambitious?

Global Negotiations

The year 1979 also saw an initiative in the UN General Assembly towards the holding of global negotiations. These negotiations —seen by many as a natural successor to the Conference on International Economic Cooperation which was held in Paris in 1977 and which had a much more restricted membership—would address the broad range of North/South issues in five agreed "baskets" of subjects, one of which is energy. The last two years of debate have not yet seen agreement on whether or when these negotiations will take place or in what institutional settings.

The Brandt Commission

The Brandt Commission in its report issued last year[7] took the position that the emerging world energy situation was one element of a developing crisis that calls for exceptional measures of international cooperation.

The Brandt Commission called for urgent action on "an international strategy on energy" as part of a broader emergency program to deal with major North/South issues, including:

- Assurances by oil-exporting countries on levels of production and agreement not to reduce supplies arbitrarily, with special arrangements to meet the oil needs of poorer developing nations.
- Commitments to more ambitious conservation targets by consuming countries in order to hold down consumption of oil and other forms of energy.
- Avoidance of sudden oil price increases thereby giving incentives for both production and conservation. Oil pricing arrangements could include price indexing related to world inflation, denomination of prices in baskets of currencies or International Monetary Fund special drawing rights, and guarantees of the value and accessibility of financial assets received by oil producers.
- Major investment in exploration and development of hydrocarbon and other conventional energy sources (oil, gas, coal, hydroelectricity) in developing countries as well as accelerated research and development of new types of energy, especially solar and other renewable forms. An expanded role for international and regional financial agencies in such activities is also recommended.

The Brandt Commission also proposed the establishment of a global energy research center under UN auspices to coordinate information and support research on new energy sources.

UN Conference on New and Renewable Sources of Energy

A major item on the calendar of international discussions with institutional implications was the UN Conference on New and Renewable Sources of Energy held in Nairobi, Kenya, in

August of 1981—the first UN conference devoted specifically to an aspect of the energy problem. The restriction of the scope of the energy issues considered to new and renewable sources of energy represented a compromise between those concerned that the UN system is not yet in a position to address energy issues and those who would like a comprehensive energy debate. The conference's objective was the elaboration of measures for concerted action to promote the development of new and renewable sources of energy. Most participants and observers agreed that the conference achieved a large measure of success. At the technical level, priorities were agreed upon and planning programs developed.[8] An air of realism prevailed.

The question of institutional follow-up, however, proved very difficult, and no decisions were made other than to refer the matter back to the Director General for Development and Cooperation at the UN for his consideration. Without trying to prejudge the Director General's proposals, it is expected that some form of secretariat, developed largely from existing resources, will emerge to undertake the task of maintaining the momentum developed in Nairobi.

Financing and investment will be key issues. The World Bank has estimated that some $60 billion will be required for new and renewable energy in the Third World in the 1980s. To achieve anything near this level of investment, the new and renewable energy industry will have to be developed to the extent that its turnover measures in the billions of dollars, not the millions of dollars as it does today. What are the institutional vehicles for this?

International Energy Development Needs

International discussions all reveal a set of common themes: indigenous energy production in oil-importing developing countries must be increased, there is a need for more and better methods of financing and investment, information flows must be encouraged, development and transfers of technology must be effected, and research and development relevant to specific regional conditions must be undertaken. All of these issues are being considered in some way in existing institutions. But it is the question of how to provide a more coherent and coordinated thrust that particularly challenges policymakers.

In October of 1981, the Cancún Summit in Mexico brought together twenty-two world leaders to discuss the range of North/South issues that have occupied so many international gatherings in recent years. Not unexpectedly, energy issues formed an important part of these deliberations, and the threads forming the fabric of the debate elsewhere were evident: they included the need for a framework for insuring an orderly transition from hydrocarbons to other sources of energy, the potential contributions of regional energy schemes, the need for stronger conservation measures (particularly in industrialized countries), the need for more public and private investment in the oil-importing developing countries, and the requirement for more and better international exchanges of information on energy. The summit group participants reaffirmed the theme of economic interdependence, with the prosperity of any single country or group of countries being dependent upon the existence of growth and stability in other countries.

It is evident that the oil-importing nations of the Third World will require large amounts of financing if their hopes for increased energy production and the management of energy demand are going to be realized. The World Bank has estimated that from 1980 to 1990, the total investment required for a significant program is $450 billion to $500 billion in 1980 dollars.[9] Most of this investment will have to come from the commercial banking community and from domestic savings and bilateral assistance programs, but there is also a major role for international institutions.

The World Bank

The World Bank is already the largest source of public funding for energy resource development. An accelerated program for fossil fuel development, including financing for exploration, started in 1979, and by 1980 funding for the generation, transmission, and distribution of electric power had doubled. Current World Bank plans call for an outlay of $13 billion for energy over the period 1981–85. Yet, in the words of the World Bank's former president, Robert McNamara, "This amount is, however, some $12 billion short of what is both desirable and feasible; the Bank's resources today are not adequate to meet the energy needs of the developing countries as well as provide for support of our developing members' other essential investment needs."[10]

The World Bank has identified basic objectives for an energy development program. They include:

- much higher levels of investment for energy production,
- the reorientation of development plans to take account of this newly scarce and expensive factor of production,
- substantial investment in energy conservation, and

• a massive effort to insure that the minimum fuel requirements of the urban and rural poor can be met in the next two decades.

Is the World Bank the most appropriate institutional vehicle to undertake such a program? Provided funding that is truly additional and substantive can be made available, a strong case can be made for a new institution.

As long as the World Bank is restricted by its present one-to-one gearing ratio, it cannot hope to meet the demand for energy development funds. Its reluctance to change this gearing ratio for projects such as integrated rural development is understandable since it is difficult to forecast accurately the rate of investment return on such projects, but it should consider changing its gearing ratio for conventional energy projects since they generally offer fixed minimum rates of return with little risk. Changing gearing ratios is, however, not going to generate the volume of funding that will be required to meet future energy financing needs. New approaches are needed.

The Energy Affiliate

Proposals for a new multilateral institution which could take the form of an affiliate of the World Bank are high on the agendas of a number of international meetings. The common elements of energy affiliate proposals include concessional lending, universality of access by oil-importing developing countries, and a management structure that reflects the interest of the investors in the facility. Funding levels suggested range up to $30 billion. Innovative financing mechanisms, such as issuing bonds to participating governments or setting variable interest rates on loans to the recipient countries with the rate fluctuating according to the actual return on investment experienced, will be necessary. However, there are significant differences among governments on the desirability of establishing such an institution. Given the magnitude of finance and investment flows required to meet the developing world's needs, the international community has a collective duty to respond, but an agreed-upon process has been elusive to date.

Regional Approaches

Regional lending institutions also have an important role to play in stimulating investment in energy projects. Regional bodies have the advantages of commonality of interests, geographic concentration, and similarity of energy problems.

Latin America offers an interesting example of regional institution building to implement common programs for energy.[11] OLADE, the Latin American energy organization [Organisaçion Latinoamericana de Energia], was formed in 1974 with a broad and ambitious mandate that includes promotion of rational and effective policies for exploration, transformation, and commercialization of energy resources; creation of a financial institution for energy development; creation of a regional energy market; and exchange of technological information. This organization is an outgrowth of the ARPEL organization [Asistençia Reciproca Petrolera Estatal Latinoamericana] that the state-owned oil companies of nine Latin American countries had established in 1965 to promote technical assistance and exchanges of information.

The energy cooperation program set up a year ago by Venezuela and Mexico to assist nine other Central American and Caribbean countries represents a significant regional approach to providing oil supplies to oil-importing countries and, at the same time, offering incentives for the development of indigenous energy production and the rationalization of national energy programs.

The Association of South East Asian Nations Committee on Petroleum, at its meeting in Manila in October of 1981, discussed support for its members' energy programs and reviewed energy investment requirements for the region. Prospects for cooperation appear to be encouraging.

Bilateral Approaches

Bilateral investment of concessional funds traditionally has been a major source of funding for energy projects. While these funds have often been tied to procurement from the donor country, it is through this channel that most of the concessional investments in the energy sector have been carried out. Donor countries have different areas of expertise, and their bilateral energy projects usually reflect their own interests. It should not be difficult to encourage governments to increase their spending in areas that have been identified as national priorities by the donors (for example, hydroelectricity is clearly of interest to Canada, so one could expect the Canadian government to look more favorably on project proposals that involve hydroelectricity). National development plans could be linked to international development spending in order to strengthen a donor nation's capability in a particular technology. Such spending need not come directly from the central government—state corporations may well be the most appropriate vehicles for these projects.

The establishment of Petro-Canada International (PCI) by the Canadian government is an example of this type of aid vehicle. Its purpose is to explore, on concessional terms, for hydrocarbons in oil-importing developing countries under conditions where private oil companies might not normally be willing to invest. It is a subsidiary of the national oil corporation, Petro-Canada, and it will use the services of its parent as well as goods and services obtained on the open market for its activities. Canada sees PCI as a novel vehicle for utilizing Canadian expertise and capabilities in the furtherance of energy development in the Third World. To expand this activity, PCI will be open to cooperative projects with other nations with similar interests.

Private Sector Approaches

The multinational corporations are also institutions by which funds are channeled from the developed countries into the Third World for industrial and resource development. However, being nonconcessional, these funds have had to be returned with interest to the corporations over the long term. To date, the interests of the energy corporations have been directed towards the export of energy from those few energy-rich developing countries to the developed world. There are signs, however, that the multinationals are now looking to the business opportunities presenting themselves in supplying energy to Third World markets. Given the pressures that keep the cost of energy more or less the same around the world, the multinationals cannot be expected to market energy at less than world prices. On the other hand, they could be induced (for example, by governments through favorable taxation arrangements, as well as by the profit motive) to make the investment necessary to provide the infrastructure for distributing energy within Third World countries as they do in the developed world.

Smaller corporations also have an important role to play. These corporations are frequently more flexible than the multinationals and represent another form of bilateral resource transfer from developed countries to the Third World. They frequently are prepared to invest in Third World markets to develop energy infrastructures that are not transnational. These companies offer a new and exciting means of developing Third World energy resources since they operate on a scale that is manageable in a national economic context and are often perceived to be more responsive to national interests and aspirations.

Energy Research and Development

Energy research and development is an area that has had little global focus. Most of the activity under way is being done in developed countries on issues of relevance to them but with little application for the developing world. This reflects research and development activities more generally; the OECD has estimated that 92 percent of all research done occurs in the developed world and only 8 percent in the Third World. The IEA has made some encouraging but modest efforts through its Committee on Research and Development to involve Third World countries and their problems. In addition, UNESCO is playing a role through funding and directing research in regional and national institutions.

As a small contribution to the promotion of energy-related research in and for developing countries, the Canadian government announced at the August 1981 Nairobi conference, a $10 million grant to the International Development Research Centre. This center, funded by Canada but directed by an international board of directors from developed and developing countries, will establish an expanded program of energy research in developing countries.

The transfer of technology is a natural consequence of investment in a country. Many developing countries believe that technology transfer is the missing link in the process of developing their economies, including their energy development. But how can technology transfer be institutionalized in an international context?

There have been several UN initiatives to try to institutionalize the transfer of technology. The United Nations has established an Intergovernmental Committee on Science and Technology for Development. The frustrated discussions on the UN Conference on Trade and Development's (UNCTAD's) code of conduct for international technology transfer are well known. The difficulties usually arise over the transfer of technical information, because it is this information which has a financial value. For a technology transfer to be effective, proprietary information should be treated in the same way as financial resources—it can be granted, loaned, or invested in a country in the expectation of achieving a return.

Conclusion

First, the foregoing review of discussions, negotiations, and proposals for international cooperation in energy illustrates how political and economic considerations must influence and indeed direct the search for international institutional arrangements in the same way that the energy crisis itself, as Symposium I concluded, is essentially political and economic. Second, it is difficult to conceive of institutional arrangements that do not address, particularly in the North/South context, a much broader range of issues and problems than simply energy.

Bold global plans and strategies can excite the imagination and provide aims to which the community of nations must aspire. At the operational level, however, their very comprehensiveness may make them inefficient or difficult if not impossible to implement. Furthermore, as the lack of response to the Portillo proposal has shown, there is little evidence of political will to consider massive global undertakings.

Indeed, in the country hosting this Symposium, there is a sizable school of thought which believes that if only the appropriate climate can be created, development will largely flow from free market forces and private investment. However, history suggests that the world might have to wait a very long time for this to happen, and the last thing mankind can afford is the luxury of time.

Politics and diplomacy have long been defined as the arts of the possible. The creation of the international institutions that flow from these processes could also be included in this definition. In a world of sovereign nation-states, each with its own array of energy problems, issues, approaches, and solutions, we must stress the interdependence of our economic futures and the national benefits that flow from international cooperation, but we must be realistic about what is attainable. Just as there is no single solution to the energy problem, there is no single solution to institutional arrangements for international cooperation in the field of energy.

NOTES

1. *Report of the* UN *Conference on New and Renewable Sources of Energy, Nairobi, 10–21 August 1981* (document A/conf. 100/11) (New York: United Nations, 1981), p. 3.

2. See R. Bohm, L. Clinard, and M. English, eds., *World Energy Production and Productivity: Proceedings of the International Energy Symposium I* (Cambridge, MA: Ballinger, 1981), pp. xiv–xv.

3. See the *Report of the High Level Group for Energy Technology Commercialization* (Paris: OECD, 1981) noted by the OECD Council on June 4, 1981, and endorsed by the Governing Board of the IEA at the Ministerial Level on June 15, 1981.

4. *1980 World Bank Atlas* (Washington, DC: World Bank, 1981).

5. *Development Cooperation 1980* (Paris: OECD, 1980), p. 127.

6. United Nations General Assembly, document A/34/PV.11 (September 27, 1979), New York, 1979.

7. *North-South: A Programme for Survival,* The Report of the Independent Commission on International Development Issues under the Chairmanship of Willy Brandt, The Independent Commission on International Development Issues, 1980.

8. See the report cited in note 1.

9. *Energy in Developing Countries* (Washington, DC: World Bank, 1980), introduction.

10. Ibid.

11. A useful presentation of the Latin American experience is contained in a paper by A. del Valle, "Regional Aspects of Energy Planning and Development," Lecture #11, UN Symposium on Energy Planning in Developing Countries, Stockholm, 1981.

Work Session on the Role of Nuclear Power

CHAPTER FOUR

The Role of Nuclear Power

Sigvard Eklund
Director General
International Atomic Energy Agency

Energy questions are very popular topics for discussions at national or international levels in meetings of different kinds. It is difficult to estimate the number of major meetings held on energy or, more specifically, nuclear energy, during 1980, but a preliminary survey indicates that around forty conferences were held in 1980 concerning energy in general and around sixty conferences on nuclear energy. These global figures certainly only represent the lower limits, since not every meeting may have produced proceedings whereby they would be registered in our [i.e., the IAEA's] information system.

The publications on energy or nuclear energy are likewise very numerous. From our INIS (International Nuclear Information System) we know that each year an average of 75,000 individual articles, reports, and so forth concerning nuclear energy are published.

Under these circumstances I hope that my readers understand that it is not possible to fully reflect the literature concerning nuclear power. I have therefore limited myself to those aspects which I have found especially important. For those interested in more extensive surveys I recommend the final report of the Committee on Nuclear and Alternative Energy Systems published in 1979 by the National Academy of Sciences under the title "Energy in Transition 1985–2010." Another report summarizes the results of a seven-year study performed at the International Institute for Applied Systems Analysis (IIASA) under the name *Energy in a Finite World: Paths to a Sustainable Future.* This report was synopsized at last year's Symposium by Professor Häfele himself.

More details are provided in a second volume, *Energy in a Finite World: A Global Systems Analysis.* [The aforementioned two works have been published as *Energy in a Finite World*, Vols. I and II (Cambridge, MA: Ballinger, 1981).]

INTRODUCTION

A fundamental task is to provide the growing world population, now 4,500 million people and by the year 2000 around 6,000 million, with food and housing. The intensive agriculture necessary to produce the needed quantities of foodstuff is based on extensive use of fertilizers which in turn need large quantities of energy for their production. We recognize here a link between a growing population and the need for energy. Another factor influencing the need for energy is the attempts being made to close the gap between developed and developing countries by industrialization, which increases energy consumption.

This position paper will be limited to a survey of the present installations and, to a certain extent, the operating experience of power reactors in the world. In addition, some estimates of nuclear growth in the future will be given. The survey will start with an introductory section giving estimates of the need for nuclear power made during the last seventeen years, estimates which are interesting when they are compared with what was actually realized. The plausible reasons for the discrepancies between what was forecast and what was actually constructed give some indication of the uncertainties associated with every attempt to forecast what will happen in the nuclear field even during the next ten years.

It is possible to foresee with a rather high degree of accuracy what will happen in the field of nuclear power during the next five years. We like to think that the future is ours and can be formed according to our wishes and intentions, but the fact is that the next few years are already committed to developments decided upon by masters who were in charge a few years ago. The succeeding five years, say 1986–90, are much more veiled in uncertainty, and after that period unforeseeable changes can occur, completely outside the forecasts based on estimates possible now. Predictions under these circumstances are then possible only by very extensive analysis of the influence which different factors may have, as in the IIASA study, *Energy in a Finite World.*

For the purpose of this Symposium, I have tried to be as factual as possible without too many speculations. Certain assumptions, however, have to be made about the future—for example, that peace will prevail and no military actions will be taken against nuclear power reactors, that no major accidents will occur releasing radio-activity in an uncontrolled way, and that no further country outside the nuclear-weapon states will test nuclear explosives.

In spite of assuming such a favorable scenario it is obvious that the situation of nuclear power is not very bright at present and that the long-term future is uncertain. For example, it is hard to see how a utility in the United States could, within the next ten years, dare to decide to construct and operate a big nuclear power plant, thereby committing an investment of a couple of billion dollars. Present tariffs do not permit the accumulation of the capital necessary to build a nuclear power station. The economic risks are very high both in the case of an accident and in the case of a change in the attitude of the general public during the construction time leading to a denial of the operating license when the reactor is ready.

We have seen several results of events of this type in the United States during the last couple of years. I am especially referring to Pilgrim 2, which is owned by Boston Edison and was well along in its construction when work on it was stopped and the project cancelled after an investment of almost 300 million US dollars. We see a similar development in the Federal Republic of Germany where a "green" political wave has caused anxiety in governmental circles and has forced the government to take a very cautious approach vis-a-vis the nuclear opponents.

In France, where during 1980–81 fifteen reactors were commissioned in fifteen months, the government has decided to decrease the number of nuclear power reactors to be ordered during 1982 and 1983 by one-third. In the socialist countries, on the contrary, extensive programs are being launched guaranteeing their nuclear industry orders for the whole decade.

Let us recognize that it is really an open question whether the nuclear industry in the West can survive under these circumstances. Depending on the answer the present nuclear power era either will represent only a short parenthesis in this history of energy or, alternatively, will survive into the next century with the prospect of being an energy source comparable to coal.

PRIOR FORECASTS

From the UN Conferences on the Peaceful Uses of Atomic Energy in Geneva up to conferences on nuclear power this year, forecasts have been made of the nuclear power capacity expected to be installed in the future, some of which are condensed in Table 4–1. It is striking how the forecasts have been modified with time and how the most recent ones made for the year 2000 are only one-fifth of those of six years ago, in spite of the fact that the price of oil during these six years has increased by a factor of at least seven.

Table 4-1. **Nuclear Power Forecasts—World Totals**

Year of estimate	Forecasts for year[a]			Source
	1980	1990	2000	
1964	150-200	--	--	The Third (UN) Conference[b]
1971	310	1,300	3,500	The Fourth (UN) Conference[b]
1974	220-250	1,300-1,900	3,600-5,300	IAEA Annual Report (July 1, 1974-June 30, 1975)
1975	220-230	1,000-1,300	2,000-2,600	IAEA Annual Report for 1975
1978	167-187	532-699	1,084-1,652	Int'l Nuclear Fuel Cycle Evaluation[c]
1981	136	389-512	725-1,096	IAEA internal data

[a]In gigawatts-electric.
[b]On the Peaceful Uses of Atomic Energy.
[c]Available through the IAEA.

The reasons for this drastic reduction are many. First, I would like to refer to the economic situation in general, leading to a much slower increase in the need for electric power in many industrialized countries, to the more efficient use of energy through conservation measures, and to the reduction of energy waste. In many countries the performance of the regulatory agencies charged with issuing operating licenses for the reactor plants under their supervision may represent another reason. Because of the agencies' requests, often during the construction period, for design changes and extra safety measures in the form of equipment or operational codes, it became difficult to estimate the final capital and operating costs for a nuclear power station. New environmental regulations in many cases also exercised an influence on decisionmakers. The attitude of the general public, or of environmental opponents, often able to intervene in courts at a late stage during the construction, has scared off many utilities. In their attitude towards nuclear power, many people have had difficulties in distinguishing between peaceful and military uses of nuclear energy.

Another feature which has strongly influenced the general

public is the question of radioactive waste. While earlier discussions concentrated on the front end of the fuel cycle—that is, the availability of uranium, enrichment services, and so on—they have now focused on the back end of the fuel cycle—the storage and retrievability or unretrievability of the spent-fuel elements or the fission products after reprocessing. The debate on these issues went off the track when it was proposed that the barriers established around the radioactive waste must withstand the changes which may occur in geological periods, for example, during the next ice age. Public debate consequently became concerned with a nonproblem but nevertheless was able to raise emotions to a level and affect the energy policies of many countries.

Within a period of five years, nuclear power lost its place as a favorite alternative and became "the energy of last resort." Politicians in many countries became wary of making decisions on such difficult problems with wide political implications and instead referred them to a referendum, as occurred in Austria in 1978 and in Sweden in 1980.

This development took place essentially towards the end of the 1970s. It was also remarkable that while nuclear power was increasingly looked upon as a threat to mankind, not a word was said about nuclear weapons and their consequences. Permission to store radioactive waste underground was withheld, or extended only with very severe limitations, while underground testing of nuclear weapons with an average frequency of once per week was not questioned at all. However, it can be noted with a certain satisfaction that in the last few months the general public has seemed to become much more concerned about an antinuclear program directed, as it should be, against the use of nuclear weapons.

PRESENT STATUS

The present situation regarding the installed nuclear capacities in different geographical areas as of September of this year, and the situation with regard to reactors under construction, is summarized in Table 4-2.

In spite of the nuclear situation in the United States referred to earlier, the installed capacity of the nuclear power plants there will almost treble by 1990 when those now under construction are completed. The only other industrialized countries which have large rates of increase are France, Japan, and the countries with centrally planned economies.

Table 4-2. **Status of Nuclear Power Reactors as of September 1981[a]**

	Reactors in operation		Reactors under construction	
	No. of units	GWe	No. of units	GWe
OECD North America	84	60	97	101
OECD Europe	103	52	65	61
OECD Pacific	24	15	11	9
CPE Europe	45	18	33	24
Asia	9	4	14	9
Latin America	1	0.3	8	6
Africa and Middle East	--	--	2	2
World total	266	149	230	212

[a]In the left column of this table, the term "OECD" refers to the twenty-four countries that are members of OECD, including Japan; OECD Pacific is composed of Australia, Japan, and New Zealand; and CPE Europe stands for European countries with centrally planned economies.

On the other hand, the developing countries have only limited plans, and we foresee no great increase in the number of developing countries committed to nuclear power in the 1980s beyond the nine countries already having nuclear power plants in operation or under construction. Some four to five additional countries, such as Egypt, are now considering nuclear power, but not all are likely to make commitments.

Table 4-3 shows that at the end of 1980, nuclear capacity in the world amounted to 136 GWe, or 7 percent of the nearly 2,000 GWe installed total capacity including all types of electric power generation. Industrialized OECD and centrally planned European countries have almost 98 percent of the nuclear capacity in the world, whereas developing countries have only around 2 percent.

Table 4-3 also shows that the total installed nuclear capacity in the world is estimated to be approximately 458 GWe by the year 1990, which would represent 13 percent of the world's estimated total electric generating capacity at that time. The nuclear share of electricity generation is slightly higher than its share of total electric capacity, since nuclear power plants are normally used for base-load generation.

The OECD and centrally planned European countries will continue to be the countries with the largest share of nuclear generation in the next ten years. Table 4-3 also shows that Asia (besides Japan) and Latin America will start to generate electricity by nuclear power on a more significant scale at the end of this decade.

Let us look for a moment at the 256 power reactors in operation at the beginning of 1981. Figure 4–1 depicts the age distribution of operational reactors older than eight years. As shown, a

Table 4-3. **Estimates of Total and Nuclear Electrical Generating Capacity by Main Country Groups**[a]

Group of countries	1980			1985			1990		
	Total	Nuclear	%	Total	Nuclear	%	Total	Nuclear	%
OECD North America	710	57	8	890	130	15	1,065	150	14
OECD Europe	440	45	10	580	105	18	735	150	20
OECD Pacific	180	15	8	255	25	10	340	50	15
Centrally planned Europe	370	16	4	545	35	6	745	75	10
Total for industrialized countries	1,700	133	8	2,270	295	13	2,885	425	15
Asia	130	3	2	235	10	4	400	20	5
Latin America	100	0.3	0.3	130	3	2	180	10	6
Africa and Middle East	65	--	--	80	2	3	120	3	3
Total for developing countries	295	3	1	445	15	3	700	33	5
World total	1,995	136	7	2,715	310	11	3,585	458	13

[a]In gigawatts-electric except where percentage is indicated.

total of sixty-five reactors have been in operation for more than ten years, including six which have been in operation for more than twenty years and thirty-two for periods between eight and ten years. One hundred fifty-nine out of the 256 reactors operating in the world are less than eight years old. Altogether around 2,400 reactor-years of experience have now been accumulated, and the technology of nuclear power has reached a state of maturity, safety, and reliability. It is appropriate to recall at this stage that there has not been a single fatal accident in a civilian nuclear power plant by exposure to radiation or radioactivity.

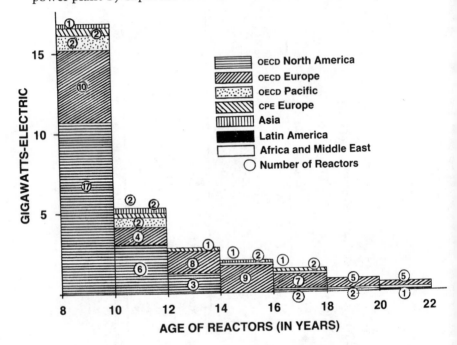

Figure 4-1. **Age Distribution of Reactors More than Eight Years Old**

How have these reactors performed? Figure 4–2 summarizes the load and operating factors between 1975 and 1979. The load factor is a performance measure in that it is the energy actually produced divided by the energy which could have been produced with operation at maximum power the whole of the time concerned. The operating factor is a measure of availability—in other words, time in operation divided by the total time.

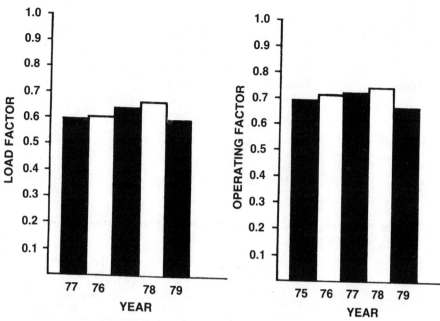

^aApplies to all plants excluding prototypes and those starting operation in the second half of the year.

Figure 4–2. **Average Load and Operating Factors for the Years 1975–79**[a]

It is interesting to note that following 1975 there was a slow increase in both factors until 1979, when both dropped significantly due to regulatory actions after the Three Mile Island accident. Since the 1978 and 1979 data are based on 156 and 176 reactor units respectively, there is no doubt about the significance of the drop.

In this context it should be mentioned that data from the last World Energy Conference indicate that the "unavailability" of nuclear units in 1979 has generally been similar to that of fossil fuel plants in a comparable size range—namely, about 30–35 percent, in other words, corresponding to the lower operating factor.

The technical improvements in proven reactor types during the 1980s will probably be only minor and based mainly on the experience gained during the last three decades nuclear power reactors have been operating and, as mentioned earlier, during which time 2,400 power-reactor-years of experience have been accumulated. It should be recognized that experience will increase rapidly during the 1980s. In the beginning of the decade 250 reactor-years are to be added each year; in the middle, about 450; and in 1990, some 600—in other words, the accumulated experience then will be some 6,000 reactor years.

PRESENT TRENDS

After this report on the present status of nuclear power it is natural to ask about the trend in new orders. Table 4–4 gives an indication of the situation.

Table 4-4. **Orders and Postponements of Nuclear Plants during 1980**

	Orders and letters of intent sent during 1980		Cancellations and postponements during 1980	
	Number of reactors	*Power (GWe)*	*Number of reactors*	*Power (GWe)*
OECD North America	--	--	12	13
OECD Europe	12	12	--	--
OECD Pacific	4	4	--	--
Centrally planned European economies	1	0.6	--	--
Asia	2	2	--	--
Latin America	--	--	--	--
Total	19	18.6	12	13

The year 1980 was again not promising in terms of new orders. Only nineteen reactors with a total capacity of 18.6 GWe were ordered in France, the Federal Republic of Germany, Japan, the Republic of Korea, Romania, and the United Kingdom. However, twelve orders for reactors—with a total capacity of 13 GWe—were either canceled or postponed in the United States. Thus, the total net capacity increase in 1980 was only 5.6 GWe.

Comparing the general nuclear situation in 1980 with the period up to 1990 as shown in Table 4–3, it seems at first view that we have now reached the lowest point. However, appearances are deceptive—a point I would like to underline with the next two figures. Figure 4–3 depicts the amount of nuclear capacity to be added annually during the period 1981–90 based on reactors under construction or fully committed for construction. In 1981, about 43 GWe will be added to nuclear capacity; for the period 1982–86,

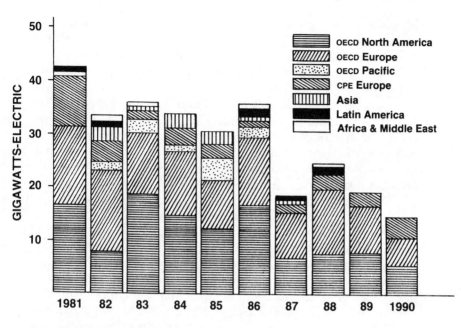

Figure 4–3. **Starting Dates of Reactor Operation**

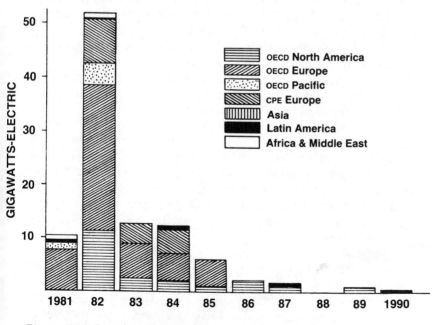

Figure 4–4. **Starting Dates of Reactor Construction**

between 30 and 35 GWe will be added annually. Beyond 1987, however, the nuclear capacity addition per year will be in the range of only 15 to 25 GWe. The picture becomes still darker when we consider plans for starting reactor construction during the same period (see Figure 4–4).

Construction of slightly more than 10-GWe capacity will begin in 1981, and about 52-GWe capacity will be begun in 1982. After 1983, construction of nuclear power plants already committed drops drastically to about 12-GWe capacity in 1983 and 1984 and to figures below 5-GWe capacity after 1985. This last figure, however, does not include the centrally planned economies.

One of the reasons for this slowing down of nuclear power construction is the long lead times from commitment to commercial operation for plants now under construction. The average lead times are as follows: for Japan (eight plants under construction), sixty-one months; for France (twenty-nine plants), sixty-three months; for the Federal Republic of Germany (seven plants), eighty-two months; and for the United States (eighty-one plants), one hundred twenty-one months. The difference may be almost exclusively due to the more or less complicated regulatory procedures for construction permits, operating licenses, and so forth. If new individual plants are not committed now, and if we take into account the long lead times I have just mentioned, a general slowdown of nuclear power programs beyond the year 1990 is likely, with serious consequences for the nuclear industry in many countries. In other countries—for example, the countries with centrally planned economies—the nuclear industry is flourishing, and in countries such as Argentina, Brazil, and Romania, it is being carefully nourished for a rapid future growth.

It is of interest to refer here to statements made at the last annual General Conference of the IAEA in Vienna five weeks ago. To look first at the brighter side, we have just noted that the countries with centrally planned economies have embarked on a very extensive nuclear power program, at first based upon the 440 MWe light-water reactor of Soviet design; later on its enlarged offspring, the 1,000-MWe reactor; and in the case of Romania, on the Candu-type reactor.

The Soviet Union has announced that its nuclear capacity in 1980 of 13,000 MWe will be increased by 1985 to a total of 38,000 MWe.

France is another example of a country which wants to decrease its dependence on imported fuel by a vigorous nuclear power program. The goal of the previous administration of adding 5,000-mwe installed capacity every year may be somewhat curtailed through the parliamentary debate which took place at the beginning of October. A number of European countries will, by 1985, be producing more than 35 percent of their electricity by nuclear means; France, about 50 percent. The Federal Republic of Germany, with a nuclear industry second to none, seems at present to suffer from an efficient working opposition against nuclear power and associated activities. The Gorleben reprocessing plant, which has been at a planning stage for a long time, will evidently not be licensed for construction, and even the plans for intermediate spent-fuel storage facilities have not been accepted, in spite of the fact that the Biblis reactors urgently need additional storage capacities for spent fuel. The Kalkar fast-reactor project under construction since 1973 in collaboration with Belgium and the Netherlands is finding it hard to fill a gap of $400 million in the project, the total cost of which is estimated to be $2,300 million. The high-temperature reactor, long a fixture of the German program, has similar problems. A second "enquête commission," composed of parliamentarians and technical specialists and comprising both pronuclear and antinuclear representatives, has been appointed to define the future role of nuclear energy in the Federal Republic of Germany.

Let me as a Swede say a few words about the Swedish nuclear program after the referendum in March 1980, when the electorate with a solid majority decided that nuclear power would ultimately be replaced by other energy sources in Sweden. It was further decided by the referendum that no more than twelve nuclear power reactors should ever be built in Sweden. In practice, nuclear power would then be phased out around 2010. At present, nine reactors are in operation representing a generating capacity of 6,400 mw, and the remaining three are scheduled to be connected to the grid during 1982–85, when in total 9,400 mw of nuclear power will be operational in Sweden and approximately 40 percent of the electric power will be generated by nuclear means. The largest opposition party, the socialists, now advocates the use of coal as a replacement for nuclear energy and has suggested the establishment of a parliamentary commission to consider the transition from nuclear to other energy sources.

In Sweden more than 40 percent of the energy consumed is used for the simple purpose of heating buildings. In the last few years, Finnish and Swedish experts have developed a concept

REQUIREMENTS

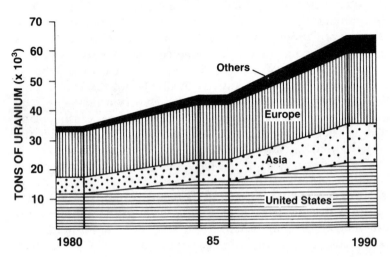

PRODUCTION

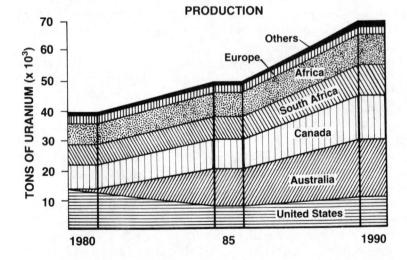

Figure 4–5. **Uranium Requirements and Production**

SECURE (Safe and Environmental Clean Urban REactor) consisting of a light-water-moderated enriched uranium reactor for district heating purposes. The safety features of the reactor, which may be located above or below ground level, are based on simple physical principles. In the case of overheating or failure of the circulation pumps, boronated water would enter into the reactor core and quench the chain reaction. Because of the limit on the number of reactors to be built in Sweden, the concept cannot legally be implemented, even if it was accepted by the public.

It was therefore of great interest to learn from the report given at the IAEA's General Conference this year by the representative of the Soviet Union that reactors for district heating are under construction in two cities in that country, namely, Gorki and Voronezh. In both cases each block consists of two reactors, each with a thermal output of 500 MW. If these pilot plants turn out to be successful, a new market has been opened for nuclear energy which may contribute to a revival of the nuclear industry.

URANIUM REQUIREMENTS AND PRODUCTION

After this outline of the nuclear power reactor situation during the 1980s, it is natural to turn to the provision of fuel during the decade and to other aspects of the fuel cycle. The challenge of the 1980s with regard to natural uranium is to reconcile a much-reduced requirement for uranium for existing and planned nuclear reactors with the present and likely future overcapacity in the uranium mining industry.

Present industry-based forecasts of uranium requirements and production, as shown in Figure 4-5, are therefore quite pessimistic, and they reflect the industry's concern about uncertainties in the rate of future additions to installed nuclear generating capacity.

Since it reached its peak in 1978 the market for uranium has more or less continuously declined. In August 1981, prices for uranium sank to US $55 per kg (1981 $) on the spot market—less than half its 1978 value of US $112 per kg (1978 $). Because of the general perception that additional uranium will be readily available from new production and from stockpiles, there is little hope for a turnaround. This trend will cause fairly drastic changes in the geographical distribution of uranium production, as shown in Figure 4–6.

During the 1980s, uranium production should grow considerably in Australia and Canada where large new mines are under development, while production from the United States and Africa

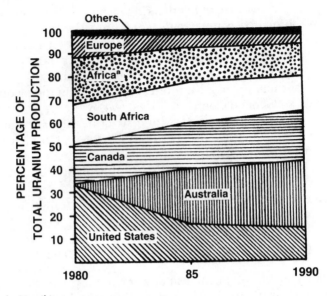

ᵃIncludes Namibia.

Figure 4–6. **Relative Participation of Various Regions in Uranium Production**

should remain static and decrease in relative importance. This also implies that developing countries will have little chance to attract capital for new uranium ventures and will be particularly affected by the economic impact of low uranium prices.

One additional point should be made with regard to the availability of assured supplies of uranium now and in the future. Table 4–5 shows an estimate of maximum, technically attainable production capabilities from the known resource base for the years 1980, 1985, and 1990. These figures are considerably higher than those for estimated uranium requirements and actual production, and they indicate the large reserve capacity built into the existing industry. But until the dismal conditions of the present uranium market are reversed, there must be considerable concern about the uranium industry's willingness to develop additional production centers for the years beyond 1990.

URANIUM ENRICHMENT

Most of the reactors in operation or planned for the decade require

Table 4-5. **Estimated Maximum Uranium Production Capability**

	1980 Number of countries	1980 Capability ktu/yr[a]	1985 Number of countries	1985 Capability ktu/yr	1990 Number of countries	1990 Capability ktu/yr
OECD North America	2	30	2	30	2	42
OECD Europe	3	4	3	5	5	7
OECD Pacific	2	2	2	14	2	21
Africa	4	14	5	18	6	23
Latin America	1	<1	3	3	4	5
Asia	2	<1	2	<1	3	2
Total	14	50	17	70	22	100

[a]Kilotons of uranium per year.

Table 4-6. **Capacities of Isotopic Enrichment Plants[a]**

	1980 Number of countries	1980 Number of plants	swu x 10³	1985 Number of countries	1985 Number of plants	swu x 10³
OECD North America	1	3	21,000	2	4-6	35,300-44,300
OECD Europe	4	7	3,800	4	9	12,800
OECD Pacific	2	1	30	2	3-4	30
Centrally planned Europe	1	1	7,100	1	1	7,100
Asia	--	--	--	--	--	--
Latin America	--	--	--	--		180
Africa and Middle East	1	1	6	1	1	200-300
World total	9	13	32,016	11	19-22	55,960-65,060

[a]A 1-GWe light-water reactor requires a capability of approximately 110 thousand separative work units (swu) per year.

enriched uranium. Table 4–6 represents the capacities of isotopic enrichment plants in 1980 and 1985. Taking into account that the estimated total nuclear capacity of around 450 GWe in 1990 requires a capacity of approximately 50 million separative work units (SWU) per year, which will be available in 1985, overproduction could be expected in the near future if all newly committed facilities were to be built.

No problems are expected in regard to the capacity of fuel fabrication plants (see Table 4–7), as the available capacity of around 9,500 tons of uranium per year in 1985 is in accordance with the requirements of the estimated 310-GWe capacity in the same year.

SPENT-FUEL STORAGE AND REPROCESSING

The forecast of spent-fuel storage capacity shows that during the 1980s no major problems are foreseeable on a worldwide and regional basis. However, it must be stressed that an overall comparison of spent fuel arisings and available storage capacity does not reflect the real situation, because the spent fuel cannot be freely distributed among the available storage locations. Therefore, some individual states and utilities will have inadequate storage capacities, and some alternative storage techniques will have to be used—transshipments to other pools, cask storage, double stacking of spent fuel, and so forth.

The major storage problems are, however, likely to occur in the following decade—1990–2000. Figure 4–7 summarizes the data available to IAEA from its International Nuclear Fuel Cycle Evaluation (INFCE) and the International Spent Fuel Management (ISFM) studies. The 1990 data suggest that the problems might be resolved on a regional basis, whereas the data for the year 2000 indicate that major alternatives for storage must be explored. Due to the lack of new reactors, the at-reactor storage capacity ceases to grow while the arisings continue to do so. This implies that the additional needs for spent-fuel management will have to be met by away-from-reactor storage as well as by reprocessing or by final disposal of spent fuel. The studies show, moreover, that even if the projected reprocessing capacities are in fact achieved on schedule, there will be a significant amount of fuel to be stored or disposed of.

For reasons of completeness only, I would like to make a few remarks on reprocessing. As shown in Table 4–8, the available

Table 4-7. **Capacities of Fuel Fabrication Plants**[a]

	1980 Number of countries	1980 Number of plants	Tons of U/yr	1985 Number of countries	1985 Number of plants	Tons of U/yr
OECD North America	1	6	2,900	1	7	3,300-3,700
OECD Europe	6	13	3,510	7	14	4,860
OECD Pacific	1	4	990	1	4	1,050
Centrally planned Europe			Figures not available			
Asia	1	1	21	1	1	21
Latin America	--	--	--	--	1	180
Africa and Middle East	--	--	--	--	--	--
World total	9	24	7,421	10	26	9,231-9,631

[a]Applicable to light-water reactors only. Approximately 25 tons of uranium are loaded annually in a 1-GWe light-water reactor.

Table 4-8. **Capacities of Reprocessing Plants**[a]

	1980 Number of countries	1980 Number of plants	Tons of U/yr	1985 Number of countries	1985 Number of plants	Tons of U/yr
OECD North America	--	--	--	1	3	2,550
OECD Europe	4	5	840	5	7	2,115
OECD Pacific	1	1	210	1	1	210
Centrally planned Europe			Figures not available			
Asia	1	1	100	1	2	200
Latin America	--	--	--	--	--	--
Africa and Middle East	--	--	--	--	--	--
World total	6	7	1,150	8	13	5,075

[a]Applicable to LWR fuel only. Approximately 25 tons of uranium are loaded annually in a 1-GWe light-water reactor.

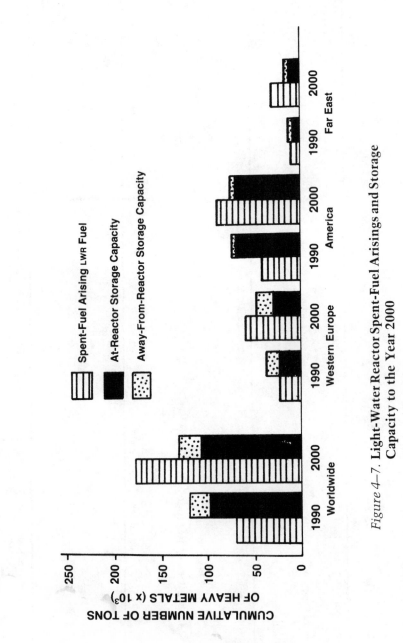

Figure 4–7. Light-Water Reactor Spent-Fuel Arisings and Storage Capacity to the Year 2000

capacity in 1980 of 1,150 tons of uranium per year represents only about 20 percent of the capacity needed to reprocess all irradiated fuel; in 1985 the theoretically available capacity of 5,075 tons of uranium per year would be enough to reprocess around 50 percent of the spent-fuel arisings. Everybody will agree that urgent decisions at the political level are needed to demonstrate first of all the technical feasibility of large-scale industrial reprocessing. This is essential in order to provide the basis for long-term commitments for international institutional arrangements and to restore confidence in this part of the nuclear fuel cycle, which is so important for the introduction of fast breeder reactors.

In this context I would like to refer again to the safe management and disposal of radioactive waste and spent fuel. Many countries have extensive programs to explore the suitability of repository sites located in geological formations of their territories and to establish national systems for the long-term management of radioactive wastes. Some countries set up special national organizations to deal with these issues. In addition, much progress has been made on the various conditioning and packaging techniques that are necessary prior to storage and disposal of all types of radioactive wastes.

During the 1980s a number of countries must define and implement appropriate waste management systems for their national nuclear power programs. This will include the management of low- and intermediate-level waste, the interim storage of high-level waste as well as the definition of sites, and, possibly, the construction of repositories for high-level and alpha-bearing waste. The method of solidifying high-level wastes will be used on an industrial scale rather than on the pilot scale previously used.

As is now the case, the storage and disposal of radioactive wastes is a matter for national control. However, regional and even international solutions, including the acceptance of wastes from other countries in national repositories, may be agreed upon for the storage and disposal of high-level wastes. Such solutions are, for obvious reasons, of special interest to smaller countries. One related prerequisite may be the basic safety requirements to be applied.

ECONOMIC ASPECTS

This review of the nuclear power situation in the 1980s would not be complete without mentioning the economic aspects of energy supply.

The dramatic rise in energy costs, driven by oil price increases since 1973–74, has played a significant role in worldwide inflation and consequent economic recession and unemployment. All countries—both industrialized and developing—are faced with the problem of finding reliable energy supplies at acceptable costs. The major issue is not a choice between mutually exclusive options, as some would have it, but of optimization of all available options by taking into account technical, economic, and environmental aspects. Without question, nuclear power is technologically and commercially ripe for an immediate and expanding contribution. It is therefore not surprising that most industrialized and several developing countries have viewed nuclear power as a promising option to meet part of their future electricity demand.

Fulfilling the economic promise of nuclear power will, however, require concerted action to overcome the public acceptance problems which have severely retarded its growth. In many countries, this has resulted in a reluctance to commit new nuclear power plants, even though the economics generally favor nuclear power over oil and coal for electricity generation. For example, nuclear-generated electricity costs only about half as much as electricity from oil-fired power stations.

In comparing costs of electricity from nuclear and coal-fired power plants, the result depends on a number of factors, and there is no single global answer. However, results from current IAEA studies indicate that in most situations large nuclear power plants can, depending on the cost of coal, produce electricity at costs as much as 12–40 percent below the costs of electricity from coal-fired power plants. In some special situations, however, such as in areas of the United States where low-cost coal is available to plants at nearby or mine-mouth locations, coal-fired power plants can deliver electricity at costs competitive with or lower than nuclear power. The key economic factor for coal-generated electricity is the cost of coal delivered to the power station. The IAEA studies indicate that coal plants have an economic advantage when coal can be delivered to the power plant at costs below $30 per ton. For nuclear power, the key factor is the plant investment cost, which is significantly increased when interest rates are high and lead times long. One of the challenges for nuclear power in the 1980s will be to achieve shorter licensing and construction times, in order to reduce investment costs.

As we know, nuclear power plants are much less affected by the costs of fuel resources than are fossil-fueled power plants. Doubling the price of uranium would increase the cost of nuclear-generated electricity by only 10 percent; doubling fossil fuel prices would lead to a 35–75 percent increase in the costs of electricity from fossil-fired power plants. (The lower value refers to coal and the higher to oil.) Thus, those countries with large commitments to nuclear power are less affected by fuel price increases.

To what extent will developing countries make use of nuclear power during the decade? It will increase elevenfold but from a small base and in half a dozen countries only. The reason is the consistent trend by nuclear power plant designers and manufacturers, because of economies of scale, towards units with a generating capacity of 1,000–1,300 MW. Units of such size require the existence of a prepared infrastructure in the receiving country—namely, an electric grid of sufficient capacity as well as the manpower and facilities necessary to cope with routine maintenance and emergency situations.

An old rule of thumb says that no generating unit should be larger than 10 percent of the total generating capacity in an electric grid. The very large stations referred to can only be incorporated in systems with capacities of at least 5,000 to 7,000 MW, which in turn means that these large reactors can only be introduced in a very few developing countries. We know that some reactor manufacturers are now studying the possibility of constructing smaller nuclear power reactors—between 200 and 400 MW—where what is lost in economy of scale is compensated by design simplification while still maintaining the same degree of safety as that of the larger units.

It is certain, however, that developing these small units, marketing them, and getting them licensed will take considerable time. In the meantime, one can only hope that the switch to nuclear energy by developed countries will ease the pressure on the crude oil market and make it possible for developing countries to expand their conventional electric systems to the size and infrastructural maturity required for the incorporation of nuclear reactors.

SAFETY QUESTIONS

Important developments in nuclear safety will take place in areas of regulation, operational safety, and safety systems design.

The primary challenge during the 1980s will be to set regulatory priorities among the outstanding safety issues in such a way that significant disturbances in old and new nuclear plants are minimized. In this regard, we must have clear regulations so that the level of understanding between regulators and applicants can be improved. Those who design, build, and operate a nuclear power plant must have a reasonable confidence that the rules will not be changed in midstream. If a change is needed it should result only from a thorough analysis with consideration given to the real cost effectiveness of any modification. It is appropriate to note the importance of the international harmonization of nuclear standards —in particular, harmonization of the level of basic criteria and approaches —and the contribution made by the Nuclear Safety Standards (NUSS) program of the IAEA. This ambitious effort to prepare internationally agreed safety standards and guides is nearing completion.

Operational safety has been improved by developments in two major areas in which efforts will have to continue—namely, evaluation of operating experience and consideration of the human factor. It is becoming difficult to identify the few significant items out of an increasing flood of event reports from national and international exchanges. It is essential that information on accidents and incidents be analyzed by highly experienced specialists and that the results reach the persons who need them. The IAEA is now taking steps to play a more vigorous role in the collection, analysis, and dissemination of information on operating experience. The broad membership of the IAEA and its access to information from East and West, developing and developed countries, provide it with unique resources in this field.

The human element has, after Three Mile Island, been recognized as a factor which influences safety and must be taken into account in the design, operation, maintenance, and management of plants. The IAEA should, in full cooperation with member states and with full consideration of different conditions prevailing in different countries, attempt to establish competence criteria for operating and maintenance staff.

In the design area, much effort has been spent on hypothetical core-melt accidents. Three Mile Island has shown the need to give appropriate emphasis to the more reasonable type of accident. In safety system design, we must always keep in mind that adding more and more devices in an effort to improve safety may make the plant so complicated that the probability of failure will increase.

It is appropriate to point out that there are bound to be failures in power reactor systems, as in every other complex technology. But it must also be recalled once again that up to now there has not been a single fatal accident caused by radiation in a nuclear power plant built for peaceful purposes, nor an emission of radiation harmful to the public. The many built-in barriers to prevent release of dangerous amounts of radioactivity to the biosphere have, up until now, fulfilled their purpose.

The increased use of nuclear power should also be followed by the acceptance of nuclear power in the minds of the public and news media as a natural part of our environment. A steam-valve leakage or turbine trip in a nuclear power station should have no more news value than similar happenings in a conventional power station. Words are misused these days. If, for example, the Three Mile Island incident can be called "a catastrophe," one would hope that any future accidents could be similar "catastrophes"!

But Three Mile Island was a catastrophe from an economic point of view. It will be essential for utilities in the future to see to it that they, by mutual arrangements, share the economic burden which an accident may impose on them.

The IAEA is organizing a Conference on Nuclear Power Experience in September 1982 in Vienna, an event which should give member states and their utilities a most valuable survey of the vast amount of experience already accumulated today over some 2,400 reactor-years of operation. With this survey, public understanding of the safe operation of power reactors should be substantially improved.

NONPROLIFERATION, SAFEGUARDS, AND TECHNOLOGY TRANSFER

The fact that nuclear reactors produce plutonium and that reprocessing and enrichment plants can be used to make weapons-grade material has, of course, profoundly influenced the role and development of nuclear power, and it is the reason for the international safeguards and nonproliferation regime (NPT) laboriously built up since 1958. The refusal of a few nonnuclear-weapon states to accept safeguards on their entire fuel cycles is seriously impeding their nuclear power programs. On the other hand, fears that IAEA safeguards and the NPT would not be adequate to stop the spread of nuclear weapons led in the late 1970s both to attempts to limit the transfer of nuclear technology and to highly restrictive export policies designed essentially to stop the development of the fast breeder reactor because of its use of plutonium fuel.

These restrictive export policies, which went far beyond the requirements of the NPT, succeeded in disrupting the structure of supply assurances vital to the health of nuclear power. Foreseeably, they had little effect in impeding the transfer of technology. They also led to serious strains with those countries, particularly in western Europe and the Far East, that regard the development of the fast breeder reactor as essential to diminish their dependence on external supplies of fossil fuels and uranium.

In short, the policies of the late 1970s further compounded the damage being done to nuclear power by the factors I have already mentioned.

A further severe blow was inflicted in June this year by the Israeli attack on the Baghdad Nuclear Center. Not only has this attack damaged the credibility of IAEA safeguards and the NPT, on which international commerce in nuclear plant and materials largely depend today; it has also ended the immunity from armed attack which nuclear plants had enjoyed since 1945, thereby opening the prospect that radiological warfare could be launched by using conventional weapons.

Further dangers lie ahead. Unsafeguarded facilities in one of the few countries I have referred to could provide a basis for further tests and a kind of race in developing and demonstrating nuclear explosive capacity in the region concerned. One hardly needs to stress the instability of the Middle East, which has again been tragically demonstrated, or of southern Africa.

There are also some other countries that have placed all their present facilities under IAEA safeguards but have not joined the NPT and are legally free to construct unsafeguarded plants in the future. In short, the grim prospect that one or more nonnuclear-weapon states outside the NPT might explode a nuclear device during this decade cannot be ruled out. The impact of this prospect on nuclear power development as well as its broader implication for international security in today's unstable world should be clearly understood. Wisdom, prudence, and interest dictate, therefore, that no further time should be lost nor effort spared in the formidable task of creating the political conditions that would encourage these countries to join the nonproliferation regime and, hopefully, make it universal.

One further point deserves mention. It is beyond the reach of most countries to use the light-water power reactor as a source of military plutonium. I profoundly hope that this remains true and that attempts to develop the technique of transforming civil reactor plutonium into a source for weapons will not be pursued. If they were, the effects on the IAEA system and on confidence in it could be far-reaching.

There is also, however, the equally grave risk that in focusing on the world's political trouble spots and the destabilizing effects that nuclear energy could have on them, we may overlook the impressive safeguards and nonproliferation achievements of the last three decades. In the 1950s many of us believed that the spread of nuclear weapons would be rapid and unavoidable. In fact, no country has openly joined the nuclear weapons club since 1964, despite the vast dissemination of nuclear technology that has taken place since then and despite the fact that today twenty to thirty countries would be technologically capable of doing so. Membership in the NPT now stands at 114 countries and continues to grow. Many of the most recent members—such as Indonesia, Bangladesh, Turkey, and, most recently Egypt—are in, or close to, areas of political tension, and their readiness to accept the constraints of the NPT is therefore particularly significant.

Every industrial country in the world except the nuclear-weapon states has now put all of its nuclear plants and materials under IAEA safeguards. The same is true of all the developing countries except the four I have mentioned. Some encouraging progress has been made in Latin America towards a truly nuclear-free zone, and all nuclear plants in that region of which the IAEA is aware are now under safeguards. In fact, throughout the world, more than 95 percent of all nuclear facilities and material outside the nuclear-weapon states are now under safeguards, and four of the nuclear-weapon states—the most recent being France—are placing part or all of their civilian plants under IAEA safeguards.

There have also been encouraging signs that we may be moving away from the restrictive export policies of the late 1970s. Australia and Canada have concluded new agreements with their main customers, and the United States has indicated that it generally will not stand in the way of reprocessing unless it poses a proliferation risk. Finally, I should mention that the IAEA last year established a Committee on Assurances of Supply with a view to rebuilding that structure of confidence in nuclear supplies which largely had been lost towards the end of the last decade. While the committee has a long and difficult task before it, it is already helping to build confidence by insuring that both supplies and the related question of safeguards are now discussed in an open international forum instead of behind closed doors.

I have spoken of the risks that lie ahead and the impressive gains we have been able to achieve. For the next decade or so it seems unlikely that any major new nonproliferation measures are in the offering. We shall have to rely on the NPT and the IAEA safeguards operation and seek to maintain and strengthen them. This task would be greatly helped if the nuclear-weapon states could make some visible progress in meeting their arms control commitments under the NPT—in particular, if they could agree on a comprehensive treaty banning all forms of nuclear explosion. Otherwise, I fear that we will be in for a very rough passage in 1985, at the next five-year conference to review the treaty, and in particular in 1995, when the NPT's twenty-five-year term runs out and vital decisions must be taken about its continuation.

The question of access to technology, especially the so-called sensitive technology, is difficult. Experience repeatedly shows that there is no way of permanently preventing other countries from mastering even the most sensitive fuel cycle processes, if they are determined to do so. This has been the lesson of nuclear energy since 1945. Knowledge once given away cannot be retracted. On the other hand, no one wishes to see an unnecessary proliferation of small enrichment and reprocessing facilities. Many solutions have been suggested, including a multinational fuel cycle center which unfortunately up to now has not received political support. The answer does not lie in a policy of denial. On the contrary, as we have recently seen, such denial is likely to achieve exactly the opposite of what it seeks and to spur the proliferation of scattered, small, national fuel cycles instead of one or more internationally interdependent, large fuel cycles. We must seek the solution at the regional as well as the international level, and perhaps the Far East, in which there are now a number of expanding nuclear power programs, could explore further the possibility of regional fuel cycle cooperation.

CONCLUSION

Based upon available documentation now in the IAEA in the form of national forecasts and plans, I have tried to give you a picture of the actual nuclear situation and to indicate what we may expect in this field in the decade which has just started.

I should like to conclude my address by sharing with you my personal appreciation of the situation. Thege caused by an interruption in oil supply or if energy-related costs drastically affected the whole economy of its people, a situation which nobody wants. It might become even more antinuclear if there were nuclear plant accidents, irrespective of whether these accidents affected the nuclear parts of the plants or not. Every effort must therefore be made to insure objective reporting by the public media so that unfair comparisons are not made between failures in nuclear plants and failures in other technical undertakings of the same complexity.

Furthermore, let us recall that thermal reactor systems only represent a temporary contribution to the energy provisions of the world, on a time scale comparable to the oil period. A long-term contribution presumes the development of fast systems, breeders, whereby nuclear energy could make a long-lasting contribution to the world's energy problems; in other words, like coal, but with much less environmental impact although involving other problems as well. Unfortunately, the dynamics of the technical development is not considered when the politicians plan for the future —a future which is seen as extending only through the next election.

Considering these different circumstances, I personally conclude that there will be a steady but slow growth of nuclear power's share of the world's electricity supply during the 1980s—a growth which ultimately will lead to the use, in advanced countries, of the commercial breeder at the end of the century.

CHAPTER FIVE

The Safety of Nuclear Power Reactors

H. G. MacPherson
Institute for Energy Analysis
Oak Ridge Associated Universities

THE NATURE OF THE PROBLEM

Although the nuclear era was introduced to the world by a devastating bomb explosion, the danger associated with nuclear power plants does not arise from any concern with explosive forces. Rather it comes solely from the possible spread of radioactivity. The core of a nuclear reactor contains fearsome amounts of radioactivity even after the reactor has ceased its chain reaction, and some of this radioactivity persists for hundreds of years. Maintaining reactor safety, then, means keeping the bulk of this radioactivity confined and away from people. To date the record is one of success, yet many are not satisfied. Why?

Compared with the safety of other industrial activities, there are major factors that, taken together, make nuclear safety something special. Some of these are as follows:

- Ionizing radiation cannot be seen, heard, or felt. It is perceived as a mysterious new threat that will strike without warning.
- There is the possibility of a catastrophic event involving the deaths of thousands of people in a single accident.
- Some of the effects of radiation are delayed in action, with cancers developing as much as several decades after the exposure and with genetic effects even longer lasting.
- Nuclear power carries the stigma of being related to the nuclear bomb, and, in fact, bombs of some degree of effectiveness could be made from the plutonium in spent fuel elements. Although this is not strictly a safety issue, it does affect how people think about nuclear power safety.

Although there are broader issues that engender some opposi-

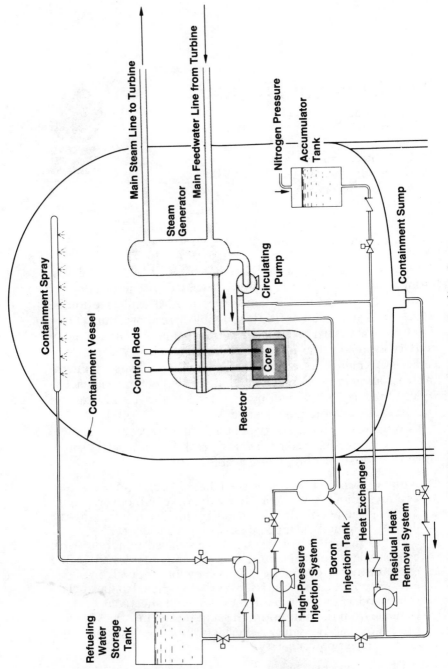

Figure 5-1. Pressurized Water Reactor Emergency Cooling Systems

tion to nuclear power, safety is certainly the issue uppermost in the minds of ordinary people. For there to be a nuclear future, the safety issue must be either resolved or accommodated.

TECHNICAL CONSIDERATIONS

The release of radioactivity from normally operating reactors has been reduced to such low levels that it is no longer a public safety issue. Thus, only accidental releases need be considered. Estimating the harm of such an accident to people involves three steps: determining the probability of the accident and its associated radioactivity release, calculating the spread of this activity among populated areas, and estimating the health effects produced by the resulting exposure to radiation. The combination of these three steps is now called Probabilistic Risk Analysis.

Probability of Accidental Release of Radioactivity

I will use the pressurized water reactor, a nearly universally used reactor type, to illustrate the points I wish to make. Figure 5–1 is a greatly oversimplified schematic drawing of such a reactor system showing some of the important features. In normal operation the fuel and its fission products are encased in fuel rods. These are long slender tubes of zirconium alloy filled with uranium dioxide pellets, some 40,000 of them in a big reactor. The reactor cooling water flows along the length of these rods, removing the heat produced in fission and transporting it to the steam generator. This cooling water is at high pressure and is contained within the primary pressure boundary, consisting of the reactor vessel, the steam generator, the pumps, and connecting piping. This primary system, along with control and other equipment, is situated within a large containment building. Thus, in normal operation, there are three physical barriers between the fission products and the free atmosphere: the fuel cladding, the reactor pressure vessel and the containment building.

Although great care is taken in the design, construction, and operation of these facilities, unusual events will occur and equipment failures can be expected. Each reactor has the built-in capability of handling many types of such events. For example, if a pipe in the primary system were to break, some of the primary cooling water would squirt out into the containment atmosphere and flow into the containment sump. There are several emergency systems available for supplying make-up water to the reactor so that the fuel rods would still remain cooled and intact. Should these systems fail, the core would become dry and overheat, releasing fission products which would come out of the broken pipe into the containment volume. Ordinarily the fission products with their radioactivity would remain there and the nongaseous fission products would be gradually purged from the containment atmosphere by a recirculating filter system and by a containment spray system. Nevertheless, the containment structure itself could fail or develop a leak, allowing radioactive fission products to escape to the atmosphere.

A significant accident, then, involves three successive steps: some unusual initiating event, one or more failures of engineered safety systems, and the failure of the containment vessel. By analysis of the details of each possible failure mode, probabilities can be assigned for each of these steps. If P_a is the probability of a certain initiating event occurring, P_s is the probability of the relevant safety system failing, and P_c is the probability of failure of the containment, and if these events are independent, then the probability of some release of activity from this sequence is the product P_a x P_s x P_c. To get the magnitude of the release of radioactivity from this postulated sequence, one must estimate in detail the amount of fission products released from the fuel rods, how much escapes into the containment vessel, how much is sequestered there, and how much is available for leakage.

The calculation of these probabilities and the associated fission product releases constitutes the first step of the elaborate process called Probabilistic Risk Analysis (PRA). The pioneering effort in applying this technique to reactors was the monumental Reactor Safety Study (WASH-1400)[1] headed by Professor Norman Rasmussen. That study required a three-year effort by a large group of talented people and was published in 1975 in a main report and eleven appendices. At first the report was criticized rather sharply, in part because it boldly announced that power reactors are much less dangerous than many other rare but well-known threats to human life.

Figure 5–2 is taken from this report and shows the probability of immediate death from one hundred reactors compared with earthquakes, tornados, and hurricanes. The use of this figure was justly criticized because it omits the probability of delayed cancer deaths. Overreacting to this criticism, the US Nuclear Regulatory Commission issued a regrettable statement on January 18, 1979, repudiating the findings and restricting future use of the study. However, the value of that study came to be recognized shortly thereafter, in part as a result of the March 28, 1979, accident at Three Mile Island, and similar assessments are now being carried out on a number of other reactors.

The studies to date have shown that the most important initiating events for accidents are moderate-sized leaks from the

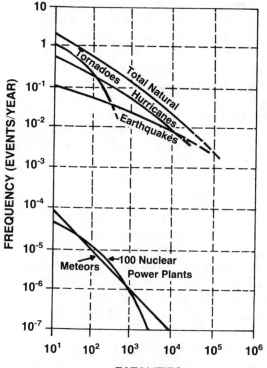

FATALITIES

Notes: 1. For natural and man-caused occurrences, the uncertainty in probability of largest recorded consequence magnitude is estimated to be represented by factors of 1/20 and 5. Smaller magnitudes have less uncertainty.
2. Approximate uncertainties for nuclear events are estimated to be represented by factors of 1/4 and 4 on consequence magnitudes and by factors of 1/5 and 5 on probabilities.

Source: Reactor Safety Study, WASH-1400 (NUREG-15/014), (US Nuclear Regulatory Commission, October 1975), Executive Summary, Figure 1-2, p. 2.

Figure 5–2. **Frequency of Fatalities Due to Natural Events**

primary system and certain transient events such as loss of offsite power. The studies also show that no appreciable escape of radioactivity will occur unless the fuel melts and the containment is breached. For the particular reactor picked for the Reactor Safety Study, the probability of a core melt turned out to be about 8×10^{-5} per reactor-year, and a significant leak from the containment occurred about 15 percent of the time.[2]

The Reactor Safety Study identified about one thousand different possible accident sequences for pressurized water reactors (PWRS) and boiling water reactors (BWRS), of which fifty-nine were determined to be the most probable. All of the sequences were classified into a few release categories. For each release category, an estimate was made of the amount of each radioactive element released and the time and altitude above ground at which it was released. This information, together with the sum of the probabilities of all accident sequences contributing to each release category, was used as input to the next step of the risk assessment.

Spread of Activity in Populated Areas

The movement of the radioactivity from the reactor accident scene is calculated using historically observed weather patterns at the site. This, combined with the known population distribution around the site, gives the probable number of people exposed to different levels of radiation. In cases of serious accidents, it is assumed that the people in heavily contaminated areas can be evacuated in accordance with experience with evacuations that have taken place frequently for other reasons, such as the wreckage of tank cars or trucks carrying hazardous materials. The art of calculation of the atmospheric dispersion of radioactivity from the scene of an accident is well developed and has been used for many years.

Health Effects of Radiation Exposure

Despite the great amount of publicity given to the dangerous effects of radiation, the evidence of the harmful effects of radiation on humans is very scarce, and is nonexistent at dose levels as low as 5 rad, the yearly allowed occupational dose. The effects of very high doses of hundreds to thousands of rad applied over a short period of time are very clear. The person suffers acute radiation sickness, and, if the dose is high enough, death is likely to occur within a few days or weeks. The Reactor Safety Study assumed that early death was unlikely for a whole body exposure of less than 320 rads but almost certain for an exposure of 750 rads, with a dose of 510 rads causing a 50-percent death rate. Although there is some difference of opinion on the exact numbers to use, these differences are small and the agreement among experts must be considered good.

The situation is quite different with regard to the cancer-causing effects of low levels of radiation. All of the known radiation-induced cancers have been caused by relatively high doses, in excess of 20 rad. In all, about five hundred cancer deaths can be attributed to radiation exposure: about one-half of them occurred at Hiroshima and Nagasaki, a few among radium dial painters, some among miners from breathing radon, and the remainder from intense therapeutic x-ray treatments. Although different organs of the body have different sensitivities, the overall effect is thought to be equivalent to about one extra cancer death among one hundred people when each is exposed to 100 rads of whole body radiation. When radiation is applied in this high dosage range, it is generally assumed that the number of extra cancer deaths will be proportional to the product of the number of people exposed and the radiation dose. This is the linear theory, and, if strictly applied, would lead to the expectation of one hundred extra cancer deaths for each million man-rem of radiation. (Some experts would use a factor higher than the 100 used here.) The controversy arises from the extrapolation of this relationship to doses less than those known to cause cancer—in other words, less than about 20 rad. Some believe that there is a threshold level below which cancers will not appear. Others point out that even one quantum of radiation can cause cell damage and dislocations in DNA chains, and that, in the absence of evidence, the safe assumption to make is to extrapolate the linear hypothesis down to zero dose. There is, however, evidence for repair mechanisms, both at the microscopic cellular level and with gross effects of high exposure. The latter arises from animal experiments, which show that if a dose of electromagnetic radiation is given in small increments the effect is much less than if given all at once. The present predominant view, as represented by the latest report of the Committee on Biological Effects of Ionizing Radiation (BEIR)[3] and by the panel assembled to provide input to the Reactor Safety Study, is to attribute fewer cancers to low-level radiation than would be predicted by the linear hypotheses.

This is an issue which will probably never be settled by experiment, because a radiation-induced cancer is no different from a naturally occurring cancer and because so many of us will die of cancer whether irradiated or not. At present, about 20 percent of the total deaths in the United States result from cancer, and, because cancer is a disease whose probability increases with age, about 25 percent of our increasingly aging population now living in the United States can be expected to die from cancer. Thus, if one were able carry out a giant experiment and give each of 1 million people a dose of 5 rem (the yearly occupationally allowed dose) and if the linear hypothesis held, then one would expect five hundred extra cancer deaths to occur. However, if the probability of natural cancer death were exactly 0.25 for each of the million people, the number actually observed to die of cancer could fall anywhere within a random distribution of numbers, centering on 250,000 but having a standard deviation of 433. Thus, if five hundred extra cancer deaths were to occur they would not be statistically observable.

In fact, the effect would be even more difficult to detect than this calculation would indicate, because no two large samples of the population would have a natural cancer incidence rate that agrees within 0.2 percent. The futility of ever resolving this issue suggests consideration of a new uncertainty principle; namely, if a postulated hazard is so small that it cannot be detected, then it may be ignored.

Mention should be made of certain epidemiological studies that purport to show cancer-causing effects of low-level radiation. The problem that these all encounter is the question of what to use as a control population. It is known that some tendency to be susceptible to cancer is an inherited trait, and thus cancer patients constitute a special population. Noncancerous persons cannot therefore be used as controls, if quantitative conclusions are to be drawn. All such epidemiological studies that have claimed to show cancerous effects of low levels of radiation have been challenged and shown to be inconclusive, at best.

My personal experience is with one of the best of such studies, the Oxford study of childhood cancer, which included a large fraction of all of the children dying of cancer in England and Wales over a period of years.[4] In this study there was a clear association between cancer and prenatal x-rays. Fifteen percent of the children who died of cancer had been subjected to x-rays while in their mother's womb, whereas only 10 percent of the noncancerous controls had been so subjected. To show that the association was causative, however, required that the a priori likelihood of being x-rayed for the children destined to die of cancer was the same as that for the control population. Unfortunately, the extensive published data showed that this could not be the case. In fact, in certain characteristics, such as the frequency of illness of the mother during pregnancy, the cancerous children showed themselves to belong to a population group about halfway between the x-rayed controls and the non-x-rayed controls. Thus there is no way of showing that the association of cancer with prenatal x-rays was a causal one.

The Reactor Safety Study panel of health consultants[5] agreed to use something less than the linear hypothesis extrapolation for low-level radiation but also chose to treat cancers of individual organs separately. It turned out that, due to inhalation of fission products, lung cancers predominated, and the net effect was to expect 104 cancer deaths per million man-rem, a very similar result to using the linear hypothesis and considering only whole body doses.

Results of the Safety Studies

Table 5–1 shows the amounts of human damage associated with their probabilities of occurrence per reactor-year reported by the Reactor Safety Study. Table 5–2 indicates the risk that an individual would be exposed to if he were one of the 15 million

Table 5–1. Single Reactor Probabilities and Consequences

Chance per reactor-year	Early fatalities	Early illnesses	Latent cancer fatalities
One in a million	1	300	5,100
One in 10 million	110	3,000	13,800
One in 100 million	900	14,000	25,800
One in a billion	3,300	45,000	45,000

Source: Reactor Safety Study, WASH-1400 (NUREG-75/014) (US Nuclear Regulatory Commission, October 1975).

Table 5–2. Yearly Risk for an Individual Living Near a Reactor[a]

Chance of	Risk per year
Early death	2 in 10 billion
Early illness	1 in 100 million
Latent cancer death	1 in 100 million

Source: Reactor Safety Study, WASH-1400 (NUREG-75/014) (US Nuclear Regulatory Commission, October 1975).
[a]Based on 15 million people living near one hundred reactors.

people living near to one of the one hundred reactors that the study assumed would be operating. Although these tables show single values, the uncertainties in both probabilities and consequences are great. The Reactor Safety Study indicates that the probabilities are thought to be within a factor of five and the consequences within a factor fo four. More recently, an uncertainty factor fo ten has veen associated with the probabilities.[6]

A study authorized by the German Federal Minister for Research and Technology and carried out by the Gesellschaft fur Reaktorsicherheit under the management of Professor Adolph Birkhofer[7] gave results that were consistent with those shown above, despite the fact that they used the linear hypothesis for the calculation of latent cancers and considered the whole population of Europe (670 million people) as the target population.

The question is frequently raised as to whether the Reactor Safety Study predicted the Three Mile Island accident with a reasonable probability. The simple answer is that it did not, for at least two reasons. The first reason is that a risk analysis made for one reactor is not necessarily valid for a different reactor. The Reactor Safety Study used as its model the Surry reactor manufactured by Westinghouse, whereas the Three Mile Island reactor was built by Babcock and Wilcox. They are different.

A sequence similar to that causing the trouble at Three Mile Island was identified for the Surry reactor. It involved the following steps: The initiating event (1) is the loss of the main feedwater pump, estimated to happen three times per year. (2) To give trouble, the auxiliary feedwater system must also be inoperative for a long time, and the probability of this was placed at 4×10^{-5} per demand. If these two events happened, then the primary reactor coolant system would overheat, increasing its pressure, and the relief valve on the pressurizer would open. Normally it would close automatically when the pressure decreased, and the probability of its sticking open (3) was placed at 10^{-2}. Even then there would be no severe consequences providing the pump injecting make-up water operated. The probability of its not operating (4) was placed at 2×10^{-3} per demand. Thus, the probability of this sequence occurring in the Surry reactor was taken as the product of the probabilities of the independent failures, or 2×10^{-9} per year, an extremely unlikely event.

However, the Babcock and Wilcox reactor at Three Mile Island has a once-through steam generator, so that the loss of the main feedwater pump was immediately felt in the primary system as an increase in temperature, producing a pressure spike that directly opened the safety relief valve.[8] Thus, the availability of the auxiliary feedwater system was irrelevant and step (2) of the Surry reactor sequence was bypassed, yielding an overall probability of about 6×10^5 per year, much higher than Surry, but still rather low. Had a risk analysis been carried out for the Babcock and Wilcox reactor before the accident, this is the number that might have been reported. There was, however a hidden factor. This is that the operators had been instructed in such a way that caused them to turn off the water injection pump that would have made up for lost inventory. Thus, there was a human error connection that made step (4) of the Surry sequence highly probable, it is important to note that human intervention also terminated the accident and made the consequences much less than would have been expected.

I should add that Babcock and Wilcox reactors have been changed, both in hardware and in operator instructions, to make the likelihood of this accident negligible.

Future Use of Probabilistic Risk Analysis

Each probabilistic risk analysis carried out to date has revealed some design peculiarity of that specific reactor that contributed significantly to the probability of an accident, and which could be changed for the better without too much difficulty. Thus, it is clear that application of the PRA technique will lead to safer reactors. Furthermore, this technique is now being used as a guide to profitable areas for safety research. Ultimately, it is intended that it also be used as a guide to test the validity of rules for reactor regulation.

INSTITUTIONAL CONSTRAINTS

There is universal agreement that the primary responsibility for the safe operation of nuclear power plants lies with the owners of the plants. In the United States this means the electric utility companies. Due to the fractured nature of the utility industry in which reactors are owned by sixty or so separate companies, few of the owners had the technical strength to provide the assurance of safe operation. Thus, great dependence had to be placed on the

Nuclear Regulatory Commission and on the reactor vendors. A welcome change now seems to be in progress.

The accident at Three Mile Island in March 1979 brought an immediate and dramatic response from the leadership of the United States utility industry.[9] The accident demonstrated that the threat of core damage was no longer a hypothetical event that could be ignored, and it also showed the extent of economic disaster that accompanies a core-damaging accident. The Edison Electric Institute immediately established an Oversight Committee chaired by Floyd Lewis of Middle South Utilities. This committee prevailed upon the Electric Power Research Institute to establish the Nuclear Safety Analysis Center (NSAC), and later, in conjunction with the Atomic Industrial Forum, it also established the Institute of Nuclear Power Operations (INPO).

The initial tasks of the Analysis Center were (1) to carry out an in-depth analysis of all that went wrong and right at Three Mile Island, (2) to screen routinely reported reactor malfunctions for important or repetitive events that might affect reactor safety, (3) to analyze the important events, and (4) to devise a way of communicating the results directly to the operating staffs of nuclear power stations. Under the leadership of E. L. Zebroski these tasks were carried out in a timely fashion. The continuing work of analysis of important malfunctions and communication with the utility staffs has now been transferred to INPO, while NSAC continues with its technical functions related to reactor safety. These include conducting a Probabilistic Risk Analysis on a specific reactor plant and using the occasion to train utility personnel in the technique. The Analysis Center has also initiated an in-depth study of degraded core phenomenology, which is directly managed by a Knoxville concern, the Technology for Energy Corporation.

Recognizing that the Three Mile Island accident was preceded by inadequate management practices and was turned into a disaster by wrong operator decisions, INPO was set up by the Oversight Committee with Admiral Wilkinson as president. According to John Selby, "Its purposes are to establish industry-wide benchmarks for excellence in nuclear operation and to conduct independent evaluations to assist utilities in meeting the benchmarks. It will determine educational and training requirements for operating personnel and will accredit training organizations."[10]

The evaluations by INPO amount to very thorough inspections of nuclear plants and their operating procedures, thus applying peer pressure for achieving improvements in practices. With regard to training, Admiral Wilkinson has said, "The most serious problem facing the industry is the lack of a trained working force in adequate numbers . . . INPO is committed to the safe operation of this nation's nuclear power plants. Not one of them can operate or should be allowed to operate without an adequate number of highly qualified, competent people."[11] With many new reactors due to come on line within the next few years, the task of locating and training suitable plant personnel will require great effort.

These measures and others initiated by forward-looking utility executives will certainly improve the outlook for the safe operation of our nuclear plants. The change is welcome, since it marks the first time since the very early days of reactor development that the utilities have taken the lead in examining the technical and operational aspects of reactor safety.

SOCIETAL CONSTRAINTS—HOW SAFE IS SAFE ENOUGH?

Regulatory authorities such as the US Nuclear Regulatory Commission have the responsibility of protecting the public from inadvertent mistakes in the construction and operation of reactors. They do this by reviewing the plant construction and operation to assure that all steps are taken that are needed to provide *reasonable assurance* that the health and safety of the public are not endangered. This *reasonable assurance* provision involves qualitative judgments on the part of the regulating authorities. This qualitative nature has been necessary until now because information has not been available to provide quantitative estimates of the degree of safety or nonsafety. The present system has worked rather well as judged by the safety record to date, although it has been a source of friction between utilities trying to build reactors and the regulatory authorities who continually demand changes in the name of improving safety.

With the increasing sophistication and use of Probabilistic Risk Analysis, it is now becoming possible to make quantitative estimates of the risk to the public, and a consensus is developing for quantitative safety goals. These are desired by the industry as a way of putting a limit on the number and cost of proposed new safety features and will be welcomed by the regulating agencies so that they can work to some standard.

There are several views from which one approaches the question: "How safe is safe enough?" The national regulator presumably should be concerned primarily with the total number of people in the whole country that might be killed or injured each year by reactor malfunctions. In contrast, people living near a nuclear plant want to know what the danger is to them. There is also another view, perhaps representative of the news media, giving special emphasis to the remote possibility of very large accidents involving deaths of thousands of people at a time. The safety goals being considered take all of these views into consideration.

For the risk to the general population, there is a tendency to set a limit for reactor accident deaths that is significantly less than the total number of deaths associated with a coal-fired plant delivering the same energy. Current proposals provide limits that range from less than 0.1 death per large-reactor-year to as much as 2 deaths per reactor-year.[12] The number of deaths would be estimated from Probabilistic Risk Assessments and would appear as sums of products of probabilities of each accident type and the number of deaths involved. The desire to avoid large accidents can be taken into consideration by giving added weight to accidents involving many deaths. One way of doing this is to raise the number of deaths to some exponent before carrying out the multiplication and summation. In this way, for example, an accident in which 1,000 people died would be considered 4,000 times as bad as one involving a single death if the exponent chosen were 1.2.

Proposals for limiting the risk to any individual living near the plant tend to be related to the normal death rate, the accident death rate, or the risk of dying of cancer. Values suggested for the allowable probability of death per year are in the range of 10^{-5} to 10^{-6} per year. These figures compare with the average mortality risk of about 10^{-2} per year, the average cancer death rate of about 2×10^{-3} per year, or the accidental death rate of 5×10^{-4} per year, all in the United States.

For both the population and individual risk proposals, there is some disagreement as to how much weight should be given to a delayed death, such as by cancer occurring several decades after exposure. There seems to be a consensus that a delayed death should count less than an immediate death, by as much as a factor of five or even more. This seems to be to be a debatable point.

Since the calculation of risk involves a considerable uncertainty, an additional constraint has been suggested, relating to the frequency of severe core damage. The logic for this is that there is no chance of large releases of radioactivity unless the core is destroyed, and the calculation of the probability of the accident to this stage avoids uncertainties regarding containment failure, fission product paths through the containment, meteorological conditions and health effects of low-level radiation. A suggested goal is a probability of core melt of less than 10^{-4} per reactor-year, equivalent to a probability of less than 1 in 300 of a core melt during a reactor lifetime. This latter constraint is met only marginally by the Surry reactor, according to the Reactor Safety Study, and other light-water reactors (LWRS) may fall short of this goal.

If the US Nuclear Regulatory Commission should embrace goals such as those outlined above, there is still the question of how to apply them. Using the medium results of the Reactor Safety Study, our light-water reactors now meet the safety goals mentioned, in some cases by a considerable margin. Does this mean that some of the safety features currently employed can be relaxed? Or should the risk analysis be required to show a wide margin of safety above and beyond these goals? One possible use of the goals is related to changes that are proposed to improve the safety of reactors. Provided that safety goals are already met, the proposed changes could be subjected to cost-benefit analyses in which one would balance additional lives saved against the cost of the improvements. Placing the value of a life under these circumstances at 1 to 2 million dollars would be consistent with practice in other industries.

EDUCATION AND INFORMATION

Although engineers and scientists designing and building even the first reactors held safety to be a prime requirement, the complexity of modern power reactors is such that no designer could foresee all of the possible malfunctions. This situation has required the nuclear power industry to go through a learning process in regard to reactor safety.

Another important aspect of nuclear education is the manner in which the public receives information on the effects of nuclear radiation. As often as not, such information is presented in ways that tend to scare people rather than educate them.

The Safety Learning Process

As in most industries, the reactor safety business appears to learn through crises. A number of such crises can readily be recalled. In the early days there was a crisis of metal fuel elements that swelled too fast under irradiation, saved by the introduction of oxide fuel elements. For the oxide fuel rods there was a short crisis of densification of the pellets and collapse of the cladding, quickly solved by better quality control. Then there was the crisis of the China Syndrome, brought on by the increase in reactor size and power density. Although each of these appeared in crisis form, there was always a background of research already available so that remedies were quick in coming and the consternation, genuine for a time, was short-lived. A current problem that is in crisis form as this paper is being written is that of the possibility of cracking a reactor vessel by cold thermal shock.

This current problem has roots as old as the beginning of reactor technology. It is one on which research has been continuous, and yet one that has taken a new turn within the last couple of years. Before the first reactor operated, Wigner predicted that neutron bombardment would alter the crystal structure of materials and change their physical properties. Experiments were initiated immediately, using accelerators to produce neutrons before reactors were available to produce them. Such research has continued to this day, and the radiation embrittlement of pressure vessels has been studied almost continuously. From the start, surveillance specimens of pressure vessel steel have been placed within reactor vessels so that the rate of radiation hardening could be measured. Early concerns were with the brittle fracture of a vessel while at pressure and temperature; here, the situation seemed under control. Recently, however, Dick Cheverton of the Oak Ridge National Laboratory made more sophisticated calculations of the effect of cold thermal shock on irradiated pressure vessels and predicted that small flaws could propagate nearly all the way through the pressure vessel wall. He has been able to confirm the reality of his calculations by experiments with thick-walled steel cylinders suddenly cooled on the inner surface by immersion in liquid nitrogen. The very cold temperature was necessary in the experiment because the steel used had not been irradiated. Calculations were made to predict the result employing the measured toughness characteristics of the steel used, and small cracks were found to propagate exactly as predicted.

In an operating reactor, fast neutrons from the reactor core cause the temperature below which the steel pressure vessel is brittle to increase. So long as the vessel is maintained above this temperature it is relatively tough, and brittle fracture or cracking will not occur. However, there are possible events that can cause the water in the vessel to become suddenly cooled. For example, when there is a turbine trip in a pressurized water reactor, the reactor will scram [shut down quickly], but to continue to remove decay heat it is necessary to bypass part of the steam from the steam generator directly to the condenser. If the amount of this bypass is not controlled, the water in the primary system can be cooled too much and too rapidly. When the cooling water is cooled, it contracts, and the pressure in the primary system is decreased. This automatically starts the high-pressure emergency core cooling pumps, and the system is brought up to full volume and pressure with cold water. An event like this has happened at least once, and the calculations indicate that, had the reactor been a few years older, the pressure vessel would probably have cracked through a large fraction of its wall thickness as a result of the cold shock and would have burst open as it was repressurized.

This state of affairs reverberated throughout the community of pressure-vessel steel experts for some time, and finally, last June, pressurized cold shock was recognized by the US Nuclear Regulatory Commission as the most important safety issue of the moment. As a result, a massive systems analysis effort was mounted, and fourteen of the older PWRS are having the state of embrittlement of their pressure vessels evaluated.

Fortunately, there are remedies available or on the way. It has been known for some years that steels with a low copper content are less susceptible to radiation hardening, and the later reactor vessels have been constructed with this in mind. Studies are being made of ways to anneal pressure vessels in place to relieve the effects of radiation. Instructions are being given to operators on how to avoid repressurizing any pressure vessels that have experienced cold thermal shock so that they will not burst open. Finally, there are hardware changes that could be made to automatically relieve pressure in the event of a cold shock incident.

Thus, this current high-priority problem of reactor safety is following the well-known path of such crises. A new danger is perceived and verified. There is consternation and some panic at first, and nuclear critics call for shutting reactors down. Most aspects of the problem are already well understood before there is public recognition of the problem, and remedies are available as a result of previous research and development.

That there has been some success in learning reactor safety is supported by a study of licensing event reports that Peter Roberts made while he was in residence at Oak Ridge Associated Universities' Institute for Energy Analysis last year.[13] Licensing event reports (LERS) are brief reports made by reactor operators and required by the US Nuclear Regulatory Commission whenever an unusual event with possible safety significance occurs. Reporting requirements have become more stringent over the years, so that a correction must be made for the continuous trend of an increasing number of reports before a smaller number of reports can be used as an indication of increased safety. After making this correction, Roberts found that the adjusted frequency of issuance of LERS decreased as the reactor station gained years of experience. He also found that this learning process was accelerated when there were two or more reactors located at the same site, indicating rather good communication among reactor staffs. This latter observation supports the views of Alvin Weinberg, who for years has been a proponent of clustering reactors and taking advantage of the size of the operation to add strength and breadth to the operating staffs.[14]

Informing the Public

All nuclear advocates of my acquaintance have complained at one time or another about the treatment of nuclear safety in the press or television. Of course, such complaints are by no means unique to the nuclear industry. Few, if any, prominent public figures are even moderately happy for very long about their treatment in the free press, and Harry Truman advised, "If you can't stand the heat, get out of the kitchen." Newspapers and television shows must have readers or viewers, or they don't make a living. To grasp the public's attention they must present the unusual, the exciting, the scandalous, or the controversial. Nevertheless, we would like to see a certain respect for the truth and some degree of fairness in presentations by the press and television.

A problem in accomplishing this is that, in a complex matter like the health effects of radiation, often neither the reporters, their editors, nor the public audience knows enough to recognize false or unsubstantiated claims. To add to the confusion, a number of persons with suitable scientific credentials become involved in advocacy positions and use their expertise to fight a cause rather than to establish truth. Thus, the newscaster can obtain a salable and sensational product by selection among apparently qualified sources.

An example of this procedure is exhibited in a television program entitled "We Are the Guinea Pigs" which has been shown on some public television stations. This is a very effective film that vividly portrays the panic at Three Mile Island. Scenes of crippled or deformed calves are shown a number of times, along with liberal use of frightened children and townspeople, concerned farmers, and testimony about cows promptly dying and cases of acute radiation sickness. Alarming statements about the effects of radiation are given by witnesses such as Drs. Gofman, Sternglass, Kaku, and Caldicott.

This program would have the effect of convincing viewers that the allegations of serious radiation effects at Three Mile Island on people and animals in the area were true. After an exhaustive study of such claims, both the US Nuclear Regulatory Commission and the US Environmental Protection Agency found conclusively that there was no relationship between the reported bizarre happenings and the radiation from Three Mile Island. The *New York Times* commented editorially on April 18, 1980:

> What is not at all reassuring is the behavior of 'experts' who have inflamed public fears by dealing recklessly with statistics . . . . Even in nuclear fables there are people who cry wolf.

A more even-handed treatment of radiation effects would be preferable, so that the public can be informed without bias. This is not easy to accomplish because of the differences of factual interpretation among scientists with acceptable credentials. One suggestion is to expose media representatives to direct experimental data rather than to theoretical interpretations and opinions of scientists. As an example of this, after Peter Groer presented a paper giving a statistical interpretation of the latest data on leukemias among atomic bomb survivors, one reporter insisted on seeing the raw numbers that told how many had died of leukemia in each radiation dosage interval.[15] Reporters are intelligent people, and, although they may not understand the fine points of whether a linear, a quadratic, or a threshold model best fits the data, or how to apply statistical tests for significance, by looking at the numbers they can gain a common-sense view of what the data mean.

A two-day meeting should suffice to present all of the data available on health effects of radiation on humans and a representative sample of the experimental data on animals. After all, as indicated previously here, the data on humans are not very voluminous. To be practical, the meeting should be organized by the press representatives, and, although scientists should be asked to supply the data, they probably should not be involved in presenting it, or it would degenerate into just another scientific meeting.

THE OPTIONS

The lack of new orders for nuclear plants and the delays in completing currently planned plants constitute a de facto nuclear moratorium, at least in the United States. The safety issue is only one component of this moratorium, the other principal components being the reduced need for electrical generating capacity below that envisioned a few years ago and the very high interest rates that militate against high capital cost projects. So far as the safety issue is concerned, there are the following options, or combinations of them:

1. Decide that present light-water reactors, with the hardware and institutional changes now being made as a result of the Three Mile Island experience, are adequately safe.
2. Decide that light-water reactors can be made acceptable by some radical technical fix, such as building them under water or under ground, or by insisting on remote siting and accepting the modest extra cost of transmission.
3. Abandon the LWR concept and concentrate on one or more other reactor types that are thought to be inherently safer.
4. Abandon all low conversion ratio reactors and concentrate all future efforts on reactors that conserve fuel, such as breeders.
5. Abandon fission energy.

Needless to say, there are advocates for each of these options. Personally, I favor the first one, although only with the proviso that utility managements maintain their present degree of safety consciousness and insist on continued improvement in the operation and maintenance of their plants. What I fear most is that any declaration that reactors are already safe enough will lead to laxness among plant personnel. The safety predicted by the risk analysis studies is there only so long as the plant is maintained in its approved condition and so long as operational mistakes are avoided.

The second option, involving a radical technical fix or remote siting, is instinctively opposed by most industry representatives as being either uneconomic or ineffective, or both. Nevertheless, such an approach has the potential of reducing risk by one to three orders of magnitude.[16, 17] My personal view is that this option could be made attractive and would prove reasonably practical if given a real try by industry. If public opposition to reactors should harden, it could become a necessary approach if we are to add to our nuclear power capacity. However, to implement this option would require a firm decision that the first option is not to be allowed. I see no mechanism for making this decision.

The third option—looking for an inherently safer reactor—has romantic appeal, especially to some of us die-hard enthusiasts for various reactor types. Studies that involve this approach are being conducted at Julich and at the Institute for Energy Analysis at Oak Ridge.

NOTES

1. *Reactor Safety Study*, WASH-1400 (NUREG-75/014) (US Nuclear Regulatory Commission, October 1975).

2. The numbers used here are revision of WASH-1400 numbers, taken from N. C. Rasmussen, "Methods of Hazard Analysis and Nuclear Safety Engineering," in *The Three Mile Island Nuclear Accident: Lessons and Implications*, T. H. Moss and D. L. Sills, eds., (*Annals of the New York Academy of Sciences*, Volume 365, 1981, pp. 20–36.

3. Committee on Biological Effects on Ionizing Radiation, "Effects on Populations of Exposure to Low Levels of Ionizing Radiation" (Washington, DC: National Acadmey of Sciences-National Research Council, 1980.)

4. J. R. Trotter and H. G. MacPherson, "Do Childhood Cancers Result from Prenatal X-rays?" *Health Physics*, Volume 40, p. 20–36.

5. See Appendix VI, Section 9 to reference cited in note 1.

6. See reference 2 cited in note 2.

7. As summarized in *Nuclear Safety*, Volume 21–1, pp. 21–24, 1980.

8. S. Levine, "Various Applications of Probabilistic Risk Assessment Techniques Related to Nuclear Power Plants," address to the Annual Meeting of the National Safety Council, Chicago, 1980.

9. J. D. Selby, "The Electric Industry's Response to Current Events," in *The Outlook for Nuclear Power* (National Academy of Engineering Annual Meeting, November 1979), pp. 45–50.

10. Ibid.

11. E. P. Wilkinson, Remarks before the EEI [Edison Electric Institute] Senior Chief Industrial Relations Conference, New Orleans, February 1981.

12. A summary of various proposals for safety goals is presented in "An Approach to Quantitative Safety Goals for Nuclear Power Plants," NUREG-0739, Advisory Committee on Reactor Safeguards, October 1980.

13. P. C. Roberts and C. C. Burwell, "The Learning Function in Nuclear Reactor Operation and Its Implications for Siting Policy," Report ORAU/IEA-81-4(M) (Institute for Energy Analysis, Oak Ridge Associated Universities, May 1981).

14. See, for example, "Gatlinburg II, An Acceptable Future Nuclear Energy System," M. W. Firebaugh and M. J. Ohanian, eds., Condensed Workshop Proceedings, ORAU/IEA-80-3(P), pp. 2–4 and 234–35 (Institute for Energy Analysis, Oak Ridge Associated Universities, March 1980).

15. Peter Groer, Institute for Energy Analysis, private communication.

16. One such approach is described in O. H. Klepper and C. G. Bell, "Under Water Containment of Large Power Reactors," ORNL-4073 (Oak Ridge National Laboratory, 1967).

17. For the effect of remote siting, a comparison of the Diablo Canyon site with Zion, Indian Point, or Limerick indicates a substantial improvement, as shown in Figure 9 of the reference cited in note 8.

CHAPTER SIX

Summary with Selected Comments

Session Chairman/Integrator:
Harvey Brooks
Benjamin Peirce Professor
of Technology and Public Policy
Harvard University

INTRODUCTION

The panel began its deliberations by agreeing on the formulation of four questions regarding the future role of nuclear power in the mix of energy sources potentially available to the world. These questions were as follows:

1. How critical is nuclear power to insuring adequate energy supplies for economic growth and world development? In other words, how essential is it in various national circumstances to maintain nuclear as a viable option among alternatives for electricity generation?

2. What are the minimum prerequisites—technical, institutional, and political—necessary to maintain nuclear as a viable option? (This question assumes a positive answer to the first question.)

3. What is the highest prudent practical rate of growth in nuclear capacity, and what additional measures beyond the minimum prerequisites would be required to achieve it?

In addition to the chairman/integrator and those presenting papers, the work session included the following assigned participants: Marcelo Alonso, Pierre Jonon, Amory B. Lovins, B. C. E. Nwosu, Allen C. Sheldon, Jean-Pierre Somdecoste, Miguel S. Ussher, and Alvin Weinberg. The chairman/integrator was assisted by Henry Piper, who served as rapporteur.

4. What is the likely course of events, and what would be the most suitable policies, if the nuclear option cannot be maintained in the leading industrial countries?

In practice the panel was able to discuss only the first two of these questions. However, in addressing the first question it found it necessary to assume this question's answer—an answer which is critically dependent on one's view of the future importance of electricity in satisfying final energy demand. There was disagreement on this, but the majority of the panel was persuaded by the arguments put forward in Professor Colombo's plenary paper that electrical demand will continue to grow in all countries considerably faster than total energy demand, and probably almost as fast as its reduced growth rate following the 1973 oil embargo and subsequent oil price rise. However, the panel did not attempt an independent evaluation of the pros and cons of Professor Colombo's arguments, feeling that such an assessment would be outside the scope of the charge to the panel, since it would depend on analysis of future demand for various forms of final energy rather than on any characteristics of nuclear power itself.

THE NATURE OF THE PROBLEM

The one point on which the panel was in unanimous agreement was that the nuclear option is in serious political and economic trouble in several of the advanced industrial democracies, and its long-term viability is now in serious question, at least in those countries. There was not agreement about whether the circumstances that have led to the difficulties of nuclear power in some countries are unique to those countries, or whether the difficulties are likely to spread to all countries eventually.

The reasons for the faltering of the nuclear option are complex and involve several interacting factors, of which public opposition and distrust are only one. The slowing of economic growth, the shortage of investment capital, and existing and prospective energy conservation in the face of rising prices have combined with the long lead times involved in electrical capacity planning to produce an extended "time window" of excess electrical generating capacity. There is great uncertainty as to how long this excess capacity will last, with at least one member of the panel arguing that it is a permanent phenomenon, dictated by the fact that the marginal cost of improvements in energy end-use efficiency is now so much less than the marginal cost of new energy supplies, particularly electrical generating and transmission capacity. Some believe that available opportunities for cost-competitive energy end-use efficiency investment will be exhausted relatively quickly, so that the window of excess generating capacity will be of relatively short duration, but it may last long enough to destroy the nuclear supplier industry in the absence of government intervention to preserve it. Such intervention would imply positive political leadership in several of the industrialized countries, which may be temporarily quite unpopular.

DEVELOPMENTS AND POLICIES FAVORING THE REVIVAL OF THE NUCLEAR OPTION

The panel discussed at some length what technical and political developments would tend to favor the revival of nuclear power's prospects in the major industrialized countries. Not all members of the panel were in agreement that every one of these developments would necessarily benefit nuclear power. Nevertheless, the majority of the panel felt that the following developments would be beneficial:

1. Restoration of economic growth rates in the OECD countries to magnitudes approximating those that were experienced prior to the 1973 crisis.

Discussion. Since there is a wide range of opinions about the causes of the growth slowdown of the 1970s, there is large room for differences about the likelihood of early restoration. Oil prices themselves may be a major factor underlying the decline in labor and total factor productivity that has occurred in all the industrialized countries. Revival of economic growth may be accompanied with increased oil demand and consequent pressure on prices which will arrest the growth. On the other hand, slow economic growth will retard conversion away from dependence on oil. In any case, faster growth would narrow the window of excess electrical capacity and by increasing oil prices may accelerate the replacement of the oil generating capacity with a nuclear generating capacity.

2. Another disruption of oil supplies as a consequence of various fairly likely political developments or policy changes in the oil-exporting countries or regions.

Discussion. It is possible to imagine a whole host of contingencies that would reduce international oil supplies for either short or long periods. Many exporting countries may decide to conserve their energy supplies for their own future development needs. A crisis could be induced by an embargo, by general political turmoil in one or more exporting countries, or simply by production policies of oil producers. It seems likely that energy shortages would dampen public opposition to nuclear power, but not all members of the panel were in agreement on this. It was pointed out that the present political difficulties of nuclear power came about largely subsequent to the 1973 crisis.

3. A "technical fix" in the nuclear industry, such as a policy change in favor of exclusively remote siting, nuclear parks, underground reactor construction, offshore nuclear plants, or an alternate reactor type such as the high-temperature gas-cooled reactor.

Discussion. The panel expressed the greatest skepticism about the probable value of technical fixes. Remote siting might be the most attractive, but it is not an option available to all industrial countries. Deployment of a new reactor type would be bound to introduce new technical uncertainties which would probably balance theoretically greater inherent safety. Much public opposition to nuclear power is based not on its technical characteristics but on more general factors such as distrust of the institutions and decisionmaking procedures associated with management of the nuclear industry. Such perceptions change slowly and are not likely to be much affected by technical improvements. Nevertheless, improvement of management structure could help, and its effect might be greater if accompanied with technical changes as well.

4. A continued good safety record for nuclear power, and, especially, continued success of the aggressive French program.

Discussion. Public perceptions of nuclear power may be more influenced by the worst possibilities than by expectation values of death or injury from accidents. Events which are minor in their consequences may nevertheless have a big impact because they remind the public of the possibility, however remote, of much more catastrophic events. Conversely, indications of competent management and the absence of incidents are reassuring even when they do not have much bearing on objective measures of risk. The avoidance of upsetting incidents, however inconsequential in effect, is essential to the viability of nuclear power and justifies the most intensive effort and vigilance on the part of industry and government.

5. Intensification of concerns about acid rain, the climatic effects of carbon dioxide, respirable particulates, and other environmental problems associated with fossil fuel power generation.

Discussion. To the extent that coal-fired electric generation is seen as the major alternative to nuclear power, public attention is likely to become increasingly focused on the comparative risk assessment of these two options. Given the fact that much more research has been done on nuclear safety and the risks of radiation than on the environmental and health risks of coal, increasing knowledge of the latter is likely to lead to more balanced consideration of alternatives. Whether this results in a more favorable attitude towards nuclear power will, of course, depend on concrete findings, but it seems likely that the outlook for coal may become more pessimistic with time.

6. Reform of the regulatory system for nuclear power to bring about more stable regulatory criteria within the planning and construction time of each plant, and to place different forms of energy production on a more comparable basis with respect to both technical criteria and administrative and legal procedures.

Discussion. Reform of the regulatory system is not easy, because instability in regulation results partly from a lack of public confidence in the technology and from evidences of past management mistakes as well as from incidents like Three Mile Island. Because safety criteria depend mainly on theoretical analysis rather than actuarial experience and engineering experience, stability in regulation is harder to achieve. The burden of proof has tended to be heavier on nuclear power than other more familiar technologies. The degree to which this is appropriately so is a matter of some debate. One panel member expressed the opinion that nuclear power was seen by the public as a symbol of the imposition of risks on people by industrial society without the people's consent, and that other industries would tend in the future to have imposed upon them safety criteria equally stringent to those of nuclear power. In the long run, the risks of alternate energy technologies ought to be regulated on a comparable basis.

7. Progress in limitations on nuclear weapons deployment,

especially by superpowers, and with special importance attached to achievement of a comprehensive test ban treaty.

Discussion. An important component of public opposition to nuclear power arises from its historical and psychological association with nuclear war and nuclear weapons development. Opposition to nuclear power has frequently been a surrogate for opposition to nuclear weapons, since nuclear power is politically more visible and sufficient information is available for informed debate. The majority of the panel believed that real progress towards meaningful limitations on nuclear weapons would create a climate in which a more informed and realistic debate about the risks of nuclear power could take place. A comprehensive test ban would greatly inhibit the ability of the nuclear powers to introduce new nuclear weapons systems and would make it ever more certain that nuclear weapons states could not have sufficient confidence in the reliability of their weapons to make the possibility of a first strike credible.

8. Demonstrable improvement in the management of nuclear power construction projects, especially in the United States.

Discussion. Widely publicized evidences of slack management and a lack of quality control and adequate training in all aspects of nuclear plant construction and operation have been important contributors to loss of public confidence. Following Three Mile Island, the industry has taken strong steps to improve the situation, but it will take several years to establish that these steps are effective and to restore the public confidence that has been lost as a result of a series of disturbing incidents. Similar problems in other industries do not receive comparable publicity; this publicity derives from public distrust which is in turn fed by it. Thus the nuclear industry must expect to be held to a much higher standard of performance than most other industries and will have to act accordingly if the nuclear option is to survive.

9. Improving the level of confidence in the nuclear nonproliferation treaty (NPT) through further adaptation of this regime, in order to make it acceptable to those who have not signed it.

Discussion. The relationship between the credibility of the nonproliferation regime and public acceptance of nuclear power is a complex one. On the one hand, its apparently discriminatory aspects, and the failure of the major powers to make progress towards nuclear disarmament, breed cynicism in other nations, with the danger that the separation of military and civilian technology will erode. On the other hand, such erosion leads the nuclear powers to attempt unilateral actions which further stimulate resentment in the nonnuclear nations. The IAEA's International Nuclear Fuel Cycle Evaluation (INFCE) exercise made an important contribution to improving the climate for further adaptation and development of nonproliferation policy on a multilateral basis. Confidence that the spread of nuclear power will not lead to the proliferation of nuclear weapons is necessary to public confidence in nuclear power in the major industrialized countries.

10. Maintenance, further improvements, and internationalization of the reporting system for nuclear operating experience and risk assessment.

Discussion. Accidents and untoward incidents related to nuclear power anywhere in the world will have a damaging effect on publicsconfidence everywhere, as the worldwide political effects of the Three Mile Island incident amply demonstrated. It is thus essential that the industry internationalize the learning process regarding operational safety and quality assurance so as to take maximum advantage of all safety-related information, whatever its source. The coordinated learning process should be extended to embrace the whole fuel cycle as soon as possible. Moves on the part of a number of foreign countries to join the safety analysis and information and training systems created by the US nuclear industry are an important positive development in this respect.

11. Development of more positive political leadership in support of the nuclear option, but without compromising honesty in the communication of bad as well as good news to the public.

Discussion. Regardless of the ultimate merits of nuclear power, it appears unlikely that it will receive fair hearing in the absence of government leadership to help solve a number of the problems now facing the nuclear industry. Given the high degree of government involvement in the past, it is unlikely that the industry by itself can solve these problems, especially those pertaining to the back end of the fuel cycle and the general management of radioactive wastes. On the other hand, the ultimate objective should be for the price of nuclear electricity to reflect the full cost of the fuel cycle, including the permanent isolation of wastes.

SELECTED TECHNICAL AND INSTITUTIONAL ISSUES

The panel reviewed a series of technical and institutional issues pertaining to nuclear power which may affect its economic viability and public acceptance.

Comparative Economics of Nuclear Power and Other Energy Sources

The panel was especially struck by the differences among countries in the degree to which support of the nuclear option was influenced by comparative costs in contrast with various nonmarket criteria. In a country such as the United States, with several fuel options for electrical generation, nuclear power will be judged primarily in terms of its economic competitiveness. Public acceptance is reflected primarily in the additional economic costs imposed on the nuclear option as a result of unusually agressive regulation and a high burden of proof with respect to safety in comparison with other generating options. In contrast, in countries having few indigenous fuels for electricity generation, particularly the developing countries which face rapid growth in electricity demand, decisions about nuclear power in the medium term tend to be based on nonmarket considerations such as improving energy independence, diversifying energy sources, developing indigenous industrial capabilities, and training people in advanced modern technologies. It was noted, for example, that the United States was the only country from which sufficient raw data was publicly available to permit an independent assessment of the relative cost of nuclear electricity, and even here there is by no means complete agreement about the results.

Even in some developing countries with a large hydro potential, for example, it is anticipated that the ultimate exhaustion of suitable sites will force a transition to nuclear power, so that it is necessary for each country to begin acquiring experience with nuclear technology now in order to be in a position to move forward with an indigenous and independent nuclear power program when the time comes. Thus, even though these countries are currently rich in traditional resources, nuclear power is seen as an important form of insurance against their future depletion.

The panel merely noted these differences in perception without being able to resolve the question of their objective validity.

With respect to nuclear power in the industrialized countries, many analysts and most utility economists believe that nuclear electricity is somewhat cheaper than coal and much cheaper than oil-fired base-load generation, and that this difference will continue and possibly grow despite the recent escalation of nuclear construction costs. With nuclear electricity, fuel is a much smaller fraction of the total cost, so that if fuel costs escalate in the future as it is generally assumed they will, the life-cycle economic advantage of nuclear power may increase in the instance of two plants —one based on nuclear, the other on fossil fuel—completed at approximately the same time. There were strong dissents from this view, however, even within our panel.

Comparative Risks

The panel agreed that there would be considerable value in wide dissemination and critical discussion of information on comparative risk assessment among different energy sources, although there was some skepticism as to the degree of credibility and legitimacy that expert assessments could achieve in the political process. There is no objective way of comparing risks on an aggregated basis, because of the differing subjective valuations of different types of risks to different people—for example, delayed cancers due to low-level radiation exposure from a nuclear accident versus immediate deaths due to the failure of a hydroelectric dam, or a steady rate of "statistical" fatalities due to air pollution from the burning of coal versus the same "actuarial" fatality rate due to very infrequent accidents potentially affecting large numbers of people in each event.

Fuel Resources

Because of the recent worldwide slowdown in nuclear plant orders, questions of uranium supply and enrichment capacity do not appear to be at all critical for the next twenty years and so did not receive much attention from the panel. For the medium term they are not critical considerations in relation to the future of nuclear power.

Small Reactors for Developing Countries

An interesting point that came up in the panel discussion was the issue of small reactors, by which was meant reactors in the 100- to 200-mwe capacity range. A problem with nuclear power for many developing countries is the fact that what are considered optimal size reactors from the standpoint of economies of scale in industrialized countries are too large for most electric grids in developing countries. In the light of experience even in the industrialized countries, some of the alleged scale economies may be illusory because of failure to take into account the political and procedural costs of siting very large facilities. In addition, the factor costs in developing countries may be quite different than in developed countries, especially as the former develop indigenous design and manufacturing capabilities. Thus, consideration may be given in the future to the development of a standardized small reactor suitable for the smaller electric grids of the developing countries. Perhaps such a reactor type could be developed cooperatively by some of the newly industrialized countries for sale in the developing world. The fact that electricity demand in newly industrializing countries is typically increasing at the rate of 10 to 11 percent per year, leading to a twentyfold increase in thirty years, may be an additional argument for the development of a reactor type specifically tailored to this market.

The panel did not resolve the question of whether such a smaller reactor development would be desirable in relation to the wider political questions of proliferation. This is something that needs to be carefully evaluated in the future.

The Back End of the Fuel Cycle

The issues of spent fuel management, fuel reprocessing, and waste management were discussed in order to gain a notion of the magnitude of this problem. It appeared that the temporary storage of spent fuel at reactor sites is a manageable problem for the next few years. Furthermore, if redistribution of spent fuel rods among the available reactor storage sites is permitted, then no problem is likely to arise until after about 1990. After that date, however, there will be a real shortage of fuel storage space unless reprocessing of spent fuel is introduced. Present reprocessing capability worldwide, if utilized, could accommodate about 20 percent of the needed capacity, but by 1985 it is expected that reprocessing plants will be in place that are capable of providing about 50 percent of the needed capacity. It is to be noted, of course, that reprocessing "solves" the on-site spent fuel storage problem only at the cost of intensifying other problems downstream, particularly the handling and final disposal of low- and intermediate-level wastes containing actinides.

Most of the panel felt that the present reprocessing activities were performing adequately, were well based technologically, and had a good prospect for cost effectiveness in the future. One panel member dissented strongly from this view, however, feeling that reprocessing technology was by no means proven. Another panel member felt that reprocessing was worth pursuing only to provide fuel for future breeder reactors and to gain experience for necessary future breeder fuel cycles. For the light-water reactor, it was estimated that in the French experience reprocessing cost would amount to about 25 percent of the total fuel cycle cost for a full LWR fuel cycle. At least one country today has offered to reprocess the spent fuel from other countries, and three additional countries are expected to enter the international reprocessing market in the near future.

Once reprocessing is in place, a permanent solution to the problem of dealing with nuclear wastes can no longer be deferred. Most of the panel felt that appropriate technology is available in principle to deal with nuclear wastes but that serious problems of integrating technical processes with political and regulatory procedures remain. International issues of transfer of wastes and spent fuel between countries are unresolved. Several nations have been engaged in evaluating disposal (or isolation) sites for processed nuclear waste. It is generally accepted that even though the time frame is long, the risk from waste isolation can be made acceptably small. One panel member expressed strong disagreement with this view, however.

Proliferation and the NPT

The panel spent a good deal of time discussing the question of proliferation and the future of the international nonproliferation regime. Several members of the panel felt that the nonproliferation treaty has effectively been dead since the preemptive strike by Israel against the Iraqi research reactor. Others expressed a fear that support for the NPT would disappear if the main nations pushing it, particularly the United States, chose to abandon nuclear power. There was a wide divergence of views among panel members as to the importance of preserving the NPT and trying to improve safeguards to include the balance of the fuel cycle and the storage of plutonium. The majority of the panel felt maintenance of the treaty was essential.

Public Participation

There was considerable divergence within the panel on the appropriate role of public participation in decisions about nuclear power, both individual siting decisions for nuclear facilities and generic policy decisions about such issues as away-from-reactor spent fuel storage, reprocessing, fast breeder development, and radioactive waste management. The predominant view was that such decisions ought to be the responsibility of elected government officials, with inputs both from technical experts and from a broader public having access to intelligible expert analysis representing a wide range of technical opinion. However, it was felt that decisions themselves could not be delegated to the general public in such mechanisms as referenda; a line should be drawn between public consultation in decisions and the actual making of decisions by the general electorate. The long time interval between decisions made today and their ultimate consequences for the energy system argued for a decisionmaking mechanism that did not put too much weight on contemporary constituency interests. The difference of opinion on public participation split within the panel more or less along lines of industrialized versus industrializing countries, with the latter arguing that only governmental leaders could be expected to have the long-term perspective on national interests necessary for sound decisions on complex technological issues.

Advanced Nuclear Technologies

The panel did not have time to discuss either the future of fast breeders or prospects for nuclear fusion in any detail, although there was one intervention from the audience on the subject of fusion. It was recognized that the future of nuclear power in the long term would depend on the development and public acceptance of advanced reactor designs capable of producing electricity economically from much less rich uranium ores than those currently being mined and processed for LWR fuel. In practice this probably means fast breeders. Fusion was seen as the principal competitive technology to the fast breeder, on the assumption that some form of nuclear energy would continue to receive public acceptance. The principal dilemma presented in the fusion program is whether to push for the earliest possible test of engineering feasibility in order to provide a credible alternative to breeders at the time that the resource situation begins to require the latter's deployment, or whether to push for the development of the best practicable reactor on a longer time scale. The theoretically best fusion reactor configurations would be the most superior to breeders in respect to problems like radioactive waste management and tritium containment, but their ultimate feasibility is in the most doubt, and the possibility of testing and assessing them furthest down the road in time. Thus, the best fusion reactors might become available only after the breeder is so well entrenched that it would be difficult for a newer alternative technology to enter the market, because the breeder would be so far down the learning curve.

One panel member from a developing country mentioned that in such countries there are many devotees of the position that fusion could be used to "leapfrog" the breeder.

CONCLUSIONS AND RECOMMENDATIONS

This section summarizes what seem to be the most important conclusions and recommendations that emerged from the panel sessions. The reader should bear in mind that all but one member of the panel came to the session with a strong pronuclear orientation. The conclusions and recommendations thus reflect the views of the majority of the panel, recognizing that those which imply a judgment about the need for and desirability of nuclear power were foreordained by the original selection of the panel. It seems superfluous to repeat the reservations of the dissenting member in this section, since references to dissenting views are included in the preceding sections.

Conclusions

1. Nuclear power is in serious trouble for both economic and political reasons in a number of the major industrialized countries, and at present it is not clear whether the nuclear supplier industry will survive long enough to meet a revived demand for nuclear plants if and when it occurs.

2. The survival of nuclear power as a viable option is likely to depend upon maintenance of a record of safety and managerial competence which is without precedent for any other industry at a similar stage of development. The challenge to the industry and the political institutions which govern it will thus be uniquely demanding and difficult.

3. Although technical progress in safety, quality assurance, and radioactive waste management will be essential to the preservation of the nuclear option, technical factors by themselves cannot assure this progress. Success will depend upon the skillful orchestration of technical, institutional, and procedural developments to provide adequate assurance to the public that the technology is under competent and responsible control. Justifiably or not, such assurance is not publicly perceived in several key countries, from the standpoint of either the public itself or the utility industry.

4. The close identification of nuclear power and nuclear weapons in the public mind, though probably greatly exaggerated, nevertheless means in fact that the survival of nuclear power may be determined by significant progress in limiting the testing and deployment of nuclear weapons.

5. The credibility of the international nonproliferation regime is important in its own right for world stability, and also for the preservation of the nuclear power option.

6. Many rapidly industrializing nations see the acquisition of indigenous capabilities to run nuclear power industries of their own as an important development objective, independent of their cost in comparison with other energy sources.

7. Dealing with the back end of the fuel cycle either by reprocessing or by storage of spent fuel away from reactor sites will become urgent in the next decade. The problem would be temporarily alleviated to the extent that some countries would be willing to accept the spent fuel of other countries for storage or reprocessing. This could also contribute to nonproliferation objectives.

Recommendations

1. Development of an international reporting, analysis, and dissemination system for nuclear reactor operating experience and risk assessment should be extended as rapidly as possible, since an incident anywhere could affect the prospects of nuclear power everywhere. Internationalizing the US nuclear industry's Nuclear Safety Analysis Center (NSAC) and Institute for Nuclear Power Operations (INPO) has already begun and is probably the best option.

2. Efforts should be made to insure wider public information and discussion on the comparative risk assessment of nuclear power relative to other energy options. In particular,

much further research is needed to define the health and environmental problems associated with fossil fuel power generation, to assure a knowledge base more nearly comparable to that for nuclear energy and radiation.

3. There is a need for greater stability and predictability in the regulatory criteria and procedures for nuclear power plants and other parts of the nuclear fuel cycle. Indeed, predictability and firm decision schedules may be more important than the stringency of regulations in determining the financial risk of nuclear power investments.

4. All countries have a major stake in preserving the international nonproliferation regime and the system of safeguards. The NPT and the IAEA deserve the strongest support, and an effort to effect reforms and improvements to make the system more effective and equitable should be undertaken.

5. Studies should be made of the feasibility and possible cost and safety characteristics of a standardized reactor design in the 100- to 200-MWe capacity range for deployment in less developed countries whose power grids cannot accept the present greater than 1,000-MWe plants being built or operating in the industrialized countries. The significance for proliferation of the availability of such a design, possibly manufactured in one of the rapidly industrializing countries, should also be assessed.

6. There should be adequate provision for both public input and a wide diversity of outside technical input into major decisions about nuclear power, but the responsibility for final decisions should rest with accountable, representative public officials.

SELECTED COMMENTS

MARCELO ALONSO

[The following is an abridged version of a report by Marcelo Alonso on Mexico's nuclear power program. The report was volunteered by Dr. Alonso at the work session on nuclear power; the abridgment is the editors'.]

That nuclear energy has to play a substantial and critical role in the global energy picture over the coming years is a fact well recognized and accepted by most of the people concerned with assuring an adequate future energy supply. A major consideration to be taken into account in nuclear power programs is the finitude of oil resources, since by the beginning of the next century oil production will probably begin to decline. Therefore, it seems imperative to stretch oil resources as long as possible, for which energy substitutes are required *now*. It is thus not surprising that a major oil-producing country, Mexico, has recently decided to go ahead with an important nuclear power program.

Assuming no new discoveries, no changes in technology, and no increase in production, the lifetime of Mexico's oil resources should be on the order of sixty years—a relatively short time. The policy that Mexico has adopted towards these resources is to conserve them, and to implement this policy, the diversification of energy production and the development of energy alternatives are essential. It is within this framework that Mexico has decided to go nuclear.

The Mexican nuclear program is not an isolated decision but part of a carefully prepared energy plan, covering a period of twenty years—a plan which considers all possible energy alternatives available to Mexico and is a clear definition of Mexico's energy policy, including its policy towards electricity generation. Mexico's current electric generating capacity is about 15,000 MW and is growing at the rate of 10 percent per year. By 1990 it will be about 40,000 MW, and by 2000 it is expected to be on the order of 80,000 MW. Hydroelectricity generation will grow rather slowly since Mexico's hydro potential is already largely attained, and the major portion of new electricity generation during the coming twenty years will be based on oil, coal, and nuclear thermal power units. Mexico's goal is that by the year 2000, nuclear generation will be on the order of 20,000 MW, accounting for about 23 percent of the total electric generating capacity (see Table 6-1).

Table 6-1. **Mexico's Projected Structure of Electric Energy Production**[a]

	1980	1990	2000
Thermal			
Oil	8.0 (53%)	20.0 (50%)	44.0 (51%)
Coal	--	4.0 (10%)	6.0 (7%)
Geothermal	1.0 (7%)	1.0 (2%)	3.6 (4%)
Hydro	6.0 (40%)	12.0 (30%)	13.0 (15%)
Nuclear	--	3.0 (8%)	20.0 (23%)
Total	15.0	40.0	86.6

Source: based upon information obtained from Mexico's Federal Electricity Commission.
[a]In gigawatts.

The authorization to go ahead with the nuclear program was given by the president of Mexico in September of 1981, and bids for this program's first nuclear plant, designated NP-2 (NP-1, the Laguna Verde BWR, is under construction), were issued in October of 1981. It is important to note that before adopting the nuclear program a study was carried out in 1980 by the Federal Electricity Commission (CFE) and others to compare the BWR, PWR, and CANDU systems from the point of view of Mexico's financial, material, and human requirements. The study's conclusion was that there are no fundamental differences in the requirements of the three systems and that there are no technical differences which might make one system preferable to the others in terms of design, construction, operation, and maintenance. At least in principle, this implies that the decision about the supplier will be based primarily on the package that the supplier is prepared to offer, covering the nuclear power plant and technology transfer and financing factors.

An aspect of the complete nuclear program that has been emphasized by the Mexican authorities is that of transfer of technology (TT), which, for good reasons, the Mexicans consider more important than just the construction of NP-2. It is the aspiration of Mexico to develop a national nuclear power industry capable of dealing with all aspects of the nuclear program. It is estimated that Mexican industry currently can be involved in only 30 percent of NP-2. The aspiration is to reach a level of approximately 80 percent involvement during the last phases of the program's development.

The TT program that CFE is requesting from the suppliers as an integral part of the NP-2 bid covers three major areas:

a. *Engineering:* CFE wishes to acquire capability in all engineering components of the nuclear power program, including design, management, supervision, operation, and construction of nuclear facilities. It is estimated that about 150 persons should be trained. Most of the training is to be on the job rather than in a formal university setting.

b. *Nuclear fuel cycle:* By law, the two Mexican entities responsible for the nuclear fuel cycle are URAMEX and ININ (respectively, the uranium company owned by the Mexican government and the national institute for nuclear research). The bid provides that assistance shall be given to URAMEX in developing a capability for exploration and exploitation of uranium resources. Eventually, enrichment and reprocessing will be included as well, with materials technology, particularly metallurgy and enrichment methods or heavy-water production, considered jointly with ININ. Similarly, reprocessing technology must be included in the scheme. Obviously, these capabilities have to be developed over a certain period — probably ten years at a minimum.

c. *Basic technology:* There are numerous specific industrial needs of a nuclear program for items such as concrete, tanks, pipes, pumps, valves, controls, and so forth—items for which Mexico has a certain manufacturing capability, but ones where its industries do not fully meet the quality controls of nuclear power plants. Therefore, the TT program must also be geared towards developing high quality production methods for such items, an effort that must be geared basically towards private industry.

Since the time horizon of a TT program is different from that of the construction of NP-2, and since subsequent plants are to be initiated as part of the total nuclear power program, it appears that the supplier accepted for NP-2 will be used at least for two or three ensuing nuclear power units.

Another element that will be considered by Mexican authorities in making their decision about the supplier is that of financing (which might be coupled with trade agreements), as well as the support that the government of the supplier's country is prepared to give during the program's implementation. This obviously introduces a political component in the decisionmaking process. Independent external advisory and consulting services to CFE are to be secured to help evaluate the suppliers' proposals, and the selection process is scheduled to take eight months. However, some people involved in the program believe that this timetable is too tight and that it will take about a year instead.

Besides Mexico, two other Latin American countries—Argentina and Brazil—have nuclear power programs. It is interesting to note that there are certain parallels in the ways the three countries have planned and developed their programs, although there are also important differences. Argentina was the first of these countries to initiate a nuclear program; at the end of the 1950s, it developed research groups in most of the aspects related to nuclear energy, and construction of its first nuclear power plant (300-MW HWR) was begun in 1968 with the collaboration of West Germany and the involvement of the Argentinian industry. Since 1968, its nuclear program has evolved slowly but steadily, with the construction of a second nuclear power plant (692-MW HWR) and a series of related facilities associated with the fuel cycle. Thus over a period of twenty years Argentina has achieved an almost total capability and autonomy in nuclear power, making possible significant participation by its domestic technical and industrial capabilities—participation which implies a successful effort in technology transfer, an objective that Mexico also considers essential in its nuclear program.

Brazil initiated its nuclear program in the 1960s, also in close association with West Germany (although its first nuclear power plant—600-MW PWR—is being built by Westinghouse), but has adopted the PWR type. As is the case with Argentina, Brazil's agreement with West Germany includes development of the whole nuclear cycle. However, in Brazil's case some people have criticized its program, because it apparently places the country in a situation of long-term heavy dependence on a particular country (West Germany), although it does call for a certain degree of technology transfer to domestic industries. The Brazilian program is still in its infancy and is behind schedule; therefore, its effectiveness cannot yet be fully evaluated. Mexico, for its part, apparently will try to avoid a long-term strong commitment with one particular country and will seek the involvement of the International Atomic Energy Agency in some of its arrangements.

D.R. PENDSE

The summary of the nuclear power work session started with the statement that in the foreseeable future, the demand for electricity is likely to increase at least at the same rate as in the past —probably something like 10 percent per year. This assumption is central to the further points and conclusions of this session's panel. I have two comments to make on this central assumption.

First, it is possible that there may be some unforeseen developments or unanticipated events which may take place and could completely upset this central assumption. For example, quite a few studies were published in the 1960s which said that the demand for crude oil would keep on increasing through the 1970s at a far higher rate than what actually occurred. Would it not be preferable, perhaps, to proceed with alternative scenarios rather than proceeding on one central assumption?

Second, in the past in relation to electricity as well as to oil and coal, many countries have used an approach of "demand accommodation"—namely, first to project the likely increase in demand with respect to these energy sources and then to consider how energy supplies can be increased to accommodate this demand. But the deeper question of energy policy is often not asked. In other words, is all of the projected increase in the demand necessary? What are the other implications of such a continuous increase in demand for energy? Will there be some adverse factors and features that may arise on the horizon as a necessary result of such an increase in demand? More importantly, if, as a result of our drive to increase energy supplies, we have abundant high-level concentrated energy—as, for example, when we achieve controlled fusion—what will we do with it? More specifically, with so much electricity available and with the consumption of electricity going up proportionately—"at least at the same rate as in the past"—will this situation create any new problems, and will they be less significant than the benefits which we hope to get? These questions may appear philosophical, but they are necessary, and, I am afraid, they have not been sufficiently stressed.

AMORY B. LOVINS

I thought that in most points Harvey Brooks's summary of our very diverse nuclear power panel was admirably fair. I should just like to change the emphasis a little on three points.

First, Professor Brooks listed a number of possible future events which *might* be helpful to nuclear power. We did not have a consensus that they *would* actually be helpful. The nuclear industry felt the oil embargo would be very helpful to nuclear power, but in fact it had precisely the opposite effect. I think most if not all of the events on our list might likewise backfire. Nobody knows.

Second, Professor Brooks very tactfully omitted a point which did emerge quite forthrightly from our discussions: namely, that the prospect of having a military nuclear option did influence the nuclear planning of at least one if not both of the developing countries represented on our panel.

Third, the relative cost of coal versus nuclear electricity sent out from central stations is irrelevant if, as some of us believe, *none* of those plants can compete economically with other alternatives for supplying the same energy services, notably efficiency improvements and appropriate renewables.

SIGVARD EKLUND

The chairman/integrator of our nuclear power group said that some members of the group have considered the nuclear nonproliferation treaty to be dead. I would like to emphasize that the nonproliferation treaty has, in my opinion, been successful in preventing proliferation during the time the treaty has existed. If we aim at preventing proliferation but regard the nonproliferation treaty as dead, then the treaty must be replaced by something else. Under the present circumstances, I don't think that the general political atmosphere would be favorable for such a renegotiation of the treaty. Therefore, I instead urge increased support for the treaty, and I wish to note that every attempt is being made to renegotiate it to eliminate its disconcerting features.

A few other points with regard to the nonproliferation treaty. [These points are in response to a prior oral comment which has been excluded at the request of its author.] First, I would like to say that although only three nuclear-weapons states so far have signed the treaty, France for a long time has taken the position that it would act in international dealings in the same way as if it was a part of the treaty. France has in the last few years so changed its policy with regard to safeguards that I wouldn't be astonished if under the new Mitterand administration, France adhered to the treaty—but this is a personal opinion only. Second, with regard to China, the situation is completely different. China is not even a member of the International Atomic Energy Agency, and therefore there is no consideration on its side of joining the treaty at present. Third, with regard to the advantages of joining the treaty for a developing country, there is in the treaty an Article IV—as has been previously mentioned—by means of which collaboration between the advanced countries in the nuclear field could be arranged with the less advanced developing countries. There has been some slip in that because of the policy followed by the previous administration in the United States. I think, however, that the situation will improve. And fourth, again with regard to Article VI—which is a very essential article—I still very much hope that the nuclear-weapons states will take steps in order to limit nuclear armament.

One final comment. I have heard remarks about the IAEA: that it should be split and that there should be two organizations, one dealing with the promotional aspects of nuclear power and the other with the safeguarding aspects. This is a matter which has been under consideration in the agency; there are pros and cons for both aspects. In my opinion, and I think in the opinion of the IAEA's executive body, its board of governors, there is very clearly a majority for maintaining the present situation.

HARVEY BROOKS

I would like to supplement what Dr. Eklund just said on behalf of his work session's panel. I think I would agree almost completely with the other Symposium participant's analysis of the reasons why the nonproliferation treaty is in trouble, although I am not sure that I would go so far as to pronounce it dead, as he did. I think the death announcement may be premature. In our discussion we did not treat the issue in great detail, but we were very impressed with Dr. Eklund's suggestion that the negotiation of a comprehensive test ban—that is to say, a ban against all nuclear explosives testing—was the highest priority item on the agenda for trying to preserve the nonproliferation regime. In the absence of an ability to continue testing nuclear devices, the rate of so-called improvement of nuclear weapons would be greatly slowed down. Perhaps, more importantly, the inability to proof-test stockpile weapons would sufficiently erode the confidence in the reliability of a first strike so that the whole search for nuclear "superiority" and the consequent incentive to increase stockpiles would be greatly reduced. However, also of importance is the fact that a test ban would enhance the probability of success of negotiations for a substantial reduction in nuclear weapons stockpiles. I think that is worth adding to what Dr. Eklund has already said in response to the other participant.

AMORY B. LOVINS

It's important to remember that the obligation in Article IV of the NPT was negotiated at a time when many of the connections with military technology were not as well understood, even technically, as they are today, and when it was widely assumed that nuclear power would be relatively cheap, easy, and essential. The time may now be ripe, in the light of what we now know about nuclear power and alternatives to it, to recall that the specifically nuclear context of Article IV was an artifact of that then-prevalent belief and of the nuclear background of the negotiators, but this context does not reflect the essential purpose of Article IV— namely, that the developing countries desired help with energy security. If we therefore reconstruct Article IV as an obligation to help provide energy security rather than nuclear technology per se, then the legitimate desires of the developing countries, particularly of NPT adherents, can be satisfied, while proliferators would have to become explicit about their intentions. This could for the first time make nonproliferation policy logically self-consistent.*

*Mr. Lovins notes that this argument is expanded in Chapter 9 of A. B. and L. H. Lovins, *Energy War: Breaking the Nuclear Link* (San Francisco: Friends of the Earth, 1980; and New York: Harper & Row, 1981).

JOSÉ GOLDEMBERG

I did not belong to the nuclear power panel, but I would like to take this opportunity to make two comments on the report given by Harvey Brooks. It's obvious from his report that it was a very difficult work session; most of his statements were followed by a number of reservations and qualifications. However, there is one point which he mentioned that I would like to single out, not to take issue with him but to reinforce his point: it deals with the role of nuclear energy in developing countries. My impression is that most of the work of the panel dealt with developed countries; in developing countries nuclear energy has a modest role to play, although it has been pushed down the throats of many less developed countries over the last twenty to twenty-five years. I think that's an interesting story, because it shows that developing countries are an easy prey for the prevailing modes, tastes, and interests of the companies of the developed countries. In consequence, a number of developing countries have become quite excited about and interested in nuclear energy, although nuclear energy does not have a real role to play in those countries. A balanced choice of energy sources that can be used in such countries is very, very important, and in many instances, adequate attention has not been paid to the other sources available. Just to give an example, what really stood out at the Nairobi conference was the fact that biomass is a very important energy source. In addition, hydroelectric energy is undergoing a revival, since, as you know, only 10 percent of the world's hydroelectric resources are being used currently. This seems absolutely incredible, because hydroelectric energy is one of the oldest energy sources in existence, and the revival occurred only in 1981. In other words, I think the panel on nuclear energy should take a more humble attitude towards the production of electric energy. After all, nuclear energy is only one of the methods—actually, a fairly complicated method—of producing electricity which always must be compared with other methods of electricity production. I don't think the panel gave due consideration to this fact.

MIGUEL S. USSHER

I will support what Professor Brooks, our chairman/integrator during the nuclear power work session, said about the role of nuclear energy in developing countries, and at the same time I will answer Professor Goldemberg's remark on the priority use of hydroelectricity.

Whereas in entirely industrialized societies the nuclear sector is developed mainly for electrical generation, the less developed countries (LDCs) consider the same subject from a very different point of view: an LDC's decisions on nuclear power tend to be based on nonmarket benefits such as diversifying energy sources, improving advanced industrial skills, training technicians in modern industrial technologies, developing the related mining sector, and so forth. Furthermore, in many LDCs that are now attempting the nuclear option, there are abundant indigenous energy resources such as oil, natural gas, and hydro that are going almost untouched, but the governments of these countries are aware that the ultimate depletion of these resources inevitably will happen, although it may take many years. Consequently, it behooves these countries to gain some nuclear experience now in order to move forward in an independent way in the future, when the political leaders realize either that the additional benefits mentioned above are critical or that the nonrenewable sources are decreasing too fast.

In cases like those I've mentioned, the primary use of hydro is thoroughly understood, and the development of nuclear energy is undertaken particularly for industrial and training purposes rather than just for electricity generation.

MARCELO ALONSO

I don't want to take issue with José Goldemberg, but he was not present during the three days of discussions of our panel, and I think our chairman/integrator has been very faithful in presenting our discussions. At no moment was it said that nuclear power had to be used by developing countries. What was emphasized throughout is that developing countries as well as developed countries have to take into account the nuclear possibility together with the other energy options. Which one is going to be used? Well, that depends on the good judgment and technical capability of the experts or technical people, and on the understanding of the elite or political decisionmakers in these countries. The trouble is that if some countries don't have the proper technical people, it is very difficult for them to make the right decisions, not only on nuclear power but on many other energy-related issues. Thus, I don't want the Symposium participants to have the impression that our panel was pushing for one particular form of energy. While analyzing the nuclear power options, the panel was very conscious that there are many energy options. I therefore emphasize that what countries need is the capability to make decisions by themselves and not to be forced or pushed by others into decisions.

HARVEY BROOKS

There were a number of points mentioned during the discussion, and I don't know whether I can remember them all. Of course, I have had to go rather quickly through some subjects that had a good deal of discussion in our panel.

To take up a point dealing with the appropriateness of referenda, there was considerable divergence of view on this subject, but I think the majority view of the panel was that responsible elected officials should make the decisions and bear the responsibility for their decisions. Of course, if the public doesn't like what the elected officials decide, then it's up to them to replace the elected officials. What I think the group felt was that referenda are not a very good way of making decisions on complex technical subjects. That is not to say that there should not be a public input. In fact, the summary emphasizes the importance of public input, but there's a difference between public *input* and public *decision-making*—a difference which is very often glossed over in discussions of public participation. It is one thing for people to make decisions without any input from the public, but it is another thing for responsible elected officials simply to abnegate their responsibilities by turning over to the public difficult decisions involving complex tradeoffs among competing "goods" or "bads."

On the question that was raised about the relative economics of coal and nuclear power, that part of the panel's discussion was necessarily very condensed in the summary. I might amplify it slightly to clarify the point about conservation.

It is, of course, quite true that there are many technological opportunities for increasing energy efficiency that are cheaper than increases in supply by a corresponding amount. I think nobody on the panel would deny the existence of these opportunities. Where I think there would be a divergence of opinion is in just how big that "window" of conservation opportunities is—in other words, how quickly the opportunities for more cost-effective efficiency improvements would be exhausted. It is really the size of the window that is the subject of debate, not the existence of *some* window. The point of view which would be on the other side from the one expressed by Mr. Lovins would be stated roughly as follows: at the present time, because many economies—especially those in the developed countries—are in extreme disequilibrium as a result of the recent price changes, there are many opportunities to make the economies more efficient at the current price levels by substituting labor or capital for energy; nevertheless, these opportunities eventually will be exhausted and the growth in demand for electric power will resume. During the work session, we simply did not have time to address this in detail, nor did we think it was our charge to explore this particular question, which really has little to do with the characteristics of nuclear power per se. And again, in answer to the person who raised the question about the projected increase in electricity demand, we simply did not examine into that; we essentially accepted for purposes of further discussion the arguments that were put forward very eloquently by Professor Colombo in the first special plenary session. That was a stipulation of the discussion of our panel and not a conclusion; I hope I made that clear. It is true, I think, that the majority of the panel members were well persuaded by Professor Colombo's analysis, but not all of them were.

With regard to the discussion about the nonproliferation treaty, I think I should make it clear that the majority of the panel, perhaps all the panel, believed that the NPT regime was important, and even those who said that they felt that the NPT regime had been undermined by the preemptive attack on the Iraqi reactor would agree that the NPT regime should be kept in place, even if it requires modification of the present treaty to provide greater technology transfer advantages to those nations who adhere to the NPT. But we simply did not get into detail on this question, which obviously is a very complex political and diplomatic issue.

Work Session on the Biomass Energy/ Nonenergy Conflict

Decentralized Integrated Systems for Biomass Production and Its Energy/ Nonenergy Utilization

José Osvaldo Beserra Carioca
Coordinator
Nucleus of Nonconventional Energy Sources
Federal University of Ceará, Brazil

Harbans Lal Arora
Vice-Coordinator
Nucleus of Nonconventional Energy Sources
Federal University of Ceará, Brazil

INTRODUCTION

Ever since the dawn of civilization, two factors which have played an important role for human progress and development are energy and agriculture. In the initial stages of its development, agriculture not only provided food; it also met the household, agricultural, and industrial energy needs of its people through fuelwood, charcoal, crop and forest residues, animal dung, and so on. However, agriculture also used energy: its progress was based on diverse energy inputs such as human and draft animal energy, solar thermal energy, and irrigation by gravitation. Thus, there existed a very intimate interaction between agriculture, industry, and noncommercial forms of energy.

With the introduction of coal usage, the steam engine became the most important means for converting stored fossil chemical energy into mechanical and electrical energy. Later on, hydro power, converted through turbines into electrical energy, brought impressive developments in transport, agriculture, and industry.

More recently, the discovery of petroleum has ushered in an unparalleled and unprecedented era of development which has overshadowed former progress based on traditional or other earlier commercial forms of energy. The versatile properties of petroleum derivatives have found wide-range applications in transportation, agriculture, and household uses, in addition to providing primary material for chemical and other industries. Furthermore, factors such as low production costs and ease of transport and use have enormously accelerated petroleum's utilization in most of the economy's sectors. The second half of the twentieth century may rightly be called the petroleum civilization.

Other resources have difficulty competing with petroleum's attributes: for example, the uranium-based energy now being gradually introduced, in spite of being a highly powerful and concentrated source of energy, is not capable of serving the multi-faceted uses to which petroleum currently is being put. Furthermore, important commercial energy resources such as hydro power, coal, and uranium can find only limited applications, mainly in electrical and mechanical devices, at their present stages of development.

A region's progress in agriculture, industry, and transport is intimately linked with its energy availability and energy consumption patterns—in other words, with the quality and quantity of energy that it consumes. Modern agriculture, industry, and transport systems have flourished on energy-intensive inputs derived from commercial energy resources (coal, petroleum, natural gas, uranium, and hydro power).

The world's commercial energy resources are very unevenly distributed and utilized:

- Coal, which is the most abundant resource, representing more than two-thirds of the fossil fuel reserves, is found in abundance in a few countries. Currently about 28 percent of the technically and economically recoverable reserves of this fuel are in the United States and about 38 percent in the communist countries.[1]
- Oil, which represents only about one-sixth of the world's energy reserves, is limited mostly to the OPEC countries in the Middle East, with considerably smaller reserves in some other industrialized or developing countries.
- The world's natural gas reserves are presently estimated at about 460 billion barrels of oil equivalent, which is about 70 percent of proven oil reserves and 15 percent of proven coal reserves.[2] As far as distribution is concerned, more than 75 percent of the gas reserves are in North America, the Middle East, and centrally planned economies, including China.[3]
- The world's hydroelectric potential, which is virtually a permanent energy resource, is shared almost equally by the developing and developed countries. Of the total world estimated

capacity of 2,343 GW,[4] only about 10 percent has been developed so far.

- World shale oil reserves match with those of coal. As in the case of coal, the United States has the largest share, about two-thirds of the total world reserves, while Brazil occupies second place with 25 percent.

The worldwide patterns of commercial fuel consumption are given in Table 7-1. It can be seen from this table that coal and oil constitute about 30 percent and 46 percent respectively of the current consumption picture, although, as remarked earlier, their reserves are more than two-thirds and less than one-sixth respectively of the total fossil fuel reserves. Due to oil's uneven reserve distribution and to its overproduction and overconsumption, it is the commercial energy resource which dominates world trade markets. Another important point which must be emphasized is the importance of the associated technologies of petroleum use. A large number of oil-importing and -exporting countries depend on imported technologies for their utilization of this premium product. As can also be seen in the table, natural gas constitutes about one-fifth of the commercial energy consumed today, with the remainder distributed among hydro, nuclear, and other resources.

Table 7-1. **Worldwide Commercial Energy Consumption**
(millions of barrels of oil equivalent)

	1980		1990	
	Quantity	*%*	*Quantity*	*%*
Oil	63.1	45.8	77.3	38.4
Coal	41.3	29.9	62.5	31.0
Gas	27.0	19.6	45.0	22.3
Hydro	3.7	2.7	6.8	3.4
Nuclear	0.9	0.7	4.2	2.1
Others[a]	1.8	1.3	5.7	2.8
Total	137.8	100	201.5	100

Source: adapted from the World Bank publication *Energy in Developing Countries* (Washington, DC: World Bank, 1980).
[a]Includes alcohol, other nonconventional energy sources, and unallocated energy resources.

THE NATURE OF THE PROBLEM

Most assessments of the future energy situation point to petroleum supply/demand imbalances by 1985[5]—imbalances which could be postponed until about 1990 with events favorable for supply such as the discovery of new reserves and with reduced demand resulting from conservation and the development of alternative, substitutable resources. In these assessments, it has been assumed, of course, that there would be no abrupt disruptions in supply due to political instability in the oil-exporting countries or to unexpected oil price hikes. The projected commercial energy consumption picture for 1990 given in Table 7-1 reflects this expected imbalance for petroleum and shows favorable trends towards other resources, although in this projection oil still maintains its position as the prime supplier of commercial energy.

Thus, the energy crisis which we face now and shall confront in the current and perhaps the following decade is mainly the petroleum crisis—a crisis caused by a series of important factors: the real imbalance between petroleum supply and demand, its extraordinarily high rate of consumption, its skyrocketing prices, and the artificial shortage occasioned by uncertainty, due to political and other reasons, about its availability. Some significant effects of this petroleum crisis, particularly on issues related to this paper's topic, are considered below.

Production and Consumption of Commercial Forms of Energy

The production and consumption picture for primary forms of commercial energy in 1980 is given in Table 7-2, from which the following important conclusions can be drawn[6]:

- The oil-exporting developing countries (OEDCS) are the net exporters of primary commercial energy totaling 8.9 million barrels per day of oil equivalent (MBDOE). Of this total, oil constitutes about 95 percent of the energy exported. The OEDCS are also net exporters of natural gas, which constitutes about 9 percent of their total energy exports. These countries import about 4 percent of their other forms of commercial energy, such as alcohol, other nonconventional primary energy sources, and unallocated forms of energy.
- Over one-half of the exported fossil energy (coal, oil, and natural gas) goes to the developed countries and less than one-half to the oil-importing developing countries (OIDCS). The situation is reversed in the case of oil alone.
- About 85 percent of the total commercial energy is produced in developed countries, whereas these countries consume approximately 88 percent of the total, with a deficit of only 3 percent. Thus, the energy production share of the developing countries is less than one-fifth of that of developed countries (including oil-importing and -exporting countries). Of this small production, ironically, more than one-fifth is exported to the developed countries.
- The OIDCS are the ones that have the largest dependence on

Table 7-2. **Estimated Daily Primary Commercial Energy Production and Consumption of the World's Countries in 1980** (millions of barrels of oil equivalent)

	World	Developed countries			Oil-exporting developing countries			Oil-importing developing countries			Total for developing countries		
	P^a	P	C^b	D^c	P	C	D	P	C	D	P	C	D
Oil	63.1	49.9	53.9	4.0	11.2	2.7	-8.5	2.0	6.5	4.5	13.2	9.2	-4.0
Coal	41.3	38.8	38.7	-0.1	0.1	0.1	0.0	2.4	2.5	0.1	2.5	2.6	0.1
Gas	27.0	24.0	24.9	0.9	1.5	0.7	-0.8	1.5	1.4	-0.1	3.0	2.1	-0.9
Hydro	3.7	1.8	1.8	0.0	0.4	0.4	0.0	1.5	1.5	0.0	1.9	1.9	0.0
Nuclear	0.9	0.8	0.8	0.0	—	—	—	0.1	0.1	0.0	0.1	0.1	0.0
Others	1.8	1.5	1.0	-0.5	—	0.4	0.4	0.3	0.4	0.1	0.3	0.8	0.5
Total	137.8	116.8	121.1	4.3	13.2	4.3	-8.9	7.8	12.4	4.6	21.0	16.7	-4.3

Source: adapted from the World Bank publication *Energy in Developing Countries* (Washington, DC: World Bank, 1980).

[a] Production.
[b] Consumption.
[c] Consumption minus production.

imported commercial energy—imported sources constitute 37 percent of their meager daily consumption of 12.4 MBDOE. The situation with oil is even worse; about 69 percent of the oil consumed by these countries is imported.

• Considering that about three-fourths of the world's population lives in the developing countries and their share of commercial energy consumption is merely one-eighth, one can easily conclude that the per capita commercial energy consumption in the developing countries is less than one-twentieth of its counterpart in developed countries. In the case of OIDCs, this fraction is obviously much smaller—on an order of one-thirtieth.

The important conclusion to be drawn from the above is that although the total as well as the per capita oil consumption by the OIDCs is comparatively very small, more than two-thirds of it is imported. Thus, sharp increases in petroleum prices weigh especially heavily on the already negative balance of payments of these countries. In order to redress this situation, the following solutions may be considered, individually or in combination: (a) reducing the acceleration of industrial growth, (b) intensifying oil exploration, (c) conserving energy, and (d) developing and substituting locally available new and renewable energy sources for petroleum derivatives.

Agriculture: Its Energy/Nonenergy Relationships

Agriculture is a producer of both energy and so-called nonenergy primary materials.

The former include traditional fuels such as fuelwood, crop and livestock residues, and charcoal; the latter include products such as food, feed, fiber, construction wood, and organic fertilizers in the form of leftover crop and livestock residues. It is this dual utilization of agriculture that has sustained agriculturally based economies and populations since the beginning of agriculture itself.

Most of the OIDCs depend heavily on traditional forms of energy, with those resources that are renewable and are available in solid form used mainly for cooking. It has been estimated that in poorer countries these traditional sources supply one-half to three-quarters of the total energy used; the proportion varies from 50-65 percent in Asia to about 70-90 percent in Africa. For the majority of the population of these continents, the traditional sources are, in fact, the only sources of energy.

These traditional sources may account for about 8-9 Mbdoe, or roughly the equivalent of the commercial energy consumed by the developing countries. However, considering that the traditional wood and charcoal stoves have combustion efficiencies which are 30-40 percent of those of kerosene and gasoline stoves, these traditional resources are equivalent to only about 3 Mbdoe— approximately one-third of the total commercial energy consumption. Although this is a comparatively small amount of energy, these traditional resources are extremely important for the majority of the rural population since they are practically the only sources of energy available to them.

The human and draft animal energy and agricultural tools and implements manufactured by using agriculturally based energy are still the main energy inputs for agriculture for most of the less developed countries. Modern mechanized agriculture, however, is based on highly energy-intensive technologies that derive their energy from commercial energy resources, primarily petroleum derivatives. The manufacture and operation of farm machinery, powered irrigation, and the production and utilization of inorganic fertilizers and other agricultural inputs all employ mainly petroleum derivatives.

The important thing to note is that modern agriculture uses only a very small fraction of the total commercial energy consumed. According to estimates by the UN Food and Agriculture Organization (FAO)[7] agricultural production during the year 1972–73 was responsible for only about 3.5 percent of the total world use of commercial energy. The FAO projections further estimate that this proportion has not changed significantly since then.

But the fact is that this small fraction has been responsible for a large-scale increase in food production, keeping production ahead of population growth and thereby bettering the per capita global food supply. Undoubtedly, agriculture benefits greatly by its use of petroleum derivatives, out of proportion to the amount used. It would be a folly, therefore, to give way to the pressures of supply shortages and high prices and deprive agriculture of its very small share of petroleum. Studies show that for each 1 percent growth in agricultural production, a more than 2 percent increase in the commercial energy input is needed.[8] Thus, agriculture prospectively will require a larger share of commercial energy —energy essential for the rapid increase in agricultural production required to meet the needs of a growing population.

The figures rated above show that 96.5 percent of the total commercial energy consumed, or about 93 percent of the petroleum consumed, is used by industry and transportation. This tremendous amount of energy serves to maintain the high living standards of the people in developed countries and of the urban and rural rich in developing countries. More than half of the world's population live in rural areas of the developing countries, and these people have been only marginally touched by industrial and transportation benefits. In fact, only a little more than 0.5 percent of total commercial energy is used by this segment of the world's population.

Agriculture's use of petroleum derivatives is extremely uneven globally. Most of this sector's commercial energy is used in the developed countries. It is estimated that the developing countries, with more than two-thirds of the world population, only use about 18 percent of the total commercial energy consumed by agriculture, whereas 82 percent is used by the developed countries.[9] One thus can conclude that the per capita energy consumption by agriculture in the developed countries is more than ten times that in the developing countries. However, the agricultural output in the developed countries is correspondingly higher, largely as a result of this energy input. For example, Japan's high rice productivity is mainly due to this factor.

It is interesting to observe that in the developed countries, the processing, transport, marketing, and preparation of food entail additional commercial energy inputs which take more than four times as much commercial energy as the food's initial production on the farm. In developing countries, on the other hand, these energy-intensive processes are practically nonexistent. Another important fact is that the primary beneficiaries of the agricultural production enhanced by commercial energy inputs are the developed countries themselves, both those with net agricultural surpluses and those with net agricultural deficits.

The above analysis shows once again that the share of commercial energy used by the OIDCS' agricultural sector is very small. Nevertheless, this share is important for meeting the increasing food needs of these countries' fast-growing populations. And lest we get lost in statistical jargon and enigmatic numbers, the stark fact is that statistics and numbers can hardly describe the darkness and drudgery of living conditions for the majority of the African, Asian, and Latin American rural populations and for the urban slum-dwellers of these regions.

The core of the whole problem is that the industrialized nations have too much energy, whereas the rest of the world has too little. Furthermore, the uneven distribution of commercial energy resources among industrial, transport, and agricultural sectors and the skewed disparities in the use of these resources between and within regions of the same country are some of the important factors which heighten the seriousness of the situation.

It is within this broad framework that we should view the petroleum crisis. Only then can alternative new, just, and sustainable solutions be boldly proposed, critically analyzed, wholeheartedly accepted, and rigorously implemented.

The Petroleum Crisis and Biomass

The main problem which we are facing and which could

become more serious with time is that of the increasing scarcity of, but more importantly the sharp price increases in, petroleum derivatives. Technically speaking, these liquid fuels can be replaced by those derived from other fossil reserves (coal, tar sands, and oil shale) as well as from biomass, a renewable resource.

As far as the potential fossil substitutes are concerned, these do not offer a short-term solution given the present state of technology for their conversion into liquid fuels. Furthermore, even though these fossil reserves are comparatively large, they are finite and would naturally become scarce in the future. Moreover, owing to their uneven distribution, liquid fuels derived from them are bound to create dependence on the countries that possess them and the technology for their utilization.

On the other hand, biomass can be a highly feasible petroleum substitute since the technology for its conversion into liquid fuel is relatively simpler and is already known, although technological improvements are possible and desirable. In Figure 7-1 are shown various processes for converting biomass into fuels and other primary feedstocks for complete substitution for petroleum. As shown therein, biomass seems to offer large possibilities as a short-run and probably even a long-run solution to the petroleum crisis.

The oil-shortage situation demands that agriculture be called upon to play an extended role to provide liquid fuels for (1) substitution for petroleum-derived, energy-intensive inputs to increase agriculture's own production of food, feed, fiber, and construction wood, as well as the traditional fuels and (2) substitution for petroleum in industry and transport. Agricultural resources have the advantage of being renewable but are limited by factors such as the availability of arable land, water resources for irrigation, fertilizers, and other inputs.

Obviously, then, a conflict is bound to arise if biomass is to produce both energy and nonenergy products. Here is a challenge and an opportunity to make intelligent use of agriculture so that it may play this dual role harmoniously, not only avoiding conflict but making these two uses complementary. Such an opportunity cannot and should not be missed since it offers avenues for the on-site production of environmentally cleaner nonfossil liquid fuels and other feedstocks and since it provides opportunities for overall rural development.

Energy and Agriculture: National Prospects

The real potential for exploiting the biomass alternative for liquid fuels substitution varies from country to country, depending principally on the following parameters:

- *Index A*, the agricultural self-sufficiency index, which is the ratio between current production and current consumption of nonenergy agricultural products (food, feed, and fiber).
- *Index E*, the energy self-sufficiency index, which is the ratio between current total production and current total consumption of commercial forms of energy.

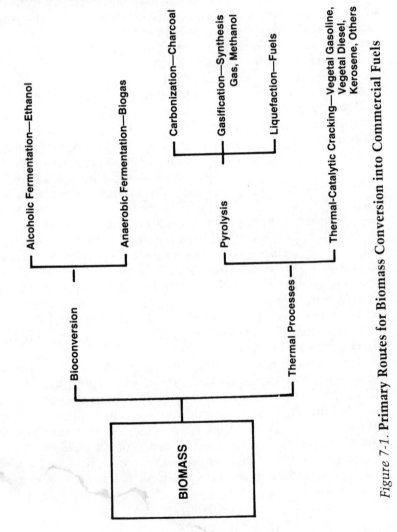

Figure 7-1. Primary Routes for Biomass Conversion into Commercial Fuels

• *Index L,* the arable land self-sufficiency index, which is the ratio of total available arable land to the land currently under cultivation.

Parameters A, E, and L and per capita consumption of food and energy are given by nation in Table 7-3. In addition, parameters A and E are plotted in Figure 7-2 as a two-dimensional agriculture/energy matrix.

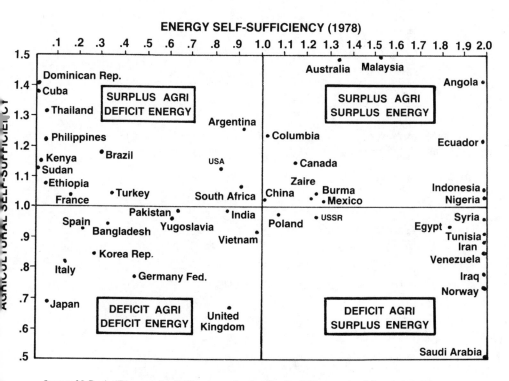

Source: N. Rask, "Biomass: It's Utilization as Food and/or Fuel," International Symposium IV on Alcohol Fuels Technology, Guarujá, Brazil, October 5–8, 1980.

Figure 7-2. **Energy and Agricultural Self-Sufficiency Matrix for Selected Countries**

As shown in Figure 7-2, the countries of the world can be grouped into four broad categories, characterized as follows:

1. A > 1, E > 1—surplus agricultural and energy production. These countries have the most favorable situations, since

Table 7-3. **Energy and Agricultural Data for Selected Countries**

	Self-sufficiency ratios[a]			*Per capita consumption*		*Population*
				Food	*Commercial*	
	A^b	E^c	L^d	*(Cal)*	*energy (kgoe[e])*	*(x 10^6)*
Africa						
Algeria	.66	7.79	1.20	2,357	468	18
Angola	1.41	9.35	12.50	2,036	131	7
Benin	1.07	0	1.96	2,153	38	3
Burundi	1.08	.15	1.20	2,260	8	4
Cameroon	1.31	.20	2.00	2,408	81	8
Central Africa	1.08	.09	4.34	2,250	30	2
Chad	1.31	0	2.43	1,793	15	4
Congo	.91	11.88	16.66	2,234	119	2
Egypt	.90	1.86	1.01	2,716	315	41
Ethiopia	1.08	.08	1.75	1,838	13	32
Gabon	.47	17.26	25.00	2,403	1,246	1
Gambia	1.44	0	1.42	2,281	73	1
Ghana	1.37	.31	2.17	2,014	112	12
Guinea	.99	.02	1.63	1,921	62	5
Ivory Coast	1.90	.02	1.23	2,563	243	8
Kenya	1.15	.04	1.69	2,060	95	15
Liberia	1.22	.05	3.57	2,374	269	2
Libya	.49	31.13	1.20	2,946	1,285	3
Madagascar	1.08	.03	7.69	2,480	53	8
Malawi	1.23	.13	1.69	2,282	35	6
Mali	1.03	.03	1.23	2,114	21	6
Mauritania	1.51	0	1.05	2,557	138	2
Mauritius	.85	.02	2.32	1,894	277	1
Morocco	.92	.19	1.14	2,568	194	20
Mozambique	1.13	.89	5.88	1,930	103	10
Niger	1.03	0	1.07	2,051	26	5
Nigeria	1.02	18.11	1.25	2,291	72	83
Rwanda	1.07	.23	1.09	2,277	12	5
Senegal	1.16	0	1.72	2,228	123	6
Sierra Leone	1.01	0	1.17	2,101	68	3
Somalia	1.03	0	1.78	2,129	67	4
South Africa	1.07	.93	1.51	2,945	2,162	28
Sudan	1.13	.02	3.70	2,247	117	21
Tanzania	1.08	.16	4.00	2,089	44	17
Togo	1.18	0	1.05	2,035	65	2
Tunisia	.88	2.02	1.05	2,657	369	6
Uganda	1.26	.14	5.55	2,070	33	13
Upper Volta	1.07	0	1.26	1,997	17	7
Zaire	1.03	1.20	16.66	2,312	47	28
Zambia	.91	.68	10.00	2,018	322	6
Zimbawe-Rhodes	1.30	.73	7.14	2,545	394	8
North America						
Canada	1.14	1.12	2.08	3,346	6,755	24
United States	1.13	.81	2.04	3,537	7,737	231
Central America						
Costa Rica	1.87	.16	2.04	2,477	384	2
Cuba	1.38	.01	1.01	2,636	794	10
Dominican Rep.	1.41	.01	1.12	2,107	316	6
El Salvador	1.64	.10	1.14	2,075	181	5
Guatemala	1.67	.05	1.38	2,166	177	7
Haiti	1.02	.10	1.20	2,040	39	6
Honduras	1.49	.06	3.12	2,074	193	4
Jamaica	.77	.01	1.14	2,663	1,240	2
Mexico	1.02	1.28	1.49	2,668	941	66
Nicaragua	1.84	.04	2.77	2,453	351	2
Panama	1.12	.02	2.94	2,357	674	2
Trinidad & Tobago	.69	3.72	1.58	2,684	3,377	1

Table 7-3—Continued

	Self-sufficiency ratios[a]			Per capita consumption		Population
	A^b	E^c	L^d	Food (Cal)	Commercial energy (kgoe[e])	(x 10⁶)
South America						
Argentina	1.24	.92	1.51	3,359	1,274	27
Bolivia	1.02	2.18	100.00	2,134	250	5
Brazil	1.18	.34	5.26	2,522	540	124
Chile	.81	.53	1.04	2,644	678	11
Columbia	1.24	1.03	5.88	2,255	476	26
Ecuador	1.22	3.71	1.63	2,109	344	8
Guyana	1.54	0	20.00	2,431	728	1
Paraguay	1.22	.12	16.66	2,779	136	3
Peru	.95	1.09	4.16	2,286	442	17
Surinam	1.04	.16	100.00	2,286	1,463	—
Uruguay	1.19	.07	5.88	3,098	717	3
Venezuela	.79	4.71	5.88	2,480	2,033	14
Asia-Middle East						
Cyprus	1.10	0	1.02	3,047	1,338	1
Iran	.86	6.72	1.01	3,193	1,230	38
Iraq	.74	24.58	0.99	2,306	430	13
Israel	.93	.01	1.66	3,145	1,607	4
Jordan	.63	0	1.01	2,067	364	3
Lebanon	.71	.03	1.13	2,495	637	3
Saudi Arabia	.47	48.18	1.14	2,472	888	8
Syria	.96	2.00	1.16	2,616	659	8
Turkey	1.04	.35	1.02	2,916	539	44
Yemen, AR	.81	0	1.12	2,179	36	2
Yemen, PDR	.65	0	1.14	1,897	356	5
Asia-Far East						
Afghanistan	1.05	4.34	1.06	1,974	32	15
Bangladesh	.94	.39	1.00	1,945	29	88
Burma	1.04	1.21	4.16	2,211	43	34
China	1.02	1.03	0.76	2,439	569	1,024
Kampuchea	.95	0	1.56	1,857	2	8
India	.99	.85	1.07	1,949	121	670
Indonesia	1.05	3.44	1.81	2,115	189	148
Japan	.69	.08	1.23	2,847	2,602	116
Korea, PDR	.97	.96	1.08	2,730	1,838	19
Korea, Rep.	.83	.36	0.93	2,682	925	40
Laos	.92	.14	25.00	1,979	41	4
Malaysia	1.87	1.53	1.85	2,594	487	13
Nepal	1.00	.12	1.17	2,070	7	14
Pakistan	.99	.65	1.38	2,255	117	80
Philippines	1.21	.06	1.51	2,155	231	47
Sri Lanka	1.12	.10	1.20	2,047	74	14
Thailand	1.31	.05	1.75	2,193	222	46
Viet Nam	.92	.99	1.29	2,032	85	53
Western Europe						
Austria	.91	.34	1.33	3,547	2,754	8
Belgium	.91	.14	1.38	3,565	4,134	10
Denmark	1.48	.02	1.36	3,432	3,689	5
Finland	.92	.09	1.85	3,130	3,541	5
France	1.05	.20	1.49	3,458	2,971	54
Germany, Fed. Rep.	.76	.45	1.25	3,362	4,092	61
Greece	1.09	.25	1.66	3,441	1,310	9
Iceland	.80	.30	1.00	2,939	3,295	—
Ireland	1.53	.24	2.43	3,519	2,239	3
Italy	.81	.14	1.53	3,462	2,197	57
Malta	.41	0	1.75	3,103	736	—
Netherlands	1.13	1.32	1.36	3,324	3,624	14

Table 7-3—Continued

	Self-sufficiency ratios[a]			Per capita consumption		Population
				Food	Commercial	
	A^b	E^c	L^d	(Cal)	energy (kgoe[e])	(x 10^6)
Norway	.72	2.28	2.38	3,126	3,790	4
Portugal	.78	.13	1.47	3,424	701	10
Spain	.93	.26	1.72	3,210	1,636	38
Sweden	.88	.18	1.63	3,168	4,050	8
Switzerland	.74	.22	1.56	3,386	2,510	6
United Kingdom	.67	.84	1.58	3,305	3,546	56
Yugoslavia	.97	.62	1.36	3,469	1,384	22
Eastern Europe						
Albania	1.04	1.91	1.13	2,624	679	3
Bulgaria	1.06	.32	1.36	3,594	3,415	9
Czechoslavakia	.90	.73	1.53	3,450	5,123	15
German Dem. Rep.	.88	.70	1.36	3,610	4,844	17
Hungary	1.11	.58	1.33	3,494	2,348	11
Poland	.97	1.09	1.26	3,647	3,806	35
Romania	1.03	.92	1.13	3,368	3,749	22
Soviet Union	.97	1.28	1.51	3,443	3,742	264
Oceania						
Australia	1.61	1.35	3.44	3,413	4,505	14
New Zealand	1.58	.57	2.77	3,443	2,579	3

Source: based on N. Rask, "Biomass: Its Utilization as Food and/or Fuel," International Symposium IV on Alcohol Fuels Technology, Guarujá, Brazil, October 5–8, 1980.
[a]See text for explanation of these ratios, or indices.
[b]Agricultural self-sufficiency index.
[c]Energy self-sufficiency index.
[d]Arable land area self-sufficiency index.
[e]Kilograms of oil equivalent.

they are net exporters of both energy and agricultural products. In this group of countries, those which have high L indices (i.e., large proportions of arable land areas) will have high prospects for producing biofuels to export or to meet their own increasing energy demands. Countries such as Colombia, Australia, and Angola, among others, fall into this subcategory. Angola, for example, has a large capacity to raise its per capita food consumption by increasing its agricultural production.

2. $A > 1$, $E < 1$—surplus agricultural production and deficit energy production. Countries in this group have prospects for converting biomass into fuels. However, their real prospects for biofuels policies will be determined by their specific locations within Figure 7-2's quadrant and by their arable land self-sufficiency indices. In Brazil, for example, the value of L is greater than 5, and the country therefore possesses strong prospects for a biofuels program by bringing large areas of land under cultivation. On the other hand, the Philippines, whose value of L is approximately 1.3, has to depend more on increasing the productivity of land areas already under cultivation and on making more efficient use of agricultural and animal residues if it is to establish a biofuels program.

3. $A < 1$, $E > 1$—deficit agricultural production and surplus energy production. These countries have few prospects for biofuels, since their tendency naturally would be to increase food production. Due to their surplus commercial energy, they have good prospects for increasing the productivity of the land now under cultivation, especially those countries (such as Egypt, Iran, and Iraq) with an L value of approximately 1. On the other hand, Venezuela, with an L value of about 6, has prospects both for increasing existing productivity and for harnessing currently unused arable lands.

4. $A < 1$, $E < 1$—deficit agricultural and energy production. These countries need land for the production of both energy and food in order to become self-sufficient. Since food production should be given preference, the prospects for energy production from biomass are very unlikely, except in the instance of using waste agricultural products. Countries like Japan, the Federal Republic of Germany, and Italy fall in this category.

AN INITIAL EVALUATION OF THE OPTIONS

General Technical and Economic Considerations

The amazing industrial development achieved by modern societies may be traced to the profoundity and vastness of the human spirit. This spirit, although at times contradictory, is at the same time entirely responsible for finding solutions to preserve mankind in the midst of the great threat now upon us—a threat that results from the industrial development which has been mankind's objective.

A rapid analysis of the genesis of this industrial civilization reveals that in eighteenth-century England, a new tendency directed towards the production of goods in large scale and at low cost appeared in the nation's economy.[10] This tendency was associated with the new and extensive use of cast iron and coal: the first steps in the evolution towards the present consumer societies, industrial centralization, mass production of capital goods, and so forth.

The need for a low-cost fuel in England brought the development of the coal industry in substitution for fuelwood, which was becoming quite scarce. In fact, wood was the basic fuel for all of Europe at that time, and with the expansion of ship construction as well as the need to house a growing population, the demand on wood was always pressing.

In primitive metallurgical processes, iron was obtained directly from iron ore. With modern large-scale energy-intensive processes, it became economically possible to explore less concentrated iron ore, by employing melting technology. Thus, the large-scale production of various articles such as locks, keys, pots, guns, and so on became a matter of common practice.

The development of the coal industry was so intensive and rapid that it brought serious environmental problems. Chimneys were installed in residences and small business houses to throw off soot and smoke which then polluted the atmosphere and caused great environmental damage, affecting even aquatic life under the river Thames.

It took approximately two hundred years for the total substitution of coal for charcoal to take place. The progressively more extensive use of coal during the end of the seventeenth and beginning of the eighteenth centuries was made feasible by deeper mines, improved drainage, and the increasing ability to transport coal in large quantities over longer distances.[11]

The increase in coal production helped stimulate the invention of the steam engine in the 1780s. About fifty years later, this engine was being used in various industries in several European countries. From then up to the present day, coal has been utilized as fuel in thermoelectric power systems, in the steel industry, and in other industries.

Furthermore, with the current high petroleum prices the utilization of coal has increased recently, since coal can serve as an alternative to some of the petroleum derivatives.

However, in spite of coal's greater abundance than petroleum, the use of coal must be restricted because of the serious consequences to the environment caused by the large quantities of carbon dioxide it produces when burned. Since carbon dioxide is already detected in high levels in the atmosphere, its further concentration might reach critical levels which would cause polar temperature variations and thus disturb the climate in many regions of the globe. Taking into serious consideration coal's environmental risks and the high costs and safety hazards of mining it, the extensive use of this concentrated and highly polluting energy source becomes unviable. Although coal is an abundant fuel, it is less valuable and much less versatile than petroleum when considered as a raw material.

About fifty years ago, 80 percent of the world's commercial energy came from coal whereas 16 percent came from petroleum and natural gas.[12] Currently, about 65 percent of the total commercial energy is derived from these last two sources. However, as petroleum replaced coal, mankind did not realize that there potentially would be serious conflicts among nations interested in that precious "black gold"—conflicts because petroleum's possession, exploration, and utilization have become a privilege of a few economically powerful and politically influential nations.

Nobody can deny the tremendous material progress achieved everywhere, especially in all modes of transport, during the last fifty years. In parallel with this progress, however, the so-called developed countries started an intense weapons program unprecedented in the recorded history of mankind. Another serious consequence of rapid material progress has been the disorganized growth of cities, making impracticable the healthy and harmonious development of human beings. Environmental pollution has achieved levels never imagined before. Psychosomatic diseases brought on by urban stresses and strains are diminishing the average useful life span of metropolitan inhabitants. Worst, this kind of material progress benefits a few while the problems created affect everyone, but that paradox is disguised under the impressive word "technology."

The development of the petroleum industry has brought innumerable goods and services to mankind, from portable liquid fuels to the fascinating and sophisticated world of plastics, synthetic fibers, and so on. Rapidly and shortsightedly, countries without petroleum reserves, attracted by these goods and services and deluded by the illusion of cheap petroleum, started to import this product unrestrainedly. Thus an indiscriminate and extensive dependence upon petroleum generated the present global economic crisis, precipitated by the sudden hike and continuous rise in petroleum prices—prices dictated by the exporting countries and by the technology dominators.

Under the impact of this crisis, mankind was suddenly awakened and realized that petroleum resources are finite and that we are on the threshold of a deficit between demand and supply of this vital product. Now, suddenly, national sovereignty is being threatened by invasions which are becoming more frequent and are motivated by the prospect of gaining domination over part of the petroleum reserves. This situation raises the following questions:

- What price should we pay for the miraculous technological development, for the centralized systems, and for the visible but highly unstable progress?
- What should we do to achieve a more just, participatory, and humane kind of progress for mankind?

Surprisingly, the reasoning of economists and development enthusiasts seems almost logical: increased population, increased demand, economicity, large systems, substitution of machines for labor, and so on. But as a result the energy-consuming curves in developed countries, especially in relation to electricity, have reached asymptotic limits. Even the large hydroelectric power plants, which are responsible for flooding vast areas of arable land, are not sufficient to supply the present level of electricity demand in large industrialized centers. It is at this juncture that nuclear energy enters the scene—but not without problems.

In fact, the future of nuclear energy appears very dangerous or at least dubious. The development and use of nuclear power will accelerate the invention of sophisticated and highly lethal weapons which will increase the tensions among the most powerful nations and probably will lead the world into a nuclear armed conflict with unpredictable consequences if not total destruction of mankind. Furthermore, the characteristics of the nuclear power systems will drive modern societies to a centralization even more serious than the one brought about by petroleum, because of the permanent risks of the high levels of radioactivity to which humans will be exposed.

The development of the nuclear system has occurred much faster than those of coal and petroleum. In spite of the vulnerability of technicians directly involved with reactor operations and of all humans indirectly through the environment, this type of industry has spread within a few decades in the developed nations. Although there currently is considerable debate in many of the developed countries about whether nuclear power is economically feasible, this form of power generation is now being extended to the developing countries, at least in part to help the large nuclear enterprises recover the high investments they have made—and this, despite the fact that the developing countries are even less equipped financially to handle nuclear power than are the developed countries.

Frequent protests have been made against the installation of new nuclear plants, seeking moratoria until the complete safety of reactors can be assured. Among these, we must emphasize the reaction of the World Council of Churches conference held in Boston in July of 1979. After detailed analysis of these problems from various aspects—ethical, moral, social, technical, religious, ecological, economical, and so on—the conference found that the development of nuclear energy is giving the governments of certain countries an extraordinary power which is extremely dangerous for the stability of mankind.[13]

The issues discussed above show, on the one hand, the human spirit's tremendous power in striving along the difficult path towards progress. However, on the other hand, they demonstrate its capacity to generate problems by its own work, thus jeopardizing the very subsistence of mankind and degrading the environment and the quality of life.

Mankind's eagerness for progress has brought enormous centralization as far as production, exploration, and utilization of natural resources are concerned, with the extent of this centralization varying according to the availability of the resource. Technology has acted as the meeting point among those who possess the natural resources, those who dominate the resource technologies, and those who are users of the resource systems. The economic struggles over these systems can be considered as responsible for their superdevelopment, centralization, and domination by a few, and the same forces which made material progress feasible have created a permanent instability, inequity, and lack of safety while at the same time keeping the members of modern societies under their domination.

The evolution of today's scientific research methods made possible the technological revolution and was responsible for the great scientific discoveries which led to the development of mechanization, fast transport systems, and rapid means of communications. Stimulated by the discovery and application of such new energy sources as petroleum, hydroelectricity, and, recently, nuclear energy, the human spirit has focused societal progress within an increasingly narrow prism, thus creating an increasingly great risk. In the name of progress and economic development, today's industrialized societies have achieved amazingly high energy consumption levels while two-thirds of the world's population still lives in subhuman conditions, consuming scarce fuelwood to meet its basic necessities. Thus, owing to the very nature of the centralized systems, a series of conflicts were and are constantly being generated, thereby risking the survival of mankind itself.

In an effort to mitigate these conflicts, various institutions —for example, the United Nations and the World Council of Churches, among others—are dedicating enormous efforts to a search for a more just and participatory society where all can collaborate on and benefit from the common good. The North/ South dialogue is yet another step in this desired direction.

The industrialized countries have nearly exhausted most of the known oil reserves within the short span of half a century, but it was only after the impact of the sudden and soaring petroleum price increases that modern society became conscious that it was consuming more energy than it should. Inadequately planned buildings and the excessive use of energy by transport vehicles and in industrial installations all have compelled the establishment of energy conservation programs. The available data reveal the enormous importance of such programs and of treating energy intensivity as a fundamental decisionmaking criterion. Thus, for example, in place of aluminum, less energy-intensive materials have been suggested. Furthermore, the employment of special materials aimed at minimizing energy losses in big buildings has been recommended. New total energy systems such as fuel cells are being improved to make possible decentralized independent systems of electricity generation that are capable of simultaneously using their residual heat for heating or cooling.

Another significant characteristic of centralized energy systems based on resources such as coal, petroleum, and nuclear power is that these systems are capital intensive rather than labor intensive, a characteristic which can be seen in industrial developments based on large mechanized devices. On the other hand, systems that are less centralized and less capital intensive may produce more job opportunities. Thus, since labor, energy, materials, and money are to an extent interchangeable, these components require careful orchestration with a view towards creating a large number of jobs and minimizing the use of scarce energy resources. And thus also, the equation of available labor with the specific needs of each nation assumes great importance, since erroneous planning in this respect is bound to generate serious social problems—especially in the developing countries, where there is a great need for a large number of new jobs at the lowest possible level of investments. Some countries such as China are already experimenting with decentralization, and others such as Japan and Brazil[14] have urgently felt the need for an industrial decentralization program.

If the use of primary materials continues to expand at the present high levels, the extraction and processing of such materials will put great strain on energy supplies, especially since the scarcity of petroleum and other fossil sources is going to become very serious over time. Recycling these materials therefore becomes quite feasible economically, particularly considering the fact that reutilizing these materials consumes much less energy[15] and minimizes their environmental effects. For example, the recycling of iron, copper, aluminum, paper, and other materials has become a real necessity for industrialized countries.

According to Gaspar,[16] if nothing unexpected happens the world's population will present the following distribution pattern in the year 2000:

Year	Population x 10^9	Distribution % Developed	Developing
1850	1.26	34	66
1975	3.97	29	71
2000	6.25	22	78

These data reveal that about 25 percent of the world's population live in industrialized countries. This minority consumes approximately 75 percent of the world's natural resources and controls approximately 88 percent of the world's total production, 80 percent of its trade and investments, 93 percent of its industrial products, and 100 percent of all scientific and technological development.[17] On the other hand, there currently are more than 2 billion people living in underdeveloped countries who suffer from hunger and malnutrition, both of which hamper their willingness to work and reduce their resistance to disease. Every year, 50 million children die from starvation or disease in these countries. A fundamental question about the utilization of natural resources may therefore be posed: How might new energy sources contribute to a reorganization of society or to a new economic and social order on a worldwide basis? Another basic question is this: How can finite natural resources such as petroleum, coal, and minerals best help to cope with the increasing demand of a world population which has few prospects of stabilizing in the near future?

Questions such as these encourage joining in the common struggle to establish a reeducation program at all levels, whose objective is to overcome the widespread ignorance responsible for the high populational growth rate in underdeveloped countries. The challenge to overcome hunger, poverty, and disease is a task not too far out of reach of the human spirit. It must be emphasized, however, that no solution to energy and food problems can be feasible and lasting without strict population growth control.

In the modern world a nation's development plan should pay particular attention to energy demand per capita. The dynamic relationship between human participation and mechanization should constitute an important variable in planning a society. While this does not mean that economic growth should be deterred, it does mean that the growth must be fair and just, with the rational and integrated use of available resources intended for the benefit of all.

Technical and Economic Considerations for Biomass-Derived Fuel

Mankind's basic needs can be satisfied with a sufficient supply of food and energy, although to attain more than a minimal quality of life, other fundamental needs must be provided for through health care, education, housing, job opportunities, and so forth. In terms of energy, which pervades all aspects of life, the present energy crisis can be indirectly characterized as scarcity of liquid fuel.

Despite the shortage of petroleum, there are still the possibilities of shale oil, coal, and biomass-derived fuel as ways to maintain the present rate of economic development. Due to the technical difficulties of producing liquid fuel from coal and shale at the same cost as petroleum, several countries have started to develop liquid fuels from biomass. This alternative, which appears to be a valid one, has raised a series of political, institutional, social, and environmental controversies which will be discussed later in this paper. But first, it is important to point out two basic questions related to the use of biomass for energy purposes:

- What is the correct size of a system which produces biomass from energy?
- What criteria should be adopted to determine the use of biomass for food or energy production?

System size. Today many countries including Brazil, forced by the present market conditions, have started their own liquid fuels production from biomass. However, in a search for economy, the industrialist often chooses a very large production plant—one which will eventually bring individual gains and collective losses. Therefore, the choice of the appropriate size for a biomass energy production system demands careful planning to avoid repeating the same mistakes made by the large centralized systems.

Let us consider the sizes of the biomass energy production systems reviewed in Table 7-4's analyses of pertinent variables. Three sizes were considered for this analysis: small, medium, and large. A small, or micro, system is defined for this paper's purposes as one producing up to 2,000 liters of ethanol per day, an amount which would be sufficient for a farm's own consumption and therefore excludes distribution costs. A medium-sized, or mini, system is defined as one producing about 20,000 liters of ethanol per day, an amount appropriate for the liquid fuel consumption of a small town. Finally, a large, or macro, system is defined as one producing about 200,000 liters of ethanol per day which would probably then be transported to large centers of consumption. The macro system is comparable to fossil fuel systems, in that they are characterized by a large concentration of capital investment.

From the table, it can easily be seen that the medium-sized system is the ideal one from a technical, economic, social, and environmental standpoint. Such mini systems appear to be a highly viable solution to the liquid fuel energy problem, especially for developing countries where distribution of income and wealth is formidably distorted. Furthermore, these mini systems could serve as a means of solving the chronic rural problems of unemployment and underemployment, thereby reducing the enormous migration to urban centers and the consequent exacerbation of urban slums.

The establishment of a decentralization policy with a view towards installing an appropriate number of such mini systems would both aid the cause of human justice and be practical, espe-

Table 7-4. **Qualitative Considerations about Capacities of Alcohol Distilleries**

	Size of distillery		
	Micro[a]	Mini[b]	Macro[c]
Agricultural sector			
Cost of transport	low	reasonable	high
Risks associated with feedstock	small	reasonable	large
Mechanization	none	some	high
Investments	small	reasonable	large
Land use diversification possibilities	large	reasonable	small
Environmental constraints	none	few	many
Industrial sector			
Level of technology	simple	simple	sophisticated
Labor requirements	semi-specialized	semi-specialized	semi-specialized
	small	small	large
Management efficiency	high	high	low
Total investments	low	reasonable	high
Resistance to technological improvement	nonexistent	small	large
Possibilities of technology diffusion	many	many	few
Environmental constraints	none	few	many
Cost per liter			
Production	high	reasonable	low
Distribution	low	low	high
Social	negligible	low	high
Total	large	reasonable	large

Source: adapted from J. O. B. Carioca, H. L. Arora, and A. S. Khan, "Technological and Socio-Economic Aspects of Cassava-based Autonomous Minidistilleries in Brazil," to be published in *BIOMASS - An International Journal.*
[a]About 2,000 liters per day.
[b]About 20,000 liters per day.
[c]About 200,000 liters per day.

cially since the investments involved would be compatible with a medium-level industrial capacity. Furthermore, the use of this type of unit within a framework of cooperatives would allow a more homogeneous distribution of wealth, unlike large undertakings which by their very nature concentrate power and profit.

Food versus fuel. It is important to consider at this point the use of land for production of energy or food. This conflict will automatically disappear if we are able to produce both commodities in large quantities.

Unfortunately, what one observes in a great number of countries, notably Brazil, is the deviation of fiscal resources towards and concentration of subsidized investments in the energy sector because of the need to produce alternative fuels. Ironically, in the very countries where the low-income population does not earn enough to feed itself, these subsidized investments benefit the high-income inhabitants of large cities by increasing the availability of alternative fuels.

In that light, let us consider for the purposes of analysis the system described in Figure 7-3, a system where both fuel and food are produced. From an economic point of view, the main concern lies in the agronomic sector, since primary feedstock is responsible for a large share of total fuel cost.

The integrated system referred to above deserves special consideration, having as it does the following attributes:

- Maximum land use with minimum investment by employing crop rotation systems.
- Use of cultures that are self-sustainable, since alcohol-producing crops usually withdraw nitrogen from the soil whereas crops used for producing esters (substitutes for diesel) fix air nitrogen in the soil.
- Incorporation of biofertilizers produced within the system to help increase the soil's agricultural productivity.
- Use of a system that recycles water and is ecologically desirable, self-sufficient from an energy standpoint, technology feasible, and economically viable.

A system developed by Professor Amaratunga[18] using the same principles is worth mentioning and is shown in Figure 7-4. In addition, other petrochemical substitution systems involving the production of biomass for alcohol and its derivatives and possessing attributes not solely focused on energy production are now at a high level of development in Brazil. There is, however, still a long way to go before a good yield can be obtained from the present technologies and before new systems such as those producing solar hydrogen can be fully developed.

Finally, decentralization should be stressed as an extremely important asset of these systems, to help promote a fair and desirable level of technological, economic, and social development in the Third World. This ideal level of development will be achieved sooner if the nations that now possess the critical technologies cooperate with those countries lacking such technologies but having usable natural resources. This should be done as a gesture of fraternity and to help attain a more participatory global society.

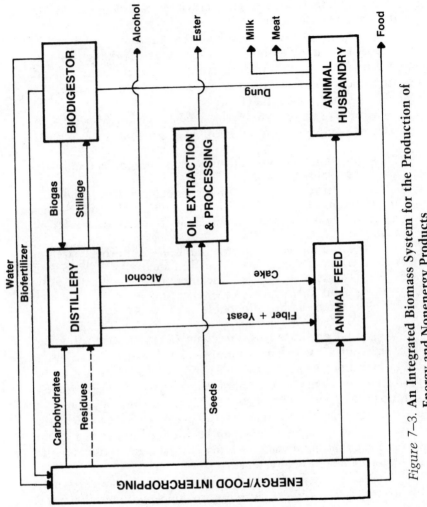

Figure 7–3. **An Integrated Biomass System for the Production of Energy and Nonenergy Products**

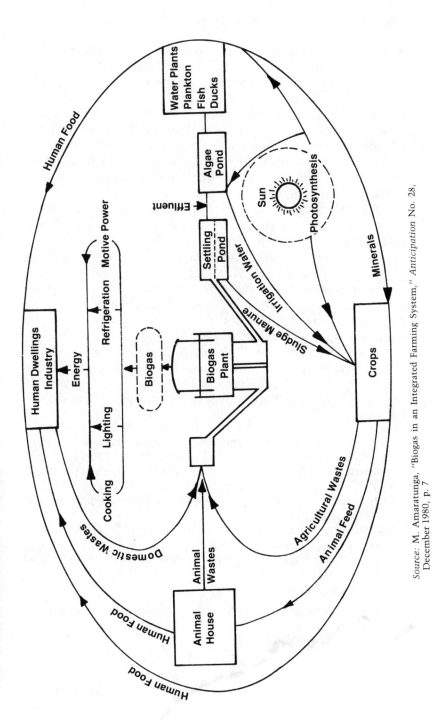

Source: M. Amaratunga, "Biogas in an Integrated Farming System," *Anticipation* No. 28, December 1980, p. 7

Figure 7–4. **Components of an Integrated Farming System**

INSTITUTIONAL CONSTRAINTS

As we have discussed, planning production factors requires a deep analysis and comprehension of all the energy supply problems of a nation and of the need to obtain the maximum social benefit from each investment in the production sector. Within this analysis, the use of biomass as an energy source and a raw material is fundamentally important. A few countries can integrate biomass as both an energy source and a primary material by using the analytical and planning framework shown in Figure 7-5. Nevertheless, it must be kept in mind that biomass constitutes only a segment of the energy problem, not a final solution.

A main difficulty also arises from the fact that in the developing countries a large part of the energy supply is derived from highly decentralized energy sources such as fuelwood, animal dung, and crop residues. Under such conditions, the problems involved in collecting data and establishing analytical methodologies become nearly insuperable. To have an idea of the difficulty, it is important to remember that about 50 percent of the world's population still uses fuelwood for cooking and one-fifth of the world's energy requirement is satisfied by fuelwood.

According to Kuhner,[19] the establishment of a biomass for energy program and the integration of this program with other alternatives has to be analyzed and executed by government, as shown in Figure 7-5. This implies that government should do the following:

- evaluate the viability, immediate prospects, longer-term potential, and environmental and socioeconomic impacts of both centralized and decentralized biomass-based energy programs;
- establish policies that make such programs operational with other energy programs and sectors within cooperative and competitive situations;
- provide the necessary financing and other incentives to create appropriate technology transfers;
- develop and improve institutions that can support these intersectoral activities; and
- provide technical assistance.

Within this framework, let us analyze some institutional constraints.

Land Use

Soil and land classifications are needed to make sound land

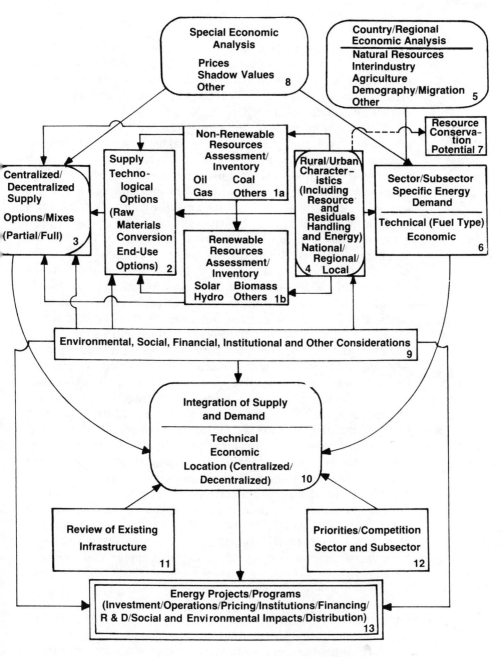

Source: J. Kuhner, "Impact of Biomass Derived Energy Programs," International Symposium IV on Alcohol Fuels Technology, Guarujá, Brazil, October 5–8, 1980.

Figure 7-5. **Program/Project Evaluation in Energy Sector Planning and Development**

use decisions, but unfortunately, data are not available in most countries. Detailed information on the supply of water for crop production (a supply which is sometimes limited or irregular in agricultural areas) as well as on the availability of human labor are important complementary data.

Cultural factors can influence negatively the establishment of new programs. Some countries like Brazil have based their agriculture on monocultures, thus leading to the establishment of very large estates (latifundium) for coffee or sugar cane crops. Although large-scale farms are still theoretically the most appropriate for the large-scale agroindustry energy systems, the necessity and clamor for a more equitable land distribution in many less developed countries may add new institutional problems to the existing social problems. In Brazil, for example, a program such as the National Alcohol Program, which receives large subsidized funds, unfortunately is absorbing and incorporating small properties to form large agroindustrial sugar cane enterprises which will eventually create conflicting situations between employers and employees. Furthermore, the incorporation of small farms usually increases the number of small agriculturists and their families who migrate to urban centers in search of other jobs. And finally, the use of land for energy purposes may compete with food production, due to a lack of central planning concerning the adequate use of land for specified ends—a problem that is occurring in many southern Brazilian regions and may occur in many other countries.

For social development, more suitable alternatives should be implemented such as the formation of agricultural communities and agro-energetic villages, in cooperative systems or otherwise. In addition, legal aspects concerning labor rights, social security, duties, and taxes must be reexamined from a more fair and just perspective. In summary, land distribution is, for many developing countries, the most difficult task to be performed, especially since there is an increasing demand on land for energy and food purposes.

Government Arrangements

The enormous delays characteristic of government bureaucracy in general are equally characteristic of government alternative fuel programs that implement policies regarding the supply of primary material, the financing of industrial units, the development and transference of various technologies, and the creation of environmental quality control. Unfortunately, in many countries the whole energy sector is under the control of state-owned enterprises, a fact which renders difficult any concrete steps towards the implementation of decentralized, independent systems—systems which require a reorientation of policies towards small-scale industries. This reorientation is a difficult task at best, for small-scale industries do not enjoy the same political prestige as do the large enterprises which hold the markets for equipment and processes and whose main concern is sophisticated capital-intensive technology. Nevertheless, government financial support and technical assistance are of primary importance for small- and medium-sized enterprises that use simpler technologies, preferably at the local level.

It is the responsibility of government organs such as national councils for scientific and technological development to assume technological leadership directly and indirectly, through the development of research within the country or through cooperative programs with other countries.

The same approach should be followed by national agronomical research institutions on the production of feedstock, with special credits issued by authorized financing agencies. The establishment of a program of rural credits to help finance agricultural activities in the food and energy sectors would provide a great stimulus to the biomass program. Unfortunately, the lack of adequate infrastructures in the developing countries hinders the fulfillment of such measures.

In any case, all these efforts should be directed towards the welfare of the people, and thus the fuel produced should be used primarily for mass transport and agricultural implements rather than for individual transport. Furthermore, biomass energy production should not affect the environment negatively, and for this reason there is a need to establish, under the coordination of national technically accredited institutions, strong programs of controlling and supervising activities affecting the environment.

To sum up, the commercialization of fuels poses specific problems which require for their solution great efforts on the part of the government at the federal and provincial levels. As part of this process, the standardization of technology and the experience gained over the years by some countries could facilitate the development and use of these fuels, if appropriate intergovernmental channels were created.

International Arrangements

In spite of the socioeconomic problems generated by the energy crisis, this crisis has awakened world consciousness to the fact that the energy problem is a world problem. Along with this, it has become apparent that food production is also a world problem. On the judicious understanding of both of these problems will depend the world's social stability and peace. Both of these intimately related problems, then, are the responsibility of the entire world community, and their solution demands collaboration at the international level.

The attention of international agencies should be focused on improving the miserable conditions of the rural poor, a group which constitutes over one-half of the world's population, by insuring that they can obtain sufficient food and energy. To achieve these objectives, there is an urgent need for adequate technological development in the Third World to alleviate the injustices caused by the existing centralized energy production and commercialization systems. The United Nations through its various organizations and agencies can play a predominant role in framing a solid program of technical cooperation in specific fields such as bioenergy technology and the production of proteins and so forth from different sources.

Pricing Policies

The involvement of private enterprise in the development of a biofuels production program implies the establishment of adequate pricing policies. Basically, these policies should address the feedstock production sector, the capital goods industrial sector, and the fuel utilization sector.

Regarding feedstocks, pricing policies should be designed to diminish the effects of price fluctuations on the international market. This policy should comprise a set of priorities at the national level in order to coordinate production with societal needs. Thus, for example, it is inadvisable for the governments of underdeveloped countries to subsidize the production of energy feedstocks if large parts of their populations suffer from a basic lack of food. What is then recommended is a more fair subsidy policy in favor of food production. In any event, however, the production of appropriate raw materials at the local level deserves full national support in order to reduce the transport of these materials and to encourage the use of local resources.

Regarding equipment production, high priority should be given to protecting small- and medium-sized enterprises and to disseminating technological innovations to them. Preferential subsidies should be awarded to those undertakings that have achieved technological improvements, especially if the improvements are without negative environmental implications. Encouragement needs to be given to labor-intensive technologies. Since the integrated systems discussed previously produce both fuel and food, and are environmentally safe and labor intensive, preferential support should be given to them in both developing and developed countries.

The pricing policies for fuel use should favor mass transport systems, agricultural implements, and so on.

Another important point to be emphasized is that the prices of both renewable fuel and fossil fuels must reflect many parameters —technical, social, and environmental as well as economic —if the biofuels programs are to achieve complete success.

SOCIETAL CONSTRAINTS

For the first time in its long history, agriculture has been called upon to produce commercial fuels to substitute for petroleum derivatives. Up until now, it has been the prerogative of agriculture to produce food, feed, fiber, construction wood, and noncommercial fuels. The so-called *green revolution* experienced during the past two decades which has kept food production ahead of population growth is now being tapped to meet the black liquid crisis by production of green liquid. And ironically, the success of the green revolution has been largely due to energy-intensive inputs derived mainly from petroleum.

The diversion of agriculture to produce commercial fuels is evidently going to impinge and put pressure on its already demanding role of producing food and other nonenergy products. There is bound to be conflict between these two divergent uses of biomass, and this conflict may loom larger and larger as increasing amounts of both energy and food are needed to adequately feed billions of people and maintain the tempo of industrial progress.

Such a conflict is a danger to the very existence of society, and societal constraints are expected to manifest themselves in various forms to curb the intensity of this conflict. The search for adequate solutions to maintain a just and sustainable society must begin before it is too late. The following sections discuss some of the complex interactions among the various societal constraints on biomass-derived food/fuel production that must be understood if these solutions are to be attained.

Evaluation of the Resource Use Conflict

The production potential of the world's terrestrial biomass resources is given in Table 7-5. It can be seen from this table that annual production of biomass grown on cultivated land accounts for less than 8 percent of total annual biomass production.

A simple calculation shows that if the biomass grown on cultivated land was converted into liquid fuels, it would produce just enough to replace the petroleum consumed in 1980, although in actuality the replacement would be only partial if the energy consumed in producing the biomass and converting it to liquid fuels was taken into account. Furthermore, it is estimated that if the world's entire 1978 production of cereals, root crops, and sugar was converted into ethanol, this would meet less than 83 percent of the world's gasoline consumption by automobiles that year. The potential of the existing agricultural crops to produce liquid fuels is thus very limited indeed.

On the other hand, the biomass produced annually on forest and shrub lands is about nine times that produced on cultivated land, and liquid fuels derived from cellulosic materials such as trees and shrubs have much greater potential than those derived from agricultural crops. Forest and shrub lands are especially appropriate for biomass-derived products since these lands are mostly marginal and therefore not usable for food crops. In fact, biomass from forested and reforested areas appears to be a sustainable solution to the liquid fuel problem, once the necessary cellulose-to-ethanol conversion technology is well established and has become economically viable. In the short run, however, sugar, cereals, grains, and oil seeds seem to be the most technologically feasible feedstock for liquid fuels.

Table 7-5. **Terrestrial Biomass Resources**

Type of ecosystem	Area (km² x 10⁶)	Biomass produced annually		Stored biomass	
		Dry matter (tons x 10⁹)	As % of total	Dry matter (tons x 10⁹)	As % of total
Forest and shrublands	57.0	79.9	68.5	1,700	92.6
Savanna and grasslands	24.0	18.9	16.2	74	4.0
Cultivated land	14.0	9.1	7.8	14	.8
Other[a]	52.0	8.8	7.5	48.5	2.6
Total biomass	147.0	116.7	100	1,836.5	100
Total energy content					
Kilocalories x 10^{15}		118.6		6,591	
Tons of oil equivalent x 10^9		40.6		639.5	

Source: adapted from H. Leith and R. H. Whittecker, Primary Productivity of the Biosphere (Berlin: Springer, 1975).
[a]Tundra, desert, swamps, etc.

The energy/nonenergy biomass resource conflict can be mitigated by the following important measures:

- Conservation in the use of liquid fuels by the industrialized world and the urban and rural rich of the Third World countries.
- Development and stimulus of less energy-intensive but more science-intensive agriculture. (Millions of small farms can greatly increase their productivity by using methods such as crop rotation, genetic improvement of seeds, interlacing of food/energy crops, and harnessing marginal lands for crops like cassava, corn, and others, for the rapid and widespread acceptance of modern efficient farming which has led to the so-called green revolution shows unmistakably that farmers possess the inherent wisdom to accept and adapt to modern agricultural developments that are within the structures of their social, cultural, and ethical values.)
- Improvements in existing technologies and the development of new less energy-intensive ones to convert biomass into liquid fuels, other fuels, and fertilizer.
- Diversification and regionalization of biomass feedstocks to reduce the risks involved in monoculture and to minimize consuming energy by transporting biofuels.
- Uses of other renewable forms of energy such as solar, wind, hydro, and geothermal in partial substitution for liquid fuels wherever possible—for example, in grain drying, industrial heat production, water pumping, and so on.
- Intensive scientific exploration of the most efficient ways to convert fuelwood, agricultural and animal residues, and industrial and urban wastes into energy and nonenergy products such as charcoal, low-calorific gas, biogas, biofertilizers, and animal feed. Residues and wastes should not be treated as waste but as feedstock for renewable resources that are compatible with biomass.

Such measures could make the production of both food and fuel from biomass a complementary rather than conflicting process.

Distribution of Income and Wealth

It is a common observation that in a free economy, large systems (agricultural, industrial, transport, etc.) normally favor enriching the already rich, thereby accumulating wealth in the hands of a few who take full advantage of subsidies such as tax reductions, easy credits, and so on.

In the United States, with its highly developed agricultural sector, there is an appalling loss of family farms throughout the nation. In fact, a large number of family farms are being taken over by big agricultural enterprises each year. In Brazil, the six-year-old Proálcool Program is already showing similar dismal consequences for the small farmers.[20] This irreversible pattern, if it remained unchecked, would extend and expand, resulting in further economic disparities within and among regions, for the big systems—more energy- and capital-intensive and less labor- and science-intensive—are not conducive to a balanced and sustainable development.

According to Brown,[21] the average per capita claim on the earth's productive resources has until recently not varied from the richest to the poorest countries by more than a factor of five. However, with the expansion of energy crops this ratio would increase dramatically. If agriculture was used to fuel automobiles, the world's affluent would be permitted to expand greatly their claim on the world's good quality land.

For a more equitable distribution of income and wealth and for a reduction of economic disparities within and among regions, small, decentralized, agro-industrial integrated systems that are integrated to produce both energy and nonenergy products seem to be the most appropriate solution, especially in the Third World countries. The technological feasibility and economic viability of such systems were demonstrated previously in this paper. The salient features of these systems which can promote all-round development are as follows:

- Creation of skilled and semiskilled job opportunities in the rural agricultural and industrial sectors to help resolve the chronic unemployment problem of rural areas.
- Exploitation of the full potential of diverse underused or unused regional agricultural resources for producing both food and fuel for the local population and surrounding areas.
- Stimulus to the development of small- and medium-sized regional agrobased industries to help obtain a sustained generation of wealth.
- Availability of cheaper, locally produced food, feed, and fuel within the economic reach of low income groups.
- More even distribution of biomass than fossil fuels among and

within the regions, enabling the judicial exploitation of biomass resources to make possible a more equitable generation
and distribution of income and a reduction in regional disparities.

• Relative cheapness of these small, decentralized food/fuel producing systems—systems which can be quickly built on small
farms.

The following concept developed by the United Nations is
especially valid for the small, decentralized system which inte-
grates food and fuel production: "Like any energy source or product, new and renewable sources of energy are themselves both an
'input' and an 'output' of the development process, not tangential
to it."[22] What this concept implies is that on the one hand, energy
is needed for development, and on the other hand, technological,
social, and economic developments are needed for production of
energy. Decentralized small systems would make this cyclic process easily manageable and adaptable to rural areas.

Environmental Constraints

As far as energy consumption is concerned, the world population can be broadly divided into two main groups: (1) a small group
consisting of the populations of industrialized countries and the
urban elite of the developing countries—a group whose energy use
is highly intensive and is based mainly on commercial fuels and (2)
a large group consisting of rural populations of developing countries—a group which consumes much less energy per capita and
which uses fuelwood, and crop and animal residues as their major
resources.

It has been estimated that about 100 million of the world's
rural poor cannot even meet their minimum energy requirements,
and about 1 billion of the world's rural population are depleting
their fuelwood supplies faster than they are being replenished,[23] in
addition to burning agricultural and animal residues and thereby
interfering with the recycling of soil nutrients. Under current
trends of population growth and increasing fuelwood demand, by
the end of the century over 2.3 billion rural people in developing
countries will need large supplies of alternate fuels.[24]

Thus, a "second energy crisis" is developing: the fuelwood
crisis. It is seemingly not very important since it involves a
relatively small quantity of energy compared with petroleum and
is very little clamored about, but it is of great significance to about
half of the masses belonging in the second group.

The doubling over the past generation of the world's food output to meet the demands of population growth has resulted in soil erosion. In fact, a large fraction of the world's crop land is now losing topsoil, a situation which is undermining its productivity. For example, the "shift" agricultural method (a method wherein a new plot is tilled when the old one is spent) being practiced in some African countries will result in irreparable loss to the soil if it is not checked. Thus, agricultural land, although a permanently renewable resource, may suffer irreversible deterioration if it is not properly managed.

It is within this broad context that we should analyze the impact of fuel/food biomass production on the environment, since the production would entail planting vast areas of fertile lands in order to meet increasing energy and nonenergy demands. The environmental consequences of biomass production include soil erosion, siltation of water bodies, deforestation, and so on.

The production of fermentation alcohol from sugar cane's amilaceous and cellulosic feedstocks simultaneously involves the production of large quantities of stillage which are highly biodegradable and consume abundant quantities of oxygen. When stillage is thrown inadvertently into rivers, lakes, or other water systems, it can destroy aquatic life and make the water system irreversibly unutilizable. Considering, for example, the ambitious Brazilian program to produce 10.7 billion liters of alcohol in 1985, the stillage produced in that year will have a polluting capacity equivalent to that produced from the sanitation waste of one and one-half times the Brazilian population. Considering further that stillage would be produced around the distillery sites, the pollution density would be much greater there. Such negative environmental effects must be eliminated if a large number of biofuel systems are to achieve sustainable success.

One possible approach is decentralization of the distilleries. Adequate treatment of stillage is practically impossible in the case of large distilleries, but if a large number of small distilleries are dispersed over a great area, the stillage would be easily manageable and could be converted into useful products such as feed, biogas, and biofertilizers. Furthermore, with the biofertilization process, the soil nutrients—which remain practically unaffected during the biofuel conversion process—could be returned to the soil, thereby preventing the soil's nutritive deterioration. Treatment of stillage wastes thus serves the double purpose of nutrient recycling and of overcoming environmental constraints.

Ethical and Moral Issues

The two groups defined above obviously both need energy and food. The first group, however, enjoys the lion's share of available resources at the cost of the second group, whose share in both energy and food is extremely minimal, resulting in standards of living that are at subsistence level or below.

The fundamental question, then, is who will benefit from these biomass-derived fuels? The answer is obvious and simple. More than 98 percent of all liquid fuels are now used by the first group. With the growing scarcity of petroleum, biomass is considered as the most viable option to meet supply shortages in the short run and satisfy the energy-intensive life styles of the first group. This biomass grown for fuel purposes would compete with that grown for food, since both need fertile lands—lands which are already largely exploited. Thus, the logical deduction is that in the absence of other viable substitutes for petroleum, the second group would be subject to ever larger energy and food deficits.

On the other hand, from an ethical and moral point of view the second group should get the *larger* share of biomass-derived food and energy products, since it would be the main producer of the biomass involved and since it currently gets a smaller share of both energy and food. Furthermore, apart from satisfying ethical and moral values, it is in the enlightened self-interest of public and private enterprises to raise the standard of living of the depressed and deprived sectors of society.

The following are two ethically and morally sound principles on which biomass production and utilization programs should be based for the overall well being of society:

- Both food and biofuels need to be produced in increasing amounts. However, if there are physical or other limitations on the production of both and a choice has to be made between the two, priority must be given to food production over energy production. The liquid fuel deficit should be met by increased conservation and reduced consumption of these fuels in the industrial and transport sectors.
- The energy/nonenergy potential resulting from agricultural and animal residues and from biofuel conversion systems should be fully and efficiently exploited with the aim of minimizing environmental damage and providing more and better forms of energy/nonenergy products for the needy.

In conclusion, it should be mentioned that attention must be paid to the methods of disposing of the effluents from biofuel conversion processes and other industries. It is unethical and unjust to indiscriminately dump large volumes of biologically degradable materials into water systems and thereby create adverse living conditions for those who probably are benefiting the least from the fuels produced.

Religious and Cultural Constraints

Societies have well-entrenched religious beliefs, cultural patterns, and customs and habits. These societal characteristics usually have become established over a long time span—many centuries, in some cases—and are not susceptible to drastic or radical changes. Therefore, any development program should conform to existing societal patterns by introducing only slow and gradual changes. Luckily, the use of biomass for energy/nonenergy applications is not new to the world. However, the production of commercial biofuels (alcohol, biogas, low-calorific gas, etc.) and nonenergy products (leaf proteins, distiller's grain, biofertilizers, etc.) would require some adaptation in the established patterns of the rural areas involved. Therefore, a well-organized and comprehensive program of education and information for diffusing new biomass technologies is necessary to dispel adverse but unfounded beliefs and taboos.

EDUCATION AND INFORMATION

Biomass as a source of food and traditional fuel is as old as agriculture. However, the demand on agriculture for the large production of commercial biofuels is relatively new. Consequently, there is an urgent need to develop new technologies, new approaches, and new attitudes to deal with the complex problem of food and commercial fuels production from biomass.

Public Participation and Decisionmaking

New means of education and disseminating information are essential to cope with the new situations described in this paper.

First of all, the stark fact that the glorious era of cheap and abundant petroleum is fast fading must be brought home to all strata of society. It is absolutely essential that this noble liquid be conserved and used for more noble ends than engine combustion. Second, it is important that the policymakers and decisionmakers, who by virtue of their power and position must choose among the various alternatives put before them by experts, become conversant with the potential and limitations of agriculture for biofuels. Although biofuels provide the most versatile substitutes for petroleum derivatives, it must be recognized that they are not a panacea for fulfilling all energy requirements.

Even in the rural areas of the developing world, there is a need to achieve a mix among traditional, renewable, and fossil-based commercial fuels. To determine the appropriate proportions of this mix, a sound knowledge and deep understanding of pertinent technological, economic, social, environmental, and political factors are essential for overall energy planning by all those involved in the policymaking process.

Once the programs have been selected by the planners and decisionmakers, their implementation should be coordinated with education and information programs that are adjusted to stay in tune with the times. For example, there is a need to reorient and modify the courses given to agronomists and soil experts in order to direct these courses from energy-intensive to more science-intensive agriculture, and there is a need for economists, scientists, and engineers to understand and appreciate new forms of available energy and their efficient utilization. More emphasis should be given to biotechnological research in areas such as crop rotation, interlacing of crops, nitrogen fixation, genetic improvements, increased photosynthetic efficiency, fermentation processes, enzymatic hydrolysis, proteins, and so on. To be effective, these scientific advances should move quickly from the laboratory phase to the phases of demonstration and diffusion.

Thus, the main objective is to change the mental attitude and outlook of the new generation involved in framing and executing a country's developmental programs, but as part of this objective, it must be remembered that the participation of the farmers—the sons of the soil—is absolutely essential for the sustained success of a biofuel/food program. In addition, appropriate initial financing is important: investments would be needed to search for technologically and economically viable as well as socially and environmentally acceptable food/fuel production systems appropriate to the region and site at hand.

In summary, the new food/fuel needs and possibilities call for earnest and realistic thinking and for the education of opinion at all levels. In all instances, a sufficiently thorough understanding of a large-scale biofuel program's precise potentials, prospects, problems, and conflicts must be attained before such a program is launched.

Technology Transfer

Technology transfer is a very touchy term and is imbued with manifold connotations. In our complex world, transferring technology from the haves to the have nots has much more complicated implications than simply loaning or donating money or an object. Technology transfer involves not only technological dependence but also equipment and accessories dependence. With vested interests, the situation can be further fraught with the dangers of possible intervention and domination by other political spheres—a domination that can be directly or indirectly related to the technology in question. There is a general awareness of this danger by the informed public, and in particular, the receiving end seems to be very sensitive to the term "technology transfer."

Fortunately, unlike fossil fuel technologies, biofuel technologies appear to have a positive edge in this respect for the following reasons:

- Biofuels production and utilization technologies are more regional and site specific than are those for fossil fuels.
- Biomass feedstocks are much more evenly distributed than are fossil fuels.
- The large-scale production and utilization of biofuels is a relatively new phenomenon.

From these important characteristics of biomass-into-fuel technologies, it appears that there would be a more genuine need for technology sharing than for technology transference with biofuels technologies. It is therefore in the mutual interest of both developing and industrialized nations to make use of the opportunity provided by biofuels to work towards the common objective of achieving a more stable and sustainable dynamic equilibrium in the development and use of new technologies.

It must be emphasized that if the new biomass technologies are to benefit both urban and rural populations, they should be directed towards the optimization of small- and medium-sized agroindustrial systems for integrated fuel/food production. It should also be observed that large systems have their due place as well and should be installed as needs dictate, but not at the sacrifice of food and at risk to the environment.

To achieve this objective of integrated food/fuel production from biomass in a short time span, scientists and researchers from both developing and industrialized countries should collaborate by sharing research results, so that these results can be transferred confidently from the laboratories to the fields. This approach needs wholehearted encouragement and stimulus.

Methods of Promoting International Communication

It is a widely recognized fact that the potential problems inherent in substituting biofuels for fossil fuels are a global concern. Their resolution requires, in our view, a regional approach to help solve national problems, leading to solutions at international levels.

There is thus an urgent need to promote communication at all possible levels: intraregional and interregional, bilateral and multilateral, through all possible means and channels. In fact, concrete and definite steps are already being taken in this direction:

- The UN conference in Nairobi in August of this year on new and renewable energy sources and the preparatory meetings held during the prior year in New York and Geneva are manifestations of the deep concern the United Nations has to bring together platforms for discussion and debate of global energy problems.
- The North/South meeting held in October in Cancún, Mexico, had on its agenda the important item of more equitable energy sharing among the twenty-seven participating countries. In the true spirit of the Brandt report's recommendations, such dialogues should be encouraged and extended to include other regions such as East/South, North/East, and South/South, for these dialogues and meetings—whose main aim is to break the existing barriers of mental resistance and reservation—are extremely important means of international communication.
- The regional, multilateral, and international organizations such as OLADE, ESCAP, and ECM [the Organisaçion Latinamericano de Energy, the Economic and Social Commission for Asia and the Pacific, and the European Common Market] can play very dynamic roles in promoting international understanding of different nations' energy problems and in finding ways to solve these problems.
- International institutions such as the UN University at Tokyo can offer excellent opportunities for the sharing of ideas and approaches to common energy problems among scientists, sociologists, and environmentalists drawn from the world's nations.

• The deep concern of the World Council of Churches (WCC) for the evolution of a more just, participatory, and sustainable society manifested itself through their organization of the international meeting on "Faith, Science, and the Future" held in Boston in July of 1979. This meeting brought together specialists from diverse walks of life and of different shades of opinion and thought; it convened them there to discuss and debate the impact of modern developments in science and energy on the future of society. The follow-up regional meetings in Madras, India; Chiang Mai, Thailand; and more recently in Lima, Peru, are a testimony to the WCC's continued dynamic concern about energy problems in the Third World countries. The slogan of this mammoth body, "Energy for My Neighbor," should echo and resonate from every nook and corner of the world.

To conclude, a pragmatic approach for promoting international communication should be encouraged through the establishment of international centers with definite goals and objectives whose results are diffused all over the world. The development of dwarf wheat varieties in Mexico, highly productive rice varieties in the Philippines, and drought-resistant cereals and oleaginous plants in India are but a few examples of how much international cooperation and collaboration can achieve. The green revolution's success with food should be extended to include green fuels.

A FINAL EVALUATION OF THE OPTIONS

In this section, criteria are defined for evaluating the available options for using biomass in energy and nonenergy applications. In conclusion, some recommendations are made.

Criteria for Evaluating the Options

There appears to be a general agreement on the following points regarding biofuels:

• They are highly viable alternatives to petroleum derivatives.
• Although they cannot offer a panacea for worldwide energy problems, they can play an important role in filling the gap between petroleum supply and demand, thereby permitting sustained development in the short run.
• They will play an increasingly important role in the subsistence and survival of more than half the world's population—those living in rural areas of the Third World.

- Although renewable, they cannot be produced in unlimited amounts, due to competition from other uses—most importantly, food.

The compelling question, then, is: Can the large-scale production of biofuels be reconciled with the increasing demand for food production? In our view, the answer to this question is affirmative, provided the energy-producing system satisfies the following criteria for harmonious and sustainable development:

- *Social:* promoting a more equitable distribution of income and wealth, thereby giving incentives for people to remain in rural areas and reducing the trend of migration to urban areas.
- *Environmental:* maintaining soils in good condition and minimizing air, ground, and water pollution in both the short run and the long run.
- *Economic:* attaining economic viability, in particular through less capital-intensive and more labor-intensive methods.
- *Technological:* employing less energy-intensive and more science-intensive approaches with an adequate mix of both sophisticated and simple technologies.

At first sight, a system which satisfies the above criteria appears to be a utopia. However, a more sober analysis would demonstrate the possibility of symbiosis among these apparently divergent criteria. Our experience[25] and that of others such as Amaratunga[26] with biofuels convinces us that such systems are fully attainable and can avoid the pitfalls and shortcomings inherent in the large, centralized systems which are landing the industrialized countries and especially the developing countries in positions of unsustainable, lopsided development with respect to biomass.

We define and defend the following position:

- A large number of small- and medium-sized integrated energy/nonenergy production systems should be installed on a decentralized, scattered basis in preference to a small number of large, centralized systems.
- In developing integrated biomass production systems, economic viability and technological sophistication should not be the sole criteria; social development and environmental safety should be given equally great consideration.

Recommendations

Recommendations concerning biomass production, conversion processes, and products utilization are given below, concluding with a section giving general recommendations.

Biomass production. Production efforts should include:

• Assessment and evaluation of biomass resources to identify and develop appropriate plant species for energy/nonenergy ends.
• Preparation of a compatibility map showing compatibility among soil types, climatic conditions, and chosen plant species.
• Development of agricultural technologies for crop rotation, crop interlacing, and so forth, with the aims of maximally using cultivated areas and reducing agricultural inputs, specifically those which are energy intensive.
• Study of how to best harness marginal lands for the production of energy crops, with the by-products to be used for nonenergy purposes.
• Research on plant genetics, to increase productivity without disturbing the ecosystem balance.
• Development of agricultural equipment and implements—especially those which can be operated on biofuels—as aids to the various phases of traditional agriculture, in order to facilitate labor conditions and to bring more areas under cultivation.

Biomass conversion. Efforts regarding biomass conversion should include:

• Preparation of a "state of the art" review of biomass energy/nonenergy conversion processes and technologies.
• Assessment and evaluation of the fuel, food, and fiber requirements for each region.
• Study to determine the compatibility between biomass potential and energy/nonenergy requirements.
• Research and development of processes for optimal biomass conversion into energy and nonenergy products.
• Design and installation of demonstration units with a view towards studying the engineering and economic aspects of technology transfer within the industrial sector.

Biomass products utilization. Efforts involving the use of biomass products should include:

• Study to enable the adoption or modification of individual, collective, and load transport systems as well as stationary systems to biofuels (alcohols, processed vegetable oils, biogas, low-calorific gas, etc.)

- Study of the adequacy of nonenergy products obtained from biomass conversion systems (animal feed, single-cell proteins, biofertilizers, chemicals, etc.)
- An integrated social, environmental, technological, and economic feasibility study of the appropriate mix of fossil and biomass-derived fuels.

General recommendations. In general, the following should be undertaken:

- Study of decentralized integrated systems for the optimal production of energy/nonenergy products.
- Definition and determination of the appropriate energy mix of fossil fuels, biofuels, and other renewable forms of energy for biomass systems.
- Monitoring a permanent program to evaluate and supervise the environmental impact of biomass products, those products' conversion into biofuels, and the fuels' utilization.
- Socioeconomic evaluation of biomass products, those products' conversion into biofuels, and the fuels' utilization.
- Reevaluation of the existing laws related to land use and ownership, including the visualization of a more just participation by the laborers in all the benefits derived from land utilization.
- Reevaluation of the existing criteria related to planning, policymaking and decisionmaking, and education and training—all of which should be geared towards the evolution of a more just, participatory, and sustainable society.
- Definition and determination of the appropriate mix of centralized and decentralized systems (renewables and nonrenewables) for the purpose of establishing biomass energy policies within the broad framework of all energy resources.
- Establishment of rural cooperatives to implement integrated systems programs and to create suitable conditions for more job opportunities and a more equitable distribution of income.
- Establishment, with the help of international agencies, of bilateral and multilateral agreements for technology transfer and training in the field of biomass production and its use for energy/nonenergy purposes.
- Establishment, with the support of the United Nations and other nonprofit organizations, of bioenergy technology centers to pursue biomass technology R&D, particularly in the Third World countries.
- Establishment of special programs to fully exploit the potential of available free or profitable urban residues for energy/nonenergy uses and for the improvement of sanitary conditions.
- Establishment and implementation of a broad-based family

planning program integrated with other welfare and development programs to make more food and energy available per capita, thereby raising living standards and reducing the trend of migration to cities.

Dedication

This work is dedicated to Dr. Cesar Cals de Oliveira Filho, the Federal Minister of Mines and Energy of Brazil. The authors express their profound gratitude to Dr. Cesar for his constantly inspiring them with his unbounded zeal and enthusiasm for renewable sources of energy through personal discussions, through his thought-provoking lectures and publications, and through his unfailing financial support. He is the father of the concept of "energy planting" originated during his tenure as Director of Coordination of ELETROBRAS. The integrated pilot plant distillery at Caucaia, a pioneering effort, is a demonstration of his broad vision and profound understanding of energy/food problems. Dr. Cesar's technical expertise and ingeniousness have provided us with innumerable challenging opportunities for introducing innovations in bioconversion technologies.

Acknowledgment

The authors are grateful to Dr. Edward Lumsdaine, Director of the Energy, Environment, and Resources Center, University of Tennessee, Knoxville, for his critical reading of the first draft of this paper and for making very useful suggestions.

NOTES

1. World Bank, *Energy in Developing Countries* (Washington, DC: World Bank, 1980), p. 84.
2. Ibid., p. 26.
3. Ibid., p. 80.
4. Ibid., p. 86.
5. J. Baguant, "Energy and Economic Development Strategies System Studies for LDCs," ph.D. Thesis, The University of Tennessee, Knoxville, June 1980, p. 82.
6. The procedure for grouping world countries according to whether they are developed or developing is a complex one, since this distinction is based on certain criteria such as the country's state of economic development, level of industrialization, gross and per capita income, and other socioeconomic parameters. In this paper, we have employed the grouping pattern used by the World Bank, as cited in note 1.

7. "Energy in Agriculture and Rural Development" in Publication of the Committee on Agriculture, Sixth Session of the Food and Agriculture Organization (FAO) of the United Nations, March 25–April 3, 1981, p. 1.

8. Ibid., p. 2.

9. Ibid., p. 1.

10. J. U. Nef, *Alicerces Culturais da Civilização Industrial*, (Rio de Janeiro: Ed. Presença, 1964). Translation from English.

11. Ibid.

12. D. Hayes, "Raio de Esperança—a transição para um mundo pós-petróleo," Editora Cultrix (1977), p. 35. Translation from English.

13. *Faith and Science in an Unjust World: Reports and Recommendations* Vol. 2, ed. P. Albrecht (Montreux: Imprimerie Corbaz, 1980), p. 104.

14. J. O. B. Carioca and H. L. Arora, "Celula a Combustível e a Geração de Energia em Regiões Remotas," *Energia* Vol. III, No. 12 (1981), p. 24.

15. See p. 186 of reference cited in note 12.

16. D. de Gaspar, "Economics and World Hunger," in *Faith and Science in an Unjust World: Plenary Presentations* Vol. 1, ed. R. L. Shinn (Geneva: World Council of Churches, 1980), p. 225.

17. Ibid., p. 226.

18. M. Amaratunga, "Biogas in an Integrated Farming System," *Anticipation* No. 28 (December 1980), p. 7.

19. J. Kuhner, "Impact of Biogas Derived Energy Programs," in the Proceedings of the International Symposium IV on Alcohol Fuels Technology, Guarujá, Brazil, October 5–8, 1980.

20. J. G. da Silva and P. G. Moraes, "O Proálcool: Uma Visão Social," *Energia* Vol. III, No. 13 (1981), p. 45.

21. L.R. Brown, "Food or Fuel: New Competition for the World's Cropland," *Worldwatch Paper* 35 (March 1980), p. 36.

22. H. Iglesias, in *United Nations New and Renewable Sources of Energy Conference News* 9.

23. C. I. Jackson, "The UN Conference on New and Renewable Sources of Energy—The Conference: Background and Prospects," *ASSET* Vol. 3, No. 7 (July/August 1981), p. 7.

24. Ibid.

25. J. O. B. Carioca, H. L. Arora, and A. S. Khan, "Technological and Socio-Economic Aspects of Cassava-Based Autonomous Minidistilleries in Brazil," to be published in *BIOMASS—An International Journal*.

26. See reference cited in note 18.

CHAPTER EIGHT

The Role of Renewable Resources in Hawaii's Energy Future

John W. Shupe
Director, Hawaii Natural Energy Institute
University of Hawaii at Manoa

INTRODUCTION

The state of Hawaii consists of the eight southernmost islands in the Hawaiian archipelago, an island chain stretching across 600 kilometers of the central Pacific and separated from the mainland United States by nearly 4,000 kilometers of ocean. These islands are the tops of shield volcanoes rising from the ocean floor and are too recent in origin to have experienced the natural cycle necessary for the formation of fossil fuels.

Energy-use patterns have played a key role in Hawaii's historical development. Through the mid-nineteenth century, Hawaii was a net exporter of energy in the form of biomass—both forest products and sugar cane. During the late 1800s, coal gradually replaced wood as the primarly energy source, and in recognition of the strategic location of Hawaii, the us Navy in 1901 located a major coaling and repair facility at Pearl Harbor—establishing early precedence for the role of these islands as a military outpost.

Hawaii's energy-use pattern paralleled the global trend of switching from coal to petroleum as the major fuel supply. In 1903 the coal-fired boilers of the Hawaiian Electric Company were converted to burn oil, and over the next seventy years there was a gradual but systematic shift to oil by essentially all segments of Hawaii's economy. At the time of the 1973 Middle East oil embargo and resulting energy shortfall, the conversion to oil was almost total.

In the intervening eight years since the oil embargo, a great deal of progress has been made in laying the groundwork for the commercialization of Hawaii's renewable energy resources as substitutes for imported oil. There is optimism in Hawaii that by the year 2000 the majority of the state's energy needs can be met by the effective utilization of a combination of its indigenous energy resources.

THE NATURE OF THE PROBLEM

Of all fifty states making up the United States, Hawaii currently is the most vulnerable to dislocations in the global oil market. It has no known fossil fuel reserves on any of its islands, and no offshore oil. There is no coal coming into the state by rail, and no natural gas by pipeline. The energy demand of its separate islands is too small to accommodate nuclear power plants, and Hawaii is not tied into a regional grid which would permit sharing electric loads with adjacent states. In fact, Hawaii's separate islands are not even connected by a common electric grid.

Current Energy Demand Data

Hawaii has a resident population of around 965,000 and a de facto population of slightly over 1,100,000 when visitors to the islands are included.[1] Since 80 percent of the resident population and most of the industrial and governmental infrastructure are concentrated on the island of Oahu, the state has both a major metropolitan area with high-density energy utilization in Honolulu and decentralized low-energy-demand communities on the outer islands.

In 1979, the total civilian energy consumption in Hawaii was 211 trillion btu, which is equivalent to about 37 million barrels of oil. Figure 8-1 compares energy utilization in Hawaii with that of the mainland United States.[2] The end uses of these various energy types are relatively fixed. Gasoline is used for lightweight vehicles; diesel fuel is used mostly for trucks and buses, with some of it consumed by heavy agricultural equipment. Residual fuel is burned in industrial boilers, while the small amount of synthetic and liquid petroleum gas goes primarily for residential and commercial water heating.

Figure 8-1 illustrates that the energy consumption pattern for Hawaii differs appreciably from that of the mainland United States. Major factors contributing to this disparity include (a) the absence of a space-heating requirement in Hawaii's subtropical climate and (b) the huge amount of aviation fuel necessary to place

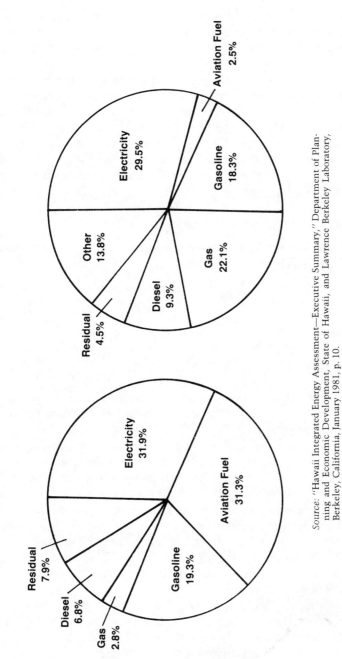

STATE OF HAWAII

UNITED STATES

Source: "Hawaii Integrated Energy Assessment—Executive Summary," Department of Planning and Economic Development, State of Hawaii, and Lawrence Berkeley Laboratory, Berkeley, California, January 1981, p. 10.

Figure 8-1. **Energy Consumption Patterns: Hawaii and the United States**

service its tourist industry. A significant point to consider in planning for Hawaii's energy future is that nearly 60 percent of its current energy demand is for liquid fuels to satisfy transportation requirements.

The increase in the cost of energy in recent years has had some influence on the level and pattern of energy utilization. Currently, the price for residential electricity in Hawaii varies from 11.4 cents per kwh in Honolulu to over 20 cents per kwh on the island of Molokai. Regular gasoline costs about 41 cents per liter ($1.55 per gallon), including local and federal taxes—slightly higher than on the mainland but still a bargain in comparison with most price figures in other parts of the world.

Current Energy Supply for Hawaii

The supply sources for the annual civilian energy consumption in Hawaii are as follows:

Supply source	Million barrels of oil equivalent	Percent of total
Imported oil	34.2	92
Bagasse (sugar cane residue)	2.5	7
Hydroelectricity	0.3	1
	37.0	100

This list illustrates Hawaii's near total dependence on imported oil, with 92 percent of its existing energy supply consisting of seaborne petroleum. This virtual dependence would be of somewhat less concern if most of this import originated within the United States. Unfortunately, the reverse is true; 62.5 percent arrives directly from foreign sources—Saudi Arabia, Oman, Indonesia, and Malaysia.[3] Alaska supplies 13.5 percent and the mainland 24 percent. However, this last figure is misleading, since products from mainland refineries may also be of foreign origin. Therefore, as much as two-thirds of the petroleum imported into Hawaii today comes from foreign crude. This was a contributing factor to the severity of the energy shortfall in Hawaii following the 1973 oil embargo.

Currently, biomass in the form of by-products from agricultural operations in the state easily represents the second largest energy supply source for Hawaii. Over 2.6 million tons of bagasse (the fibrous residue remaining after sucrose is squeezed from sugar cane) are burned each year in the boilers of the sugar companies. Much of the electricity generated from bagasse is used internally by the sugar industry for irrigation and can processing, but sizeable quantities of surplus electricity are also sold for distribution through the utility grids. On the islands of Hawaii and Kauai, for example, where there is sufficient rainfall so that little irrigation is required, the surplus electricity from burning bagasse provides nearly 50 percent of the power distributed by the utilities for public use. In addition to bagasse, greater utilization for boiler fuel is beginning to be made of other forms of biomass: cane trash (leaves and tops), pineapple waste, wood chips, macadamia nut shells, and hay.

The third traditional energy supply source in Hawaii, which also was developed entirely by the sugar industry, is hydroelectric power. The overall contribution of hydroelectricity to the resolution of Hawaii's energy supply problem is minimal. Currently, it meets less than 1 percent of total energy demand, and it is unlikely that hydroelectric power's contribution will ever approach 5 percent of the state's energy need.

AN INITIAL EVALUATION OF THE OPTIONS

Hawaii's near total dependence on seaborne petroleum is a paradox, since the state is blessed with a variety and abundance of renewable energy resources: (a) direct solar radiation, (b) steady trade winds, (c) a semitropical climate and sufficient variation in rainfall to grow many types of biomass, (d) a high thermal gradient between the sun-warmed ocean surface and the cold deep-ocean water, and (e) high-temperature geothermal resources.

Research was initiated over a decade ago on the utilization of these resources as alternatives or substitutes for imported oil.[4] The energy shortfall following the 1973 oil embargo provided additional impetus to this effort, which has been reinforced as each new Middle East crisis causes Hawaii's primary energy supply to become more expensive and less secure.

Resource Base for Hawaii's Renewable Energy Alternatives

In the early seventies, the state began a systematic program to develop an inventory of the nature and extent of its indigenous energy alternatives. A preliminary resource inventory was established for each of the major alternatives—direct solar, wind, ocean thermal energy conversion, biomass, and geothermal. These resource base studies are being expanded and refined as each of the alternatives approaches commercialization.

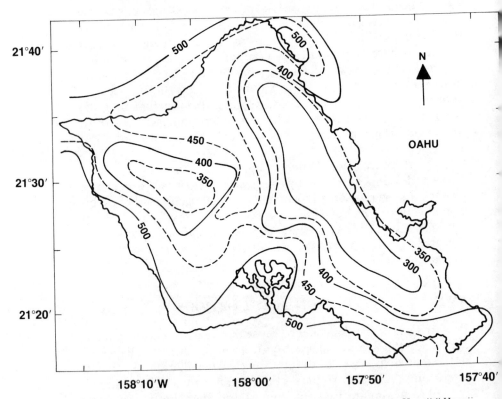

Source: T. Yoshihara and P. C. Ekern, "Solar Radiation Measurements in Hawaii," Hawaii Natural Energy Institute, University of Hawaii at Manoa, September 1977, p. 26.
[a]Gram-calories per square centimeter of irradiated space per day; i.e., langleys per day.

Figure 8-2. **Isolines of Mean Annual Isolation** (cal/cm^2/day[a])

Insolation—incoming solar radiation. The major Hawaiian islands lie between 19 and 22 degrees north latitude. For Honolulu, the midsummer sun is 3 degrees north of vertical, while the winter low is 45 degrees above the southern horizon. Therefore, Hawaii enjoys both a higher average insolation rate and a smaller seasonal variation than the mainland United States.

Although the seasonal variation is relatively small, the geographic variation in insolation is surprisingly large— particularly between the windward and leeward sides of the islands. A fairly comprehensive assessment program has been under way for seven years to obtain an inventory of insolation values throughout the state. Figure 8-2 illustrates the variation in insolation for the island of Oahu.[5] As to be expected, for most areas there is an inverse correlation between solar insolation and amount of rainfall.

The insolation data shown in Figure 8-2 are utilized in the design of various solar systems. Average annual insolation levels in excess of 500 langleys (gram-calories per square centimeter of irradiated surface) are not unusual. Many regions also have extremely clear skies much of the time and a high fraction of focusable direct-beam radiation. Consequently, Hawaii is well suited for the utilization of all types of direct solar radiation conversion systems: flat-plate and concentrating collectors, solar ponds, solar power towers, and photovoltaic cells.

Wind energy resources. Figure 8-3 represents a composite inventory of three of Hawaii's potential energy resources: wind, geothermal, and ocean temperature gradients. The northeast tradewinds, which blow across Hawaii approximately three-quarters of the time, constitute one of the most consistent and reliable wind patterns in the world.[6] This favorable wind regime is further reinforced by the terrain formation of the islands and a fairly stable inversion layer at an altitude of around 2,000 meters. These factors combine to give a channeling or funneling effect, so that as the tradewinds are forced up and over the mountain ranges and around the corners of the islands, the velocity of the normal tradewinds is doubled. Since the work that can be captured from a wind stream increases theoretically as the cube of the velocity, doubling the wind speed raises the power potential of a windmill by a factor of eight.

Figure 8-4 represents a refinement of the wind energy data from Figure 8-3 for the islands of Molokai and Maui, both of which have excellent wind regimes.[7] In fact, all of the major islands in the state have high potential for wind-power development, with average annual wind velocities of over 20 miles per hour (9 meters per second) measured 10 meters above the ground. In addition to the favorable wind patterns, there is no icing in Hawaii to contribute to blade overloading and vibration, and hurricanes and high-gusting winds are extremely rare in this area of the Pacific. Except for the corrosive effects of moist salt-laden air, which can be handled through proper design, Hawaii enjoys an ideal combination of physical and climatic conditions conducive to the introduction of wind power.

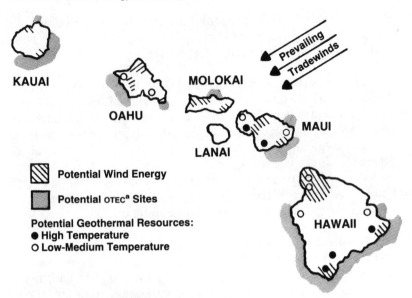

Source: adapted from "Hawaii Integrated Energy Assessment—Executive Summary," Department of Planning and Economic Development, State of Hawaii, and Lawrence Berkeley Laboratory, Berkeley, California, January 1981, p. 8.
[a]Ocean thermal energy conversion.

Figure 8-3. **Wind, OTEC, and Geothermal Resources in Hawaii**

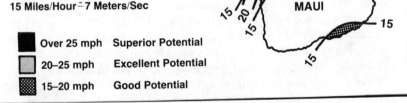

Source: W. Falicoff, G. Koide, and P. Takahashi, "Solar/Wind Handbook for Hawaii: Technical Applications for Hawaii, the Pacific Basin and Sites Worldwide with Similar Climatic Conditions," Hawaii Natural Energy Institute, University of Hawaii at Manoa, May 1979, pp. 3–13.

Figure 8-4. **Average Annual Wind Speeds for Maui and Molokai**

Ocean energy systems. The ocean itself is an excellent solar collector and storage system. The resulting temperature difference between the warm surface and the cold deep-ocean water can be used to generate electricity. This concept—ocean thermal energy conversion (OTEC)—was discovered a century ago by the French scientist d'Arsonval, but, with other inexpensive energy supplies available, there was little incentive to develop this capital-intensive resource. Now that the escalating cost of oil has provided the motivation for secure cost-competitive energy supplies, efforts are under way to develop economic ocean systems for capturing this enormous source of solar power. Unlike many other renewable resources which provide only an intermittent energy supply, there is negligible cooling of the ocean at night, so OTEC can be utilized for twenty-four-hour base-load power—a decided advantage for a utility grid.

Figure 8-3 includes a preliminary inventory of those locations throughout the state that have some potential for OTEC systems. Not all of the shaded ocean areas shown in Figure 8-3 are necessarily suitable for OTEC power plants. Additional bathymetric and ocean current studies would be required to verify that these sites are satisfactory. What this preliminary inventory illustrates are those shoreline areas where the ocean floor drops off very abruptly, so that there is access to deep-ocean water close by.

Deep-ocean water throughout the world is uniformly cold, so the maximum ocean temperature differentials will occur in equatorial regions where the ocean surface is warm. A temperature difference of 20 to 22°C (36 to 40°F) is desirable for generating OTEC power, so depths of 800 to 1,000 meters (2,600 to 3,300 feet) are required. There is an economic advantage if these sites are located near to shore, since the umbilical cable for transmitting electricity from a floating OTEC power plant to a shore-based utility grid would be both shorter and simpler in design. In addition, there is some interest in constructing shore-based OTEC plants, with the cold-water pipe going out through the surf zone and reaching the desired depth within 2 kilometers (1.24 miles) of shore. Hawaii is fortunate to have a number of sites that satisfy criteria for both near-shore floating and shore-based OTEC power plants.

In addition to OTEC power plants feeding electricity into an island utility grid, both floating, anchored platforms and un-tethered, grazing OTEC plants can be utilized to produce ammonia or hydrogen as a synthetic fuel. It has been estimated by Craven[8] that within the 200-mile zone of the islands constituting the Hawaiian archipelago, it is possible to produce 10 to 15 quads (quadrillion BTU) per year of OTEC power—an amount of energy greater than the current annual import of oil by the United States. Craven also estimates that all of the Pacific islands under US jurisdiction could produce 70 quads of OTEC energy a year within 200 miles of their shores—nearly the equivalent to the total annual US energy use.

There are many other potential ocean energy systems: tides, waves, currents, salinity gradients, sail propulsion. Since there is a low tidal diurnal variation in Hawaii and few bays or estuaries appropriate for harnessing ocean tidal surges, no significant contribution is anticipated from tidal power. Hawaii has monitored the development of wave energy systems with some interest. The state funded a demonstration of John Isaacs's wave generator in Hawaiian waters and is following progress on Lockheed's Damatoll and other wave energy conversion systems. The same is true with ocean currents and salinity gradients. Both Japan and the United States are developing sail-assisted propulsion systems which show good potential for replacing appreciable amounts of diesel fuel with wind power. However, there is strong consensus within Hawaii that the one ocean energy system with by far the greatest potential for making a significant impact on the state's energy future is OTEC. This provides justification for the fact that to date OTEC has received a greater amount of state funding than any of the other renewable alternatives.

Geothermal resources. Interest in a geothermal exploration program originated in the University of Hawaii in 1970, and joint funding was obtained from the Hawaii state legislature and the federal government to initiate this assessment program in 1973. Early geological and geophysical surveys were conducted on the island of Hawaii (commonly called the Big Island)—the home of two active volcanoes and a complex rift zone system. These surveys then proceeded up the island chain.

The first experimental well was drilled on the Big Island in 1976 to a depth of over 2,000 meters. Preliminary flashing and flowing of the well were quite encouraging.[9] Subsequent well testing verified that this is one of the hottest geothermal wells in the world (358° C downhole temperature), and the quality of the fluid is excellent. Preliminary analyses of the geothermal reservoir associated with this initial well indicated that it could provide as much as 1,000 MW (megawatts) of electricity for fifty years. There is some indication that there may be a natural recharge to the reservoir, both of surface water and of heat from the molten magma of the nearby Kilauea volcano so that no depletion of the resource will take place.

Additional joint state and federal funding was identified to undertake a broader resource assessment program, both on the Big Island and throughout the island chain. This preliminary assessment was made through the compilation and evaluation of readily accessible geological, geochemical, and geophysical data. A much more comprehensive site-specific survey program and analysis of the promising areas should be made before additional risk capital is identified and "wildcat" test drilling proceeds.

Based upon this preliminary assessment, there are twenty regions in thirteen general areas throughout the state with some potential for geothermal development.[10] These thirteen areas are shown in Figure 8-3, with the solid dots representing areas of probable geothermal resources with temperatures sufficiently high to generate electricity and the open dots representing probable lower temperature resources which could be used as process heat in refining sugar cane and for other industrial applications. Unfortunately, there are no high-temperature geothermal resources located on Oahu, where over 80 percent of the state population base and electricity market currently resides.

Biomass resources. The total land area of the state of Hawaii is 16,642 square kilometers (6,425 square miles), with nearly 63 percent of this land mass concentrated in one island— the Big Island of Hawaii.[11] The heavily populated island of Oahu constitutes only 9 percent of the total land area. Land use classification for the state is as follows:[12]

Cultivated land	9%
Pasture and range	25%
Forest land	40%
Urban, industrial, defense, transportation, and park land	20%
Nonproductive land (bare volcanic rock, marshes)	6%
	100%

Sugar cane. There are nearly 220,000 acres (89,000 hectares) of sugar cane grown in Hawaii today.[13] The primary product from this acreage is about one million short tons (0.9 million metric tons) per year of raw sugar. Major by-products are molasses and bagasse—the fibrous residue from cane processing.

Bagasse is an excellent, clean-burning boiler fuel. The moisture content of bagasse runs around 48 percent, with each wet metric ton providing about the same amount of heat during combustion as a barrel of oil. Approximately 2.5 million wet metric tons of bagasse are produced per year, most of which are burned in sugar factory boilers, replacing nearly 2.5 million barrels of oil. The heat generated provides steam and electricity for internal use by the sugar industry, as well as electricity for distribution through the utility grids.

The majority of the 280,000 metric tons of molasses produced annually is marketed as cattle feed. Although the energy balance of converting molasses to ethanol is relatively good, the economics remain marginal at best, particularly with the large fluctuation in the market value of molasses. Between 1977 and 1981 the price of this base-stock material for ethanol increased by a factor of three. Even if all of the molasses produced in the state were converted to ethanol and utilized as transportation fuel, it would not have a major impact on the amount of gasoline required. The ultimate potential for ethanol from molasses is around 23 million gallons (87 million liters) per year—about 8 percent of the current fuel use for automobiles in Hawaii.

Forest products. Most of the 1.6 million acres (650,000 hectares) of forest lands in Hawaii are natural stands, with less than 1 percent of the area planted by man for lumber, watersheds, or recreation. None of the existing stands was planted as an energy crop, although a number of small test plots of different varieties of trees have been planted in recent years. In addition, forestry thinning operations and removal of trees that have fallen from natural causes are providing a limited source of wood chips for sugar company boilers.

Different varieties of eucalyptus and the giant koa haole (leucaena) seem to show good potential for tree farms in Hawaii, both as a source of boiler fuel and as feedstock for liquid fuels. An extensive study[14] was conducted on the cultivation and harvesting requirements of several of these varieties to determine the amount and location of land appropriate for tree farming with each of these strains. Factors considered included climatic and soil conditions, elevation, topography, rainfall, land use, and ownership.

Figure 8–5 shows the potential growth area identified for Eucalyptus grandis on the Big Island of Hawaii. Limiting factors included 160 to 1,000 meters elevation, 75 to 750 centimeters yearly rainfall, and deep well-drained soils. Similar figures were completed on promising varieties of trees for each of the major islands, resulting in an overall inventory of land in the state with good potential for tree farming.

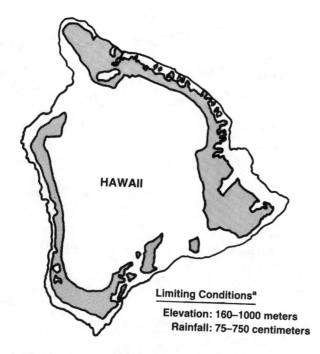

HAWAII

Limiting Conditions[a]

Elevation: 160–1000 meters
Rainfall: 75–750 centimeters

Source: A. Seki and P. C. Yuen, "Biomass Resource Potential for Selected Crops in Hawaii," Hawaii Natural Energy Institute, University of Hawaii at Manoa, in preparation.
[a]The limiting conditions define the parameters of the potential resource areas. These areas are shown as shaded portions on the map.

Figure 8–5. **Potential Resource Areas for Eucalyptus Grandis**

Other biomass sources. A number of the long-stemmed grasses were also investigated[15] for their energy potential. Currently there are over 9,000 acres (3,600 hectares) of land on the Molokai Ranch planted in guinea grass, providing about 17,000 metric tons of hay per year for the Molokai Electric boiler. This yield could probably be increased with the liberal use of nitrogen fertilizers. Figure 8–6 illustrates the potential growth areas for guinea and napier (elephant) grasses on Molokai.

There are 16,000 hectares of pineapple grown in Hawaii, yielding about 450,000 wet metric tons of pineapple chop and field waste annually. Currently, most of this pineapple waste is returned directly to the soil as humus and fertilizer, either by open-fill burning or by cultivation. Consideration is being given to

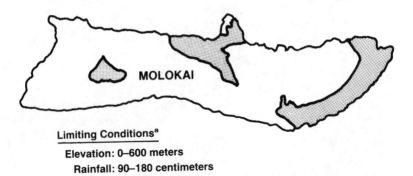

Limiting Conditions[a]

Elevation: 0–600 meters

Rainfall: 90–180 centimeters

Source: A. Seki and P. C. Yuen, "Biomass Resource Potential for Selected Crops in Hawaii," Hawaii Natural Energy Institute, University of Hawaii at Manoa, in preparation.
[a]The limiting conditions define the parameters of the potential resource areas. These areas are shown as shaded portions on the map.

Figure 8–6. **Potential Resource Areas for Guinea and Napier Grasses**

burning this material in boilers for generating electricity, particularly on Molokai. If this practice gains acceptance, it will be necessary to recover and return the potash from the boiler to the fields for soil replenishment.

Ocean crops may also be grown for their biomass yield. Studies have been made on three seaweeds with good potential yield: Gragilaria, the red seaweed; Macrocystis pyriferea, the giant California kelp; and Sargassum, the brown Hawaiian kelp. The University of Hawaii has an experimental project under way for growing single-cell algae in seawater.[16] An obvious advantage of ocean crops for an island state where land is scarce is the vast area of the sea available for cultivation. In Hawaii, there also are huge expanses of barren volcanic rock close to the sea which have potential for the growth of aquatic biomass—possibly in conjunction with nutrient-rich, deep-ocean water as a by-product of an OTEC power plant.

Resource base summary. It is evident from this review of renewable energy resources in Hawaii that the development of many of these alternatives will in no way be limited by the magnitude of the resource base. However, the availability of suitable land area could be a limiting factor for the extensive utilization of direct solar radiation (either photovoltaics or solar thermal) and for biomass (either boiler fuel farms or liquid fuel feedstock). It has been estimated[17] that in order to produce sufficient biomass to supply the total energy demand of Hawaii, 54 percent of the state's total land area would be required, at a net efficiency rate of 0.4 percent, to convert sunlight to liquid fuels—the approximate conversion rate of solar energy to sugar cane to molasses to ethanol. A photosynthetic efficiency conversion rate of 10 percent, obtained in laboratory-scale experiments with single-cell algae, would require only 2.2 percent of the total land area. Successful utilization of ocean farming could increase the biomass potential of the state significantly.

Preliminary estimates indicate that during tradewind conditions, the potential of the wind resource of Hawaii easily exceeds by a factor of ten the total electrical requirements of the state.[18] The probable limiting factor for the commercialization of wind energy—until and unless an economic method for storing electricity is developed—is the total amount of intermittent wind power that can be accommodated by a utility grid. When no storage is involved, present estimates for the upper limit on the amount of wind power that can be accepted range up to 25 percent of the lowest electrical load on the grid.

The capacity of all of the potential OTEC power plant sites also greatly exceeds the existing electrical demands of the state. Although OTEC will provide base-load power, the technical reliability and economic feasibility of full-scale OTEC systems have yet to be proven. The long-range potential for OTEC in Hawaii, however, appears excellent.

On the basis of preliminary surveys, there should also be sufficient geothermal base-load power on the Big Island of Hawaii to provide all of the state's electrical energy needs. It has been estimated[19] that the Puna area of the Big Island alone may have as much as 3,000 MW-centuries of generating capacity in its geothermal reservoirs, and there are nineteen other areas of the state that have potential geothermal resources both for electricity and for direct heat applications.

In summary, it would appear that the ultimate energy supply mix for Hawaii will include a variety of indigenous energy resources, each maximized to satisfy an energy end use of direct heat, electricity, or fuel. Since no single renewable resource is likely to emerge as the dominant source of supply for all energy end use needs, it is unlikely that the limit of the resource base for any of the alternatives will be approached.

In fact, the other extreme will probably be the case. As cost-competitive energy systems are developed for wind, biomass, geothermal, OTEC, photovoltaics, and solar thermal, additional markets will need to be identified to effectively utilize Hawaii's renewable energy resources. Eventually, Hawaii should be able to completely reverse its role of near-total dependence on imported oil to that of a net exporter of energy.

Status of Renewable Energy Alternatives in Hawaii

To date, the impact of renewable energy resources on reducing Hawaii's oil imports has been slight—except for the 8 percent input from the sugar industry's bagasse and hydroelectric power plants. However, the great amount of research, development, and demonstration (RD&D) that has gone into our indigenous energy sources in recent years is beginning to show positive results.

Table 8–1 lists both the origin of the $65 million spent on renewable energy RD&D in Hawaii and the general categories in which these funds were expended. This section will describe the progress that has been made in moving each of Hawaii's major energy alternatives toward commercialization.

Table 8-1. **Funding of Renewable Energy RD&D in Hawaii: 1971–80**

	Sources of funding ($ x 10^3)			
Program	Federal	State/county	Private	Total
Ocean thermal	22,006	3,246	1,712	26,964
Geothermal	12,336	2,173	4,334	18,843
Insolation, wind, and hydro	9,437	1,494	423	11,354
Bioconversion	1,848	1,498	766	4,112
Other	1,026	2,674	256	3,956
Total	46,653	11,085	7,491	65,229

Source: "State Energy Resources Coordinator 1980 Annual Report," Department of Planning and Economic Development, State of Hawaii.

Direct solar radiation. Almost 40 percent of the residential energy consumed in the state is used for heating water. This represents a good target for solar water heaters, and currently there are about 19,000 flat-plate solar collectors installed in Hawaii, resulting in annual savings in imported oil of around $6 million. Depending upon a family's hot water requirements and the location of their dwelling relative to insolation levels, their investment for an average solar water heater can be amortized in from three to seven years—with the current rate of state and federal tax incentives.

An optimistic projection for the replacement of the 200,000 residential electric and gas water heaters in Hawaii with solar collectors is around 65 percent. If each solar heater displaced 85 percent of the energy formerly required for heating water, this would result in an overall savings in oil imports of only 3 percent. While this is not a very impressive statistic in itself, it does illustrate the finite contribution that one of a multitude of renewable energy resources can make in reducing the amount of oil imported into the state.

In addition, flat-plate collectors are finding broad utilization for multiple-family dwellings and for commercial applications requiring water temperatures no higher than 150°F (66°C). Solar drying of agricultural products (coffee beans, macadamia nuts, tropical fruit) is also being accomplished with flat-plate collectors.

Solar cells—photovoltaic power systems utilizing solid-state semiconductor devices to convert solar radiation into electricity—are also beginning to find limited application in Hawaii. A photovoltaic power system which will generate 35 kW of electricity for the Wilcox General Hospital on Kauai will be completed in December 1981. This concentrating dual system consists of a series of parabolic troughs which focus radiation on the solar cells. Water, which is warmed as it circulates through the system to cool the cells, provides much of the hospital's hot-water needs.

Three small photovoltaic systems—2 to 3.5 kW—have also been installed on homes in Hawaii as demonstration projects, funded by the US Department of Energy (DOE) to provide operational data on the performance of photovoltaic units for meeting the requirements of three average families in Hawaii. These roof-mounted systems are metered into the utility grid, and records are kept of (a) the amount of electricity generated by the solar cells and used by the homeowner, (b) the amount of electricity provided by the utility to the homeowner, and (c) the amount of surplus electricity from the photovoltaic system that is fed back into the utility grid and which serves as a credit against utility electricity charges. In addition to these demonstration projects, there are more than two hundred fifty homeowners in Hawaii who have installed photovoltaic systems—both with and without storage—to help meet their electrical needs.

A number of design studies have been conducted on various types of solar thermal systems, both for providing process heat and for generating electricity: (1) Molokai Electric Company was selected by DOE as one of six finalists for a small community solar thermal power experiment, utilizing small-dish concentrating collectors; (2) Amfac Sugar Company and Bechtel received a $475,000 grant from DOE for studying the feasibility of a solar power tower for providing process heat for Amfac's Pioneer Sugar Mill; and (3) Brewer received DOE support to investigate parabolic concentrating collectors to supply steam for processing sugar cane. None of these projects have extended beyond the design phase as yet. With the current attitude of DOE toward funding additional demonstration projects, there will probably be some delay in the actual construction of solar thermal systems in Hawaii.

Wind energy. There are more than thirty small wind generators—1 to 40 kW—dotting the Hawaiian landscape, many funded privately. For the most part, the performance of those machines which were designed or modified to operate in Hawaii's strong wind regime and to resist corrosion has been quite satisfactory.

The Hawaiian Electric Company (HECO) was one of four utilities selected by DOE to receive the Mod-OA wind turbine generator. This is a 200-kW windmill located at Kahuku on the north shore of Oahu—one of the prime wind sites in the state. During its first year of operation ending July 3, 1981, this windmill generated 870,970 kWh—more electricity than the Clayton, New Mexico, Mod-OA had generated in over three years of operation.

HECO is very optimistic about the future of wind power for Hawaii and has entered into a major contract with Windfarms, Inc., for the purchase of electricity generated by 80 MW of wind turbine generators to be located near the 200-kW unit at Kahuku. Windfarms, Inc., will finance these huge wind turbines—the total cost of the project is estimated at $350 million—taking full advantage of all tax incentives and write-offs. HECO will not have to identify investment capital for purchase of the generating equipment but will buy the electricity from Windfarms, Inc., for distribution through the utility grid. The system should be completed by the end of 1984, at which time it will meet nearly 10 percent of Oahu's electricity needs, at a saving of around 800,000 barrels of imported oil per year.

OTEC projects. Three of the most significant salt-water OTEC projects conducted throughout the world are located in Hawaii. Mini-OTEC was the first fully operational closed-cycle OTEC system to generate electricity from the ocean thermal gradient. It was a joint venture of the Lockheed Missile and Space Company, the Dillingham Corporation, and the state of Hawaii to demonstrate the technical feasibility of an OTEC power plant. The heat exchangers and other components of Mini-OTEC were located on a barge, with a 24-inch (61-cm) diameter polyethylene pipe suspended from a buoy to bring up cold water from a 2,150-foot (655-meter) depth. During its 4-month deployment off Hawaii in late 1979, Mini-OTEC generated slightly over 50 kW of electricity, of which 40 kW were used internally to drive the system, and 10 to 15 kW were net power.

The second project to be located in Hawaii was the Ocean Energy Converter—the major effort of DOE to obtain operational data on the components of a 1-MW OTEC system. Nearly $50 million was spent in converting a Navy tanker to accommodate the massive heat exchangers and other components, and to develop deep-ocean pipes for circulating 65,000 gallons per minute of cold water through the system. The Ocean Converter was deployed 25 kilometers off the coast of Hawaii in early 1981 and obtained three months of excellent operational data before budget limitations caused premature termination of the program.

The third major OTEC project is the Seacoast Test Facility (STF). This is a $15 million project, jointly funded by DOE and the state of Hawaii, to provide a permanent shoreline facility from which to conduct reliable, economical, long-range test programs on biofouling, corrosion, and component performance. The advantage of this particular site is that the ocean falls off rapidly from the shoreline, so that a depth of 2,100 feet (640 meters) can be reached by extending a pipeline 6,100 feet (1,860 meters) out from the laboratory, through the surf zone, and along the ocean floor.

Phase I of the STF has been completed, and limited testing using warm surface water has begun. Phase II, the cold-water pipe construction, has been postponed pending receipt of DOE matching funds for installing a 28-inch (71-cm) pipe. Rather than accept an indefinite delay, the state has allocated an additional $500,000 to install a 12-inch (30-cm) cold-water pipe. Successful installation of this pipe, which has a flow capacity of 1,500 gallons per minute, was completed in November 1981. Hawaii's justification for making a heavy investment in OTEC RD&D is based not only on its interest in base-load power but also for the probable spin-off in aquaculture and desalination from the cold, nutrient-rich, deep-ocean water.

Geothermal development. When the 3-MW geothermal well-head generator on the Big Island of Hawaii first began to provide electricity for the utility grid in July 1981, Hawaii gained the distinction of becoming the second state in the Union to generate electricity from its geothermal resources. Compared with California's 900 MW of installed geothermal capacity, this 3-MW power plant is a very modest beginning. However, with twenty potential geothermal regions identified throughout the state and the possibility of an interisland electric cable, geothermal power in Hawaii could approach the current California capacity by the end of this century.

A total of $13.4 million was expended during the eight years required for exploratory surveys, well drilling, well testing, generator design, and construction leading to this 3-MW plant. Federal sources supplied $10.8 million, with the state and county providing most of the remainder. However, once this high-risk energy source was identified using public funds, the private sector took over the commercialization. Three drilling consortia have been established, and the first successful geothermal well drilled entirely with private capital was flashed on October 23, 1981. The Hawaii Electric Light Company has revised its earlier plans for constructing more fossil-fired power plants on the Big Island and has issued a Request for Proposal on a 25-MW geothermal power plant.

Direct heat applications of geothermal energy are also under review. Interest has been evidenced in utilizing process heat (1) for the refining of sugar from cane, thereby releasing bagasse steam for electricity generation; (2) in the manufacture of ethanol or methanol from cellulosic material, such as cane stalk or eucalyptus; (3) in the refining of manganese nodules or bauxite; (4) in preserving tropical fruit and in drying coffee beans; and (5) in aquaculture applications. With the projected demand for both base-load electricity and direct heat, the likelihood of rapid expansion of geothermal energy in Hawaii looks excellent.

Biomass programs. There are numerous programs and projects under way for satisfying a higher percentage of Hawaii's energy demand from a wide variety of biomass sources. The flexibility in utilization of biomass is a decided asset in meeting different energy end uses. Two promising areas for biomass usage in Hawaii are (1) standby boiler fuel for intermittent energy sources, such as wind, and (2) feedstock for different types of liquid or gaseous fuels.

In addition to greater utilization of cane leaves, tops, and field trash, the sugar industry has a task force investigating the possibility of growing cane as a total energy crop, with sugar either as a by-product or included as feedstock material in the energy processes. Row and stalk spacing, growth period, and the use of irrigation, fertilizers, and dessicants are all under review for increasing biomass yield. In addition, after decades of genetic breeding to achieve optimum sucrose production, plant strains are now being isolated for maximizing the bulk yield of the biomass, rather than the sugar.

Gasohol—a 10-percent blend of ethanol with gasoline—was introduced into the Hawaiian market in 1979 by Pacific Resources, Inc. (PRI). Although Del Monte Corporation has been producing a small amount of ethanol from pineapple juice for a number of years, all of it has gone for the production of vinegar. Consequently, the ethanol used for gasohol in Hawaii today is imported into the state, the same as petroleum is. However, the commercial production of ethanol may become a reality soon, as three ethanol-from-molasses plants are presently under design. By the end of this calendar year, Maui Distillers will have converted an unused rum plant to produce 10 to 15 thousand gallons (38 to 57 thousand liters) a month of anhydrous alcohol. Most of this alcohol will be utilized for human consumption, but a small amount will be diverted for gasohol production.

A much larger supply source is two ethanol-from-molasses plants under design, one by PRI and the other by C. Brewer and Company, Ltd. A DOE award of $900,000 for a synthetic fuel feasibility study was made to Brewer to perform the process design and preliminary engineering work for a plant to convert 50 percent of the state's total molasses production of 280,000 metric tons per year into 11.4 million gallons (43 million liters) of ethanol. This study has verified both the technical and the economic feasibility of the ethanol plant. If an assured market can be identified for the ethanol, Brewer will begin negotiations for acquiring 50 percent of Hawaii's molasses production and proceed with the construction of the plant on the Big Island of Hawaii.

Tree farming, both for boiler fuel and for liquid fuel feedstock, is also receiving much attention. The Hawaii Division of Forestry was granted a continuing annual appropriation of $500,000 from the state legislature to provide for the planting of 1 million seedlings per year for energy tree farms throughout the state. Optimum cultivation and survival characteristics are also being studied for a variety of tree crops.

The US Department of Energy is providing the major funding for a five-year demonstration project conducted by Brewer's Bio-

Energy Corporation on fast-growing strains of eucalyptus. Presently the project is in its second year with a third of the full 850 acres (344 hectares) already planted. Once these trees reach a harvestable size of from 15 to 20 meters, which will take from five to seven years, they will be cut and chipped for boiler fuel.

Another variety of tree under study by the University of Hawaii is the leucaena (giant koa haole). A recent study by Dr. James Brewbaker [20] suggests that leucaena could be grown profitably on the island of Molokai—primarily as a fuel source, but with the leaves also providing cattle feed. A model farm of 1,000 acres (405 hectares), on which approximately 4.1 million leucaena trees are grown, is estimated to be able to provide enough wood chips each year to replace 22,000 barrels of diesel oil—about one-third of Molokai's annual energy need.

There is a fifteen-month study under way with $330,000 of DOE support on the feasibility of commercial production of liquid or gaseous fuels from cellulosic biomass materials. Partners in this study are PRI, the Institute of Gas Technology (IGT), and the University of Hawaii's Natural Energy Institute. Under review [12] is a hydropyrolysis process developed by IGT to produce liquid fuels. It has worked successfully with coal, and this study will determine if it works equally well with biomass feedstocks such as eucalyptus. The process requires slightly less than 1 bone-dry metric ton of wood to produce 1 barrel of liquid fuel. Assuming a conservative estimate for biomass yield of 10 bone-dry metric tons per acre per year, 330 work days per year, and an additional 500 dry metric tons per day to produce the energy for plant operations, around 45,000 acres (18,000 hectares) of land would be required for a production level of 1,000 barrels per day or 330,000 barrels per year.

There are also projects under way at the University of Hawaii on some of the more exotic areas of biomass. Dr. Joseph Tu of the University's Food Science Department is investigating the mechanisms by which termite enzymes convert cellulose to sugar. If this enzyme can be identified and synthesized, it possibly could lead to the development of a process for converting cellulosic materials to liquid fuels.

The Solar Energy Research Institute and the state of Hawaii are jointly supporting a research project on the oil-growing potential of a single-cell algae—Phaeodactylum tricornutum—in sea water. Initial laboratory experiments with this algae suggested annual yields of around 150 barrels of oil per acre (6 liters per square meter). Subsequent testing of the process in an outdoor scaled-up 50-square-meter algal raceway system has not yet verified this high conversion rate.[21] However, the generation of both lipids and protein by this process still looks sufficiently encouraging to justify increasing the size of the project to a 1-acre (40,500 square meter) system.

BENEFITS OF ACHIEVING ENERGY SELF-SUFFICIENCY

There is strong support from the general public for the development of Hawaii's indigenous energy resources. The vulnerability of this island state has been painfully demonstrated both by severe maritime strikes, which have tied up essentially all shipping from the mainland, and by the energy shortfall following the Middle East oil embargo. Although Hawaii's citizens have consistently responded positively in accepting sacrifices associated with these shortages, each incident reinforces their resolve to strive toward greater independence from imported commodities.

The obvious benefits to Hawaii of achieving full or near-total self-sufficiency with its renewable energy resources include the following considerations:

- A secured indigenous energy supply provides protection from both short- and long-term fluctuations in the global energy market.
- This stability has a positive effect on the general economy by assuring reliable, reasonably priced energy to the local consumer, and it assists in attracting and retaining energy-intensive industry.
- Hawaii's renewable alternatives are less polluting than conventional energy supplies (oil, coal, nuclear power) and are compatible with the state's concern for the quality of the environment.
- The development of an indigenous energy supply represents a major potential growth industry for Hawaii—one which could help retain a portion of the billion dollars per year currently spent on imported oil and could help broaden the job and tax bases. In addition to the construction of energy facilities and the conversion and distribution of the renewable energy supply sources, related industries for fabricating components of renewable systems, such as windmills and solar collectors, could also be established.
- Current estimates are that during the mid 1990s, which should be the peak period for constructing renewable energy systems in Hawaii, over 10,000 workers would be required for construction and indirect employment sectors—manufacturing, professional services, wholesale and retail trade.[22] This does not

include any manpower allowance for the secondary related industries for fabricating components of renewable energy systems.

ENVIRONMENTAL, SOCIETAL, AND INSTITUTIONAL CONSTRAINTS

There are a number of obstacles to the introduction of energy alternatives that are common to many of the renewable energy options. Some of the more crucial obstacles include the following:

- **Accepted conventional wisdom.** A strong carryover feeling still prevails among many of the current energy leaders (planners, economists, managers) that renewable energy alternatives are exotic twenty-first century technologies with little potential for helping to meet the drastic shortfall of liquid fuels projected over the next two decades. Unfortunately, this prophecy tends to be self-fulfilling, for without adequate support and incentives, these energy alternatives will have only limited impact by the year 2000.
- **The national energy policy.** This accepted "conventional wisdom" is strongly reflected in the current federal energy policy. The initial Reagan administration budget for the US Department of Energy called for a reduction of 68 percent in funding for *Solar and Other Renewables* from 1980 to 1982; i.e., a decrease from $751 million to $241 million. In spite of congressional efforts to reinstate some of this support, it now appears likely that further budget cuts will be imposed. This drastic reduction in RD&D support, guaranteed loans, and other incentives will have a very negative impact on the development of renewable energy resources.
- **Financial institutions.** Most renewable energy systems are capital intensive. Although the lower operating costs of these renewable systems (since the fuel is essentially "free") may more than compensate for the financial carrying charges when life-cycle cost accounting is considered, it is still difficult to identify the front-end money needed to develop the renewable energy alternatives—whether for a decentralized photovoltaic system for a single dwelling or a 100-MW OTEC base-load power plant. In the past, public funding—both federal and local—has assisted with demonstration projects to verify the technical and economic feasibility of a renewable system, prior to major private capital involvement. Lending institutions have not been overly eager to invest in developing energy technologies—

particularly with a relatively small market potential and no guaranteed loan program from the federal government.
- **Uncertainties of the resource or the technology.** For many of Hawaii's energy alternatives, either the extent of the resource, the reliability of the technology, or both are unproven. To date, there has been only one successful geothermal well drilled in Hawaii with confirmed commercial generating capacity. Where a known potential resource exists—as with wind or ocean temperature differentials —either the technical feasibility or the economics of a commercial system for converting the resource to useful energy are not generally accepted.
- **Legal and regulatory issues.** The ownership and regulation of many of the renewable energy resources are, at best, poorly defined. Because of the confusion that exists in Hawaii on the legal definition of geothermal resources and the validity of native Hawaiian claims, the ultimate issue of who owns the geothermal resources on various types of public and private lands—the state, the Hawaiian people, the landowner—will probably have to be resolved in court. And the same may be true for access to sunlight or to the wind. The preparation and approval of permits are also a deterrent to renewable energy development, as evidenced by the extensive list of environmental and regulatory agencies involved in siting and operating an OTEC power plant in a coastal zone.
- **The market.** The bottom line on commercialization is an adequate market demand to drive development of the resource. Unfortunately, there will be no clear-cut match-up between energy demand and potential supply in Hawaii until an inter-island submarine electric cable, or some other mechanism for readily transporting energy, is in place. It's still something of a chicken and egg proposition. Until a major market develops on the outer islands, there is little incentive to develop the supply. Conversely, there is understandable hesitancy on the part of a potential energy-intensive industry, such as a manganese nodule processor, to make a major investment in a manufacturing plant until a reliable, secured, cost-competitive energy source is firmly established.

A comprehensive outline of the major social constraints concerning priorities in energy development, the environment, conservation, and the economy is presented as Volume VI, "Perceptions, Barriers and Strategies Pertaining to the Development of Alternate Energy Sources in the State of Hawaii," of the *Hawaii Integrated Energy Assessment Study.*[23] A very brief summary, primarily from this report, on the constraints for development of each of Hawaii's major renewable energy resources follows:

Technology	Major environmental, legal, social, and institutional constraints on implementation
Geothermal	Air pollution, noise, and ground water contamination; adverse industrial use and development of agricultural lands and Hawaiian Home Lands; controversy concerning ownership rights of geothermal resources; deleterious influx of drilling rigs and crews, resulting from interisland export of power; potential destruction of facilities by volcanic lava flows.
OTEC	Construction stage requirements of large land areas near beaches and marine facilities that are already overloaded; operating stage interference with beaching and surfing sites, underwater fuel lines, and underwater cables; water pollution from accidental discharge of refrigerant; negative visual impact offshore.
Wind	Interference with TV reception and flight operations; subsonic harmonic noise and vibrations, disturbing both humans and animals; bird kill; potential of thrown blades; visual pollution of large wind farm arrays.
Photovoltaics and Solar Thermal	Site disturbance and issue of land use appropriateness in the instance of central systems; possible misdirection of high-temperature radiation; uncertainties of solar rights; glare interference; environmental concerns relating to manufacturing and decommissioning of toxic semiconductor materials.
Biomass	Competing land uses; loss of recreational forest and open lands; potential for erosion; competing markets for biomass products; toxic stillage discharge; visual, noise, and air pollution.
Hydroelectric	Competing land uses; danger of downstream damage if dam fails; disturbance of impoundment site; legal issue of water rights ownership.
Municipal solid waste	Air and water pollution; increased noise and traffic disturbance from collecting operation and delivery to power plant.

To date, no major energy/nonenergy conflict has surfaced in Hawaii for biomass or for any other renewable resource. In general, all segments of Hawaiian society seem to favor the development of energy alternatives as substitutes for imported oil—as long as these projects do not impinge on their own lifestyles or individual conveniences. There have been only two renewable energy projects that have found varying levels of opposition. The most dramatic of these was the blocking of a major municipal solid waste project, which would have generated 40 MW of power and saved Honolulu $600 million over a 20-year period, by a small group of concerned citizens who felt that the presence of a power plant burning 1,800 tons per day of municipal solid waste would have a negative environmental impact on their area.

The other project that has received some objection is geothermal development in the Puna area of the Big Island of Hawaii. If a 400-MW submarine cable is completed from the Big Island to Honolulu, much of the geothermal development would be concentrated in Puna and would affect the relaxed, rural atmosphere of that area. Consequently, the rallying cry of those who oppose massive geothermal development is, "Don't make Puna the Pittsburgh of the Pacific."

The probable reason that the ethical question of growing biomass for food versus fuel has not been a major issue in Hawaii is that the state's primary crops—sugar cane, pineapple, coffee, tropical fruits, nuts, and flowers—do not represent subsistence agriculture. Few food staples such as grain are produced, and the majority of Hawaii's agricultural products—as with its energy—are imported. Consequently, it would appear that the use of agricultural land for energy would have at least as great an overall societal benefit in Hawaii as using the land for specialized export-oriented agriculture.

HAWAII'S ENERGY FUTURE

It is inevitable that Hawaii will become a net exporter of energy through the development of its renewable energy resources. The question no longer is "if" but "when." The next two decades, as the gradual transition begins to take place from near-total dependence on imported oil toward energy self-sufficiency with indigenous resources, will be crucial in planning for the state's economic future.

It is difficult, however, for Hawaii to plan effectively for its energy programs in the years ahead, since many of the key factors influencing future energy usage and the rate of penetration of the renewables into the energy market are entirely independent of any actions and decisions that will take place within this state. Included among these factors are (1) the continuing availability and pricing of Middle East oil, (2) the discovery and accessibility of new oil reserves throughout the world—including the success of the US offshore drilling program, (3) how effectively coal can substitute for current oil utilization in the face of the acid rain and the greenhouse effect controversies, (4) the future acceptance of nuclear power—including the breeder reactor, (5) the national energy policy on support and incentives for developing renewable energy alternatives, and (6) breakthroughs that may occur in the technical or economic feasibility of new energy alternatives to accelerate these alternatives' market acceptability.

Extensive analyses and projections on Hawaii's energy future have been conducted over the past decade by groups and organizations from both within and without the state. Easily the most comprehensive and definitive study was a three-year, $550,000 joint effort by DOE and the state to develop data bases and the integrated energy analyses necessary to develop policy planning for facilitating the transition from oil to renewables. This study, the Hawaii Integrated Energy Assessment (HIEA), was conducted through a collaborative effort of the Lawrence Berkeley Laboratory and Hawaii's Department of Planning and Economic Development (DPED), and resulted in a six-volume publication covering the state's total energy picture. Much of the information for the next two sections was obtained from this report.

Projected Energy Demand

The Hawaii Energy Demand Forecasting Model (HEDFM) was developed by DPED and subsequently modified in the HIEA. This tool is an econometric-based simulation model, designed to generate forecasts of the annual consumption of various fuel types for each of the four counties in Hawaii through the next twenty-five years. The model utilizes a set of equations that relate energy demand to the price of energy, to personal income, and to other economic and demographic variables. Using the projected values of these variables, the HEDFM first forecasts energy consumption using the assumption that the coefficients in the equations remain constant throughout the forecast period. The projected demands are then refined to take into consideration such conservation measures as the federally mandated improvement in car mileage and increased efficiency in appliances, lighting, water heating, and space conditioning.

Three scenarios for energy demand in Hawaii were devised in the HIEA[24] All three were based on the state's "most likely" projection of population and personal income, and each scenario also assumed that the mandated automobile mileage standards would be implemented. It should be kept in mind that all three of these demand forecasts were predicated on oil continuing to be the predominant energy supply. Therefore, these forecasts serve only as a starting point for an analysis on the role of the renewables in Hawaii's energy future and how they will affect the projected demand, both for oil and for total energy.

The first scenario, the Baseline Case, assumed a 3 percent per year escalation in the world oil price above the general inflation rate. The second scenario, the Savings Case, was similar to the Baseline Case but, in addition to improved automobile mileage, assumed that some additional relatively mild conservation practices would be instituted. The third scenario, the High Oil Price Case, was based on a 10 percent per year increase in oil price above the inflation rate—a condition that could occur if there are major disruptions in future global oil production.

The economic and demographic forecasts for Hawaii that were used in this analysis predict that in 1990 there will be a de facto population of 1,230,000 with a total personal income of $9.2 billion, based on 1980 dollars. By the year 2000 these figures will climb to 1,395,000 people earning $17.8 billion.

The average price for oil in Hawaii in 1980 was slightly less than $30 per barrel—due in part to some favorable long-range contracts of the utilities. As existing contracts expire, future oil costs in Hawaii will be governed by the world market. The HIEA assumed $30 per barrel as the standard 1980 oil price. With a 3 percent annual escalation, as used in the Baseline and Savings Cases, the cost of oil in 1980 dollars would be $40 per barrel in 1990 and $54 per barrel in 2000. The 10 percent escalation figure for the High Price Case would result in $80 per barrel of oil in 1990 and $207 per barrel in 2000.

Using these general assumptions, the HEDFM forecasts for the three cases show that the total civilian energy consumption for Hawaii will increase from 211 trillion Btu in 1979 to the following levels:

Cases	Energy consumption (вtu x 10^{12})	
	1990	2000
Baseline	262	324
Savings	245	300
High price	232	244

This tabulation illustrates that in comparison with the Baseline Case, there was a slight decrease in energy consumption for the modest conservation programs included in the Savings Case. There was an appreciable decline in energy consumption, however, when the real cost of oil increased by 10 percent annually. For the High Price Case, the increase in energy consumption from 1979 to 2000 was only 16 percent, in contrast to 54 percent for the Baseline Case.

The HIEA is an excellent study which provides a sound data base and a dynamic econometric model that will contribute significantly in planning Hawaii's energy future. This computerized model can be modified as conditions and input parameters change, in order to project energy futures based upon the best current information. Although I am very favorably impressed with the HIEA study, some of the initial assumptions and limits may be subject to question.

For example, the $40 (in 1980 dollars) projection for the price of per barrel of oil in 1990 seems low. This is just about equal to the cost of low-sulfur oil in Hawaii today—$43 per barrel. Although there is much publicity regarding the current oil glut and the possibility of the price stabilizing, or even dropping, it is highly unlikely that the global oil market will remain essentially constant in terms of real costs through the remainder of this decade. Any world crisis affecting an oil-producing nation could precipitate another cost spiral.

A second difference of opinion is on the demand for jet fuel. Even if the number of tourists increases from 4,133,000 in 1980 to 7,820,000 in 2000 (which I feel may be optimistic), there should be some increase in efficiency in transporting people to and from Hawaii—through improved technology, improved scheduling, or both. There also may be some shift in transportation energy use from liquid fuel to electricity, through electric vehicles. Having driven an electric car for the past year, I feel there is excellent potential for the next generation of electrical vehicles in Hawaii, particularly with so many of our energy alternatives—geothermal, OTEC, wind—leading more readily to electricity than to liquid fuels.

It also would appear that the Baseline Case, and even the

Savings Case, do not provide adequate consideration for effective conservation measures. The rather remarkable decline in oil imports into the United States this past year illustrates that conservation, triggered by higher energy costs, can have a significant impact. Most of my variance with HIEA assumptions tend to decrease the energy demand forecast. Therefore, although I do not anticipate a real-cost annual increase in the price of oil of 10 percent for the remainder of this century, my energy use consumption projections are reasonably close to those for the High Price Case. I predict a total civilian energy consumption demand in 1990 of 235 trillion BTU, and in 2000 of 255 trillion BTU (44 million barrels of oil equivalent).

Projected Energy Supply Sources

Table 8–2, based primarily on information from the HIEA,[25] summarizes parameters relating to the potential of the various renewable energy resources for satisfying a portion of Hawaii's electrical energy demand. Included in this table are estimates of when commercialization of the technology is expected in Hawaii; whether the resource is suitable for base (24-hour), intermediate, or peak (2-3 hour) load; the capital cost range from 1985 to 2000 in 1980 dollars per installed kilowatt; and the maximum potential of the resource by the year 2000. It should be recognized that Table 8–2 summarizes electrical energy potential only. Although there may be a slight shift by the end of the century from liquid fuels to electricity, the major supply problem for Hawaii over the next two decades will be transportation fuel—both for surface and for air travel.

A convenient method for obtaining a rough estimate of how high the price of oil must go before the various renewable alternatives become cost competitive on the basis of fuel replacement alone has been developed by Weingart.[26] Figure 8–7 illustrates the fuel replacement worth of five of Hawaii's renewable resources. Biomass would show up on this figure about the same as geothermal. The assumptions that went into this figure represent my current best estimates and do not totally agree either with the HIEA values shown in Table 8–2 for costs per installed kilowatt or with Weingart's earlier assumptions. The family of curves plotted in Figure 8–7 represent a fixed charge rate of 18 percent and a heat conversion rate of 11,000 BTU per kilowatt-hour. The capacity factor estimates are based on Hawaiian conditions, which explains

Table 8-2. **Electrical Energy Potential of Renewable Resources in Hawaii**

Technology	Commercialization	Base, intermediate, or peak load	Capital costs in 1980 $/kw		Resource potential by year 2000
			1985	2000	
Geothermal	Near term	Base	2,000	1,200	1,000 mw
OTEC	1990–95	Base	8,000	2,600	1,600 mw
Solar thermal and photovoltaics	1990–95	Intermediate (interrupted)	3,000	2,000	1,800 mw
		All three (interrupted)	8,000	2,000	450 mw
Wind	Near term	All three	1,500	700	160 mw
Biomass	Near term	All three	1,500	1,500	100 mw
Hydroelectric	Near term	All three	800	800	45 mw
Municipal solid waste	1985		2,200	2,200	

Source: Adapted from "Hawaii Integrated Energy Assessment—Executive Summary," Department of Planning and Economic Development, State of Hawaii, and Lawrence Berkeley Laboratory, Berkeley, California, January 1981, p. 24.

the optimistic projection that windmills will generate power from 50 to 65 percent of the time.

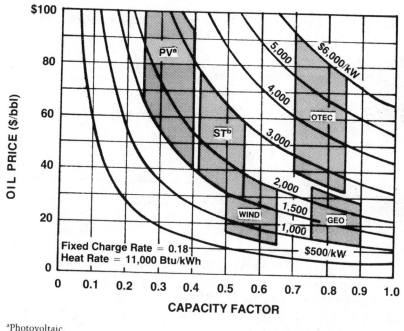

^aPhotovoltaic.
^bSolar thermal.

Figure 8–7. **Fuel Displacement by Renewable Energy Systems**

Even as a general approximation, it is apparent from Figure 8–7 that geothermal and wind are already cost competitive as fuel replacements. If oil prices continue to climb and exceed $40 per barrel, the other alternatives also show economic benefit. In addition, for those resources such as geothermal and OTEC which can provide base-load power and satisfy capacity requirements of the utility, a greater allowance should be made for replacing new capacity at the margin.

The HIEA includes a very comprehensive series of future forecasts for both energy supply and energy demand for the next twenty-five years. One such projection is shown in Table 8–3,[27] which lists the contribution which each of the major energy supply sources will make in meeting the total electricity demand for the state in 2005. Forecasts are presented for the Baseline Case, the Savings Case (additional conservation), and the High Cost Case (10 percent annual increase in oil price).

Table 8-3. **Electricity Supply for Hawaii in 2005** (kwh x 10^6)

		Projected futures	
Source	Baseline	Savings	High oil cost
Geothermal	5,684	5,339	5,306
OTEC	3,568	2,521	2,758
Wind	2,314	2,074	2,016
Solar thermal	1,120	974	1,967
Bagasse	860	860	860
Municipal waste	276	276	276
Hydroelectric	98	98	98
Photovoltaics	negl.	negl.	204
Oil	3,621	1,058	923
Total	17,541	13,200	14,408

Source: "Hawaii Integrated Energy Assessment—Executive Summary," Department of Planning and Economic Development, State of Hawaii, and Lawrence Berkeley Laboratory, Berkeley, California, January 1981, p. 14.

Table 8–3 forecasts a relatively high level of penetration of the renewables for electricity generation by 2005, with the Baseline Case showing a 79 percent penetration and the High Cost Case a 94 percent penetration. However, since the assumptions of the HIEA regarding liquid fuels are that only 10 percent of the gasoline requirement will be produced from indigenous resources and that jet fuel demand will increase by 68 percent over the next twenty-five years, the amount of the total energy provided by the renewables in 2005 is only 37 percent as forecasted for the Baseline Case and 46 percent as forecasted for the High Cost Case.

My projections for the rate and level of penetration of the renewables is somewhat more optimistic—for the reasons and assumptions I have listed previously. Last year I predicted that a reasonable energy goal for Hawaii would be to achieve 50 percent electrical energy self-sufficiency by 1990 and 50 percent total energy self-sufficiency, including jet fuel, by 2000.[28] This still appears to be an achievable goal.

CONCLUSION

Hawaii has both the incentive and the opportunity to become energy self-sufficient through effective conservation practices and the systematic development of its indigenous resources. Its near-total dependence on seaborne petroleum and its resulting vulnerability to dislocations in the global oil market provide the incentive. Its outstanding renewable energy resources and the broad commitment of Hawaii's people provide the opportunity. In addition, the comparatively benign environmental characteristics of the renewables make them compatible with Hawaii's concern for the quality of life.

Much has been accomplished to date in laying the groundwork for conversion to renewable resources—both in energy research and development as well as in public education and the establishment of a public and private infrastructure supportive of energy alternatives. There is probably a higher concentration in Hawaii of renewable energy demonstration projects—geothermal, solar collectors, photovoltaics, biomass, OTEC, large and small wind generators—than any other spot in the world. This was a determining factor in the selection of Honolulu as the site for the *National Conference on Renewable Energy Technologies,*[29] held in December 1980 and cosponsored by DOE.

Hawaii readily recognizes, however, that much remains to be accomplished before any significant inroads can be made in reducing the amount of imported oil. It also recognizes that it will be operating under a new set of guidelines in the future development of its renewable energy resources. For nearly a decade, the major energy RD&D projects which provided the foundation for Hawaii's statewide energy programs were supported primarily by federal funding. This type of support will be extremely difficult to identify under the drastically reduced DOE budget for the renewables. Compounding the problem is the fact that Hawaii is the only state in the sunbelt to vote democratic in the most recent presidential race, and all four of our senators and representatives were elected as democrats. Increasingly, the private and local government sectors will be called upon to furnish the support and incentives for the development of this state's indigenous resources.

Hawaii's energy program is fortunate that this new federal energy policy was not invoked following the 1976 election. During the intervening four years a great deal of progress was made, with much assistance coming from federal support. Geothermal, wind, flat-plate solar collectors, and various forms of biomass are already cost competitive with imported oil in Hawaii, and other energy alternatives are fast approaching commercialization. The utilities, the sugar industry, the lending institutions, and other elements of the private sector have become active participants in the development of these alternatives to imported oil. Both state and county governments also are dedicated to achieving energy self-sufficiency—ultimately.

Therefore, the appreciable momentum which has been building up over the past decade, in conjunction with the increased costs of conventional energy due to price deregulation, should assure continuing progress in the development of indigenous renewable resources for Hawaii's energy future.

NOTES

1. Department of Planning and Economic Development, "The State of Hawaii Data Book, 1980: A Statistical Abstract" (Honolulu, HI: November 1980).

2. "Hawaii Integrated Energy Assessment—Executive Summary," Department of Planning and Economic Development, State of Hawaii, and Lawrence Berkeley Laboratory, Berkeley, California (Honolulu, HI: January 1981).

3. Hawaii Integrated Energy Assessment—Volume I: (title to be determined at a later date), Department of Planning and Economic Development, State of Hawaii, and Lawrence Berkeley Laboratory, Berkeley, California (Honolulu, HI: to be published).

4. J. W. Shupe et al. "Report of the Committee on Alternate Energy Sources for Hawaii of the State Advisory Task Force on Energy Policy," published by the Hawaii Natural Energy Institute, University of Hawaii, and the Department of Planning and Economic Development, State of Hawaii (Honolulu, HI: February 1975).

5. T. Yoshihara and P. C. Ekern, "Solar Radiation Measurements in Hawaii," Hawaii Natural Energy Institute Technical Report, University of Hawaii at Manoa (Honolulu, HI: September 1977).

6. E. D. H. Cheng, "A Study of Wind Energy Conversion for Oahu — Phase II, "Hawaii Natural Energy Institute Technical Report, University of Hawaii at Manoa (Honolulu, HI: June 1977).

7. T. A. Shroeder et al., "Wind Energy Resource Atlas; Volume 11—Hawaii and the Pacific Island Region," Department of Meteorology, University of Hawaii at Manoa (Honolulu, HI: to be published September 1981).

8. J. P. Craven, in Hearings before the Senate Committee on Commerce, Science, and Transportation, 96th Congress, Second Session on S.2492, The Ocean Thermal Energy Conversion Act of 1980, April 10, 1980.

9. J. W. Shupe and P. C. Yuen, "Geothermal Energy in Hawaii — Present and Future," presented to the Circum-Pacific Energy and Mineral Resources Conference, Honolulu (Honolulu, HI: August 1978).

10. D. Thomas et al., "Geothermal Exploration and Evaluation in Hawaii," in *The Geothermal Commercialization Project Report* (Jim Woodruff, Project Manager), Department of Planning and Economic Development, State of Hawaii (Honolulu, HI: in preparation).

11. "Hawaii Integrated Energy Assessment—Volume IV: Energy Data Handbook," Department of Planning and Economic Development, State of Hawaii, and Lawrence Berkeley Laboratory, Berkeley, California (Honolulu, HI: 1981).

12. See reference cited in note 1.

13. A. Seki and P. C. Yuen, "Biomass Resource Potential for Selected Crops in Hawaii," Hawaii Natural Energy Institute Technical Report, University of Hawaii at Manoa (Honolulu, HI: in preparation).

14. Ibid.

15. Ibid.

16. E. A. Laws, "Year One Progress Report on Algal Production Raceway Project," Oceanography Department, University of Hawaii at Manoa, submitted to the Solar Energy Research Institute (Honolulu, HI: July 1981).

17. J. W. Shupe and J. M. Weingart, "Emerging Energy Technologies in an Island Environment: Hawaii," in *The Annual Review of Energy,* J. M. Hollander, ed., Vol. 5, pp. 293–333 (Palo Alto, CA: 1980).

18. See reference cited in note 7.

19. See reference cited in note 2.

20. J. Brewbaker et al., "Giant Leucaena (Koa Haole) Energy Tree Farm: An Economic Feasibility Analysis for the Island of Molokai, Hawaii," Hawaii Natural Energy Institute Technical Report, HNEI-80-06 (Honolulu, HI: September 1980).

21. See reference cited in note 16.

22. See reference cited in note 3.

23. "Hawaii Integrated Energy Assessment—Volume VI: Perceptions, Barriers, and Strategies Pertaining to the Development of Alternate Energy Sources in the State of Hawaii," Department of Planning and Economic Development, State of Hawaii, and Lawrence Berkeley Laboratory, Berkeley, California (Honolulu, HI: 1980).

24. See reference cited in note 2.

25. See reference cited in note 2.

26. See reference cited in note 17.

27. See reference cited in note 2.

28. See reference cited in note 17.

29. J. W. Shupe, ed., Proceedings of "The National Conference on Renewable Energy Technologies" (Honolulu, HI: December 1980).

CHAPTER NINE

Summary with Selected Comments

Session Chairman/Integrator:
Philip H. Abelson
Editor, Science

For the majority of the earth's people, biomass is by far the most important source of energy. Many people obtain essentially all of their energy from this source. Bioenergy constitutes about 15 percent of the world's total energy production. As fossil fuels become more expensive or unavailable, dependence on biomass will increase substantially, and this resource will ultimately have a major role in meeting world needs for food, fuel, clothing, shelter, chemicals, and gaseous and liquid energy.

Conversion of a world economy dominated by oil, natural gas, and coal to one dominated by bioenergy will proceed at different rates in different places. For example, the United States is likely to move slowly, for it has abundant coal and moderate resources of natural gas and oil supplemented by nuclear energy. By attaining greater efficiency in the use of energy, the United States can and will decrease its oil imports and will be able to pay for these imports. In many other countries, the situation is drastically different. They have little or no oil, natural gas, or coal. Their ability to pay for imported oil is limited or vanishing. They must find means of expanding their bioenergy and of using it efficiently.

Following the 1973–74 oil price rise, there was little movement to make the changes that would permit a reduction in oil imports. But the 1979 price increases have goaded people everywhere to seek means of cutting oil usage. Because biomass is

In addition to the chairman/integrator and those presenting papers, the work session included the following assigned participants: Richard L. Grant, Enrique V. Iglesias, L. Hunter Lovins, Amulya Kumar N. Reddy, and Gerald W. Thomas. The chairman/integrator was assisted by Andrew Loebl, who served as rapporteur.

widely distributed around the world, it is drawing broad attention. A sound scientific and technical basis for its expanded use is being established. Many symposia are being devoted to the topic. Thousands of research studies relating to biomass have been initiated. Some 1,700 of these are described in a bioenergy directory, published by the Bio-energy Council in Washington, DC. Important results have been obtained from these studies, but the best are yet to come.

The most casual glance at world biomass reveals great differences among the countries or regions. Differences in climate and soil quality make for contrasts in the nature of the vegetation and the productivity of the land. Agricultural systems differ. The per capita amount of land available varies. For example, the population density in Southeast Asia is much greater than that in parts of Latin America. On the other hand, urbanization has gone farther towards an extreme in Latin America than it has in India.

Despite relative differences in the quantity and kinds of biomass in different regions, there are nevertheless many common problems and opportunities for improvement both of yields of vegetation and of their conversion into more useful forms of energy. This is especially true of the conversion processes. For example, there is already a widespread and increasing use of biogas (methane + carbon dioxide), which is derived from anaerobic microbiological digestion of biomass. Improved chemical or biological processes aimed at producing methanol or ethyl alcohol will also have universal application. Until recently, scientists and engineers paid little attention to devising means for achieving efficiency in the production and use of energy. But now efforts to achieve such efficiencies have a high priority in the advanced countries, and results of these efforts will ultimately be beneficial worldwide. This will be especially true in the Third World.

From time immemorial, biomass has been the principal energy source for most of the world's people, and their practices for using it are little changed from those of ancient times. Their principal use of energy is in the form of firewood for cooking. With increasing populations exacerbating the demand for fuel, severe shortages of firewood have been leading towards deforestation and subsequent soil erosion. In some areas, dried cow dung is used for cooking fuel. Burning this material deprives the soil of both fixed nitrogen and humus.

Part of the problem of cooking fuel lies in the way wood is burned. Professor Reddy of our panel told us that in India, stoves have only a 5 percent efficiency level. One obvious solution would be to improve the efficiency of stoves. This would be a development with widespread applications. But another approach is to probe more deeply into the problem and to ask how best to furnish heat for cooking. Professor Reddy has supplied an answer for his country: one turns to generating biogas from cow dung and other biomatter. For many villages in India, there is sufficient raw material to produce needed cooking gas plus an extra amount that could be used to run motors to produce electricity. Were such biogenerators to be broadly available, the trees could be used for other energy purposes, and it would become possible to plant faster growing trees. The biodigesters' effluents, which contain fixed nitrogen and humus, are useful fertilizers and soil conditioners.

Biogas is clean burning and nonpolluting. It is a versatile energy source with many potential applications in engines and fuel cells. Biogas can be derived from almost any form of biomass, but some feedstocks are better than others. For example, vegetation that is free of lignocellulose is almost entirely converted to gas. Woody material which contains lignocellulose is only partly converted. Cattle manure, which often contains lignocellulose, accordingly is only partially gasified. Swine, fowl, and human feces are good sources of biogas, as are cellulosic wastes. Because of the value and versatility of biogas and because of the broad availability of feedstocks, this energy source will surely have major worldwide applications in the future. A good start has been made. There are about 7 million biogas generators in the People's Republic of China and several hundred thousand in India. Successful operations have been reported in the Philippines, Taiwan, and the United States. Professor Reddy of our panel reported on some of the conditions that must be maintained in the generators. He stated that the fermentation involves mixed cultures of microorganisms and that these could be adversely affected by sudden changes in the composition of feedstocks. The percentage of solids in the liquid mass must be maintained at no more than 8–10 percent when cow manure alone is used. Control of pH to a range of 6–8.5 is also desirable. The optimum temperature for the production of gas is 35°C; production stops at temperatures below 15°C. Because of this minimum, installations have hitherto been confined to tropical regions. However, Professor Reddy told of an Indian design that uses solar heat and permits operation of generators in temperate areas. There is considerable room for further research and development work which will increase the utility and versatility of the generators.

Professor Reddy's proposal for supplying cooking heat was part of a larger proposal aimed at freeing India from the crushing burden of paying for imported oil. His overall proposal is too complex to be covered in detail here.* However, the striking feature of it is that he has examined the functions being served by different energy sources, including oil and its products, and has imaginatively suggested the best local means for filling needs. He has been able to visualize how through the best use of its biomass, India might free itself of the need to import oil.

The present situation in China, India, and other parts of Southeast Asia is one in which population pressure is great and there is little biomass that is not already being exploited or overexploited. The low efficiency of stoves in India is an object lesson pointing to the need to seek means of using energy more efficiently. This need is true, of course, of both developing and developed countries.

There is a contrast in the availability of biomass resources in Latin America and some countries of Africa when compared with Southeast Asia. In Brazil, where land is abundant, some of it can be devoted to growing sugar cane and cassava for processing to make alcohol. A Brazilian program to obtain increasing amounts of alcohol as a liquid fuel is progressing, and it is anticipated that production of 200,000 barrels per day will be achieved by 1985. A quick source of biomass is annual or perennial plant vegetation. Professors Carioca and Arora of our panel described their alcohol-producing facility in Brazil, which uses cassava as a feedstock. Cassava is a hardy plant that produces high yields in good soils but also fair yields in poor soils. It can be grown in places where culture of other crops would not be feasible. Brazil is deeply in debt to international bankers and is under considerable pressure to reduce its oil imports. This could be done by obtaining a large fraction of its liquid fuels from biomass. A number of options are open, including the use of vegetable oils and methanol derived from gasification of wood. However, cassava is likely to have a major role in both Brazil and many other tropical countries.

Professors Carioca and Arora discussed some of the factors that should be considered if and when large amounts of alcohol were to be produced from cassava. They concluded that many desirable social, agronomic, and economic purposes can be served

*For a fuller explanation of this proposal, see A. K. N. Reddy, "A Biomass-Based Strategy for Resolving India's Oil Crisis" (Bangladore: Centre for the Application of Science and Technology to Rural Areas, Indian Institute of Science, 1981)—a revised and expanded version of a paper published in *Current Science* 50 (1981), pp. 50–53, and reprinted in *New Scientist*, July 9, 1981.

by the choice of medium-sized cassava-processing plants having a production capacity of about 20,000 liters of alcohol per day. Installations of this size would provide rural employment, minimize transportation costs, facilitate the return of nutrients to the soil, and minimize environmental pollution.

One of the issues that will arise about the use of biomass as an energy source is the competition between energy and food. In the United States, if a substantial amount of grain were used to make alcohol, the price of grain would increase and exports would diminish. A similar situation could arise elsewhere, leading to food shortages and more human misery. In the United States, another factor which tends to lessen grain production is arising. Monoculture agriculture has resulted in excessive soil erosion. The matter is drawing increasing attention and concern and may well lead to crop rotation practices that result in less production of grain but in preservation of soil. Our group of participants was unanimous in the opinion that cultivated lands cannot be counted on as a substantial means for producing grains for alcohol.

For the most part, one must seek biomass for energy elsewhere, rather than on cultivated lands—lands which represent only a small part of the earth's land surface. Through experimentation and selection, agricultural productivity has been greatly increased. However, little attention has been paid to improving the yields of trees, shrubs, and plants. This work is now going on but has far to go. One approach is to seek vegetation that is optimized for yield of biomass rather than food. For example, our panelist Dr. Shupe believes that in Hawaii a superior sugar cane could produce much more biomass while still yielding a substantial amount of sugar. Another approach is to select fast-growing trees such as poplars that can be coppiced. That is, after the trees are harvested, new shoots come up from the roots. Harvest cycles of as short as four years have been proposed and are being employed. Another possibility is the growth of marine and nonmarine algae. Tests of this possibility are under way. Particularly attractive is the growth of algae on land in plastic pipes in which an atmosphere high in carbon dioxide can be maintained and water loss avoided. Such growth would be particularly suited to otherwise barren arid regions with high solar inputs. The potentials for increasing the world yield of biomass are enormous, but much research and development will be necessary. Achievement of ultimates in improvement will doubtless require the passage of decades.

In choosing the optimal vegetation for biomass, an important consideration is its subsequent role as a feedstock. For example, cassava is relatively well suited as raw material for the production of ethyl alcohol. Cassava is rich in starch, which can readily be converted to glucose for fermentation to alcohol. However, many plants that are to be used for energy consist largely of cellulose or, worse, of lignocellulose. Obtaining good yields of alcohol from plants having such compositions has not been very feasible and is today a target of many studies. It is likely that the contents of various plants and trees will differ in their ease of processing to obtain alcohol. In any event, the science and technology of converting woody materials to liquid fuels is a matter of great importance. A variety of schemes can be employed. One path goes via the action of microorganisms to convert cellulose and hemicellulose to ethyl alcohol. A second method is to treat the wood with chemicals such as acids to convert it into fermentable substances. Another set of methods involves pyrolysis. Depending on conditions, a variety of products can be obtained including ethylene, carbon monoxide, and hydrogen. Carbon monoxide and hydrogen can be combined to form hundreds of compounds, including methane, methanol, ethyl alcohol, and a mixture of compounds that make up gasoline. In many countries the principal petroleum product burned is diesel oil. Neither methanol nor ethyl alcohol is suitable for this application. Recent work in Brazil, described by Professor Carioca, has provided an alternative use for modified vegetable fats such as oils derived from seeds or palm tree nuts. The crude fatty oil is relatively nonvolatile, but it can be readily modified by esterification with methanol or ethyl alcohol to form an excellent diesel fuel.

The concept of obtaining products from plants that are closely related chemically to petroleum is a new one that merits further exploration. In some instances the existence of hydrocarbons in trees has been well known. Latex from the rubber tree has been widely used for more than a hundred years. But other potentialities have not been much explored or emphasized. This deficiency is now being remedied. An inventory of chemicals in plants and trees has been initiated by a group in Brazil. The task is enormous, and were the inventory to include vegetation elsewhere, the efforts of many scientists would be needed. In the United States much talk and some effort has been devoted to expanding growth of guayule, a plant that produces a natural rubber. Related plants have been studied by Melvin Calvin, who has been enthusiastic about so-called gasoline farms. These would involve growth of hydrocarbon-containing plants on semiarid lands. Calvin is also enthusiastic about a Brazilian tree that contains a hydrocarbon that can be readily tapped, much as latex is obtained from the rubber tree.

As a nation, the United States is better situated than most with respect to energy resources. Even after oil and natural gas resources are depleted, large reserves of coal will be available. However, what is true for the country as a whole is not entirely true for all its parts. For example, the New England states are devoid of coal. There energy costs for heating are high. Thus, people in those states are increasingly turning to wood for fuel. The same is true of many of the other conterminous states. The energy problem is particularly acute in Hawaii, which has no fossil fuels and is remote from supplies of them. As energy costs rise, Hawaii must depend on its own internal renewable resources. These include geothermal, ocean thermal, wind, and solar resources as well as bioenergy. Hawaii has a vigorous program for exploiting its many sources and is likely to become a leader in many aspects of energy development.

For all the United States, a useful means of decreasing oil imports is through enhanced energy efficiency and through use of rural and urban wastes. Greater efficiency is, of course, the greatest potential source of energy savings, but utilization of wastes could make a substantial contribution. Some installations for conversion of manure to biogas are in operation, but their number could be expanded. The feedstocks employed could also utilize other agricultural wastes such as straw.

At the moment the principal users of bioenergy in the United States are the forest product companies. They have been obtaining increasing fractions of their energy needs from wood wastes and from process wastes that would otherwise pollute streams. Some companies state that they plan to obtain all their process energy needs from biomass. The energy involved is substantial, and when implemented this approach will involve a substitution of biomass for the total equivalent of about 1 million barrels of oil per day.

In other parts of the world, the immediately urgent problem is to utilize limited resources more effectively. If the experience and knowledge now being obtained in Brazil, China, and India were generally available and were applied, a considerable improvement in the living standards of a majority of the world's people would result. Beyond the present are the great opportunities and needs of the future. One opportunity is to increase yields of biomass by as much as a factor of 5 to 10. This would require development of superior strains of vegetation through selection and genetic engineering. An assessment or inventory of tropical plants might highlight additional useful sources of hydrocarbons. More efficient means of converting biomass to high-energy fuels are already under development, and advances are being made.

As the technologies become well established, they should be transferred to other countries. But technology cannot be effectively transferred unless the recipient country has a minimal level of technical capacity. Moreover, in such matters as selection of vegetation, local conditions will dictate the best choices. Again there is need for local competence.

Increased use of biomass will bring with it the need to preserve the environment. Experience has taught that in many places nature is fragile. Mistakes of the past have led to depletion and soil erosion. The goal should be to achieve a sustainable steady state. One aspect of this is the need to return essential nutrients to the soil after processing biomass.

The task of managing the lands well is enormous. Many scientific and technical considerations will be involved. In a number of countries the necessary competence to bring about and to manage a countrywide bioenergy program may be found or will soon be in place. For example, the state of science and technology in Brazil, China, and India is such that in-country competence is available. These countries could greatly benefit from exchange of knowledge and know-how. They would also benefit from interchanges with scientists and engineers in the developed countries, many of whom are now active in bioenergy developments.

However, in perhaps as many as one hundred countries, sufficient local competence in bioenergy does not exist. If these countries are to develop and manage their indigenous sources of bioenergy, they need to obtain a working access to the world's knowledge and relevant biomass and bioenergy technology. Competence in this matter is highly important for the survival of their people and improvement of their living standards. Thus, our group of participants favored creation of an international institute for bioenergy. This institute should facilitate interaction among scientists and technologists of the various countries, have an educational function, and engage in research and development projects. It should serve as a center and provide mechanisms for conducting inventories of bioenergy development opportunities. If this institute is established, efforts should be made to avoid creating an enclave of special privileges in the host countries.

Were bioenergy programs to be successful in substituting biomass-derived fuels for petroleum, important worldwide benefits would be realized. The price of oil would likely decrease, and oil reserves would last for a longer period. For the developed countries, a reduction in price of $5 per barrel would lead to a savings of roughly $35 billion per year and to a higher level of prosperity, since the Third World would provide better markets for goods. Thus, the developed countries should seriously consider in their own self-interest, if for no other reason, means of encouraging the development of bioenergy everywhere.

In its initial phases of development, financial support for bioenergy probably must come from governments. But after demonstrations of technical and financial feasibility, it should be possible to attract private capital. The matter of financial support for bioenergy deserves special attention and careful analysis. Funds provided to developing countries should not be used merely to buy and burn oil. Rather, priority should be devoted to projects that increase energy supplies.

Development of bioenergy and solar energy could have profound social effects, particularly on the rural poor. With an improved standard of living, they would feel less compelled to go to the cities. The universality of sunshine and the broad occurrence of biomass also tend to favor decentralization. This Symposium has been devoted to energy and necessarily has not dealt with population growth. However, there will be no end to human misery unless the fast growth in population abates.

We have lived in an era of profligate waste of an endowment of fossil fuels. It seems clear that substantial numbers of people now realize that we are moving towards an era in which renewable resources will be the predominant energy source and an era in which great efforts will be made to achieve energy efficiency. We will also be considering more carefully why we wish to use energy and in what form. There will continue to be awareness of the need to preserve the environment. It is hoped that there will be good cooperation among countries as each works out its own particular energy destiny.

CONCLUSIONS AND RECOMMENDATIONS

In view of both short-term and long-term opportunities and benefits, international programs designed to foster the use of bioenergy should have a high priority for implementation. Programs should include:

- R&D aimed at improving the energy efficiency of biomass utilization.
- R&D aimed at better use of nonagricultural areas.
- R&D devoted to the more effective use of woody materials.
- Inventories of biomass resources of the various countries.
- Creation of an international institute of bioenergy to facilitate rapid dissemination of knowledge and know-how.
- Financial support for a substantial number of bioenergy demonstration facilities.

SELECTED COMMENTS

HARBANS LAL ARORA

Professor Carioca and I belonged to the work session on the biomass energy/nonenergy conflict, and we recommend the following:

1. Establishment of an international bioenergy technology center (or centers) for research, development, and demonstration of various aspects in developing countries of biomass production and conversion into fuels and other products, with the support of the United Nations and other nonprofit organizations and with a concrete mechanism for World Bank support for specific projects and programs.

2. Initiation of country-specific programs to develop decentralized, integrated systems scattered through various regions of each country—programs whose purposes are the diversification and regionalization of primary materials and the achievement of more equitable development in rural areas, in order to provide desolate, neglected, and deprived rural populations with sufficient energy, food, and employment on a sustainable basis.

As far as the second recommendation is concerned, I will explain a little bit. Experience has shown that centralized energy systems—whether they are petroleum, hydro, nuclear, coal, or, for that matter, biofuels production systems (big alcohol distilleries, for example)—have mostly enriched the wealthier sections of society. On the other hand, we have demonstrated by our work— not just theoretical but experimental and practical demonstration work—that small decentralized biofuel production systems which make use of both agricultural and industrial effluents and by-products are, even though small, technologically and economically viable and also environmentally safe and socially acceptable. We thus would like to make a strong point that in any planning by any government, especially those of the developing countries, decentralized, integrated agro-industrial systems which produce not only fuel but also food, feed, and fertilizers should be given preferred consideration.

L. HUNTER LOVINS

I would like to add to Dr. Abelson's summary several points which

I think reflect the unanimous emphasis of the biomass panel. The first is that the biomass for energy option is already technically and economically viable. The three papers that were presented [i.e., the position paper by Professors Carioca and Arora, the special paper by Dr. Shupe, and the paper volunteered by Professor Reddy, as cited above] demonstrated this conclusively. Appropriate use of the various modern biomass-based energy technologies can reduce both oil dependence and the devastation caused by overuse of firewood or dung for cooking. In the instances both of a country with no more land available for biomass production—namely, India—and of a nation with perhaps a great deal of land available—namely, Brazil—biomass conversion technologies are up and running. They have been demonstrated, and they seem to offer the most hope for rural and, indeed, national development.

Because of this present viability of biomass-based energy technology, the panel recommended the creation of an international biomass center to speed the implementation of present technologies and the development of even better ones. It is crucial, however, that the work of such a center be as decentralized as possible. It is not our intent to create an international bureaucracy, an international elite, even if the center was placed in a developing country. Perhaps an "international biomass network" might be a more apt description of what is needed. What came most clearly out of Professor Reddy's paper is that there are no global solutions in the areas of biomass or of rural development. One can only see what works when something is analyzed on a local scale and implemented at a local level, with comprehensive integration with the rest of the agricultural sector and the development process.

In any biomass-based energy project, it is vital to emphasize the efficient use of the product. This was a unanimous conclusion of our panel: that without very efficient use of the biofuels (or any other kind of fuel), no energy solution is going to make sense, in either the developed or the developing countries. In the case of the United States, an efficient vehicle fleet would so greatly decrease the amount of biofuel required that just the present waste streams could supply enough liquid fuels to run the US transport sector at present activity levels. Professor Reddy had analogous figures for India.

In conclusion, the panel's strongest finding was that biomass is here—it's been technically demonstrated, and it's been shown to be economic in a variety of settings around the world. What is needed now is not so much research as implementation.

PHILIP H. ABELSON

The comments about the biomass work session were useful, and I have no contradictions to them. However, I would like to take this moment to advertise the contribution of Dr. Shupe of Hawaii. In Hawaii there are abundant sources of various forms of renewable energy, including biomass, geothermal, wind, and ocean thermal energy. The continuing development of technology for those renewable resources in Hawaii is likely to make important world contributions.

Work Session on Energy for Rural Development

Technical and Social Aspects of Energy for Rural Development in Third World Countries

Edward Lumsdaine
Director
Energy, Environment, and Resources Center
The University of Tennessee

Salah Arafa
Professor of Physics
and Solid-State Science
American University at Cairo

INTRODUCTION

The rapid escalation in the price of petroleum since 1973 (an 80 percent increase in the last two years alone) has presented the world with a new set of problems and an awareness of the importance of energy in human activities and societies. Even though the industrialized countries differ greatly from the developing countries in their economic and political structures, all count on having access to a reliable supply of affordable energy. The economies of industrialized countries have become heavily dependent on petroleum; thus, these countries will have to deal with the attendant problems of technological changes and appropriate internal and international policy as well as with environmental implications (local as well as global) of their energy resources by managing the supply side as well as the demand side. However, because the industrialized countries possess large amounts of resources and not only capital but also technical knowledge, appropriate response and policies theoretically should be relatively easy.

In contrast, the energy problem poses an altogether different kind of challenge to developing countries. Oil-exporting Third World countries have time and money on their side in planning for a post-petroleum future. Oil-importing Third World countries, however, are faced with immediate and pressing needs, of which the provision of energy is but one of many interrelated problems which must be solved with strained and meager resources. Therefore, any solution in a particular area must take into account its effect on all other social and economic factors in order to achieve lasting and optimum results. Also, the suppositions underlying any long-range planning will need to be clarified. Although the immediate problem may appear to be to find alternate renewable energy sources to replace expensive imported oil, policies may evolve differently if the high price of oil is perceived to be caused by (1) real shortages and diminishing world supplies or (2) supply uncertainties and interruptions due to political problems. Policies will also evolve differently if the high price of oil is established as an incentive in response to environmental concerns or if the country must deal with a fuelwood crisis.

The report will attempt to give an overview of the energy problems and related problems pertaining to rural development that confront oil-importing Third World countries. Available options and their technical, economic, and social aspects and consequences will be discussed in general, and a framework will be indicated which can aid in evaluating programs that will best suit specific regions or countries. There is a tremendous need for information and education, not only in developing countries about implementation of appropriate new technology but also in industrialized countries about the problems and struggles that beset the millions of rural poor in Africa, Asia, and Latin America. Since developing countries vary greatly in their cultures and resources, energy projects and policies must be tailored to particular conditions in specific regions, taking local traditions, political structure, and resources into account.[1] This paper will include a brief discussion of an Egyptian village to illustrate the type of barriers that must be faced if rural development projects are to succeed not only in providing reliable energy but also in increasing the quality of life for these people.

THE NATURE OF THE PROBLEM

It is generally acknowledged that the energy problem is not just a technical problem; it is perhaps primarily and foremost a political problem. This is true especially in developing countries that have very limited resources to either seek out conventional energy sources or develop new sources, and where in the near term, economic realities leave very little room for experimentation or inaction and no space at all for mistakes or inappropriate policies. If it can be said that the poor exhaust their resources to survive and the rich to preserve their affluence,[2] then developing countries not only must compete externally with rich countries for resources but also may face internally strong vested interests against the allocation of resources for rural development. Even though there are a number of technological solutions available, the infrastructures to implement these widely do not yet exist.

In developing countries, reduced oil imports due to high cost have led to reduced economic growth and decreased food production; the high oil prices have also resulted in unmanageably high debts to pay for imported energy, sometimes even at the cost of necessary food imports. In some areas, traditional noncommercial fuels such as wood, dung, and crop residue are also in short supply, and the increased pressure on these renewable resources is causing serious deforestation, soil degradation and erosion, and lessened productivity. In the last few decades, it has been the substitution of oil for the traditional fuels which constituted a first step up from subsistence towards modern production and development and a modest rise in the standard of living. Therefore, the developing countries with the lowest incomes are the most afflicted by the high oil prices and their consequences. Rural development suffers when high costs decrease the availability of fertilizers and fuels for irrigation pumps, both being important components of the green revolution. In economic development terms, increased energy basically means increased ability to produce the necessities and amenities of life—food, shelter, clothing, education, health care, communication, transportation, consumer goods, and so forth. Hence, there is a fundamental link between energy and rural development such that any efforts at harnessing increasing amounts of energy must be directed towards creating opportunities for productive purposes.

There is a growing consensus that successful development requires a firm agricultural foundation and that the quality of life must be improved for the poor majority of people living in rural areas. Any programs must be done not *for* these people but *with* their active participation. If this can be done (and it may not be done easily or quickly), then the rural poor may have reason and ability to reduce their birth rates; may increase their food production as well as consumption; may improve their health, life span, and education; and may no longer be forced to flee to already overcrowded towns and cities in search of employment. Carefully and persistently pursued, a fully integrated rural development program can provide a sound basis for the manufacturing and service sectors of a self-reliant and thriving national economy. Increasing energy supplies and efficiency in energy use will be very important aspects of any such comprehensive rural development strategy. Even just meeting the existing fuel needs in rural areas is a considerable task.

Energy Consumption in Developing Countries

From a number of studies, a basic figure of about 300 to 400 kilograms of coal equivalent (kgce)* per capita per year has appeared as the minimum energy input for rural subsistence, with triple that amount needed to provide a level of living that includes adequate food, shelter, safe water, health care, education, and other comforts considered the rights of all human beings. The questions that must be asked and answered are thus those not only of how energy sources are going to be found to supply these basic demands (including alternatives for oil) but also of exactly what this energy is expected to do. Energy is not consumed per se—it provides a service, removes drudgery, increases comfort. Thus, the goal should be to use energy for the specific service at its highest efficiency and least cost, taking into account environmental factors and the value of the various fuels in alternate (present as well as future) applications. Additional factors that will also influence the choice of energy sources are available capital, labor and its skill levels, and the entire sociopolitical framework in the country as well as in the individual community.

Conventional commercial fuels used in developing countries are kerosene, coal, gas, and electricity. Kerosene is especially popular in African countries. Electricity provides less than 5 percent of rural energy in Africa, 15 percent in Asia, and about 25 percent in Latin America, although accurate data about energy supplies and consumption are very scarce and difficult to obtain.[3] In rural areas, noncommercial fuels may constitute as much as 100 percent of domestic consumption in Africa and about 90 percent in India. As income rises, consumption shifts from noncommercial energy to commercial energy, which is more efficient than the former but also more costly. In many developing countries, households consume about 50 percent of all energy used. Despite a rather low per capita consumption of commercial energy in developing countries, there is a large dependence on oil (61 percent) and natural gas (15 percent). However, 75 percent of all African, 80 percent of all Latin American, and 50 percent of all Asian developing countries depend on liquid petroleum fuels for over 90 percent of their commercial energy. Table 10-1 summarizes worldwide energy consumption and population figures. Not

*1 kgce = 8 kwh = 6,880 kcal = 27,300 btu = 0.0047 barrels of oil equivalent.

only do industrialized countries consume a disproportionate amount of energy per capita (e.g., 11,000 kgce per year in the United States); there is also a large disparity in per capita consumption between different Third World countries (e.g., 1,750 kgce in Argentina, 1,200 kgce in Mexico, 500 kgce in Nicaragua and Zambia, 200 kgce in India, 100 kgce in Cameroon, and only 10 kgce in Nepal, Rwanda, and Burundi).[4]

Table 10-1. Distribution of Population and Energy Production and Consumption among the Three Worlds[a]

	First World (Australia, Canada, Japan, New Zealand, Western Europe, and United States)		Second World (China, Eastern Europe, Soviet Union, and Mongolia)		Third (Over 100 developing countries in Africa, Asia, and Latin America)		Global Total	
Population, 1981	17%		31%		52%		4.44 billion	
Energy consumption	65%		25%		10%		140 million bpd of oil equivalent	
	P^b	C^c	P	C	P	C	1980 consumption	As % of total
Crude oil	22%	65%	22%	21%	56%	14%	23×10^9 bbl	45%
Coal	40%	41%	55%	54%	$5\%^d$	5%	3×10^9 t	25%
Natural gas	58%	60%	32%	30%	10%	10%	1.5×10^9 m^3 [e]	18%
Water power	2/3		1/6		1/6		5×10^6 kwh[f]	6%
Other renewables	1/3		1/3		1/3		2×10^9 bbl[f]	4%
Nuclear power	2/3		1/3		negligible		2×10^6 kwh[f]	2%

Source: F. S. Patton, "Energy—A Global Summation," briefing of us International Communication Agency, Washington, DC, March 11, 1980, with updating and interpolation by the authors using data from 1981 *Britannica Book of the Year* (Chicago: Encyclopedia Britannica) and estimates.
[a]All percentages and fractions represent portions of the global totals.
[b]Production.
[c]Consumption.
[d]Primarily India and South Africa.
[e]Not including almost 0.2×10^9 cubic meters flared at wellheads.
[f]Rough estimates.

Income Levels in Developing Countries

Table 10-2 summarizes population and income figures for most of the world's developing countries. The table's GNP per

Table 10-2. Economic Indicators of Developing Countries

	Area (km² x 10³)	Population (x 10⁶)	Rural population (as % of total)	GNP per capita (US $)[a]	Income level[b]	Food supply[c]	Import/export ratio	Major exports	Type of economy
Africa									
Algeria	2,380	19.1	N/A[d]	900	H	3	1.1	Oil and petroproducts	OPEC mem.
Angola	1,250	6.8	N/A	550	M	4	0.5	Oil, coffee	Oil exp.
Benin	115	3.3	87	150	L	3	5.3	Cotton, cocoa	Agric.
Botswana	575	0.8	88	300	M	3	0.7	Diamonds, meat, ore	Raw mat. expm
Burundi	30	4.1	98	100	L	N/A	1.4	Coffee, cotton	Agric.
Cameroon	465	8.4	74	250	M	1	1.1	Coffee, cocoa, timber	Agric.
Cent. Afr. Rep.	625	2.3	64	150	L	1	0.8	Coffee, diamonds, timber	Agric.
Chad	1,280	4.5	86	100	L	4	1.9	Cotton, meat	Agric.
Comoros	2	0.3	N/A	200	L	2	1.8	Vanilla	Agric.
Congo	340	1.5	38	450	M	1	1.9	Oil, timber, food	Oil exp.
Egypt	1,000	41.1	44	280	M	N/A	2.1	Oil, cotton	Oil exp.
Eq. Guinea	30	0.3	45	300	M	3	0.3	Cocoa, coffee	Agric.
Ethiopia	1,220	31.1	11	100	L	3	1.3	Coffee, hides, skins	Agric.
Gabon	270	1.3	N/A	4,000	V	2	0.4	Oil, manganese ore	OPEC mem.
Gambia	10	0.6	14	200	L	3	2.4	Peanuts and by-products	Agric. exp.
Ghana	240	11.4	31	450	M	1	1.1	Cocoa, aluminum, timber	Agric.
Guinea	245	5	20	150	L	3	1.0	Bauxite, alumina	Raw mat. exp.
Guinea-Bissau	35	0.8	N/A	N/A	L	N/A	6.8	Peanuts, fish, copra	Agric.
Ivory Coast	320	8.2	34	500	M	1	1.0	Coffee, cocoa, timber	Agric. exp.
Kenya	580	15.3	11	200	L	3	1.5	Coffee, petroproducts	Agric.
Lesotho	30	1.3	3	200	L	3	16.4	Wool, mohair, diamonds	Agric.
Liberia	98	1.9	28	400	M	3	0.9	Iron ore, rubber	Raw mat. exp.
Libya	1,750	3.3	N/A	6,000	V	1	0.5	Oil	OPEC mem.
Madagascar	585	8.7	15	200	L	1	1.8	Coffee, cloves	Agric.
Malawi	120	5.8	5	150	L	1	1.7	Tobacco, tea, sugar	Agric.
Mali	1,240	6.6	N/A	100	L	4	2.1	Cotton, livestock, nuts	Agric.

Mauritania	1,030	1.6	22	300	M	4	1.8	Iron ore, fish	Raw mat. exp.
Mauritius	2	1	48	400	M	1	1.5	Sugar, clothing	Agric.
Morocco	460	19.5	38	450	M	1	2.0	Phosphates, oranges	Raw mat. exp.
Mozambique	800	12.4	55	250	M	4	2.2	Cashew nuts, textiles	Agric.
Namibia	825	~1	N/A	N/A	N/A	2	0.6	Diamonds, uranium	Raw mat. exp.
Niger	1,190	5.3	9	150	L	4	1.6	Uranium, livestock	Agric.
Nigeria	925	~90	N/A	400	M	3	0.6	Oil	OPEC mem.
Rwanda	25	5.1	4	100	L	3	1.5	Coffee, tea	Agric.
São Tomé & Principe	1	0.1	N/A	N/A	M	3	1.6	Cocoa	Agric.
Senegal	195	5.7	29	350	M	3	1.2	Peanut oil, phosphates	Agric. exp.
Sierra Leone	70	3.5	15	200	L	2	1.9	Diamonds, coffee, cocoa	Raw mat. exp.
Somalia	640	3.6	28	100	L	4	3.7	Livestock, bananas	Agric.
South Africa	1,130	23.8	N/A	150	L	1	0.5	Gold, ores, food, diamonds	Raw mat. exp.
Sudan	2,500	18.3	13	300	M	3	2.0	Cotton, gum arabic	Agric.
Swaziland	15	0.6	8	400	M	3	1.3	Sugar, wood, asbestos	Agric.
Tanzania	945	18.0	8	150	L	4	2.0	Coffee, cotton, fruits	Agric.
Togo	55	2.7	15	200	L	2	1.9	Phosphates, cocoa	Raw mat. exp.
Tunisia	155	6.4	48	900	H	2	1.6	Oil, clothing, phosphates	Oil exp.
Uganda	240	12.6	7	200	L	3	1.6	Coffee	Agric.
Upper Volta	275	6.9	11	100	L	4	0.6	Cotton, livestock	Agric.
West Sahara	265	0.1	N/A	N/A	L	N/A	5.4	N/A	
Zaire	2,345	25.6	26	150	M	4	0.45	Cobalt, copper, oil	Raw mat. exp.
Zambia	755	6.0	34	450	M	4	2.0	Copper	Raw mat. exp.
Zimbabwe	390	7.4	N/A	450	M	1	0.8	Chrome, ore, gold	Raw mat. exp.
Asia and South Pacific									
Afghanistan	655	15.5	12	150	L	4	0.7	Fruits, nuts, carpets	Agric.
Bangladesh	145	88.7	7	100	L	4	3.2	Jute, leather	Agric.
Bhutan	45	1.3	N/A	50	L	3	2.1	Timber, fruits	Agric.
Burma	675	35.3	22	100	L	2	0.8	Rice, teak	Agric.
Kampuchea	180	4.8	N/A	N/A	N/A	2	6.9	Rubber	Agric.
Cyprus	10	0.6	42	N/A	N/A	1	2.2	Clothing, potatoes	Bal. growth
Fiji	20	0.6	39	1,000	H	1	1.8	Sugar, petroprod., fish	Agric.
India	3,290	651	21	150	L	3	1.3	Textiles, yarns	Bal. growth
Indonesia	1,920	142	N/A	250	M	2	0.5	Oil, timber, nat. gas	OPEC mem.
Iran	1,650	37.7	N/A	1,500	H	2	0.5	Oil	OPEC mem.

Table 10-2—Continued

	Area (km² x 10³)	Population (x 10⁶)	Rural population (as % of total)	GNP per capita (US $)ᵃ	Income levelᵇ	Food supplyᶜ	Import/ export ratio	Major exports	Type of economy
Iraq	440	12.8	N/A	1,200	H	2	0.35	Oil	OPEC mem.
Israel	21	3.9	N/A	N/A	N/A	1	1.9	Diamonds, chemicals	Industrialized
Jordan	95	2.2	43	600	M	3	4.8	Phosphates, vegetables	Agric.
Laos	235	3.7	N/A	100	L	2	3.7	Timber, tin	Agric.
Lebanon	10	3.2	58	1,200	H	1	3.1	Paper products, food	Agric.
Malaysia	330	13.9	27	850	H	1	0.7	Rubber, oil, timber	Oil exp.
Maldives	300	0.1	N/A	N/A	L	4	3.0	Fish	Agric.
Nepal	145	14	4	100	L	3	2.3	Jute	Agric.
North Korea	120	17.9	N/A	400	M	1	0.95	Metal ores	Raw mat. exp.
Pakistan	800	82.4	26	150	L	3	2.0	Rice, cotton, carpets	Bal. growth
Papua New Guinea	465	3.1	11	450	M	2	1.0	Copper, coffee	Agric.
Philippines	300	48.4	29	400	M	3	1.5	Coconut oil, metal ores	Bal. growth
Saudi Arabia	2,240	8.4	N/A	4,000	V	1	0.4	Oil	OPEC mem.
Solomon Islands	30	0.2	N/A	N/A	M	3	1.0	Copra, timber, fish	Agric.
South Korea	100	38.2	47	550	M	1	1.3	Clothing, textiles	Industrialized
Sri Lanka	65	14.9	22	150	L	2	1.5	Tea, rubber, coconuts	Agric. exp.
Syria	185	8.6	46	650	M	1	2.0	Oil	Oil exp.
Taiwan	36	17.7	51	1,200	H	1	0.95	Machinery, clothing	Industrialized
Thailand	540	43.1	17	350	M	1	1.4	Rice, rubber	Agric. exp.
Turkey	780	45.2	42	800	H	1	2.3	Fruits, cotton, textiles	Bal. growth
Vietnam	330	52.3	N/A	150	L	1	3.0	Clothing, fish, rubber	Agric.
Western Samoa	3	0.2	N/A	250	M	1	4.1	Copra, cocoa, fruits	Agric.
Yemen, AR	200	5.2	5	300	M	4	N/A	Cotton, coffee, hides	Agric.
Yemen, PDR	340	1.9	N/A	250	L	4	2.5	Petroproducts, fish	Agric.
Latin America									
Argentina	2,760	27.1	80	1,500	H	1	0.6	Meat, corn, nuts	Industry
Bahamas	15	0.2	N/A	2,700	V	1	1.1	Oil	Oil exp.
Barbados	0.4	0.2	N/A	1,500	H	1	2.4	Sugar, clothing, petro.	Agric.

Country	Area	Population	Urbanization (%)	Per capita income ($)[a]	Income[b]	Food supply[c]	Food ratio	Principal imports/exports	Economy type
Belize	25	0.1	N/A	N/A	N/A	1	1.4	Sugar, clothing	Agric.
Bolivia	1,100	5.6	37	350	M	4	1.3	Tin, gas, silver, oil	Oil exp.
Brazil	8,510	123	59	1,000	H	1	1.3	Coffee, beans, machinery	Industrialized
Chile	755	11.1	83	800	H	1	1.1	Copper, metal ores	Industrialized
Colombia	1,140	27.3	63	500	M	3	1.0	Coffee, cotton	Bal. growth
Costa Rica	50	2.2	41	900	H	1	1.5	Coffee, cotton	Agric. exp.
Cuba	110	9.7	N/A	800	H	1	1.0	Coffee, bananas, beef	Agric. exp.
Dominican Rep.	50	5.4	42	700	M	2	1.4	Sugar, metal ores	Agric. exp.
Ecuador	280	8.4	N/A	550	M	3	1.0	Sugar, coffee, gold	OPEC mem.
El Salvador	20	4.8	39	450	M	2	0.9	Oil, coffee, cocoa	Agric.
French Guinea	90	0.1	N/A	N/A	N/A	N/A	1.5	Coffee, cotton	Agric.
Grenada	0.3	0.1	N/A	400	M	3	2.1	Timber, shrimp	Agric.
Guatemala	110	7.3	34	550	M	1	1.2	Cocoa, nutmeg, bananas	Agric.
Guyana	215	0.8	40	550	M	2	0.9	Coffee, cotton, chem.	Agric. exp.
Haiti	30	4.9	20	200	L	3	1.4	Bauxite, sugar, rice	Raw mat. exp.
Honduras	110	3.7	31	350	M	1	1.1	Coffee, bauxite	Agric.
Jamaica	10	2.2	37	1,100	H	1	1.3	Bananas, coffee, beef	Agric. exp.
Mexico	1,975	67.4	63	1,100	H	1	1.3	Alumina, bauxite, sugar	Raw mat. exp.
Nicaragua	130	2.7	49	650	M	1	0.9	Oil, machinery, coffee	Oil exp.
Panama	75	1.8	50	1,150	H	1	4.1	Coffee, cotton, meat	Agric.
Paraguay	410	3.1	36	550	M	2	1.7	Petroproducts, bananas	Bal. growth
Peru	1,285	17.3	55	800	H	1	0.6	Cotton, soybeans, timber	Agric.
Puerto Rico	10	3.1	N/A	2,400	V	1	1.4	Copper, lead	Bal. growth
Surinam	180	0.4	N/A	N/A	N/A	N/A	1.3	Chemicals, petroproducts	Industrialized
Trinidad & Tobago	5	1.2	25	2,000	V	1	0.8	Alumina, bauxite	Raw mat. exp.
Uruguay	110	2.9	81	1,300	H	1	1.5	Oil, clothing, phosphates	Oil exp.
Venezuela	900	13.9	N/A	2,500	V	2	0.8	Wool, clothing, meat	Industrialized
								Oil and petroproducts	OPEC mem.
People's Rep. of China[e]	9,560	971	N/A	350	M	2	1.1	Industrial products	Bal. growth
United States[e]	9,360	223	N/A	7,500	V	1	1.2	Machinery, chemicals	Industrialized

Source: geographic, import/export, and food supply date—*1981 Book of the Year* (Chicago: Encyclopedia Britannica); urbanization and economy type date—P. F. Palmedo et al., "Energy Needs, Uses, and Resources in Developing Countries," BNL 50784 (Upton, NY: Brookhaven National Laboratory, March 1978), Appendix A, with some supplementary information by E. Lumsdaine and S. Arafa for those countries not listed in the source.

[a] Predominantly 1975 data, averaged from a number of sources.
[b] L = < $250; M = $250-$750; H = $750-$1,750; V = > $1,750.
[c] 1 = adequate, 2 = inadequate, 3 = serious shortage, 4 = critical shortage.
[d] Data not available or unreliable.
[e] The economic indicators of the People's Republic of China and the United States are given for comparison purposes.

capita figures are approximations given for purposes of comparison. A large number of the countries shown (about 40 percent) have annual GNP per capita levels of less than $250, and 85 percent have levels below $750 (1975 US dollars). Only those countries with exportable oil resources and those that have achieved considerable industrialization have higher income levels. Since these numbers are averages, a large segment of the rural population may have annual incomes of less than $100 per capita. According to World Bank definitions, one-third of the total population of the less developed countries have annual per capita incomes of less than $50, not including the value of rural subsistence resources such as gathered wood and crops for household consumption.[5] Many developing countries exhibit dual economies; for instance, in Brazil, the top 20 percent of the population earns thirty-six times as much as the bottom 20 percent.[6]

Table 10-2 also gives 1980 import/export and food supply data by country (although political instability, war, and natural disasters result in atypical values for some countries), and it categorizes the countries by economic type. Those countries where industrial output constitutes the largest percentage of GNP have been designated as "industrialized." However, such countries still have large undeveloped rural areas which have not been integrated into the economic growth of the urban sectors. This is also true for oil-exporting countries, including some OPEC countries. Oil-exporting developing countries face the problem of deciding how their oil should be used for development and at what rate. Balanced-growth economies are those which show some industrial activity and contribution to GNP but still have basically agrarian societies with severe unemployment problems. Developing countries that export minerals that account for major contributions to GNP have been classified as "raw material exporters." Energy consumption in these countries is considerably lower than in industrialized and balanced-growth countries, and hydroelectricity is an important source of energy. The agricultural economies have mostly subsistence levels of production, with the bulk of their energy supplied by noncommercial fuels. In about 20 percent of such nations, agricultural exports are a significant portion of GNP; market price fluctuations, however, have not allowed steady growth and development.[7] As a whole, centrally planned Asian countries use slightly less commercial energy per capita than noncommunist Asian developing economies.[8]

Future Energy Demands of Developing Countries

A number of studies have made projections of future energy

demands for developing countries. Because accurate consumption figures are scarce and a number of assumptions about price elasticity, population growth, GNP growth rate, future energy supply mix, and so forth must be made, it is very difficult to project energy demand on a worldwide basis. By the year 2000, an annual consumption of 2,000 kgce per capita should be typical for industrialized and oil-exporting developing countries (this is about 15 percent of current US per capita consumption), whereas the remaining less developed countries are expected to have a subsistence average of about 700 kgce. To raise all of the developing countries to 15 percent of US per capita energy consumption levels would require doubling the energy input.

Supply projections for the developing countries as a group are even more uncertain than demand projections. Energy supplies require not only a proven reserve but also capital, skilled manpower, and appropriate infrastructures for production and distribution; thus, assumptions about allocations to different developing countries are at this time practically meaningless. On the other hand, since these countries have relatively small needs, it may not be extremely difficult to exploit smaller petroleum fields that were uneconomical while world prices were low. Less developed countries so far have only averaged about 1 percent of the drilling density of the United States.

In the near term, developing countries will continue to need oil to fuel their struggling economies, which are primarily based on oil, and to prevent serious depletion of noncommercial sources. Some data suggest that a lowering of even the current subsistence standards of living will occur in many areas if petroleum becomes too costly and no noncommercial fuels are available. Therefore, energy is important to developing countries, not only to satisfy economic and social needs at present standards of living but also to provide growth and improvement in all areas of human activity. The consequences of stagnant or failing developing economies are political instability, distressed international monetary systems, increased power for OPEC, and deterioration in the quality of life.[9]

Related Social Needs and Policies

Developing countries not only face the problem, under conditions of rising energy costs, of providing sufficient energy to sustain their current standards of living and to carry out plans and programs for growth and development; they also must directly attack the problems of absolute poverty in their countries. These requirements often conflict under conditions of very limited resources and capital. Some of these countries already have such large debts that the major share of additional borrowing will have to be used for interest payments and amortization, leaving only 15 to 20 percent available for development projects.[10]

The most critical areas are sub-Saharan African countries, which have the least prospect of growth, the lowest levels of literacy (less than 25 percent), and life expectancies of only thirty to forty years; and South Asia, which contains half of the world's poor. It has been found that primary education, including education for women, must go hand-in-hand with programs that improve health and nutrition. Historically, countries with higher literacy rates have grown faster, even when all other factors have been discounted. With a program of health and education, an emphasis on family planning must be included also, since population growth puts tremendous pressures on land resources as well as on employment opportunities. In general, if family planning information is available, increased income will in time bring reduced fertility, which in turn fosters raised income levels. Without social programs, development assistance will do little to help the very poor. During the past period of easy energy availability, poverty in several Asian nations was alleviated when growth and development policies were well structured. Such policies are even more crucial now. Thus, developing countries will need to determine how to allocate resources between strictly energy-related projects and human-assistance programs for education, health, nutrition, and family planning. In particular, energy development must not be undertaken at the expense of these important social concerns. This implies that energy development must be accompanied with an appropriate infrastructure that will further social goals. Where defense priorities place strong pressures on national budgets, social programs especially may suffer. Past development efforts have too often neglected the need to obtain strong input and cooperation from rural inhabitants and have failed to insure an equitable distribution of the benefits accruing from the development. In other words, growth in the modern sector of a developing country has not necessarily meant any improvement in the conditions of its very poor. On the other hand, an exclusive devotion to rural social programs can lead to overall stagnation of the country's economy. In the end, mutually supportive programs will reap the greatest success in terms of economic and social development while at the same time preserving the environment and the community's cultural cohesion.

AN INITIAL EVALUATION OF OPTIONS: TECHNICAL AND ECONOMIC CONSIDERATIONS

Rich, industrialized countries face the choice of how much to develop renewable technologies versus centralized systems, especially nuclear power. The oil-importing developing countries are faced with a dilemma of a very different sort: if they continue to shift to oil, their troubles will greatly increase in future years; if they interrupt the shift, they will be in trouble now.[11]

There is by no means a consensus that the Third World's salvation lies in small-scale renewable energy technologies.[12] It has often been said that most of the developing countries are in areas of abundant sunshine and thus have great potential for solar utilization, particularly when a variety of technologies are available to exploit this limitless energy resource. But just one example will give an appreciation of the complexities of the rural development problem. Increasing the efficiency of rural cookstoves from their present levels of less than 10 percent to a 20 percent efficiency level in India, Egypt, and other nations of the Third World is technically feasible and would greatly relieve the danger of deforestation, yet this improvement has not met with any degree of success, primarily for the following reason: high-efficiency cookstoves such as the Indonesian *Singer*, the Philippino *Ipa*, the Kenyan *Jika*, and the Indian smokeless stove cost about $10 to $20 and are thus being used almost exclusively in urban areas. There simply are few alternative energy resources that are within the economic reach of the rural poor.

However, centralized energy sources such as electricity and commercial sources such as coal, oil, and gas are expensive, and the poor must compete with industrialized and urban sectors for these resources. The advantages and disadvantages of centralized and decentralized energy systems must be considered by each country and region in order to determine the most appropriate local application.

Centralized Alternatives

In most rural areas at present, a centralized power plant (nuclear-, coal-, oil-, or gas-powered) would not be matched in scale to development needs. First of all, such large installations are very capital intensive. Rural areas have very inefficient load factors. Electricity in villages is too expensive to be used for cooking and is unsuitable for draft and transport. On the other hand, a smaller centralized system may be advantageous in an area which is already industrialized, and a larger electric grid system could possibly grow from originally small, decentralized networks, to be fueled later by a centralized source when demand has grown sufficiently for efficient utilization. Building a system in reverse order is wasteful. Small-scale hydro in remote locations can also be more economical than extending a grid.

Even where grids have been extended to rural villages, only about 20 percent of the households can afford hook-ups. Thus, electricity is seen as secondary to supplying the basic needs of the poor. Centralized electricity has proved not to be the expected stimulus for increased agricultural production, even where it is heavily subsidized, contrary to the common expectation that links rural electrification automatically with progress and growth. In most developing countries, even when electricity is available the utilization level is low, mainly because of cost but also because the supply is unreliable, with extended power outages occurring almost daily. In India, the unreliable supply of electricity to power coal-mining equipment has been an important factor in the low productivity of the coal industry, although a shortage of rail hopper cars, labor problems, and corruption of labor officials are also hampering coal development.[13] For application in rural areas, coal for direct use has the disadvantage of high transportation cost, and it is less versatile than oil or natural gas in many applications, in addition to its unfavorable impact on health and environment. Therefore, centralized systems do not appear well suited for helping rural communities to become self-reliant and to develop local industries and commerce that could constitute a strong incentive against population migration to the cities.

Decentralized Renewable Energy Systems

If small industries are to be established in rural areas in order to counteract unemployment and provide income-producing job options, then not only must a reliable source of energy be available (not necessarily electricity), but the industries must also have access to investment capital, transportation to and from markets for raw materials and finished products, and a pool of manpower trained in such areas as accounting, marketing, and management.

In addition, a reliable source of energy (not necessarily electricity) must be available—a source that might or might not be based on a renewable resource.

It is not sufficient to have a renewable energy source available; the technology must be adequately developed and integrated into the existing culture and economy. The problems with the introduction of biogas and solar cookers in India and the failure of higher efficiency cookstove projects in Africa are due to economic as well as cultural factors. Although many developing countries have experimented for years with renewable energy sources for application in rural areas, only China and to a much lesser extent India have been reasonably successful in implementing them. Prices of alternative energy systems vary a great deal depending on local conditions, buying power, type of technology used, and so forth. For example, a biogas digester for a single family costs, excluding labor, about $50 in China (Chinese design) but over $500 in Brazil (Indian design). Nonetheless, it is useful to briefly compare the present status of development and implementation of alternate energy technologies.

Small-scale hydro systems. This technology is universally well developed and understood. For example, since 1953 China has built more than 80,000 small-scale hydro systems with a total capacity of nearly 6,000 MW. This is 20 percent of that country's total electric capacity and 50 percent of its rural electric capacity. The system sizes range between a few kw and 500 kw. The small-scale hydro systems are commune-owned, standardized, efficient, simple, durable, and reliable. China appears to be eager to help other developing countries to realize their hydro potential.

India's hydro development is left to the individual states with the assistance of the central government, and several hundred installations are in operation. However, India has so far realized only a portion of its small-scale hydro potential. Similarly, a large number of developing countries have developed only a small fraction of this resource: for example, Bangladesh has not developed its small-scale hydro potential at all; Burma currently utilizes about 5 percent of its total potential small-scale hydro capacity; and Pakistan, the Philippines, Colombia, and others currently have developed less than 10 percent of their total known potential capacity.

It should be noted, however, that even electricity from small-scale hydro systems is still rather expensive for the rural poor. In China, for example, it costs about 1-3 cents per kwh (compared to

278 • Energy for Rural Development

6-7 cents per kwh for electricity generated by diesel fuel). In addition, small-scale hydro systems are site-specific and require trained local personnel for maintenance. Such systems, depending on their installation, offer the additional benefit of water regulation and flood control; but power generation may be secondary if impounded water is needed for irrigation, and silting can sometimes be a serious problem. In areas of high population densities especially, the environmental effects of these systems could pose difficulties. In general, small-scale hydro shows the greatest promise for small industries development and agricultural processing with some additional household use.

Wind. The use of wind power dates back thousands of years. Many of today's industrialized countries have used wind turbines (for electricity) and windmills extensively in the past, especially until the advent of cheap, easily available fossil fuel. Although windmills for pumping water are well developed technologically, their cost is still relatively high for standard systems, and thus they have not found extensive use in rural areas that still rely primarily on animal and human power for irrigation, such as the rural areas of India and Egypt. (India has developed a unidirectional wood-sailcloth type windmill which is relatively cheap but has not yet found extensive use.) With few exceptions, wind turbines are expensive, and it is unlikely that this technology will find its way into the rural areas of the developing countries in the near future, since other options will most likely be more attractive economically and socially. Furthermore, there are very few wind turbines that are durable and reliable enough to be used in rural areas. By the same token, windmills are not well matched to rural needs. For potable water pumping, their intermittent operation may be acceptable, but in some areas like India, the windmill is most productive during the rainy season, when it is needed the least.

Photovoltaics. Although photovoltaic systems generate electricity without moving parts and are thus mechanically reliable and quiet, the cost of such systems is still too high even for affluent nations to use, with the exception of some very remote applications. The us Department of Energy cost goals of $2 per peak watt by 1982 and $0.70 by 1986 will not be achieved. At present, it appears that their goals will be off by 100 to 200 percent. However, further cost reductions from the present average of $10 per peak watt appear to be forthcoming with the polycrystalline, ribbon, and other techniques of producing photovoltaic cells. It is thus unlikely that photovoltaics will be used in rural areas of developing countries except for demonstration projects, although very small systems may find application for specific purposes such as powering a communal educational television set or providing refrigeration in remote rural health centers.

Biomass. Biomass covers a wide variety of organic materials such as trees, crops, manure, seaweed, algae, and so forth. The cellulosic material which comes from agricultural by-products and industrial and municipal wastes is the most abundant organic material on earth, and biomass today is the major source of energy for half of the world's population living in rural areas in developing countries. Major efforts are under way to increase the energy yields from biomass and to convert these yields to more versatile energy forms. The traditional rural practice of using wood as cooking fuel is a very wasteful process, with an energy efficiency of 5 to 10 percent at most. In India, animal dung represents about 10 percent of the total fuel consumed (as much as 50 percent in rural areas, mostly for cooking).

Of the two technologies (biological and thermochemical), the biological process involving enzymatic breakdown of the biomass by microorganisms is the most popular for rural application. Two methods are distinguished: one involves the anaerobic digestion of biomass (mostly waste products), yielding methane gas, carbon dioxide, and small amounts of hydrogen sulfide gas; the other operates by fermenting biomass materials (primarily those that are carbohydrate-rich) and distilling the result to produce ethanol.

Biodigesters—gas. In China there are reports of more than 7 million family-size biodigesters constructed during the last twenty years.[14] However, during one of the author's recent visits to some communes in China outside of Shanghai,[15] only a few communes reported having digesters, and some were not operating at all or only at reduced capacity. (In cooler climates, digesters frequently are operated only during the warm season; this may have been a factor here.) The commune people seem to prefer using coal-cake stoves and are more interested in the fertilizer that is produced in the digesters than the gas. Of course, an important advantage of the biogas-generating process is the fertilizer yield; in many areas fertilizers are in critically short supply for food production. Another advantage of this process is improved sanitation at the rural village level, an important health benefit.

In India, it has been reported that more than 50,000 biodigester plants are in operation with more planned. The program is of somewhat limited success, in that it is not yet self-sustaining but is still heavily subsidized (from 20 to 50 percent) by the government, although it has received generous cooperation from banks. Cost and cultural problems have been hindering the widespread use of the Indian-designed gobar plants. The methane gas has been used mostly for cooking, although a number of pumpsets being run by biogas are being field-tested. Korea has reported the completion of more than 25,000 biodigesters and Taiwan about 10,000. However, the red mud plastic biogas bags used in parts of Taiwan and the Philippines are still very expensive. Brazil has a program to develop biogas plants for rural areas, and many other countries such as Egypt are experimenting with this technology but are not yet actively pursuing a biogas production program in their rural areas.

Although biogas plants are relatively cheap, rugged, and simple to construct, operate, and maintain, problems such as hydrogen sulfide corrosion and increasing mosquito and wood-boring insect infestation (such as that found in Nepal) must be dealt with carefully. There is a lack of data on user acceptance of biogas digesters as well as on their long-term equitable operation. Ownership problems may occur since plants may be shared among several poor families for efficient operation. Distribution may be a problem also, as well as operation, maintenance, and initial financing. Biodigesters may be easiest to introduce into societies that already have a tradition of cooperative behavior (e.g., China).

Biodigesters—electricity. Small electric power stations using the gas from sewage treatment plants can be a source of electricity for populated rural areas. Such plants were built in some areas in the 1930s (for example, in the city of João Pessoa, Brazil) but fell into disuse. Many such plants are operating in China. One Chinese plant in Foshan, Guangdon, has been operating since 1974 with an output of 90 kw of electricity and 60 tons per day of fertilizer.[16] Although not without technical problems, the plant has been running almost continuously. The obvious advantage of such a plant for a larger rural community is that it acts not only as a sewage treatment facility but also as a fertilizer and power-production plant, all of which are important factors for rural development.

Biodigester—liquid fuel. The production of ethanol in very small plants by biomass fermentation and subsequent distillation is being tested in such countries as Brazil. Ethanol is attractive as an alternative to petroleum since it can be used as a transportation fuel in farm machinery. Especially given present technological limitations, each country must address its own food/fuel competition issue when food products rather than cellulose materials are used as feedstock in ethanol production. Brazil (a grain-importing country) is using its alcohol fuels program to end its dependence on foreign oil, bring unused land into production, and create jobs for the rural unemployed. However, there has been an increase in its food prices due to these energy crops—an increase which has imposed additional hardships (including malnutrition) on the poor. Furthermore, damage to topsoil can occur by excessive demands on it and by inadequate crop practices.[17] Any alcohol fuels program must also consider its net energy yield: sugar cane as feedstock is attractive because the stalks can be used after crushing as fuel for the distillation cycle, but when corn is used as feedstock and fossil fuels are used in the distillation cycle, the net energy yield may be negative.

Fuelwood. Direct combustion of fuelwood for cooking and heating is still the principal source of energy for nearly 2 billion people living in rural areas of Third World countries. If the present use trend continues, most of these people will have to turn to a replacement fuel by the year 2000. Already, the removal of tree cover has resulted in the destruction of the fragile ecosystem in the Sahel and erosion and flooding in the Himalayas, and millions of hectares lie bare yearly in Brazil and India. Thus, afforestation programs and proper forest management techniques as well as efficient fuelwood use techniques are urgently needed if this vital resource is to continue to be an energy source for rural people without severe damage to the environment. Evidence suggests that forested areas are decreasing by 25 percent every twenty years under present trends. In Pakistan, 25 percent of the nation's commercial wood fuel is imported. In Nepal, it now takes one day to collect the wood that formerly could be gathered in two hours.[18] Uncontrolled harvesting damages forests; tree plantations, however, are not yet economically competitive against "free" wood, especially since traditionally women and children are the wood gatherers.

Charcoal. Charcoal making using inexpensive, primitive kilns is a well-established but low-efficiency technology. New kilns that permit more control in burning and new ways of utilizing the process's waste products (mostly gases) must be developed to improve charcoal-making efficiency. Charcoal is a valuable commercial commodity and is preferred over fuelwood by many of the urban poor. It has, of course, the advantage of being less bulky in transportation and more efficient in use than fuelwood. Some research is currently aimed at making charcoal briquettes out of agricultural waste (nut shells, corn cobs, bark, etc.).

Direct solar. Direct solar methods for crop drying, cooking, and distillation are feasible technologically but are still very expensive. Despite much research, testing, and especially rhetoric, in no country are solar devices for cooking, crop drying, or distillation of drinking water widely used. A large gap still exists between the technologies tried in testing labs and at universities and the actual application of these technologies in the field. Although there are reports of several hundred solar cookstoves and a few crop dryers in countries such as China, India, and Upper Volta, the $30–$50 price of a solar cooker simply cannot be afforded by the average rural family. Also, the outdoor location and the amount of supervision required by a solar cooker, together with the fact that the solar cooker can best be used around noon does not match the traditional customs of many rural societies. In India, government-subsidized solar cookers are being tried out, but with limited success due to both cultural and cost reasons.

Since the Third World rural poor are for the most part simply trying to meet survival needs, such solar techniques as passive solar heating, while they may be cost effective and applicable in some developing countries (depending on the climate), are not important in rural areas where people deal with the elements as best they can. For example, in Egyptian villages, rudimentary structures built of adobe bricks, wood beams, and straw roofing—all local materials—serve as shelter. In Bhavnagar, India, the use of a solar distiller and wind/photovoltaic system for pumping provides rural people with some sanitary drinking water but at a cost that would be prohibitive if the system had not been given free to the people by the government. Crop dryers of the plastic bag type are being field tested in Brazil and are exhibiting some potential. However, the ultraviolet degradation of these crop dryers will limit their use to only one or two drying seasons.

Animal power. The Indian Institute of Management, Bangalore, has estimated that the 80 million work animals in India provide power equivalent to 30,000 MW. India's most recent five-year development plan (its Sixth Plan) aims at better utilizing this vast source of animal power in farm operations and transportation by improving the health of the animals through adequate nutritive fodder and by modernizing the animal-drawn vehicles. There are a number of ways to calculate the energy efficiency of animals where the input is composed of feed, maintenance, health care, supervision, and the output of milk, dung, and useful work. Results vary from 0.5 x 10³ to 5 x 10³ kilocalories per hour for an average of eight hours per day. In a rural economy employing no mechanization, it can be said that animals, like humans, convert roughly one-third of their food energy input into useful work output.

Cost and Implementation of Technology in Rural Areas

Table 10-3 gives a rough estimate of the price ranges of different energy technologies. Although there are some changes in these prices occurring in various places, basically this table shows that costs vary so much from country to country that it is difficult to state prices with any degree of accuracy. Thus, decisions about selecting one technology over others must be based on local price structures and other conditions. For example, kerosene is at present the only fuel source for many urban poor for cooking and lighting, and diesel generators are often the only source of electricity in remote areas. Diesel motors are also frequently used to drive pumpsets for irrigation.

General recommendations can be made about the proper

Table 10-3. **Cost Estimates for Various Energy Technologies**

Technology	Cost estimate
Heat	
Biogas (methane)	$0.07 - $3 per GJ
Fuelwood (commercial)	$0.40 - $2 per GJ
Kerosene	$5 - $10 per GJ
Solar cookers	$7 - $35 per unit
Electricity[a]	
Photovoltaic cells (@ $10/peak watt)	$1 - $3 per kwh
Mini hydro (excluding dam and penstock)	$0.03 - $0.10 per kwh
Wind generators	$0.04 - $0.20 per kwh
Diesel generators	$0.20 - $0.50 per kwh
Central station (excluding construction)	$0.01 - $0.15 per kwh

Source: E. Celeski et al., *Household Energy and the Poor in the Third World* (Washington, DC: Resources for the Future, 1979), p. 54, with updating by E. Lumsdaine and S. Arafa.
[a]Bert Sørensen in *American Scientist* Vol. 69, No. 5, September/October 1981, p. 502, gives the following generating costs (including construction costs) for Danish conditions: small wind generator, 2.5¢/kwh; large wind generator, 3.5¢/kwh; large coal-fired plant, 4.8¢/kwh; light-water fission reactor, 3.9¢/kwh.

sequence of introducing appropriate renewable technologies into Third World rural areas: subject to local needs and conditions, biomass, wind-driven pumps, and wind generators should come first, followed by solar hot water systems, solar dryers, small-scale hydro, and solar stills. However, practically all of these technologies are still too expensive for the areas in question and require too much education to enable their local manufacture, installation, operation, and maintenance.

In the context of integrated rural development, R. Ramakumas gives the following prioritized list of appropriate applications for renewable energy sources:[19]

1. Safe water supply for drinking
   Reliable energy supply for irrigation
   Education (including television)

2. Expansion of irrigated acreage
   Crop drying (possibly also cold storage)
   Fertilizer production

3. Home heating (if required by climate)
   Hot water for schools, community showers, health centers
   Energy centers for small industries

To be successful, these technologies must be introduced gradually and must include government financing and local participation. Because of limited local and countrywide resources, choices must be made and priorities established on development. Simultaneously, an institutional structure for implementation must be developed, and local research and development programs must be set up and evaluated. In the long run, it may be best to have little or no subsidization of fossil fuels. However, for the widespread adoption of renewable energy technologies, a considerable amount of capital will be required even for the most economical systems.[20] Energy centers for small industries are best set up in areas easily reached by a number of small villages; to increase the potential of economic benefits to a larger area, they preferably should not be located in one village.

The risk factor involved in applying a new technology is a deterrent even when the system is relatively economical. For example, when compared with diesel fuel, wind for irrigation is potentially economical at present, but not for the small farmer for whom the risk of a wind system would be too great because of its intermittent operating characteristics. In many places, cooperative wells have not worked out well due to village power structures; responsibility for maintenance is an especial problem. Thus, renewable technology is not only a technical and economic problem, but also a managerial problem.

INSTITUTIONAL BARRIERS AND SOCIETAL CONSTRAINTS

Set against both (1) the tremendous needs of Third World rural areas for energy sources and associated human services and (2) a variety of technological options to at least partially alleviate these problems are a number of very strong institutional ae to topsoil can occur by excessive demands on it and by inadequate crop practices.[17] Any alcohol fuels program must also each country. The basic institutional problem is one of conflict between the usual existing strong centralized authority and the need for a decentralized structure to implement reforms. But apart from the political/institutional framework, some countries have social constraints that are of predominant importance. These constraints most often take the form of a need for food that supersedes that for energy. In such countries, especially in sub-Saharan Africa, the foremost concern is food self-sufficiency.

In addition, in the Third World countries there is at present no general agreement on the manner in which development should occur, although job creation is a common goal. Many economists advocate capital-intensive solutions as the most effective way of raising GNP and hence the general development level, whereas others believe in the creation of more low-cost work places. The latter view is the one which most closely corresponds to the "appropriate technology" approach to development.

Institutional Barriers

Eighty to ninety percent of almost 400 million people in sub-Saharan Africa live in rural areas, subsisting on meager crops and livestock. At the same time, through their agricultural production they provide the major resources to meet urban consumption, and they provide up to 40 percent of exported GNP. Taxes on this agricultural production are the major source of government revenues (e.g., there was a 35 to 70 percent taxation on cotton in Mali during 1974–78). African countries have a poor record of plowing resources back into agriculture, even though this sector is the mainstay of their economies. Government intervention is unbalanced in that it places heavy restrictions on marketing yet provides no support for research, extension, and manpower training or other improvements in the agricultural sector. Much of traditional rural Africa lacks even the range of farm implements, ox plows, and animal-driven transport vehicles used, for instance, in Asia.[21]

Annual aid in the form of grants and low-interest loans during the period 1976–78 has been between $10 and $20 per capita in Sudan, Kenya, Burundi, Tanzania, Ivory Coast, Mali, Cameroon, Zambia, and Malawi, and as much as $30–$70 in Botswana, Lesotho, and Swaziland. In some countries, this constituted over one-quarter of the total annual agricultural investment or one-half of the support for agricultural/rural development. By comparison, Bangladesh annually received $9 per year during the same period. Aid has been relatively ineffective because bureaucracies are inadequately staffed, and, despite worthwhile goals, results have been meager because there was a lack of coordination on objectives and strategy among the various aid agencies and because aid was directed towards specific projects without a knowledge of national policies on rural development.

Therefore, the first step towards success in rural development programs is a strong national commitment by the developing country with budget support for the agricultural sector, coupled with strong administrative support at all levels and with the involvement of the local beneficiary population. Second, consistent foreign and domestic trade policies are important, with appropriate technical inputs supported by international financial and technical assistance. However, aid is required not only for the original capital investment but even more importantly for recurrent operating and maintenance costs on a long-term basis, perhaps as long as ten or even fifteen years.

Imported technical assistance is usually not effective; it must lead to long-term training of nationals, to the point of their self-sufficiency. Unfortunately, vested interests in the developing countries have also been a strong barrier to effective program results and support. An example of this type of barrier can be seen in India, where, through excessive government controls on foreign companies, politics is delaying oil development. (In general, oil-producing developing countries are very sensitive about any control or pressure from industrialized countries or their corporations.) Also in India, the coal production industry is nationalized and stagnated; output per manshift is the lowest anywhere because of unfavorable working conditions and the self-interests and exploitive methods of contractors and union bosses. The poor suffer when the wealthy are in control of energy sources, including appropriate technologies. Furthermore, since the wealthy prefer "modern" fuels, they influence national policy strongly in this direction. The poor suffer additionally in that formerly "free" fuels become commercialized or polluted with development. The latter holds true for some commodities such as drinking water: distilled water is too expensive to be affordable by the poor and its health benefits may not be obvious to them, but meanwhile ditch water— their primary source of drinking water—is becoming increasingly polluted by development.

National policy planning must rank and integrate human assistance needs and the need of the energy sector,[22] and integrated energy planning must both clearly define national objectives and outline appropriate physical controls, technical methods, education and information needs, and pricing factors.[23] Integrated energy planning can be established in less developed countries even with inadequate manpower resources.[24] Energy planning not only must include the least cost approach to meeting future energy needs but must consider also reducing dependence on foreign oil, supplying the basic needs of the poor, reducing the trade and foreign exchange deficit, setting priorities on development of special regions or economic sectors, insuring continuous energy supplies at stable prices, preserving the environment, and so forth. A related area that is interwoven with energy production and delivery as well as development in general is the transportation system; some countries' systems may need extensive rehabilitation and extension. In addition, countries with extreme shortages of capital and a surplus of labor due to widespread unemployment may need to favor policies in energy development that are labor intensive rather than capital or energy intensive, contrary to present tendencies.[25] Another barrier that may impede effective policy choice is that appropriate small-scale technology is sometimes perceived to be "second class," even though it can offer many advantages for developing countries. Industrialized countries can do much to dispel these ideas by vigorously developing and applying these technologies where appropriate, especially in their own rural communities. However, these technologies not only must be technically, economically, and politically suitable; they may also have to overcome cultural barriers. Democratic political systems face an additional handicap in that officials running for election will not be anxious to advocate energy conservation and the application of less popular energy technologies, unless even the illiterate poor are educated about the advantages of such fuels and methods.

Societal Constraints

In India especially, development has helped upper income groups without any appreciable benefit to the low income rural population. Lack of skilled labor is not a problem in India as it is in some countries—India has a surplus available for export. Mahatma Gandhi supported the concept of self-sufficient villages using appropriate technology and decentralized means of production. However, in 1970, India's Fuel Policy Committee recommended that the energy sector be directed towards development of fossil fuels and that renewable resources be discouraged. It is not usually recognized that the choice of technology is not independent of social structure. Local involvement must be encouraged and not prevented, both in planning and implementing the technology. It is difficult to achieve this local involvement in dual economy, caste system, or other severely class-conscious societies. Some societies have no traditional ways of cooperation or of including women in their decisionmaking processes. There is a lack of data about the cultural barriers encountered in appropriate technology applications. Biomass digesters appear to face resistance since their use involves overcoming traditional taboos in the disposal of human and certain animal (especially pig) wastes. In general, political structures and ownership patterns make the wealthy the primary beneficiaries of newly introduced, expensive technologies. The greater consumption of resources by the wealthy may affect the poor's quality of life adversely, where quality of life is defined as a system of strong cultural values and a sense of belonging together with the preservation of natural resources. Therefore, development programs must aim not only at increasing the rural poor's standard of living but also at preserving and improving their quality of life.

Another factor that may pose a barrier to the introduction of new methods and technologies for rural development is the religious framework of village life. The effects of some different religiocultural systems is reviewed below.

Confucianism's high regard for education and its value system—a system which places the group over the individual's interests—provides a context which is quite conducive to innovations that can benefit communities and that rely on strong leadership and education for introduction. Confucianism has influence over

1.2 billion people in East Asia, and this cultural tradition is credited as a factor in the development of Japan, South Korea, Taiwan, and China.[26] On the other hand, Hinduism with its caste system is embraced by about 900 million people, primarily in India; it presents a considerable barrier to economic development because low-caste landless or almost landless field workers constitute about 60 percent of all families in India's 600,000 villages (or over one-half of the entire population). Higher caste landowners account for only 15 percent of the village families but own 75 percent of the land, and traditionally they have controlled the resources because their agrarian system is based on exchanging labor for a share of the harvest. Therefore, land reform and other types of income distribution restructuring will be needed before the Indian agricultural sector can be developed to its full potential.[27]

Another system that is resistant to change (especially to that coming from Western ideas) is Islam, which strongly influences at least 300 million people, mainly in Africa but also in the Middle East, Malaysia, and Indonesia. In the Islamic faith, there is an ingrained suspicion of scientific progress together with the belief that mankind has neither the right nor the power to change things from the way they have been ordained by Allah. Coupled with this is the Muslim treatment of women as second-class citizens denied access to education and cash-productive work (that is, work for pay as opposed to domestic labor).

The Roman Catholic opposition to contraceptives is another barrier, not because it prevents technological progress as such but because it contributes to the population growth which keeps people in perennial poverty and nullifies any advances made in agricultural production. In Latin America, a strong class system and economic inequality between the classes is an associated barrier. In tribal Africa, the barriers to technology and progress are not so much cultural as environmental. No viable, more productive alternative has yet been found to their environmentally destructive slash-and-burn agriculture, which produces mainly millet and sorghum.[28]

It will take long-term education and effective information dissemination to change present political and cultural attitudes. It will be an enormous task worldwide to make lifestyles based on renewables desirable. Since world energy consumption will almost certainly increase within the next decade at least, there will be increased pressure on industrialized countries to drastically reduce their oil consumption in order to make petroleum stocks available to alleviate poverty in the Third World. The role of education will be crucial in this task also.

INFORMATION AND EDUCATION

Education plays a strong role in aiding the adoption of new technologies. Literacy, primary education, and secondary education are important in their impact on raising rural standards of living. New methods can be effectively introduced only with the full cooperation and participation of a large part of the population, and this requires an understanding of the issues involved as well as a vehicle for communication, such as written material and teachers who are intimately familiar with local conditions and are personally interested in working with local people to achieve improvement. There is an extreme lack of teaching material and general information about renewable resources in Third World countries. In particular, education about the fuelwood crisis is of crucial importance in its consequences to the very poor. In addition, trained labor will be an important factor in the ability of rural areas to develop or attract small industries.

The Need for Information at National and International Levels

Education and information will be very important in carrying out the developing countries' energy policies and in establishing strong, rational, working energy plans. Some countries will need to have their libraries supplied with basic informational tools for policy planning, resource assessments, and energy development. Assistance from industrialized countries may be required, but the initiative and commitment must come from the developing countries. Conversely, the populations of the industrialized countries must have factual information about their responsibilities in the wise use of resources and about energy conservation goals, as well as about the pressing needs and conditions in developing countries. As shown by a recent poll,[29] the need for aid to Third World countries is generally acknowledged by the public in many industrialized countries, but there are differing opinions about the appropriate extent of such aid. An impediment to generous aid seems to be the apparent unwillingness of some developing countries to solve their problems. Leaders in Third World countries have a great responsibility to mobilize their well-to-do ruling classes in order to initiate integrated energy planning as soon as possible, following up on the recommended programs with full political and economic support. The areas of cultural and religious traditions and patterns will in many cases be very difficult to deal with or even to discuss; again, it will be important that any proposed changes in this area come with input from the people that will be affected. Although it may appear that revolution is an effective way of bringing about change, this by no means applies in the area of rural development. A revolution may bring about the beginning of change, but historically, very intensive education and indoctrination programs have been used to achieve major and lasting changes in cultural patterns, and even then, the new system usually had at least some roots in the traditional societal fabric.

Information Gathering at the Community Level

The energy needs and resources of rural communities are complex and interrelated and therefore must be viewed in a holistic way, taking into account the environmental setting, population distribution, sociocultural context, and changing economic system. The social science methodologies needed to obtain data on behavior, attitudes, and social organization in a community are less difficult to implement than are the "software" components of some energy projects—components which often involve taboos or matters of a highly personal nature. Conventional surveying does not take into account the complexity of the relationships involved.[30] The social science techniques that have been found most useful in determining existing attitudes and practices and in designing more acceptable and effective projects are those in which the local people have been most involved in the identification of the community's felt needs and priorities.

Face-to-face communication raises awareness of present practices and alternative opportunities and defines problems and priorities. Ultimately, joint analysis by the community members and project team members of the practices and alternatives leads to a greater understanding of needs, resources, practices, and objectives. Nearly all data-gathering techniques are based on a combination of three main approaches:[31] observing, asking questions, and listening (the single most important phase for establishing a dialogue between community residents and field workers). If residents of a community do not participate in a project's design and planning, the project risks being inappropriate to local needs; if the residents do not participate in implementation, the project probably will not survive; if they do not receive benefits from the project, it is meaningless. Residents will get involved in a project voluntarily only if they can clearly see that they personally have something at stake in its success. Energy projects will interest participants only insofar as they also address other perceived needs such as increased income, improved crop production, and added food or services. Unfortunately, the present rigorous curricula in science and engineering have given many technologists of the developing countries a sense of élitism, and their training seldom includes courses in sociology, community development, or extension methods. Therefore, each project budget may have to set aside about 20 percent or more of its funds to educate personnel and develop appropriate informational material. It is both necessary and more fruitful to conduct a large number of small experimental projects rather than one large-scale project if the benefits of rural information dissemination and training are to be maximized. As part of this, projects must be kept rather flexible in order to accommodate the different needs and desires of different communities.

Research Priorities

It is interesting to note that returns in agricultural research have been high (as much as 40 percent in India, for example). In the developing countries, there are a number of areas in agrarian energy research that urgently need exploration and that hold great promise of improving these countries' energy futures. Although the priorities of individual countries and their local needs may dictate a different ranking than the one given below, these research areas may be enumerated as follows:

1. Inventory of the developing country's energy uses, needs, and resources.
2. Establishment of regional and national energy plans.
3. Systematic search for conventional energy sources (with the assistance of industrialized countries).
4. Collection of data on renewable energy resources (wind, possible hydro sites, biomass, waste); matching these resources with seasonal energy needs.
5. Improvements in forest management to prevent soil erosion, silting, desertification, flooding.
6. Increasing wood-burning efficiency and forest yields; planting suitable trees on marginal land; improving the design and materials of hand tools, animal harnesses, and so forth.
7. Incorporating women users in the process of designing

more efficient hardware, especially stoves, to facilitate future acceptance.

8. Development of more efficient charcoal conversion techniques; design of efficient low-cost kilns that can process waste materials.
9. Development of biogas digesters with improved efficiency, reduced cost, and simplified maintenance.
10. Development of transportation fuels based on renewable resources that do not compete with food production. (These could be very valuable in the mid-term.)
11. Development of intensive programs of operator training and popular education in the use of biogas digesters and other appropriate technologies.

Not all of the above areas are pertinent in each country, and countries with similar problems and conditions may be able to benefit from cooperative programs and information exchange. Some of the research will need to originate with technical help from more advanced countries, but work should be transferred as rapidly as possible to the developing country and, where feasible, to local institutions.

Application: Example of an Energy Analysis of an Egyptian Village

To illustrate the interrelationship of a village's social structure with its economic patterns and to show some of the barriers that need to be overcome when renewable technology is introduced, a brief discussion is included here of energy use in an Egyptian village. Basaisa, a typical small village in the Nile delta consisting of forty-two homes and two hundred eighty people, has been investigated in detail as to its energy resources and energy consumption patterns in connection with a cooperative research project between the American University in Cairo and the University of Tennessee, Knoxville. The project was especially concerned with the social ramifications of introducing appropriate energy technology to a developing rural community.[32]

Life in Basaisa is organized around a rhythmic agricultural cycle in which men, women, and children all have a vital role to play in the production and processing of the crops that provide their livelihood. The main agricultural activities are (a) preparing the soil with organic manure; (b) plowing the land consecutively at three different depths, followed by furrowing and sowing; (c) cultivating the crop, which includes weeding, removing larger vegetation, applying fertilizer (depending on the particular crop), irrigating every ten to fifteen days (depending on the season and crop), and picking cottonworms in early summer for a period of forty-five days; culminating in (d) the harvest. A tractor is used occasionally for plowing; donkeys or camels are used for transporting fertilizer to the fields; oxen or water buffaloes are used to turn the irrigation pump (water wheel) with a few diesel pumps available to the village on a rental basis. The other tasks are done by human labor, with the children responsible for the cottonworm picking. The latter activity is shared among three villages: all the children are organized into work gangs of from ten to twenty children under the leadership of an older boy, and the gangs then rotate among the three villages, spending three days at each village. Working hours are from 6:30 in the morning until dusk, with one and a half hours off for lunch and rest.

Activities associated with agriculture (and overlapping the domestic tasks) entail processing the harvest, including threshing, winnowing, roasting, and grinding grain and drying food; slaughtering animals; producing milk, butter, and cheese; cooking and baking; preserving food and storing grain; caring for clothing; building and repairing houses; gathering, processing, and storing fuel; and marketing and purchasing. The adobe buildings are heavily roofed with the dried stalks of cotton, rice, and corn, thereby conveniently storing this material for its later use as fuel for cooking and baking. The irrigation canal on the village's eastern border provides its main source of water for cleaning cooking utensils, washing, bathing, and irrigation. Potable water is supplied by three public and four private water pumps; pools of stagnant water accumulate around the troughs of these pumps and create active breeding places for diseases such as malaria and eye infections. (The canal water, on the other hand, transmits bilharziasis and encolostroma.) A few houses possess a rudimentary system of sewage disposal. The nearest primary school is 5 km away; the health center and market are both 3 km away. Village land holdings total about 35 hectares divided into very small parcels. About forty to fifty villagers tend the fields as a full-time occupation, with some women (especially widows) also having land cultivation as their major responsibility.

Agricultural practices and energy use. One-third to one-half of the acreage is planted in cotton (the allotment is determined by the government), and the cotton crop takes approximately eight months from soil preparation to harvest (April to October). The remaining land is planted with rice and clover or rice and wheat (or corn), making up a one-year cycle (eight months for the rice). There is also a land pattern which lets one-third of the land rest (with or without clover). The regulation of the cotton crop by the government poses some problems for the farmer, because this produces a crop that does not provide food for the family. The same land does not have cotton two years in a row.

The most intense demand on human labor is during the wheat harvest (May) and cotton/rice cultivation (May through July), followed by the cotton and rice harvest in September/ October. It is estimated that the fellah spends one hundred eighty days at ten hours per day and one hundred twenty-five days at eight hours per day working in cultivation. Occasionally, beans and other vegetables are planted also. December through mid January is the least active period in the cycle and is devoted to irrigation and harvesting clover. Clover constitutes an important crop for animal fodder. It is estimated that 20 percent of the plowing is done by tractor (the remainder by plow and buffaloes); 20 percent of the irrigation by tractor or diesel pump, (the remainder by the traditional water wheel) and only 10 percent of the threshing and winnowing by tractor (the remainder by traditional methods). It takes one day with a water buffalo to plow a 1-acre (0.4-hectare) field, including six men to furrow the soil. With a tractor, this task of plowing would take about four to five hours. Cotton-seed planting is done by hand with a small pointed stake; this task is frequently done by younger adolescent girls.

The village has seven water wheels; water runs through ditches to fields that are as much as 1 kilometer away. Water rights and responsibilities are tied to land ownership: the water wheels are owned and maintained by all whose land lies along the ditches extending from the wheels; maintenance expenses are divided proportionately; the water is apportioned according to need—a system which gives rise to conflict. In general, each wheel runs an average of six hours per day in the winter and ten hours per day in the summer. Each farmer is responsible for providing his own locomotion for the water wheel—usually water buffaloes but sometimes cows, donkeys, or even camels. One family member is needed to keep the animal walking around the wheel by gentle prodding. The farmer then follows the ditch to open it so the water is directed to his field. Care has to be taken not to allow water to run onto neighboring fields which may be ready for harvest. It takes about one to three hours to irrigate 1 acre, depending on the field's distance from the water wheel. Rice paddies demand the greatest attention and must be irrigated every other day during the first month of the rice growing season. However, cotton requires chemical fertilizer whereas rice does not.

The government cooperative is heavily involved in cotton cultivation, providing fertilizers and seed, paying the government wage to the cottonworm pickers, and then accepting the crop. These government costs are then deducted from the proceeds of the crop. Occasionally, the government will also spray with pesticides. A camel driver and his beast provide transport of the crops from field to village or government cooperative; he is paid with a share of the crop. The entire cotton crop must be sold to the government. Fifteen hundred kilograms of rice per acre are sold to the government cooperative; if the farmers' yields are higher, the difference is marketed by them as they wish. Corn and wheat are marketed only after home needs are met. The cooperative also extends some credit to farmers—credit which has to be paid back when the crop is sold.

Five tractors exist in the surrounding area and are available for hire. Handpowered threshers are traditional; affluent villagers rent a larger thresher powered by a rubber belt run off a tractor. A handpowered winnower is required to separate the chopped particles into grain and chaff. The chaff is a valuable product, since it is used for animal fodder, for mixing with animal dung to make fuel for baking, and for mixing with mud to make adobe bricks. But despite the presence of the tractor, animals are still the farmer's most precious energy resource. They not only provide labor but also produce considerable quantities of organic waste that are essential for crop cultivation. Waste production is related to food intake; in the winter when the farmer relies primarily on clover for animal nutrition, waste is available in larger quantities. A cow or water buffalo can consume approximately 30 kg of clover per month.

The use of animal manure is institutionalized in a closed cycle. Manure is spread on the soil at the time of soil preparation. After the harvest, the land is plowed for the next crop and then irrigated. When the land is dry, the farmer collects the old manure residue from the land using a special rake; mixes the residue with dry silt, ashes, and some thin straw; and spreads this mixture in the animal shed under the livestock. The new waste will seep into it, and the renewed manure mixture will be collected to be spread yet again on the field. The manure is transported to and from the field by donkeys.

Energy use in households. The work of rural women includes child bearing and rearing and household provisioning and management (baking, cooking, washing clothes, manufacturing and re-pairing household goods, and gathering fuel and water). It also includes aspects of agricultural production and processing, live-stock raising, artisan production, trade and marketing, and some other income-generating activities. These varied tasks are intrin-sically related, but an attempt has been made to estimate the time required for each.

Domestic water use. Since domestic water usually must be carried some distance, consumption is estimated at only about 20 liters per day per capita. Canal water and ground water are from different sources; their composition, properties, and taste differ, and villagers discriminate between these two kinds of water and use them for different needs. Ground water is used primarily for drinking, bathing, and personal hygiene and for prayer ablutions and making tea. Canal water is used for washing clothes and cooking certain vegetables like okra, lentils, and cereals. The canal water is ½ km away from the village center, whereas the hand pumps for the ground water are no further than 150 meters at most. A large shaded clay pot in the home is filled approximately ten times per day with water carried from a hand pump. Canal water is drawn for specific tasks and not stored. The women even prefer different pumps for different tasks; one pump is preferred for its taste in drinking water whereas another is preferred for its con-venience in watering livestock because it has a trough (although in general, most livestock are watered at the main canal).

Canal water is used for washing clothes since it is softer; clothes washing is done at the canal or, usually during the winter months, at home. In the latter instance, the women transport water from the ditches to their homes where it is then heated on a kerosene stove or on the traditional clay-brick stove fueled by animal or agricultural wastes. Animal fuel is usually preferred for this task as it has a higher combustion quality. Clothes are washed twice, rinsed and bleached, and then stretched over the dry agricul-tural waste piled on the roof for drying. If the clothes are washed at the canal (an occasion for socializing), they are still bleached and dried at home. Rinsing is easier at the flowing canal and water quantity there is not limited, but the heated water at home makes washing easier. Four dung cakes or 1 liter of kerosene are needed to heat 10 to 15 liters of water.

Although older boys and men use the communal shower at the mosque, women and young children usually bathe in a large round shallow pan at home. Bathing at home involves the following tasks: water is brought from one of the pumps and heated on the kerosene stove (which requires about thirty to forty-five minutes and consumes 1 liter of kerosene); thorough soaping is then followed by rinsing off with warm water. When several children are washed in a row the wash water is recycled, but clean rinsing water is used. Including the time to heat the water, bathing four small children will consume about two to two and one-half hours; this task is done at least once a week, usually the evening before or on the morning of Friday, the Muslim holy day.

Tea hospitality. One of the most deeply rooted traditions of rural life is the expression of hospitality symbolized in offering a small glass of sweetened tea. Tea is also the basic household drink following every meal, and it is the housewife's task to prepare it. The water for the tea is heated on the kerosene stove; when boiling, tea and sugar are added and the brew is boiled again until the desired taste is reached. If many guests appear during the day, the time used for tea preparation may be over one hour, and 1/3 liter of kerosene may be consumed. The tea is almost never prepared on the brick stove.

Cooking. Most cooking, however, is done on the traditional clay-brick stove that is fueled by renewable sources which burn at about 20 percent efficiency. Each household has at least one kerosene stove (every married woman brings one as part of her dowry). Even though cooking consumes about two to three hours on an average day for a family with four to five children, cooking is not done every day but usually every other day or so, and leftovers are consumed with bread, milk products, and molasses. Rice is cooked every cooking day (requiring about 1½ hours), as are stuffed vegetables such as tomatoes and onions (3 hours); chicken and soup are made about once per week. Baking is usually done only for feast days and special occasions, and potatoes and tomatoes are baked during Ramadan [the ninth month of the Muslim year]. Cabbages stuffed with a mixture of rice, butter, onions, garlic, tomatoes, and herbs, including mint, are also popular and are done in quantities to assure leftovers for the next day.

Fuel preparation. Dung cakes are prepared two or three times per year by women and girls from the dung which has been spread by the men in an open place in the village for this purpose. The fresh dung is mixed with chaff and ashes and formed into 15-cm-diameter cakes which are left to dry in the sun and then stored on the roof for later use. The fuel used in an average day requires about 1 person-hour of preparation. Fuel from vegetative wastes is used more frequently, since animal waste is also used for fertilizer. Cotton stalks are burned directly, but rice and corn wastes are twisted into knots before burning. A dung cake burns for about fifteen minutes, and its ash residue is used for patching the home's floor. Cotton waste gives good heat and little smoke or ashes, but it is clumsy to store and handle. Rice and corn stalks give only average heat and make much unpleasant smoke in the house, besides leaving useless ash; however, rice residue is plentiful and so far has no other uses.

A few households also have butagas stoves or small one-ring burners; however, these status symbols are not used every day since butane is not always available, its cost is high, and it must be transported from a village 3 km away. The average consumption of kerosene per household for cooking and lighting in wicklamps is about 3 to 4 liters per week. Those households that do not own or cultivate land can purchase agricultural wastes from those that have a surplus. Cotton residue is preferred for baking. Among girls attending school or those involved in handcraft production, there is increasing resistance to preparing dung cakes. Also, because dung is valuable as fertilizer, there is pressure to consume less as fuel. The value of the energy consumed in this village per day per person is about one US cent.

Women's energy role. The study of this village points out the tremendously significant role that women play both as energy resources and as managers of energy, human and renewable. Table 10-4 summarizes the time expended on tasks performed by women in the village. This table does not include the tasks of child bearing, child rearing, bringing lunch to husbands in the field, cutting clover for animal food and the other participation by women in crop cultivation; nor does it include the time spent in visiting and socializing. On the average, however, women spend ten to twelve hours per day performing domestic tasks, and girls are needed for assistance at an early age. Thus, a minimum of about 3,600 kilocalories of human energy are expended by women daily per household. It therefore is imperative that strategies be pursued that can lighten the food processing and diverse portage tasks of women, in order to create time in which women can benefit from educational opportunities and increase their chances for participation in income-generating activities.

Table 10-4. **Female Household Labor in Basaisa, Egypt[a]**

Task	Time per task per household	Frequency	Average hours daily per person
Fetching water	1-1½ hours	daily	1
Cooking	2-3 hours	4-5/week	2
Baking	5-7 hours	1/week	1
Washing clothes	3-5 hours	2/week	1
Making tea	1 hour	daily	1
Milking cows	½ hour	daily	½
Cleaning animal shed	½ hour	daily	½
Leading livestock to field	1 hour	daily	1
Feeding poultry	½ hour	daily	½
Cleaning & grinding grain[b,c]	½ day	1/month	½
Bathing children	2-3 hours	1/week	⅓
Marketing	4-5 hours	1/week	⅔
Milk processing	1-2 hours	1/week	1½
Purchasing	1-2 hours	daily	1½
Dung cake preparation[c]	several days	3/year	1

[a]Childbearing and infant care, crop production and harvesting, and recreation (visiting) are not included in this list. Daughters are required to help with tasks at an early age.
[b]120 kg per month.
[c]These tasks involve male assistance, especially for dung cake preparation.

Introduction of appropriate energy technology. Another result of this study was to discover the relevant potential of some small-scale renewable energy technologies to meet basic human needs within the community. The photovoltaic system powering the mosque's loudspeaker and a communal television set was highly acclaimed in this project, since it provided more reliable power than the electric grid (the latter having been brought to the village for political prestige reasons after the start of the project). The water from the solar distiller has been accepted only for babies' drinking water and for medicine. The solar water heater for the mosque's communal shower is beginning to be quite popular as a viable system, whereas solar cookers, solar ovens, and solar food driers are still in the working model stage and it is too early to tell how widely they will be accepted and utilized by the village. The educational and awareness process in introducing a biogas digester has only just begun.

This socially-oriented project is now in its third year, and one thing it has pointed out so far is the tremendous amount of time required to work with the villagers in preparing them to be active participants in the development of renewable energy technologies (a process which began five years ago, two years before the actual start of the project). It has also pointed out that the cost of the renewable technologies is still far beyond the means of the average developing village.

Example: An Organization Providing Training/Teaching in Developing Countries

The US Peace Corps, with about six thousand volunteers, has been quite successful in providing developing countries with project help that is well matched to communitywide or country-wide needs. From the original people-to-people approach, the Peace Corps has grown into a more comprehensive development assistance program that encourages task sharing by the host country. More recently, emphasis has been placed on appropriate technology. In 1980, 25 percent of the projects were related to health care, including the provision of safe drinking water, family planning, and so forth; 20 percent were related to food production, including horticultural and irrigation projects and especially freshwater fish culture, which has been extremely successful in providing protein to widespread protein-deficient grain diets; 40 percent were in primary, secondary, and vocational/technical education and teacher training; 10 percent concerned facets of economic development such as credit unions, farm cooperatives, and small-business marketing (with special assistance to women); and 5 percent were specifically related to energy and conservation, including forestry projects, the development and construction of efficient mud stoves, and biogas digester programs.

In Africa, projects have concentrated especially on agricultural extension, fisheries, cooperatives and credit unions, health care, and education. These programs are strictly tailored to local needs—thus, one project involved rodent control in Niger; another the development of primary and secondary textbooks for schools in Togo. Projects in Latin America have focused on nutrition, with, for example, assistance on animal husbandry in Ecuador and vegetable gardening in Costa Rica. As this range of projects indicates, the Peace Corps has been accepted and has achieved success because its organization and the people and projects involved are highly adaptable to the wide variety of problems and conditions found in different rural areas, because they take local cultural practices and behavior into account, because they encourage or require extensive local participation, and because they have emphasized decentralized assistance at the village level.

FINAL EVALUATION OF OPTIONS

There is such a diversity among developing countries in size, climate, vegetation, soils, water resources, raw materials, agricultural practices, population densities, cultural traditions, and economic and political structures that it is not possible to recommend solutions to their rural energy problems that would be applicable to even a majority of these countries. Each country must analyze its own needs and resources. In general, however, decentralized energy development shows the greatest promise for rural areas and regional market towns, bringing with it economic opportunities as well as environmental benefits and the preservation of the quality of life.

Criteria for Evaluating a Renewable Energy Project

Figure 10-1 illustrates the relationships that are involved in the activities of a rural community. In a quasi-stable, traditional village, a crisis or imbalance occurs when one factor (capital, natural resource, or human resource) undergoes rapid change without corresponding compensations either at the same level or along inner levels. Pressures in developing countries originate from rapidly increasing populations, droughts or floods, deforestation, high imported-oil prices, war or other political upheavals, lack of local or national capital, and so forth. However, if a foreign assistance program only provides capital, failure of the program can almost be guaranteed, because development programs must be responsive to and work with the local people, addressing their specific needs and resources (social as well as material). Development programs, including the introduction of renewable technologies, must be passed, through the sieve of culture, tradition, and societal organization in the community. Figure 10-1 can be used as a tool to evaluate past projects (successful as well as unsuccessful), and the resultant insights should then be applied to new or continuing projects and options, since change or shortcoming in one resource factor at the community level can to a great degree be counterbalanced by adjustments in the other factors.

An application of Figure 10-1 is illustrated by analyzing alternate cooking fuels in the village of Basaisa, described above. Table 10-5 lists the advantages and disadvantages of the fuels considered, using some of the criteria given in Figure 10-1. There is at present no urgent need to change fuels in Basaisa, although a certain degree of dissatisfaction with the traditional use of dung and agricultural waste can be discerned. Dung is recognized as valuable for fertilizing crops, and the more educated girls dislike the task of dung cake preparation. Rice stalks would not be used for cooking fuel because of their excessive smoke (a health hazard), if another use could be

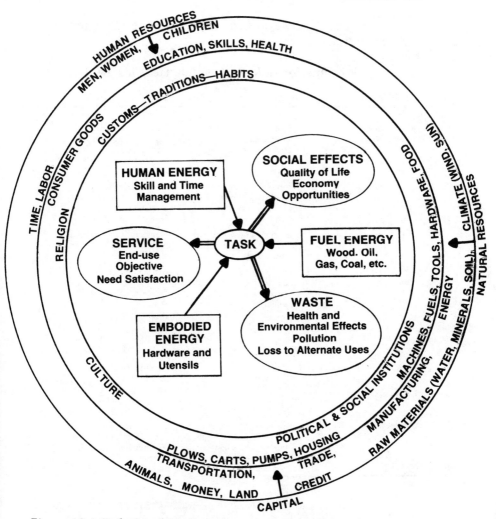

Figure 10-1. **Relationship of Energy to Community Resources**

found for them in this frugal economy. Charcoal is not available in this village; it would have little advantage over the fuels used at present and the current fuel efficiency levels. Conservation as it is understood in this context (i.e., changing to a more efficient clay-brick stove) would not be appropriate in this fairly advanced village, since stove efficiency is already estimated at about 20 percent, since kerosene and butane stoves have already been widely accepted and deeply ingrained into the local tradition of marriage dowries, and since activities such as tea making which consume a considerable amount of kerosene are a very important cultural factor.

Table 10-5. **Evaluation of Rural Cooking Fuels for Basaisa, Egypt**

Cooking fuel	Advantages	Disadvantages
Wood, dung, crop waste	Free Traditional Uses existing utensils Good-quality heat (dung)	Labor-consuming Smoky, unhealthy Decreases soil fertility Low efficiency Deforestation possible
Charcoal	Somewhat efficient Employment possibilities May use waste products	More costly Wood used inefficiently Smoke-producing Not available everywhere
Conservation	Disadvantages of wood, waste, and charcoal mitigated	New hardware required New habits required May be too expensive
Kerosene or butane gas	Labor-saving Easy to use Socially accepted, preferred	Expensive Difficult to obtain Oil-base, shortages
Electricity	Clean Labor-saving	Very expensive New stove required Unreliable supply Requires training in use
Solar	Nonpolluting Renewable Free fuel	Expensive hardware Intermittent Time-consuming New habits required Special location required
Biogas	Fertilizer production Advantages of kerosene, butane, and solar Better sanitation Adaptable in scale	Large or cooperative capital investment needed Maintenance training needed Some hazards

For cooking, electricity has too many disadvantages, since it would entail acquiring a new stove and learning how to use it. Also, the supply of electricity from the grid is unreliable and much too expensive. Solar cookers were felt to be promising by villagers and investigators who wanted to experiment with this technology, but the lack of rugged, cheap, developed hardware has been a handicap, new techniques of cooking must be learned, and it may be difficult to fit the new cookers into the established household routine. On the other hand, since cooking is traditionally not done every day, this may be an advantage for solar cookers, although it may be a matter of some time before this method is accepted to the point where a solar cooker becomes a status symbol or even part of a girl's dowry. The present analysis shows that, after biogas has become fully established in the village, it may supersede solar because of its adaptability to cooking in this village and because it will require fewer cultural changes. Gas cooking is already a status symbol in the community; fertilizer would be available for the fields (an important factor); and biogas would be much less expensive than butane or kerosene. Cooperative agreements will have to be worked out in the village for the acquisition, operation, and maintenance of the biodigester, and training will have to be obtained for its installation and operation over the annual seasons and to insure the establishment of safety procedures to prevent the plant's explosion. The long-range prospects of the biodigester look very promising, and animal dung and human waste collection should not be difficult since animals are mostly kept in sheds already and since the community already has an established communal sanitary facility in the mosque.

In this type of analysis, each community (together with the project leader or adviser) will have to evaluate the relative importance of each resource factor to determine (1) the most suitable solution to the community's needs and (2) how each change should be implemented (including consideration of financing; possible disruption of work routines for men, women, and children; job opportunities; training required; political pressures; level of community cooperation needed; effects on the environment, health, etc.; and local raw materials and labor skills available to manufacture the required hardware).

Any evaluation of planned rural development projects must also answer these questions: How is the local project related to national goals? What are its social benefits, economic benefits, and potential for proliferation? What are the local priorities for development? Are there other goals that can be integrated with the particular task under consideration? If the objective is to find a substitute for unavailable kerosene, the solution may be different than if a change is desired to achieve better health (e.g., through less smoky cooking methods, better nutrition, better sanitation, or safe drinking water). The time frame from introduction to complete community acceptance of such changes will most likely be quite long—perhaps as much as a decade or more.

International Assistance

In the past, few governments of developing countries have paid much attention to their rural problems, although these problems have become aggravated as people move to already overcrowded cities in search of employment and a better standard of living. Unfortunately, programs to help rural people have not been well planned and organized, and the time and educational tasks involved have been seriously underestimated. Instead of establishing and working with long-range rural development plans, governments have been in a position of having to respond to crises. With increased population pressures, future resources will become even more limited.

For the above reasons, the following tasks are of great urgency:

- the assessment of energy resources, followed by energy planning and policies (including funding) for
- the development of promising resources (conservation and conventional fuels; then renewables, depending on local conditions), together with
- education in policy planning and alternate energy technology, research, and application (for people in government, teachers, farmers, local artisans, etc.).

As part of this, relieving the fuelwood crisis must receive priority with programs that increase use efficiency, the substitution of agricultural wastes, and forest productivity.

Third World oil-importing developing countries will need aid—both technical and financial—to establish such programs. Aid by the United States to Third World countries is expected to remain at about 8 billion dollars per year, with a major emphasis on efforts by developing countries to help themselves and with encouragement to private business to invest in energy development. Rural development will be difficult and will require hard work, serious commitment, and considerable reform. The World Bank is planning to lend 13 billion dollars for energy programs in the period 1981–85, but it is estimated that twice this amount is needed to reduce Third World oil imports without draining money from critical areas in health, agriculture, water supply, and education. The World Bank's fund allocations have a built-in bias toward the larger, more affluent developing countries.

If, in this task of energy and rural development, the developing and the developed countries cooperate by conservation methods (which contribute to the financial strength of the individual country as well as to world security) and by speeding the transition from oil to renewable energy sources, then there is a better chance to preserve the world's ecosystems and improve the world's economy. Ultimately, the survival of the human race depends on this.

Acknowledgments

Mary Celeste Gleason has been very helpful in searching the University of Tennessee library system for books and reports (often on very vague instructions). Special thanks are offered to Monika Lumsdaine, who has done a tremendous job in abstracting material, editing the manuscript, and contributing many valuable ideas during hours of discussion. The authors gratefully acknowledge the support of the National Science Foundation for the Basaisa, Egypt, project (NSF-INT-78-01127).

NOTES

1. Detailed studies, such as A. K. N. Reddy's "Alternative Energy Policies for Developing Countries: A Case Study of India," in *World Energy Production and Productivity,* Proceedings of the International Energy Symposium I, ed. R. A. Bohm, L. A. Clinard, and M. R. English, (Cambridge, MA: Ballinger, 1981) are a necessity for these countries.

2. A. K. N. Reddy, op. cit., p. 350.

3. E. Cecelski, J. Dunkerley, and W. Ramsay, *Household Energy and the Poor in the Third World* (Washington, DC: Resources for the Future, 1979).

4. P. F. Palmedo, et al., *Energy Needs, Uses and Resources in Developing Countries,* BNL 50784, Brookhaven National Laboratory, Upton, NY (March 1978), Appendix A.

5. Ibid., p. 13.

6. Ibid., p. 128.

7. Ibid., pp. 15–21.

8. Ibid., p. 22.

9. P. F. Palmedo et al., *Energy Needs, Uses and Resources in Developing Countries and the Implications for US Assistance* Vol. I (draft), Brookhaven National Laboratory, Upton, NY (February 1978).

10. World Bank, *World Development Report, 1980*—an Overview (New York: Oxford University Press), p. 7.

11. J. Foster et al., *Energy for Development: An International Challenge* (New York: Praeger), published for the 1981 North/South Roundtable, Society for International Development, p. xxi.

12. R. Critchfield, "Most of Earth's Nations Seek to Save Forests, Use Renewable Fuels," and M. Strong, "Nairobi: Energy Politics," *The Christian Science Monitor,* August 13, 1981.

13. P. Singh, "India's New Energy Monarch," *National Herald,* New Delhi, November 20, 1980, editorial page.

14. Chen Ru-Chen (translation by Li Nian-Guo), "The Development of Biogas Utilization in China," prepared for the UN Conference on New and Renewable Sources of Energy (Guangzhou Institute of Energy Conversion, Chinese Academy of Sciences, September 1980), p. 2.

15. E. Lumsdaine, March 1981.

16. Chen Ru-Chen, Huang Cong, and Xiao Zhi-Ping, "A Biogas

Power Station in Foshan—Energy from Nightsoil," (Guangzhou Institute of Energy Sources, Chinese Academy of Sciences, 1978).

17. A. B. and L. H. Lovins, "If I Had A Hammer," in Bohm et al., op. cit., p. 141.

18. Cecelski et al., op. cit., pp. 50–51.

19. R. Ramakumar, "Technical and Socio-Economic Aspects of Solar Energy and Rural Development in Developing Countries," Sharing the Sun Conference, Oklahoma State University, Stillwater, Oklahoma, 1976, pp. 164-165.

20. World Bank, *Energy in the Developing Countries* (Washington, DC: World Bank, August 1980).

21. U. Lele, "Rural Africa: Modernization, Equity and Long-Term Development," *Science* Vol. 21, February 6, 1981, pp. 547–553.

22. M. Munasinghe, "Integral National Energy Planning in Developing Countries," *National Resources Forum* Vol. 4, 1980, pp. 359–373.

23. ————— "An Integrated Framework for Energy Pricing in Developing Countries," *The Energy Journal* Vol. 1, No. 3, July 1980, pp. 1–30.

24. D. A. Rondinelli, "Administration of Integrated Rural Development Policy," *World Politics* (Princeton, NJ: Princeton University Press, April 1979).

25. World Bank, *Employment and Development of Small Enterprises*, Sector Policy Paper, February 1978, p. 5.

26. Thirty years of forceful leadership were required in China to achieve successful implementation of a new social structure and rural development.

27. R. Critchfield, "Is the Food Gap Closing?" *International Wildlife*, Vol. II, No. 5, September/October 1981, pp. 29–35.

28. Ibid.

29. G. Gallup, "Opinion Divided on Aid to Third World," *The Knoxville News Sentinel*, August 13, 1981, p. 22.

30. Cecelski et al., op. cit., p. 27.

31. A. Whyte, "Guidelines for Field Studies in Environmental Perception," MAB Technical Notes (UNESCO: Paris, 1977).

32. S. Arafa, C. Nelson, and E. Lumsdaine, "Utilization of Solar Energy and the Development of an Egyptian Village: An Integrated Field Project," National Science Foundation Project #NSF-INT-78-01127.

CHAPTER ELEVEN

Energy in Rural China

Vaclav Smil
Professor of Geography
The University of Manitoba

Studying Chinese affairs has always been interesting, often intriguing, and, until most recently, very frustrating, owing to the absence of things one takes for granted elsewhere: a variety of scholarly journals, regular statistics, public policy debates concerning key issues of the country's development. As for China's energetics, it was still possible to put together critical accounts of the nation's resource endowment and of its production and utilization of fossil fuels and electricity from information widely scattered in Chinese newspapers, radio broadcasts, and news agency releases.[1]

But only since 1978, the second year after Mao's death, the flow of critical information from China has started to grow, and now we have a relative richness of surveys, analyses, and outlooks dealing with various aspects of the Chinese economy. In consequence, the following look at China's rural energy is based entirely on original Chinese sources of the past four years: on papers in renewed scientific journals, on articles in leading national newspapers, on domestic radio broadcasts, and on Xinhua (Chinese News Agency) releases. And many more details can be found in my other recent writings based on the same sources.[2]

CHINA'S ENERGY

In absolute terms China is an energy giant: it has the world's largest hydro energy potential, and it ranks third and thirteenth respectively in coal and crude oil reserves. In total primary energy production it is now fourth behind the United States, the Soviet Union, and Saudi Arabia, but as soon as the Saudis stop producing above their usual capacity China will be third again; in total consumption of fossil fuels and electricity the country is now ahead of Japan, though still much behind the two superpowers.

In relative terms, China is a poor although rapidly industrializing country whose per capita consumption of modern energy resources (fossil fuels and hydro power) is almost exactly 600 kg of coal equivalent, or 17.6 GJ, per year—roughly ten times below the European and Japanese level and twenty times below North American per capita consumption. More significantly—since a comparison with the world's richest countries implies that a poor nation has not made it unless it wastes energy as they do—China's per capita commercial energy consumption is only about half of Brazil's value, although it is approximately twice India's per capita usage, and when ranked globally China fits somewhere around one hundredth place in the array of one hundred fifty countries for which commercial energy statistics are regularly available.[3] Shortages of fossil fuels and electricity are so severe that one-third of all industrial capacity cannot be used owing to the lack of fuel; industries suffer from repeated interruptions of electricity supply; and the most efficient fuels—liquids and gases—are used in much lower proportions in China than in any other large country, rich or poor, in the world. (Coal now supplies 71 percent of all primary energy in China.)

Moreover, the country's energy use is extremely wasteful: its nationwide average of fuel conversion efficiency is a mere 28 percent, compared with the rates of 40–50 percent in North America and Western Europe and nearly 60 percent in Japan.[4] And since China's industrial production is using nearly two-thirds of all its commercial energy and since preconversion losses and transportation (railways, trucks, planes) each account for one-tenth of the total, the modern energy resources left for household use and for the countryside are even scarcer than the national average would indicate.

RURAL ENERGETICS

Indeed, China's rural areas, where four-fifths—or 800 million—of the world's most populous nation live, suffer from acute and widespread shortages of *any* energy, and, again, the recent Chinese surveys and writings enable a quantitative appraisal of this everyday hardship affecting hundreds of millions of poor peasant families.

Modern Energy Resources

Direct consumption of fossil fuels and electricity for household uses is very limited, and even the total use of modern energy resources, including both direct and indirect energy inputs into farming and rural industries, remains relatively small, although it is incomparably higher than a generation ago.

Only some 25 million tons of raw coal are burned in villages for cooking and heating;[5] in terms of the standard coal equivalent this is no more than 16 million tons, or (assuming 29 MJ per kg of standard coal) 580 MJ per year per capita. Household consumption of liquid fuels (mainly kerosene) is altogether negligible, and although there is no breakdown available for the use of electricity—that more than four-fifths of it is used for water pumping and for local small industries is, however, certain—per capita consumption of all electric power in rural China is less than 40 kWh per year.[6] Even if all of this energy were used in households, it would be sufficient only to keep one 50-watt light bulb per person lit for two hours each day of a year.

Energy subsidies into farming rose very sharply during the past decade, and no less than 60 million tce are now used to produce and to fuel field and crop-processing machinery (small walking tractors, wheel tractors, water pumps, and threshers, above all), to synthesize nitrogenous fertilizers, and to provide other nutrients and pesticides as well as a variety of other inputs (pipes, hand tools, plastics).

An official breakdown of energy flows for the year 1978, the first such information published in 1980 after twenty-two years of statistical blackout, puts the total direct agricultural consumption of all fossil fuels and hydroelectricity at a mere 4.8 percent of the nationwide total of 586 million tce.[7] Of these 28.12 million tce (or about 815 PJ), about 14 million came from coal, 10 million from liquid fuels, and 4 million from hydroelectricity. The efficiency levels both of small rural industries burning raw coal and of agricultural machinery using gasoline and diesel fuel are rather low, and hence the official estimate of a 75 percent conversion waste may be still too low.

Shortages of modern energy resources in Chinese farming production perhaps can be best illustrated by the fact that for each of the 160 million hp potentially available from diesel-powered machinery, no more than a little over 50 kg of diesel oil is available annually, sufficient for only fifty days of normal operation.[8] As in the urban industries, rural productivities are thus seriously depressed by the unavailability of high-quality energy, even for the tasks where it is not easily substitutable. Not surprisingly then, household cooking and heating remain as dependent as ever on traditional biomass energy resources.

Traditional Fuels

According to the best available estimates, the heat content of all traditional biomass fuels burned in China's villages in 1979 was equivalent to 290 million tce, or about 360 kg of standard coal (10.5 GJ) per capita.[9] However, the low combustion efficiencies—on the average, no more than ten percent—made only about one-tenth of this total available as useful heat while, at the same time, contributing significantly to the rapid deterioration of China's already seriously degraded environment.

The extent of environmental degradation in China is of truly critical proportions, and virtually all of this degradation's most worrisome demonstrations—continuing large-scale deforestation, worsening erosion, desertification, soil quality deterioration, and local climatic changes—are directly tied to rural energy scarcities, leading to often desperate searches for fuel and to the removal of any trees, stalks, straws, roots, weeds, and dung for domestic cooking and home heating.[10]

Crop residues—largely cereal straws, corn stover, and millet stalks; but also legume straws, potato vines, cotton stalks, and roots—are by far the most important source of phytomass fuels in China's deforested farmlands. Annual production of crop residues naturally fluctuates with the harvests, and for the past several years it clustered around 450 million tons annually. At least 220 million tons are needed for animal feed and as raw material for thatching and numerous household utensils, leaving about 230 million tons (or an equivalent of 120 million tce) for fuel.

This is a far from adequate amount, as a typical household (4.7 people per family) needs 10-12.5 kg of straw per day for cooking and water heating, but 230 million tons per year would provide an average of a mere 3.7 kg of crop residues daily.[11] Even when 70 million cubic meters of firewood (equivalent to about 28 million tce) are added, the average daily amount of fuel that is readily and sustainably available is still less than 2.5 kg of standard fuel per day whereas 5 to 6.5 kg is needed, and about 4.7 kg of standard coal equivalent is estimated to be actually consumed by an average rural household each day.

The difference between that sustainably available and that actually consumed—more than 2 kg of standard coal equivalent per household per day—comes literally from every accessible source which conceivably can burn. Villagers illegally fell trees in protected forests and along roads, strip the bark off living trees, dig out roots and stumps, carve out pieces of sod, rake leaves and debris—thereby stripping the soil of its protective cover, fostering rapid erosion and nutrient loss, and rendering the denuded sites unfit for eventual revegetation. When all accessible natural cover is gone, peasants turn to burning animal dung, depriving their crops of the only fertilizer available—a fuel-food dilemma, so often described in Indian context,[12] which is now an everyday problem for several hundred million Chinese peasants, some of whom, when all other possibilities seem to vanish, grow sweet potatoes and use dried tubers for fuel, and this in a country where provinces the size of large European nations suffer from famine. As Wang Menkui puts it—dramatically but without exaggeration—all these practices will "bequeath infinite calamities and misfortunes to posterity."[13]

Since 1978, Chinese newspapers and journals have carried numerous descriptions of these environmental degradations occurring all over the country. In the arid northwest, the region of China's worst erosion and desertification problems, roads are frequented by animal-drawn carts loaded with illegally cut tree trunks, branches, and roots,[14] and the situation is no different in the rainy south: in Guangdong province, stealing even from the protected national forests is flagrant, and a large proportion of roadside trees are being destroyed by peasants breaking off branches, peeling off bark, and digging up roots.[15]

And the shortages of fuel in rural areas appear to be worsening. Official estimates acknowledge that of 800 million rural inhabitants, 500 million suffer from a serious fuel shortage for at least three to five months per year: in the best-off provinces, 25 percent of all families are so affected; in the worst-off regions, 70 percent of peasants lack fuel for up to six months of every year.[16]

One of the obvious ways to alleviate the situation—the planting of small family fuelwood lots—was until recently discouraged or even outright prohibited. In Shanxi's Yan'an-Yulin region, a dry and heavily eroded loess area drastically short of fuel, the private planting of trees for fuelwood was repeatedly prohibited, most recently by the provincial administration's order in October 1979, as a clear sign of "promoting capitalism."[17] That no relief can come from a wider distribution of liquid fuels is all too clear: even if they were more available, their transportation to hundreds of thousands of villages in a country with a dismal transportation network would be impossible, as the Chinese are only too well aware.

What are needed, then, are programs capable of both harvesting local resources and initiating the modernization of the Chinese countryside. Despite counterproductive dicta like that mentioned above, extensive efforts have been devoted to precisely some such programs for the past generation, especially during the 1970s.

SMALL-SCALE SOURCES

The three outstanding approaches adopted have been the opening of small coal mines, small hydroelectric stations, and family biogas digesters. All of them are commendable and have been made to work in numerous instances. However, as the following reviews will show, their implementation has not been without some serious problems, and the Chinese are now, after a decade of rapid expansion, reevaluating the place and the scale of these rural energy systems.[18]

Small Coal Mines

Small outcrop mines were first opened on a huge scale during the Great Leap Forward years (1958–60) when more than 100,000 native pits engaging over 20 million people went into operation, mostly to supply the thousands of primitive iron furnaces. When this wasteful and costly Maoist experiment collapsed, output from small mines fell sharply and started to recover substantially only in the early 1970s. By the mid 1970s every province acquired a number of small mines, which came to account for about one-third of China's raw coal output. In general, these mines were much larger than their predecessors from the 1950s; in fact, according to the Chinese usage every mine not run directly by the Ministry of Coal Industry came to be categorized as a small mine, though some of these enterprises may be producing hundreds of thousands tons of coal per year and even have a degree of mechanization.

In some southern provinces, traditionally coal-deficit areas, small mines were to be the main solution to cut dependence on imported northern coal. However, with the retreat from Maoist economic orthodoxy there came the inevitable reevaluation of the small mines' role. The reasons are obvious. Most small mines produce fuel of very low quality; sorting and washing is done with only 17 percent of China's coal output, so even the large mines produce largely substandard coal; the life span of many outcrop mines where extraction rests on pickaxes and shovels is ephemeral; labor productivity is dismal (again, even large mines have very low productivities); and costs are very high.

Also, as the example from Shanxi, China's leading coal-producing province, shows, small coal mines may become an outright nuisance. More than 3,200 small pits operated in Shanxi by December 1980, an increase of 1,300 in just two years, but since over 800 of them were opened without permission and most of the mines were located within boundaries of large state-run coal mines, thus reducing their extractable reserves, the net effect of the small enterprises has been largely negative.[19]

Moreover, the output of these mines has always been absorbed by various local industries, leaving very little for household consumption (as already mentioned, a mere 24 million tons per year), and so these mines have never been so helpful in easing rural energy shortages as it might have appeared. That their importance will decline in the future is clear; on the other hand, the building of small hydro stations will continue, although not without changes.

Small Hydroelectric Stations

In 1949 the country had only two large hydro stations (built during the Japanese occupation of the northeast) and only fifty small plants; in total, China's hydro capacity was then 5.6 MW. The 1950s brought ambitious Soviet-inspired plans for many large hydro projects, but the number of small stations had been increasing only slowly until 1956 when the National Program for Agricultural Development set a target of 1,000 small stations with a total installed capacity of 30 MW to be built in a year. This campaign was a failure, and the construction of small hydro stations started again on a large scale in 1958.

Ambitious plans envisaged 900 MW within a few years, 1 GW in 1962, and 2.5 GW of small hydros by 1967. Actually, during the first year of the Leap only some 130 MW were installed, and another 200 MW were added in 1959. Much less was accomplished in 1960; and afterwards, as the Leap collapsed and the country was plunged into a deep economic crisis, the construction of small hydros was virtually abandoned until 1965–66; it increased considerably in the late 1960s and has been going strong ever since. The available nationwide figures show the total number of small hydros growing from about 6,000 in 1965 to 35,000 in 1971, 50,000 in 1973, more than 60,000 by September 1975, nearly 87,000 by the end of 1978, and 90,000 at the beginning of 1981.

The greatest possible dependence on local resources, maximum thrift, and speedy construction have been hallmarks of the program. Stations are financed predominantly with locally accumulated funds, and the state's investment is destined only for the necessary design, equipment manufacturing, or operator-training assistance. Most of the dams, canals, and dykes for small hydros are built in the same way as countless water projects in China's long history: large numbers of peasants congregate on a site —usually during the winter and spring slack season, although many places now have permanent construction teams—equipped with shovels, picks, chisels, steel rods, locally made explosives, shoulderpoles, carrying baskets, wheelbarrows, and pull carts and build the small earth-filled or rock-filled structures in a matter of tens of days or a few months.[20]

Although some stations operate under heads exceeding 100 meters (water is usually brought through steel or concrete penstocks down a mountainside from a high-lying reservoir), most of the small plants have generating heads below 10 meters, and their construction is characterized by great simplicity. Cement, steel, and timber—three commodities which have always been in short supply in China—are consumed in the minimum amounts, and much of the generating and transmission equipment—small turbines ranging from rather primitive wooden devices to reinforced concrete and metal propellers (fixed blade Kaplan or Francis turbines seem to be used most often), generators, transformers, cement poles, wires, and switches—is now manufactured on a county or regional level. The layout and the equipment of generating halls is simple: belt drives, large flywheels, no turbine brakes, and minimum instrumentation are common.

However, some larger or suburban stations are rather modern and are synchronized to a local main grid. Fairly sleek, mass-produced versions of small, compact hydro-generating sets with power ratings between 0.418 and 28 kW are even available for export; the sets larger than 3 kW can be also coupled directly to working machinery by means of a pulley on the generator shaft.

Most small hydro stations are small indeed, although their average size has gone up from 32 kW in 1970 to 79 kW in 1980; the weighted mean calculated from the available provincial figures was about 60 kW for the years 1977–80. I have been also collecting hundreds of regional and county figures which indicate the highly skewed size distributions and significant local variations. Modal station size appears to be less than 25 kW and higher averages result from frequent inclusion of a handful of much larger medium-sized plants in the published totals.

For example, Fengkai county in Guangdong had 337 stations with 11,966 kW as of the end of 1977; but 4 county-run backbone plants had 8,660 installed kW resulting in an average of just 9.9 kW for the remaining 333 commune- and brigade-run stations. A typical station in the rainy south tends to be smaller than an average installation in the north: the extremes on my file show Liping county in Guizhou with just an 11-kW average and Hui county in Henan with a 512-kW mean. Extreme turbogenerator sizes range from miniature units of 400-watt capacity manufactured by the Tianjin Electro-Driving Research Institute for use on springs and small creeks to sets of several hundred kilowatts.[21]

The contribution of small hydro stations to China's rural electricity supply is essential. In 1980, 90,000 small stations had a total capacity of 7.1 million kW or 36 percent of the country's total hydro-generating capacity, which in turn is about one-third of the total power plant capacity. As for the generation during 1980 of a total of 300 billion kWh, 50 billion kWh came from hydro stations, with the small units contributing 12.7 billion kWh.[22] But the importance of small hydro plants goes beyond the fact that they have been producing roughly every twentieth kilowatt-hour in China: in most cases these stations have been the first source of power for the villages and have served as the foundation of the rudimentary electrification of China's countryside. In 1980 their output met 34 percent of all agricultural electricity consumption.[23]

And besides providing electricity for a wide variety of uses —the most frequent applications are to run irrigation and drainage pumps, to operate food- and fodder-processing equipment, and to power local small industries, with just a minimum household consumption for weak lighting—reservoirs of the larger small hydro stations are helpful in regulating water supplies and preventing floods, always an important consideration in China, and are almost invariably used to breed fish and other aquatic products. This multiple benefit accruing from small hydro projects is the main reason why they will continue to be built in the 1980s.

Recently released official cost figures also indicate that the small stations are worthwhile economically. Average investment (including the cost of transmission equipment) is about 1,200 yuan (about US $700) per kW identical with large plant needs, and the average cost per generated kWh (assuming a forty-year operation) is 0.2–0.3 yuan, or about 12–20 cents.[24] These advantageous costs are, of course, largely a result of cheap labor and local materials minimizing the state investment.

Small hydro stations, however, are not without disadvantages. In the country where 15 percent of the soil is affected by serious erosion, thousands of small reservoirs are filling up with silt very quickly, forcing the abandonment of the stations. This silting is now pronounced not only in the Huang He basin with its easily erodible loess deposits, but also in the Changjiang (Yangzi) basin, where many small reservoirs are now losing their storage capacity to serious silting in a mere two or three years![25]

Small reservoirs are also subject to pronounced seasonal fluctuations, and, owing to small storages, load factors are rather low even during the years with normal precipitation. Available figures indicate a mean annual generation time of 1,789 hours nationwide in 1980, with many county averages below 1,500 hours; values around 2,000 hours appear most typical, implying the usual load factor of only 20–25 percent, a very low value in comparison with large hydro projects or thermal stations.

The main disadvantage is the drastic lowering of water storage or the complete desiccation of small reservoirs during the frequent severe droughts afflicting the densely populated eastern part of China. Thus, the record drought of 1977–78 virtually wiped out small-scale hydro generation from large parts of Anhui, Zhejiang, Jiangxi, Henan, and Hubei provinces, and crops there had to be irrigated and drinking and industrial water had to be secured by emergency well drilling and water pumping from distant lakes and rivers. Moreover, most small hydros have only simple equipment and are indifferently managed and unreliable, and inefficient generation is thus common. As a result, the Chinese, while certainly having no intention of abandoning their small hydro station program, are putting greater stress on building large dams with great storages and developing huge pumping stations.

And they are making small stations bigger: the average size of 4,000 new stations put in operation during 1980 was 200 kW, about three times the 1979 mean. The very fact that 4,000 new stations were added in 1980 but the 1979 total has not changed means that another 4,000 small hydros abandoned. The January 1980 meeting on small hydro stations organized in Chengdu by the Ministry of Water Conservancy, the Bank of China, and the Farmers' Bank of China recommended that in all suitable places (i.e., in locations with sufficiently large water resources, and some investment funds, equipment, materials, and skills), the traditional small-station limit of 12,000 kW should be lifted, and loans should be made available to build larger small stations. China's new-found appreciation of costs and rational economic management is surely responsible for this move. In the long run, the Ministry of Water Conservancy would want to gradually eliminate all small stations with capacities of less than 500 kW, a move which would mean abandonment of most of the existing units.[26]

These considerations will also affect the choices of new locations. The total capital costs of a small hydro station at an existing nonpower dam are typically three to four times the equipment cost, but the costs at completely undeveloped but accessible sites are as much as seven times the value of the machinery. In remote sites the transportation costs can increase this multiple up to 25.[27] The Chinese are also having some second thoughts on power distribution from small hydros. Rural transmission losses are running, according to an official admission, as high as 25 percent, and more effort will thus be made to build bigger stations rationally integrated into local or regional networks.

The trend is clear: There is a transition taking place from microstations (under 100 kW), which are still the most frequent existing units; to ministations (100–1,000 kW), which are the preferred new units today; to larger small stations of up to a few megawatts, integrated in rural networks of about a 10-MW capacity.

Biogas Digesters

Spectacular diffusion of biogas technology started in the countryside of two counties in Sichuan province in the early 1970s: in the spring of 1974, the province had 30,000 units; in the fall of the same year, 120,000. In July 1975, the number reached 410,000; in the summer of 1976, 17 million Sichuanese peasants were using biogas for cooking and lighting; and by the end of 1977, the province had 4.3 million digesters providing 20 million villagers with fuel. Nearly 1.5 million digesters were at that time in other provinces, and the nationwide total reached 7 million in August 1978, 15 times the 1975 total.[28]

International comparisons show best the considerable lead which China attained in rural biogas technology. In India, only 8,500 digesters were built between 1962 and 1973, and some 15,000 were in operation by 1978. In relative terms this translated to just 1 digester per 10,000 rural families, and comparative figures were 24 for the Republic of China (Taiwan) and 36 for the Republic of Korea (South Korea)—while China's ratio in 1978 was over 400 digesters per 10,000 rural families.[29]

The rapid spread of biogas digesters in some parts of rural China was made possible largely owing to the relative simplicity of their construction. The typical Chinese family digester is a round or rectangular structure consisting of loading, fermentation, and sludge chambers. Fermentation materials pass from a funnel-shaped loading compartment into the main chamber, where the generated gas displaces the liquid in the pool as it gathers below the rigid cover plate. Spent material is removed from a narrow sludge compartment.

As long as the digester is tightly sealed, a variety of construction materials can be used: stone slabs, irregular stones, bricks, concrete, so-called triple concrete (a traditional Chinese building material usually comprising a mixture of lime, sand, and clay with water); digesters can also be carved in sandstone or shale.[30] Perfect sealing is, of course, essential for anaerobic fermentation, and the task is made more difficult by considerable outward pressure in the tank: the total force acting against the top and walls of even the smallest digester surpasses 10 tons, and leaks in improperly built and sealed tanks are a common occurrence.

In general, the stress has been on using the cheapest local materials and practices, with as few accessories coming from outside a village as possible. Preparation of materials takes three workdays; digging the opening, three to four days; laying bricks and casting concrete each one day. Coating cement is most time consuming, taking ten days, while refilling the soil and miscellaneous tasks take three more days. This adds up to a full month of work, with about two-thirds of the total time requiring an experienced laborer present to cast the concrete, build the brick dome, and put on a perfect coating. Yet in their digester cost estimates, the Chinese do not include any labor charge, a dubious assumption imparting a picture of unrealistic cheapness.

Besides the millions of family digesters, mostly of 4- to 10-cubic-meter capacity, the Chinese have built some 36,000 larger village-scale tanks. The latter are four to ten times more voluminous than the private units, and their gas has been used to run crop-processing machinery and also to generate electricity.

Pig manure is, naturally, the most voluminous waste to be fed into Chinese digesters, and since its carbon-to-nitrogen ratio is about 13:1 (and, when mixed with human excrements, even lower), crop residues or wild grasses are often added to raise the ratio to the optimum range above 20:1. In general, the Chinese try to avoid fermenting a single material and almost invariably compost all plant matter before feeding it into a digester.

The Chinese are rather flexible as far as the recommended loading ratios are concerned. The Sichuanese biogas manual gives just one example of a "fairly satisfactory" mixture (10 percent human excreta, 40 percent animal manure with crop stalks of grass, the rest water) and stresses that a suitable mixture must be devised to fit local conditions. Water is always half of the load; human and animal wastes contain only 4–20 percent solids and residues between 10 and 20 percent solids, so the total water content of the digesting material should be 90 percent or more (a level of 7–9 percent solids is considered best). In case of uncertainty about the actual water share, the Chinese advise making the content too diluted rather than too thick, except in winter when a higher concentration of solids is actually advantageous.

The typical daily generation rate of a well-run family digester is 0.15 cubic meters of gas per cubic meter of tank volume. In a south Chinese rural setting, 1 cubic meter of biogas per day—this is just 24 MJ—is considered to be "more or less" enough for cooking three meals and boiling water for a family of five or six people, making a 6- to 7-cubic-meter digester about the smallest practical size.[31] In some larger digesters operated with closer supervision the generating rates are higher, but winter rates are still at best only half of the summer performance.

The advantages of family-scale anaerobic fermentation in rural China are numerous and undisputable. The Chinese have summarized these advantages into ten points: family-scale digesters save fossil fuels, require a smaller labor force, save fuelwood and grasses, save straw and other crop residues for animal fodder and bedding, reduce expenditures for fuel, lighten household labor, eliminate many insect pests and diseases and markedly improve the hygienic conditions of rural areas, protect forests and timber, enable mechanization of some rural crop-processing tasks and local electricity generation, and narrow the gap between the standard of living in cities and that in villages.[32]

Curiously enough, this list leaves out what in many cases is undoubtedly the second most important reason for anaerobic fermentation in China: the production of an excellent organic fertilizer. In comparison with aerobic fermentation, the anaerobic process results in a superior product: it increases the ammonia content by 120 percent and the amount of quick-acting phosphorus by 150 percent, and each family digester in warm climates can yield 100 kg of fertilizer twice per year.[33] The sludge contains about 650 ppm of nitrogen, 40 ppm of phosphorous, and 9,400 ppm of potassium. Effluent which is regularly removed with added new material is produced in volumes about 1.5 times greater than the sludge, and this effluent is either applied directly to land with irrigation water or stored in tanks for later use. Its nutrient content (about 500 ppm of N, 15 ppm of P, and 2,000 ppm of K) is significantly lower than the concentrations in sludge.

Another important nonenergy benefit with well-run digesters is a significant improvement in rural sanitation. Sichuanese studies found substantially lower counts of Escherichia coli (50 percent lower after three months in one case; a 43-percent difference between input and removal in another), complete destruction of Shigella and spirochetes (tularemia, hemorrhagic fever, autumn fever) after about thirty hours, a 50-percent reduction of Ascaris lumbricoides eggs in one hundred days, and complete elimination of hookworm eggs after sixty to eighty days.[34]

Perhaps most significantly, anaerobic fermentation sharply reduces and after forty to forty-three days completely destroys Schistosoma japonicum (blood fluke) eggs. In spite of repeated massive eradication campaigns, crippling schistosomiasis is still far from eliminated from China's thirteen southern provinces, and any method contributing to its control is of great human and economic benefit.

These undisputable advantages and the spectacular rise of numbers were enough for many foreign observers to conclude that China's biogas program has been an unqualified success worthy of extensive emulation abroad. Yet a closer look reveals not a few problems and puzzles surrounding China's biogas program, and it also raises some essential doubts about its transferability. Certainly the most obvious, although casually overlooked, puzzle is the spatial distribution of the digesters. Five million units, or over 70 percent of China's total, are in the Sichuan Basin, while relatively few digesters are in the provinces climatically more suitable for fermentation than Sichuan.

Mesothermic methanogenesis is strongly affected by temperature decreases below 25°-30°C and, consequently, the best conditions for virtually year-round and fairly efficient fermentation are in the completely frost-free provinces of Fujian, Guangdong, Guangxi, and Yunnan (the southern part). However, Guangdong had only 30,700 biogas digesters in December 1979; Fujian, 22,000 units in April 1980; and Yunnan, just 10,000 digesters in May

1979. According to the Nationwide Methane Production Leadership Group in the summer of 1980, 63 percent of the 7 million digesters were in Sichuan, 9.4 percent in Jiangsu, 5.4 percent in Zhejiang, and 5 percent in Shandong, leaving less than one-fifth of the total for the rest of the country.

As for future expansion, the bold figures of the summer of 1978—when China's second national biogas conference heard about plans for 20 million digesters in 1980 and 70 million units by 1985[35]—have been forgotten. In June 1979 another national biogas gathering affirmed the commitment to the spread of technology, and the delegates urged "to speed up the use of methane in rural China"—but no target totals were cited. And the nationwide total in July 1980 was the same as in June 1979 and only marginally higher than in December 1978.[36] An obvious question here is: has not the overdue relaxation of the heavy hand of "command farming" since 1978 (a relaxation which led to more meat, eggs, and vegetal oil) also resulted in appreciable lessening of rural mass-construction campaigns, including the "enthusiastic" building of digesters?

I strongly feel this is the case, because in many instances a small family digester is simply of no great benefit compared with the time, effort, and investment put into producing woven baskets, tobacco, eggs, or pork (and, as usual, composting the animal wastes in other traditional ways). But even if I am quite wrong, the post-1978 cessation of the nationwide growth of digester numbers, coming as it did after the announcement of ambitious future goals, remains a significant fact.

More importantly, it means that a good many digesters were abandoned since 1978. Latest available claims show that during 1978–80, 330,000 new digesters were built in Shandong, 400,000 in Hunan, and another 400,000 in Zhejiang. Additions in just these three provinces would have pushed the national total to 8.2 million instead of the stagnating 7 million. Assuming that the official nationwide total and the provincial claims are equally credible (we may yet learn that all the digester figures are quite erroneous), one has to conclude that, as a bare minimum, 15 percent of the old digesters disappeared between 1978 and 1980. Is this the result of exercising a bit freer choice, or is it just natural attrition? The former would testify to a less than enthusiastic and less than voluntary acceptance of digesters (and one would not be surprised at that, reading about a provincial party committee's "demanding that 400,000 tanks be built before the end of the year"), while the latter would make an average digester last just two to three years, a not too appealing proposition.

Such rapid obsolescence is especially not too appealing in view of the real cost of a typical small family-size digester unit. Available 1973–78 figures quoted just 20–40 yuan per unit, while the 1980 figures are 50–80 yuan according to Yang Chao of the National Leading Group on Biogas Development in China or 60–100 yuan according to Wu Wen of the Institute of Energy Conversion of the Chinese Academy of Sciences.[37] This is a difference that could be explained in Latin America by rates of inflation, but as such inflation rates still do not exist in China, one must then seriously contemplate the fraudulence of pre–1978 quotations.

Even more important, the official statistics from the National Methane Production Leadership Group released in the summer of 1980 show that of all digesters constructed by 1979 only about 55 percent could be used normally, and that even among these working digesters "not too many can often be used to cook rice three times a day, still fewer every day for four seasons."[38] This means, in effect, that of China's 7 million digesters, at best one-half but much more likely around one-third should be counted as reliable energy generators, quite a comedown from heady plans of 20 million working units by 1980–and from the now entrenched but so misleading total of 7 million.

Local descriptions have a bitter accent. For example, a *Guangming Ribao* article in the summer of 1980 described the situation in an Anhui provice county (hardly a unique case in China) where most of the hastily built digesters turned out to be useless while costing large sums of money and taking up several hundred thousands of workdays. Predictable response of peasants: "The great marsh gas projects have not only failed to solve our firewood problem, they have even affected our livelihood."[39]

And even should everything go well, the reliable operating period of digesters averages seven to eight months in the warmest south and just three to five months north of the Chang Jiang, necessitating the provision of additional winter fuel supplies. Usually no gas at all is generated at ambient temperatures lower than 10°C. Taking this as the norm, 19 percent of China's counties could obtain some production for more than ten months a year, 25 percent for eight to nine months, 34 percent for six to seven months, and 22 percent for less than five months.[40]

This makes the use of biogas as the sole household fuel impracticable throughout China. Difficulties and irritations with the existing digesters are hardly an encouragement to build more. Indigenous "triple concrete" digesters proved largely useless.

Operating units commonly have water and gas leaks; are inconvenient to use and especially to clean; and since not much attention is paid to proper carbon-to-nitrogen ratios, correct pH levels, and the need for frequent loading and removal, many digesters gradually turn into just waste pits and are abandoned. Thus, looking for quick success and instant benefits caused poorly maintained units to fail, usually in their second year of operation. Furthermore, the gas generated from them is burned with low efficiencies, so that at least half of it is wasted. The simple truth may be that most of the digesters have been a burden to their builders, a costly loss, and a discouraging attempt to "modernize."

And the inherent limitations to the performance and expansion of biogas generation will turn out to keep the output and potential contribution surprisingly low. Even if all 7 million digesters (with an average volume of 8 cubic meters and an average generation rate of 0.2 cubic meters of biogas per cubic meter of volume) would operate for seven to eight months per year, they would produce only about 2.5 billion cubic meters of biogas—just 2.2 million tce or less than 1 percent of China's current rural energy use! Even the tenfold expansion of biogas generation envisaged by those unrealistic 1978 plans would cover no more than 5–7 percent of the rural energy consumption in 1985.

CONCLUSIONS

The image which emerges is often in sharp contrast with hagiographic writings of uncritical and uninformed observers which predominated in western coverage of China during the 1970s. This image shows China's inadequate output of modern energy resources and its highly wasteful use resulting in the immobilization of a substantial share of industrial capacity; in chronic and deep shortages of any fuels in rural areas, leading to tremendous pressure on the country's already seriously degraded environment as the peasants destroy vegetation and burn animal dung; and in the much less than admirable performance of the small energy technologies promoted to ease these difficulties, a fact which has led the Chinese to a critical reevaluation of the effort and investment to be put into these approaches.

The magnitude of the problem is staggering: 500 million people in the Chinese countryside do not have enough fuel to cook their meals and heat their water and homes. And there are no simple solutions. Undoubtedly, the problems will persist for a long time, critically affecting China's modernization effort. The only significant relief can come from a combination of sustainable (surely no frenzied mass campaigns of the past are worth repeating) development of any available resources.

Establishment of a very large number of village and family woodlots, building of larger small hydro stations within integrated regional networks, gradual extension of better designed biogas digesters, opening of small coal mines where the resources and economics are favorable, widespread conservation of currently wasted fossil fuels, introduction of solar water heaters and tapping of the country's large geothermal potential (simply for hot water rather than for the much more difficult electricity generation)—all of these ways have their place in China's strategy of rural energy supply.

The fate of these endeavors will be, inevitably, one of the key determinants of the success or failure of the broadly based economic modernization which is to carry the world's most populous nation from its current poverty to a modicum of prosperity within a generation. On the global energy scene few developments of the next two decades will be more critical to watch than the achievements of Chinese rural energetics.

NOTES

1. V. Smil, *China's Energy* (New York: Praeger, 1976), 246 pp.; V. Smil, "China's Energetics: A System Analysis," in *Chinese Economy Post-Mao* (Washington, D.C.: US Government Printing Office, 1978), pp. 323–69.

2. V. Smil, "Energy Development in China," *Energy Policy* 9 (June 1981), pp. 113–26; V. Smil, "China's Agro-Ecosystem," *Agro-Ecosystems* 7 (June 1981), pp. 27–46; V. Smil, *China's Environment: A Critical Modernization Constraint*, report prepared for the World Bank (September 1981), 86 pp.

3. United Nations, *World Energy Supplies* (New York: UNO, annually).

4. Chen Xi, Huang Zhijie, and Xu Jinzhang, "Effective Use of Energy Sources Is Very Important in Developing the National Economy," *Jingji Yanjiu* 5, May 20, 1979, pp. 20–24.

5. Li Shaozhong, "Developing Utilization of Organic Energy Resources to Contribute to Agricultural Modernization," *Renmin Ribao* 13, (June 1979), p. 2.

6. Shangguan Changjin, "Ways Must Be Found to Solve Energy Problems in Rural Areas," *Nongye Jingji Wenti* 4, 1980, pp. IV-56 - IV-58.

7. Wu Zhonghua, "Solving the Energy Crisis from the Viewpoint of

Energy Science and Technology," *Hongqi* 17, September 1, 1980, pp. 32–43.

8. See note 6 above.

9. Anonymous, "Tackling Rural Energy Shortage," *Beijing Review* 32, August 10, 1981, pp. 6–7.

10. See V. Smil, *China's Environment*, note 2 above.

11. See note 4 above.

12. V. Smil and W. E. Knowland, eds., *Energy in the Developing World* (New York: Oxford University Press, 1980), 386 pp.

13. Wang, Mengkui, "Pay Attention to Solving the Rural Village Energy Problem," *Guangming Ribao*, July 19, 1980, p. 4.

14. Jiang Shaogao, "Why Ban a Good Thing Helpful to Peasants?" *Renmin Ribao*, November 13, 1979, p. 2.

15. Guangdong provincial broadcast, January 28, 1978.

16. See notes 6 and 13 above.

17. See note 14 above.

18. V. Smil, *China's Energy Wonders: A Closer Look*. Paper prepared for us Agency for International Aid Workshop on Chinese Development Experience, Washington, DC, December 19, 1980, 25 pp.

19. Xinhua news release in Chinese, December 19, 1980.

20. V. Smil, "Energy," in L.A. Orleans, ed., *Science in Contemporary China* (Stanford: Stanford University Press, 1980), pp. 407–34.

21. See note 20 above.

22. Jing Hua, "Small Hydropower Stations," *Beijing Review* 32, August 10, 1981, pp. 22–27.

23. Ibid.

24. Ibid.

25. Guo Tingfu, "Soil and Water Conservation Workers Appeal," *Guangming Ribao*, November 10, 1980, p. 2.

26. See note 22 above.

27. S. A. Feder, "Small Hydro Plants," *Energy International* 17 (1) (January 1980), pp. 21–23.

28. V. Smil, "Energy Solution in China," *Environment* 19 (7) (October 1977), pp. 27–31.

29. See note 18 above.

30. A. van Buren, ed., *A Chinese Biogas Manual* (London: Intermediate Technology Publications, 1979), 160 pp.

31. Chen Ruchen, Xiao Zhiping, and Li Nianguo, *Digesters for Developing Countries* (Guangzhou: Guangzhou Institute of Energy Conversion, 1979).

32. See note 28 above.

33. Ta Meitian, "Homemade Gas for China's Countryside," *China Reconstructs* 26 (5), p. 27.

34. M. McGary and J. Stainforth, eds., *Compost, Fertilizer, and Biogas Production from Human and Animal Wastes in the PRC* (Ottawa: International Development Research Centre, 1978), 94 pp.

35. Xinhua news release in English, August 22, 1978.

36. Xinhua news release in English, June 4, 1979.

37. Wu Wen, in *Bio-Energy '80 Proceedings*, The Bio-Energy Council, Washington, DC, 1980.

38. Huang Zhijie and Zhang Zhengmin, "The Development of

Methane Is an Important Task in Solving the Rural Energy Problem,"
Hongqi 21, November 1, 1980, pp. 39–41.

39. Anonymous, "What's Holding Up the Development of Marsh
Gas?" *Guangming Ribao,* June 23, 1980, p. 2.

40. See note 38 above.

CHAPTER TWELVE

Summary with Selected Comments

Session Chairman/Integrator:
Ishrat H. Usmani
Inter-Regional Rural Energy Adviser
Department of Technical Co-operation
for Development, United Nations

It is a great pleasure and a privilege to present to you the gist of the findings and conclusions of the work session on energy for rural development which I had the honor to chair during Symposium II.

First of all, let me say that whereas the work of other sessions revolved around world regions or around energy sources like nuclear power, coal, gas, oil, biomass, and so forth, our session dealt not with any special group or energy source but with a particular topic; namely, "energy for rural development." The Symposia Series' program committee decided to consider this topic in Symposium II because it was realized that 75 to 80 percent of Asia's population, about 90 percent of Africa's, and 50 to 55 percent of Latin America's—totaling some 2 billion people—live in rural areas, of whom about 800 million to 1 billion people, a number almost equal to the combined populations of all the developed countries, live in total isolation, far removed from the centers of economic activity and, particularly, from the electric power grids and energy supply lines which invariably in these three continents converge on the urban areas.

Our session not only had the advantage of an excellent position paper by Drs. Lumsdaine and Arafa and a very scholarly case study of China by our distinguished colleague Professor Smil; it also had the benefit of inputs from several participants on their experiences in such countries as India, Sri Lanka, and Pakistan in

In addition to the chairman/integrator and those presenting papers, the work session included the following assigned participants: S. David Freeman, Alejandro D. Melchor, D. R. Pendse, and David J. Rose. The chairman/integrator was assisted by Tom Wilbanks, who served as rapporteur.

Asia; Egypt, Kenya, and Sengal in Africa; and Brazil, Mexico, Colombia, Ecuador, and Peru in Latin America. We also benefited from a joint session with the group dealing with the energy problems of energy-deficit industrializing nations. The joint session provided us with a forum for cross-fertilization of some of the ideas ventilated at our session and their session and it helped us to formulate our views on the importance of reducing oil consumption by using hydro power instead of oil for power generation where possible and also by pooling locally available renewable energy sources wherever feasible.

There was very broad agreement that the problem of rural development is a complex issue which not only involves energy inputs in different forms and from different sources, including those of humans and animals in rural areas; this issue also entails the need to remove constraints and barriers that hinder the whole process of socioeconomic development. Such barriers include the following:

- Lack of education.
- Inavailability of technical skill.
- Resistance to change in established and traditional ways of life because of societal stratification, cultural values, and religious beliefs.
- Lack of capital and poverty of resources in general.
- Lack of knowledge about the existence of technologies that could be employed to improve the rural areas' quality of life and increase their agricultural and industrial productivity.
- Absence of a governmental and administrative set-up at the village level to respond to the calls of national plans for the implementation of development projects.
- General apathy of the ruling elite towards the problems of people in rural areas.
- The tendency of the developing countries' governments to favor big, prestigious projects that bring visibility to the regimes in power rather than small, productive projects at the rural level that help achieve national goals by helping to build a solid base for socioeconomic development.

Despite complicating factors and constraints, discussions in our session and the available literature by some distinguished energy experts tended to confirm our view that energy, when all is said and done, was the one single common denominator which stimulated the whole process of socioeconomic development for all sectors, thereby improving people's quality of life and improving the human environment in general. This is borne out by the UN Energy Statistics for 1979, according to which the world average per capita energy consumption was 1.4 tons of oil equivalent; but of this, the developed countries, with a total 1979 population of about 785 million, had a per capita energy consumption level of 4.3 tons of oil equivalent, whereas the developing countries, with a population of 2.14 billion, had a per capita consumption of only 0.3 tons of oil equivalent. Unfortunately, figures for the rural areas of the developing countries are not precisely known, but their average per capita consumption of commercial energy is likely to be about one-tenth of the developing countries' figure as a whole; that is, roughly 0.03 tons of oil equivalent per capita.

Now, broadly speaking there are three sectors where energy is consumed in rural areas: first, the domestic and social sector; second, the productivity (agricultural and industrial) sector; and third, the transport sector. Of these, estimates made by distinguished energy analysts, some of whom attended our session, show that the bulk of the rural areas' total energy consumption is in the domestic sector.

The rural domestic sector uses energy primarily for heat —principally for cooking food—and this energy is usually provided by fuelwood obtained from cutting trees or removing bushes and other vegetation cover or from animal dung. With practically no means available for scientific afforestation, felling trees leads to soil erosion, first causing the loss of fertility and eventually leading to desertification. The practice of burning animal dung for fuel destroys the dung's valuable nitrogen content and deprives the soil of a very valuable organic fertilizer. (Our session considered the option of using crop residues in solid form by compaction or densification—forms acceptable as a substitute for fuelwood after the crop residues' requirements as fodder for animals and soil conditioners for fields have been met.) The rural domestic sector has practically no input of energy from an external source except for some kerosene for illumination, and that only in those rural areas that are accessible by road to an oil depot and within a reasonable distance of it.

As for the productivity sector, which includes agriculture and, in some cases, small, traditional village industries such as grain milling, pottery, carpentry, textile weaving, and so forth, the rural areas depend almost entirely on human and animal energy. Since this source of energy is limited, total stagnation and a diminished scope of socioeconomic development result. The same sources of human and animal energy are used to meet the needs of the transport sector.

The participants in our session had different views on the introduction of energy in forms that could help the development process, but based on the experience of rural electrification in both the developed and the developing countries, it appeared reasonable to assume that rural electrification on a decentralized, village-to-village basis—electricity produced by harnessing locally available renewable sources of energy such as solar, wind, biomass, and, in some cases, mini hydro—would meet practically all the basic power needs of the rural communities. These power needs include (a) pumping water for irrigation to improve agriculture; (b) running agro-based village industries and thereby adding to the overall level of village production; (c) lighting the village's homes, streets, school, and community center; (d) running an educational television; and (e) operating a cold-storage facility for the preservation of perishables, particularly a medical refrigerator for storing life-saving drugs and medicines.

We in the United Nations have been conducting and monitoring the development of rural electrification in Sri Lanka with the help of alternate energy sources, and I am very pleased to report that we have succeeded in demonstrating that the electric power output from solar, wind, and biomass sources can be integrated and supplied to a rural community to meet the basic needs outlined above. Our experience with solar thermal devices in Sri Lanka has not been very happy, but solar photovoltaic cells, wind generators, and biogas plants have been trouble-free.

In light of our experience with UN-sponsored projects and projects implemented under other bilateral, multilateral, and national programs, I believe that "energy packages" consisting of fuel and power sources harnessed from locally available renewable sources seem to offer a very viable solution to the problem of energy supplies for rural areas. Fuel could come from densification of crop residues and power from solar photovoltaic (PV) cells, since the latter require no engines or generators for conversion and need very little maintenance. In the course of time, PV cells could be produced within the developing countries at prices that would compete with oil-fired generators. In fact, there is every chance that the price of PV cells will come down exponentially in the next five to ten years from the present price of $10 per peak watt to about $2.5 per peak watt. Such an expected drop is not the imagination of US Department of Energy officials but is based on forecasts communicated to us in the United Nations by some of the leading manufacturers of PV cells.

The exploitation of renewable energy sources in the developing countries' rural areas in order to produce "energy packages" for fuel and power raises the question of the attitudes of the industrially advanced countries' public and private sectors towards the solution of the developing countries' energy problems. Assuming that the developing countries' ruling elite give top priority to energy conservation and energy planning in their overall economic development plans, and assuming further that they approach the industrialized countries for technical and financial assistance, it would be in the interest of the industrialized countries themselves to go all out to provide such assistance. If they do not, and if all the nearly 2 billion people in the Third World's rural areas were to switch to oil as the energy source to meet their fuel and power needs, these rural people's consumption would amount to several million barrels of oil per day. This would bring an unbearable pressure on the world's depletable and finite oil reserves. The choice, therefore, is between confrontation and international cooperation in the spirit of interdependence.

The session emphasized the need to promote intensive research and development programs on those problems of harnessing renewable energy sources that are of special significance to the developing countries' rural areas. These programs could first be carried out in the already well-equipped and staffed laboratories and centers in the advanced countries and later taken up in the developing countries. A regular exchange of scientific and technical experts would materially help to arrive at short-term and long-term solutions to energy problems.

Finally, a point was made in our session that the countries which have huge coal deposits may help the developing countries in the short term by offering coal under the same type of aid which was started by the government of the United States under its PL 480 food program for poor countries. If mankind is one family consisting of different colors of the same spectrum, then the challenge of fighting poverty through energy supplies based on coal and renewable energy sources can pull us through the so-called energy crisis.

SELECTED COMMENTS

MARCELO ALONSO

[The following is a report entitled "The Challenge of Expanding Electricity Supplies" that was volunteered by Marcelo Alonso at the work session on energy for rural development.]

In dealing with energy alternatives, the possible contribution of nonconventional energy sources to electricity generation is a very important consideration. For that reason, I would like to report briefly, as an illustration, on the electricity supply situation in four Latin American countries: Colombia, Ecuador, Peru, and Mexico. (See Table 12-1 for reference figures on these four countries. The situation is very similar in the rest of Latin America with the possible exception of Argentina and Chile, both of which are extensively electrified.) The governments of the four countries mentioned have established as a top priority of their energy policies the provision of electricity to as many people as possible, recognizing that electricity is an important factor in improving the people's quality of life and taking advantage where possible of the extensive hydroelectric potential not yet developed in those countries. The bulk of the electricity generation capabilities in those countries are owned and operated by their governments, and their policies favoring the provision of electricity have resulted in ambitious, capital-intensive plans to increase the capabilities of their electricity systems—systems that have been growing at rates between 7 and 11 percent. This demanding situation is aggravated by the fact that a substantial fraction of their populations—more than 60 percent, in some cases—has no electricity. The population lacking electrification is divided into two major groups: those living in isolated areas, villages, and small towns; and those living in shanty towns around big cities and urban centers.

Table 12-1. **Reference Figures for Four Latin American Countries**

	Colombia	Ecuador	Peru	Mexico
Total elec. gen. capacity (GW)	4.3	1.0	3.2	15
Growth rate (%)	10	11	7	10
Energy/capita (kw)	0.247	0.247	0.181	0.34
Pop. without electricity (%)	50	58	60	30

While Mexico's population is more or less uniformly distributed (except for the huge concentration in Mexico City), the populations of Colombia, Ecuador, and Peru are concentrated on the Andean plateau and the Pacific coast, whereas, in the vast eastern regions around the Amazon basin, the populations are very sparsely distributed in relatively small towns and villages. In consequence, the existing national electric grids cover most of the western regions, while population centers in the eastern regions are more or less isolated from the main electric power systems.

Accordingly, expanding the electricity supplies in those countries by using mostly hydroelectric power involves a combination of two options: namely, centralized and decentralized electric energy production.

For those people within reasonable reach of existing grids, the best solution is to connect them to those grids by expanding transmission and distribution lines and by building large hydroelectric stations. This is expensive; for example, Colombia will expend $150 million in the next four years to hook up 150 thousand houses to the national grids (at $1,000 per house); even so, that will cover only 10 percent of the houses without electricity. However, the long-term benefits of this effort fully justify the investment.

In those regions where the size of the population centers or their location and distribution do not justify extension of the transmission lines, a decentralized solution is preferred, using mini hydroelectric and small diesel units for electricity generation. In fact, all those countries have strong programs for increasing the number of small hydro and thermal units, some locally designed and built. The decentralized solution also has some problems, such as maintenance and (in the case of the diesel units) fuel supply.

Of course this approach to the supply of electricity does not mean ignoring other energy sources that are not directly related to electricity generation but that can effectively satisfy some aspects of energy demand such as heating (water and homes), drying, and cooking. Clearly, the actual mix of energy sources depends on the specific features (size, location, economic activity, etc.) of the population region or center under consideration.

One last remark is that the effort required to expand electricity supplies—in terms of financial, material, and human resources—is beyond the capability of most countries, and international collaboration is essential. In the case of Latin America, OLADE [the Latin American energy organization] provides an excellent framework for this cooperation. Also, both the International Development Bank and the World Bank are providing substantial support for the expansion of electricity generation.

MACAULEY WHITING

I am impressed with the reports of the work sessions on the industrializing nations, on biomass, and on energy for rural development. The technology is rapidly coming into place to solve many of the world's energy problems. I wonder if any of those sessions addressed the issues of the diffusion of this technology into rural communities. For instance, for 50 percent penetration of minimum electrification into remote communities, are we looking at a decade, a century, a millennium?—With present methods of diffusion, what kind of time span are we looking at? Also, are there means available for the people who are going to be subjected to this electrification to decide (1) at what rate they want to become subjected and (2) whether they really want to be exposed at all? Some of these issues seem appropriate for consideration at the present stage.

EDWARD LUMSDAINE

[The following comment was submitted in written form after the close of the Symposium.]

In his oral presentation, Dr. Usmani did not summarize or report the conclusions of the work session debates by the participants. Instead, he chose to use this forum to present his point of view on the issues—a viewpoint which differs significantly from those of the authors and participants. Dr. Usmani, being the chairman/integrator, can, of course, speak on whatever he wishes, provided that he first state that he is presenting his own opinions. However, he should not have neglected to present at least a brief summary of the authors' and participants' points of view, since this was the objective of the forum. At any rate, the fact that the session participants were unable to reach a consensus should have been mentioned.

Work Session on Industrialized Market-Economy Nations

CHAPTER THIRTEEN

Reducing Oil Dependence in the Industrialized Market-Economy Nations

Ulf Lantzke
Executive Director
International Energy Agency

Frederick W. Gorbet
Director, Office of Long-Term Cooperation
International Energy Agency

INTRODUCTION

The industrialized market economies are in the process of transition away from energy systems dominated by oil and towards more balanced, resilient, and sustainable systems. Ultimately, this transition must lead to new systems that are not fossil-fuel dominated, but this is unlikely to occur until well into the next century. The more immediate challenges are related to reducing dependence on oil.

THE NATURE OF THE PROBLEM

Increasing the productivity and production of energy worldwide is a necessary element in an overall approach to the energy problem. Nevertheless, the problem itself is more complex and multifaceted than would appear by simply characterizing it as one of steeply rising costs or inadequate availability at what might be deemed reasonable or acceptable prices.

The Role of Oil: Excessive Dependence

The energy problem is and will remain, at least over the next two decades or so, fundamentally a problem of oil. The economic

339

and social strains that have emerged over the past eight years have been greatly influenced by the need to absorb and adjust to a fifteenfold to twentyfold increase in oil prices. But, within that period, the timing and underlying conditions leading to those price increases, as well as their explosive character, not only have exacerbated the adjustment process but also have demonstrated clearly the political dimension of the problem.

In 1973, the OECD countries[1] used 1,960 million tons of oil equivalent (Mtoe) of oil*—53 percent of total primary energy (TPE)—and imported 68 percent of their requirements. Of total net imports, over 85 percent came from OPEC, which then accounted for 65 percent of free world oil production. In 1980, despite the dramatic fall in oil consumption from 1979 levels which has captured so much attention recently, the OECD countries still used 1,865 Mtoe of oil—50 percent of TPE—and net oil imports still accounted for 62 percent of total requirements. Of total net imports, over 85 percent came from OPEC, which then accounted for 65 percent of free world oil production. In 1980, despite the dramatic fall in oil consumption from 1979 levels which has captured so much attention recently, the OECD countries still used 1,865 Mtoe of oil—50 percent of TPE— and net oil imports still accounted for 62 percent of total requirements. Dependence upon OPEC has fallen, but OPEC still accounts for almost three-quarters of OECD net imports and 59 percent of free world oil production.

These statistics mask a number of important changes that have taken place over the past decade in the structure and organization of the world oil market and through the adjustments to higher prices that have been made in the industrialized countries. Some of these changes and their implications will be discussed in subsequent sections of this paper. In this context, however, the critical point is that dependence upon oil imports from a relatively small number of countries* in a politically uncertain, if not unstable, area of the world remains uncomfortably high. This dependence carries with it vulnerability to short-term disruptions in supply and/or increases in oil prices. Over the longer term, its continuation could also lead to a limitation of options available to deal with broader political and strategic issues.

It would certainly be incorrect, and indeed misleading, to

*Including bunkering requirements, which accounted for 69 Mtoe in 1973 and 72 Mtoe in 1980.

**Indeed, the relative dependence is increasingly becoming concentrated on Saudi Arabia, which now accounts for more than 40 opercent of OPEC production compared with less than 25 percent in 1973.

attribute all of the oil price changes recorded over the last decade to the political and economic manifestations of a producers' cartel. The basis for increasing real oil prices is to be found in geology and economics, and in particular in the decline in world reserves during the early 1970s, the peaking of US oil production at about the same time, and the sharply higher real costs of subsequent reserve additions and alternative energy sources. Yet the timing and the extent of price increases have been shaped not by these factors but ny a major war in the Middle East and as a result of internal revolution, neither of which was predictable or controllable by either the industrialized countries or, indeed, by OPEC.

If there is one major lesson to be drawn from the 1970s, it is that we must continue to expect the unexpected. Increasing the resilience and flexibility of energy systems, particularly in the short to medium term, is a desideratum that ranks with attaining minimum cost and sustainable energy systems.

The Overall Context

OECD energy growth has slowed dramatically since the 1973 oil shock, from 5.2 percent per year from 1960 to 1973 to 1.3 percent per year from 1973 to 1979. In 1980, OECD energy use fell by 3.2 percent while GDP grew by 1.2 percent. These trends reveal a significant increase in the productivity of energy in the industrialized world: energy use per unit of output has fallen by 12 percent over the past seven years. Oil use is now lower than it was in 1973, although until 1979 it grew rather slowly but with marked variations through the period depending upon economic growth. Relative to GDP, oil use has decreased by almost 20 percent since 1973.

Although further increases in efficiency can be expected, overall energy use, even in the industrialized countries, is likely to keep on growing. In the developing world, which now accounts for only about 10 percent of all energy use and about 50 percent of global population, energy use (and probably oil use as well) must also continue to grow, and at rates substantially higher than those in the industrialized world, if development aspirations are to be met. The global demand for energy will therefore continue to rise, though more slowly than before. It may, according to a recent study of the International Institute for Applied Systems Analysis (IIASA)[2], increase from 1975 levels by 65–100 percent by the end of this century, and perhaps again by a similiar magnitude in the subsequent thirty years.

Against this background, it is imperative that every effort be made to increase the productivity of energy in order to reduce the risk that energy availability will become a constraint on economic growth. But efforts will also have to be made to develop new sources of supply and to facilitate a smooth transition to better balanced and more sustainable energy systems.

In approaching the general problem, there are three sets of issues that can be distinguished according to the time frame chosen for analysis:

1. Over the longer term, say through the early to middle part of the next century, the problem is to move towards sustainable energy systems that are not based on fossil fuels. Potential candidates appear to be solar, hydrogen, fusion, and possibly fission based on breeder reactors. Technologies are still not developed to the point where any of these options can be relied on as the mainstay of a future energy system, and within the overall structure of such a system particular attention will have to be given to the problems of satisfying liquid fuel demands. This issue can be dealt with only in the most speculative terms at this point in time.

2. Over the next twenty years or so, the major challenge will be to put in place more diversified energy systems, even though these might not be sustainable over the very long term. In this time frame, the options are well known: increasing use efficiency and greater use of coal and nuclear energy. The constraints by and large are not technical, nor are they primarily economic at current oil prices. They are in essence social, institutional, and political constraints which reflect the complex and pervasive nature of the energy problem and the fact that its resolution will require tradeoffs to be made among many competing objectives, including national security, economic performance, environmental and safety concerns, and alternative views about desirable institutional structures.[3]

3. Finally, over the next ten years at least, it will be critical to develop supporting mechanisms that can safeguard the process of transition by minimizing the risks (and the consequences) of unanticipated disruptions in oil supply. In all three areas, but in this more than others, international approaches to the problem are necessary, since imbalances in the world oil market—regardless of how caused—have the potential to affect all countries adversely and, indeed, to

restrain economic growth possibilities through higher oil prices to the point where the structural changes required in energy systems become much more difficult to implement.

Each of these issues is important; all are interdependent. However, the balance of this paper will focus primarily on the next twenty years, where the options are better defined and where more effective policies in the short term can substantially affect the pace of change.

A Digression on National Differences

In what has preceeded, and much of what will follow, the OECD or IEA countries are lumped together as a group. It must be borne in mind throughout the at the industrialized market economies are not homogeneous. There are important differences of perspective and interest bearing on energy questions, even within individual countries as well as between the. Resource endowments differ, as do states of development, systems of government, and even basic philosophies about the role of government. As a result, prospects differ—particularly on a regional basis—and so do the strategies and instruments of energy policy. Nevertheless, despite these differences in circumstances and approach, the industrialized market economies do share a common desire to reduce dependence on oil and a common view that achieving this objective requires international cooperation among themselves and with other countries.

AN INITIAL EVALUATION OF THE OPTIONS: TECHNICAL AND ECONOMIC CONSTRAINTS

The pace of the transition away from oil, and the manner in which this transition is managed, will have profound implications for the attainment of economic and social objectives and, indeed, for broader security concerns. This section briefly reviews the progress that has been made since 1973 and sets out one view of what could be achieved over the next twenty years.

Recent Performance: The Beginning of Adjustment

Although overall dependence on oil has not decreased substantially over the past eight years, the aggregate figures mask a number of developments that need to be noted briefly as a point of departure for discussing future prospects.

Table 13-1. **Key Indicators: Energy and Oil Use**

	OECD total	Regional breakdown		
		N.America	Europe	Pacific
Index (1973 = 100)				
TPE/GDP ratio[a]				
1975	97	96	94	101
1979	91	91	94	90
1980	88	89	90	85
Oil/GDP ratio[a]				
1975	92	96	86	90
1979	88	92	84	81
1980	80	85	77	69
Oil use (Mtoe)				
Total[b]				
1973	1,892(53)	898(45)	704(59)	⟋290(71)
1975	1,735(51)	849(46)	621(56)	265(66)
1979	1,951(51)	958(46)	692(52)	301(65)
1980	1,793(48)	869(43)	648(52)	276(60)
Industrial[c]				
1973	467(43)	150(29)	222(55)	95(57)
1975	401(42)	138(31)	182(51)	80(53)
1979	518(47)	203(39)	206(51)	110(59)
1980	n.a.[f]	n.a.	n.a.	n.a.
Residential/commercial[d]				
1973	407(47)	179(40)	188(57)	39(62)
1975	353(43)	145(32)	165(53)	42(62)
1979	363(40)	142(30)	180(50)	41(57)
1980	n.a.	n.a.	n.a.	n.a.
Transport				
1973	664	444	163	57
1975	669	449	163	57
1979	769	502	196	71
1980	n.a.	n.a.	n.a.	n.a.
Electricity generation[e]				
1973	251(24)	90(15)	89(27)	72(56)
1975	228(22)	81(14)	75(23)	73(51)
1979	229(19)	87(13)	81(20)	61(39)
1980	n.a.	n.a.	n.a.	n.a.
Net oil imports (Mtoe)				
1973	1,324	295	735	293
1975	1,191	313	616	262
1979	1,343	426	623	294
1980	1,116	335	560	266

Source: Data for 1973–79 are drawn from annual publications of *Energy Balances of OECD Countries,* Paris (OECD). 1980 data are from preliminary unpublished IEA Secretariat estimates.

[a]TPE and oil use measured in Mtoe; GDP in 1975 US dollars.

[b]Excludes marine bunkers: 69 Mtoe in 1973 and 72 Mtoe in 1980. Parentheses show percentage share of oil use in TPE.

[c]Includes nonenergy use. Parentheses show percentage share of oil use in total industrial energy use.

[d]Includes agriculture and public administration. Parentheses show percentage share of oil use in total sectoral energy use.

[e]Percentage share of oil in total fuel inputs to electricity generation shown in parentheses.

[f]Not available.

As Table 13–1 indicates:

• Energy and oil intensity, relative to GDP, have decreased steadily since 1973. Progress has been most notable in the Pacific region and most limited in North America, where the share of oil in total energy use actually increased slightly from 1973 to 1979. The North American experience reflects both price controls and some specific factors related to natural gas availability in the 1976–77 period.

• On a sectoral basis, progress was made in actually reducing oil use substantially in the residential sector and for electricity generation. In the residential sector, the greatest displacement of oil was made in North America, where electricity use grew by 25 percent from 1973 to 1979 and increased its market share from 20 percent to 26 percent over the same period. Industrial use of oil fell in Europe but increased markedly in North America, again reflecting to some degree substitution of oil for natural gas in the mid 1970s.

• Oil use in transportation increased in all regions, particularly after 1975 when economic growth recovered. Price increases for motor gasoline have also tended to be lower than increases faced by consumers of other petroleum products, because of the relatively large specific tax component in the price of gasoline.[4]

On balance, the experience of the past seven years is somewhat mixed. Up until 1979, some modest progress had been made in reducing dependence on oil and in increasing the development of alternatives.* Progress was also made in putting in place comprehensive energy strategies in OECD countries, with the result that today nearly all countries have well-defined energy policies with supporting measures aimed at diversifying their energy economies. The process of structural change takes time, and it therefore would be unrealistic to be unduly pessimistic about the limited statistical gains registered through that period.

In the same vein, however, it would be equally unrealistic to be overly optimistic about the performance in 1980. Although the recent decline in oil use has been remarkable, it is still the case that OECD oil use in 1980 was more than 1 million barrels per day (Mbd) above the level recorded in 1975, at the trough of the last recession. During the 1976–78 economic recovery, oil use grew by 4 percent per year, setting the stage for the price explosion of 1979. There are obviously many differences between the current experi-

*OECD domestic production of energy increased by 10 percent from 1973 to 1979, but 58 percent of the absolute increase occurred in 1979, reflecting the long lead times associated with energy development. The increase in production amounted to the equivalent of 4.7 Mbd of oil. Nuclear power and coal each accounted for about 35 percent of the increase.

346 • Industrialized Market-Economy Nations

ence and the earlier recession, but it is still too early to judge confidently how much of the recent decline reflects structural factors as opposed to shorter term cyclical conditions. As well, variation in climate and changes in consumer psychology that have had an impact on consumer stocking behavior also appear to have played an important role in the 1979–81 results.

The Next Twenty Years: A Reference Scenario

At the IEA we have recently developed a reference case scenario for 1900 and 2000.* This has been constructed in the form of complete and consistent energy balances, on a regional basis and for the major energy economies; it thus provides a look at one possible future pattern of production, trade, conversion, and use. The methodology employed was to begin with sets of national projections for 1990 and adjust and extend them with a view to indicating what might be achievable in reducing oil use, taking into account technological, economic, and institutional factors and allowing for the continued development and implementation of energy policies.

The purpose of the scenario is not to present a projection of what is likely to happen, but rather to provide a quantitative and internally consistent framework within which issues of energy policy can be addressed and discussed. The numbers themselves are secondary to the directional trends and their implications.

The main features of the scenario are set out in Table 13–2. A critical assumption is healthy economic recovery over the 1982–85 period and sustained economic growth thereafter at about 3.2 percent per year. Total primary energy requirements grow at an average annual rate of 1.6 percent, with final energy consumption growing at about 1.2 percent per year over the twenty-year period. The overall efficiency of energy, as reflected by the TPE/GDP ration,** increases by about 25 percent by the end of the century.

*The scenario has been developed only for the IEA countries, which include all OECD members except France, Finland, and Iceland. For reference, oil consumption in these three countries amounted to 139 Mtoe in 1979. The scenario represents a "bottom-up" and essentially judgmental view of possible future developments. It is more fully elaborated in *Energy Policies and Programmes of IEA Countries, 1980 Review*, OECD, Paris, July, 1981. A parallel exercise, looking at global energy developments over the same time frame and using a "top-down" approach, with greater emphasis on econometric modeling techniques, is under way within the IEA, and results will be published in late 1981 or early 1982.

**The TPE/GDP ratio is a very rough measure of overall energy efficiency. It suffers from many drawbacks, however, including an inability to take account of structural changes in GDP over long periods of time and the inclusion of conversion losses in TPE, which can be particularly important if, for example, the share of electricity in total energy use grows markedly over time as it does in the reference scenarios. To overcome the latter difficulty, total final consumption (TFC) provides a better measure of energy use. The TFC/GDP ratio falls by about 32 percent in the reference scenario (1970–2000), compared with a decrease of almost 10 percent from 1973 to 1979.

Table 13-2. **IEA Reference Scenario** (Mtoe)

	Reference case		
	1979	*1990*	*2000*
Total primary energy	3,612	4,236	5,100
Nonoil energy consumption	1,794	2,666	3,780
Oil consumption	1,818	1,570	1,320
of which: *Net oil imports*[a]	1,206	974	730
Domestic energy production	2,486	3,142	4,205
Coal	727	1,100	1,770
Oil	707	678	680
Gas	695	713	750
Nuclear	123	336	555
Hydro	232	285	350
Other	2	30	100
Net nonoil imports			
Coal	11	60	40
Gas	30	142	215
Total final consumption	2,617	2,918	3,369
Industry (incl. nonenergy use)	1,040	1,270	1,667
Transport	737	709	680
Residential/commercial	840	939	1,022
Oil use			
Industry (incl. nonenergy use)	475	375	420
Transport	732	707	670
Residential/commercial	325	265	135
Electricity generation	215	136	50
Memorandum items			
Net oil imports (Mbd)[a]	24.5	19.8	14.8
Oil consumption as percent of TPE	50.3	37.1	25.9
TPE/GDP ratio (1973 = 100)	91.2	77.7	68.3
Oil/GDP ratio (1973 = 100)	88.4	55.4	34.0
Electricity consumption (Mtoe)	358	494	655
Electricity consumption as percent of TPE	13.6	16.9	19.7
Share of oil in total sectoral energy use (%)			
Industry	45.6	29.5	25.2
Residential/commercial	38.7	28.2	19.8
Electricity generation[b]	18.6	8.6	2.4

Source: IEA Secretariat unpublished estimates, Spring 1981.
[a]Includes marine bunkers. Conversion at 49.2 Mtoe/year = 1 Mbd.
[b]Measured on the basis of fuel inputs.

Oil consumption declines throughout the period. In 1990 it is lower than 1979 levels by 248 Mtoe (about 5 Mbd) and falls by a further 250 Mtoe through the 1990s. Relative to GDP, oil consumption falls by 37 percent in the 1980s and a further 38 percent in the 1990s, a rough measure of both increasing efficiency and substitution.

The pattern of declining oil use, on both a sectoral and regional basis, is shown in Table 13–3. Over the next decade, virtually all of the absolute decline in oil use is expected to occur in North America, reflecting the possibilities for larger efficiency gains as well as more easily available options for substitution. In Europe, the decline in oil use accelerates in the 1990s, and through that decade the absolute reduction amounts to almost half of the total IEA reduction. This again reflects growing opportunities for substitution through coal, natural gas, and nuclear power. In the Pacific region, oil use remains at about 1979 levels through the next twenty years but declines substantially in relative terms, from 66 percent of TPE in 1979 to 35 percent by 2000. Overall, oil use is

Table 13-3. **Changing Patterns of Oil Use, 1979–2000** (Mtoe)

	IEA total	Regional breakdown N.America	Europe	Pacific
1979 levels				
Total oil use	1,818	958	559	301
Industry[a]	474	203	161	110
Transport	732	502	159	71
Residential/commercial	325	142	142	41
Elec. generation	216	87	67	61
Net oil imports[b]	1,206	426	486	294
1979–90 change				
Total oil use	-248	-226	-23	1
Industry[a]	-99	-70	-10	-19
Transport	-25	-62	21	16
Residential/commercial	-60	-32	-34	6
Elec. generation	-80	-54	-17	-8
Net oil imports[b]	-232	-167	-64	-1
1990–2000 change				
Total oil use	-250	-132	-116	-2
Industry[a]	45	27	-16	34
Transport	-37	-50	5	8
Residential/commercial	-130	-70	-53	-7
Elec. generation	-86	-18	-40	-28
Net oil imports[b]	-244	-159	-77	-8

Source: IEA Secretariat estimates.
[a]Including nonenergy use.
[b]Including marine bunkers.

reduced to about 26 percent of TPE by the end of the century, accounting for 28 percent in Europe and only 22 percent in North America. On a sectoral basis, the strong performance witnessed in the residential/commercial sector is anticipated to continue and indeed accelerate, in response to higher prices, government conservation measures, and increasing substitution of natural gas and electricity. Oil use could fall from almost 40 percent of total energy use (1979) to less than 20 percent by 2000. The decline would be concentrated in North America, where the share of oil in total residential/commercial energy use falls to less than 10 percent by the end of the century.

Similarly, the progress made in phasing oil out of electricity generation continues, with the result that by the end of the century only about 1 Mbd is used in this sector. Coal and nuclear are the major substitutes, together accounting for almost 80 percent of electricity generation in 2000.

The industrial and transportation sectors are more complex. Oil use in industry is expected to decline substantially over the next ten years, reflecting primarily the substitution of coal. But in this process, oil use will become increasingly confined to premium use industries and, at some point, will begin to grow again as total requirements of these industries expand. Thus, total industrial oil use is shown as growing through the 1990s, even though oil's share continues to decline to about 25 percent by the end of the century. There is a modest decrease assumed for oil use in transport over the next twenty years. This is entirely concentrated in North America and reflects the potential for greater efficiency, primarily through the design of more efficient automobiles and the turnover of the car stock.

The general picture, therefore, is one of transition to a much more energy-efficient economy, and one in which oil is increasingly reserved for premium industrial use and transportation.

On the supply side, the major options for the next twenty years are coal and nuclear power. Coal use would grow by 150 percent and account for 35 percent of TPE by the end of the century, compared with 20 percent today. Although some coal increasingly would be used for production of synthetic liquids and gases towards the end of the century, the major markets would continue to be for electricity and industrial use, where its share would grow from 40 percent to 52 percent and from 17 percent to 29 percent respectively. Nuclear power would increase by 170 percent in the next ten years and by a further 65 percent in the 1990s. Despite this growth, its share of total energy would still be less than 11 percent in 2000, though it would account for about 25 percent of electricity generation, a level now reached only in Belgium, France, Sweden, and Switzerland. Natural gas could maintain its current share of TPE (about 20 percent), but this would require substantial growth in natural gas imports.

Uncertainties and Risks

One need only list the options considered above—massive expansion of coal; substantial increases in nuclear energy; greater natural gas imports—to recognize that each is surrounded by much uncertainty and carries with it considerable risks. In general, however, these uncertainties and risks are not technical, nor with current levels of oil prices are they related to economics. In some instances they are deeply embedded in the social and political structures of energy decision making and reflect the inherent uncertainty about how tradeoffs among conflicting objectives are likely to be resolved in democratic societies. In other instances, they are more analytically based and reflect primarily a lack of information about available options, their costs, and how to implement them.

Efficiency gains and demand growth. It is in this area more than any other that the uncertainties derive from lack of specific information. The rational use of energy is generally viewed as favorable by all segments of society and consistent with virtually all social goals. As well, most analysis suggests that the technical and economic potential for further significant efficiency gains is far from exhausted. Yet it is difficult to reach very hard conclusions on how much of this potential is likely to be realized in a given time frame. In part, this is because inadequate information exists to allow a detailed overall assessment of what gains have actually been made over the past decade. In part, it is because results in this area, both past and future, are achieved through the diverse actions of millions of individual decision makers, each in different circumstances and likely to react to the same overall influences in different ways. As a result, this analysis is more prudent than some in its outlook for future energy demand, even though considerable increases in overall energy efficiency have been assumed.

Electricity demand. In the context of the reference scenario, electricity demand is itself a major source of uncertainty. The scenario indicates an average growth in electricity production of 2.7 percent per year, considerably lower than historical growth rates but still much higher than total energy growth. Market share would increase marginally in the industrial sector but quite markedly in the residential/commercial sector.

Relatively rapid electricity growth is critical if coal and nuclear power are to continue to play major roles in meeting incremental energy demand. Yet in most countries substantial excess capacity now exists, load growth projections are being revised downward, and utilities are delaying expansion plans— in part because of demand uncertainty but also because of high interest rates and, in some cases, regulatory difficulties. Greater attention needs to be given to these issues and, in particular, to assessing the following:

- How much do the recent reductions in load growth reflect cyclical, as opposed to structural, factors? For example, the countries' electricity consumption profile was flat from 1973 to 1975 but grew by 4.6 percent per year from 1975 to 1979 before declining in 1980.

- For any given load-growth estimate, what are the risks of overbuilding as opposed to underbuilding, and how asymmetrical are the consequences? In particular, will overbuilding provide opportunities to retire existing oil-fired plants prematurely at lower overall system cost, or will it facilitate greater substitution of electricity for oil in end use, taking into account additional efficiency gains that can be captured with heat pumps?

- Are the structure of electricity tariffs and the regulatory regimes governing tariff-setting appropriate to encourage efficient use of electricity and to provide utilities with adequate revenues to undertake new investments?

Coal. The renaissance of coal has been a long time coming, but there are increasing indications that it is now under way. Several recent reports have confirmed the potential of coal to supply a major portion of total energy growth in the industrialized countries[5], and over the past two years OECD coal production has increased by almost 18 percent, representing more than 2 Mbd of oil equivalent. Nevertheless, there are still a number of uncertainties that will affect the pace of coal expansion. The most important is the general uncertainty about electricity capacity growth discussed above. But prospects in the industrial and synthetics markets are also unclear. Real possibilities to expand coal use in industry are not well known, and only the cement industry, which has a natural advantage, appears to be moving towards coal in a determined way. Uncertainties center on costs of conversion, payback periods, infrastructure availability, technology of industrial coal combustion, and the ability of industry to finance conversions in current economic circumstances.*

As well, environmental concerns are affecting coal use. Key areas of substantive uncertainty relate to (1) the impacts of sulfur and carbon dioxide emissions, and (2) for sulfur dioxide, particularly, the impact of the cost of appropriate abatement technologies on the overall competitive position of coal. But there are also procedural uncertainties, or in some cases simply lack of information about necessary procedural requirements related to environmental protection, which can constrain small- and medium-sized industrial users, in particular, from considering conversion to coal.

Finally, there continues to be a degree of uncertainty about whether future international trade in coal will remain free and competitive and whether its growth will be supported by adequate and timely infrastructure development.

Nuclear. Nuclear power is surrounded by the greatest uncertainties, and these have their roots in social and political attitudes towards the deployment of nuclear technology. As noted earlier, nuclear power almost tripled in OECD countries from 1973 to 1979, but since 1979 little has happened. In 1980, only four units were put into operation, and although sixteen new units were ordered in Europe and Japan, there was an equal number of cancellations in the United States. The anticipated growth in nuclear power foreseen for the 1980s in the reference scenario could well materialize, given the reactors that are now under construction and on order, although the possibility of future cancellations and continuing increases in lead times must be recognized. For the 1990s, there is much greater uncertainty, and the possibility exists that, without substantial changes in the overall climate affecting nuclear power development, a contribution on the scale assumed will not materialize. The consequences of such a shortfall could be far reaching. For example, if only half the foreseen increase in nuclear power occurs over the next twenty years, an additional 4.5 Mbd of oil equivalent would have to be "supplied" in 2000—through greater use of other fuels, through greater energy efficiency, or perhaps

*The IEA Coal Industry Advisory Board has recently completed a major study of the potential for coal use in OECD industry. The results will be published by the IEA in 1982.

through lower economic growth. This example illustrates the integrated nature of the scenario and the need to consider all the options in a comprehensive and consistent framework.

Natural gas. Unlike coal and nuclear power, natural gas is not constrained by uncertainties about its future use. It is generally recognized as a premium fuel, and the key questions surrounding its future role have more to do with availability, price, and the impact on overall security of imports from potentially insecure sources. Exploration for natural gas is still a relatively young industry, and recent oil price increases, together with technological advances, have led to a considerable expansion of effort with notable success, particularly in Canada and the North Sea and, to a lesser degree, recently in the United States. It is still too early to assess the overall potential of recent finds, but their development will take considerable time and be very costly. Over the next twenty years, gas markets, particularly in Europe, increasingly will be supplied by imports—primarily from North Africa and the Soviet Union. The smooth growth of this trade will require resolution of current uncertainties about gas prices and minimization of the risks of possible supply interruptions.

The balance of uncertainties and risks is hard to read. The scenario may be optimistic on the supply side, particularly for nuclear, but it is not necessarily unrealistic and there may well be scope for even greater contributions from natural gas, possibly from coal, and from other energy sources (mainly solar and biomass), which in the scenario are assumed to provide about 2 Mbd of oil equivalent by the end of the century. At the same time, the potential for increasing efficiency may well be understated, although it must be cautioned that the bulk of potentially achievable efficiency gains will occur only slowly, as existing capital stock is upgraded or replaced. Overall, it appears that an outcome like that presented in the reference scenario can be achieved, but its attainment will require supportive energy policies in the industrialized countries —energy policies that aim to provide a greater degree of certainty and reduce risks, thereby creating a climate within which consumers, industry, and government can take the measures necessary to increase energy efficiency and develop alternatives to oil.

Implications

Achieving such results would make a major contribution to maintaining overall flexibility in the world oil market and "creating room" for growing oil use by the nonindustrialized countries. Over the next ten years, assuming OPEC production levels of about 30 Mbd, oil use in the developing countries could increase by about 4.5 percent per year without putting undue strain on the oil supply/demand balance. This would constitute a major element in creating the overall conditions for balanced, noninflationary growth.

Over the longer term, the increasing reservation of oil to premium use industries and transportation will ease the further transition to sustainable energy systems by providing more time to deal with the liquid fuel problem.

An interesting implication of the reference scenario is the marked regional difference in oil import dependence that could emerge towards the end of the century. In North America, net oil imports are reduced to about 2 Mbd. In the Pacific region they continue at about 5-6 Mbd, close to current levels, and in Europe they fall from about 10 Mbd to 7 Mbd. At the same time as overall dependence on the Middle East is declining and relative dependencies shifting, new interdependencies will emerge among the industrialized countries, based upon growing world trade in coal and the increasing export of hydrocarbons from the North Sea and possibly from Canada and Australia. This changing situation is likely to pose new challenges and opportunities for international cooperation in energy matters and in other areas as well.

Continuing international cooperation will also be required to protect against the possibility of further oil price shocks arising from unforeseen and unforeseeable disruptions that might occur in oil supplies. A repeat of the 1973-74 or 1979-80 experience could lead once again to reduced economic growth and increased inflation, both of which could seriously affect the prospects for making the transition described in the reference scenario in a smooth and timely manner.

INSTITUTIONAL CONSTRAINTS

Institutional constraints are in general a reflection of the societal constraints discussed in the next section. They derive from the conflicts that may arise from time to time in basic objectives: adequate energy supplies, a healthy environment, noninflationary economic growth, equitable income and wealth distribution, and national and international security. Typically, they find their expression in institutional interventions, usually though not always at the government level, designed to effect a politically determined balance among such objectives. Examples are price control mechanisms, leasing policies, licensing and permitting procedures, regulatory regimes, barriers to trade, and so forth. Generally, but somewhat simplistically, they can be characterized as "market interventions." In this sense, such interventions are constraints only from one point of view, for from other perspectives they represent checks and balances that are often seen to be desirable in the political context of individual countries. Constraints of this type in some instances can be eliminated but, more typically, can only be reduced as the underlying basis of consensus shifts to accord relatively greater emphasis to energy objectives.

There are, however, two other types of institutional constraints that can and should be more directly addressed. The first is essentially procedural and has to do with the bureaucratic apparatus surrounding, in particular, licensing and permitting procedures. This "hidden cost of regulation" can be substantial. For example, at a recent Parliamentary Symposium on Energy sponsored by the OECD, one participant remarked that among the inputs required to build a new nuclear plant in Germany was "400 tons of paperwork."

Total lead times for nuclear power now average from seven to nine years in OECD countries (except France), compared with only five years in the early 1970s. In the United States, lead times average eleven years, and some reactors are now scheduled to enter commercial operation more than fifteen years after being ordered. Increasing lead times, and even more important, the increasing uncertainty surrounding the planning process for new investments, can act as an effective constraint not only in the nuclear area but in other energy areas as well. In many cases the cause is lack of effective coordination of process rather than substantive difficulties. It is questionable whether such lengthy procedural delays serve a useful purpose or are indeed necessary, and it would be desirable to consolidate and streamline procedures so that lead-time uncertainty is reduced and process does not become a substitute for substantive decision.

The second type of constraint that can be addressed directly includes those institutional barriers to the effective transmission of market signals—barriers which thereby can lead to inefficient energy use. Common examples include the lack of incentive for owners to insulate rental dwellings, the difficulties often encountered by industries wishing to develop their own supplies of electricity in reaching satisfactory agreements with utilities, bulk metering practices for electricity and natural gas in multiple-unit dwellings, and, in some countries, the lack of organizational structures to promote district heating. In areas such as these, there is a strong case for government action to ensure that market signals are transmitted to those whose actions should be affected by them and that institutional barriers to the effective response to such signals are overcome.

The remainder of this section discusses in more detail the difficult issue of striking an appropriate balance between market forces and government action.

Pricing Policies and Market Forces

Although the IEA countries do share a common view that primary emphasis should be placed on market forces in dealing with the energy problem, there are distinct differences in approach. For example, of the twenty-one member countries of the IEA, only six countries (West Germany, Japan, Sweden, Switzerland, the United Kingdom, and the United States) do not control oil prices. And it is only within the last two to three years that such controls were removed in Japan, Sweden, the United Kingdom, and the United States. Canada is the only country in which crude oil prices continue to be controlled at levels below the world price, but in other countries the operation of product price controls can affect, and indeed on occasion has affected, the availability of oil, particularly when domestic prices do not keep up with rapid changes in the international price.

Price controls on oil have been an issue for discussion in many of the annual reviews of IEA countries' policies and programs[6], and the general view within the agency is that controls on product prices, where they cannot be removed for political reasons, should be operated in a way that insures that the price of crude oil is passed through with minimum delay. This is particularly important to avoid imbalances in supplies among countries if the price of crude oil is rising rapidly. It is also generally agreed that oil prices to domestic consumers should reflect international levels, although the difficulties in moving to international levels in Canada in the short term are recognized as being unique.

More recently, at their June 1981 meeting, IEA ministers discussed proposals for the economic pricing of all energy sources and agreed that this should be given priority in future IEA work. This is a difficult issue for a number of reasons. In the first place, once one considers fuels other than oil, an unambiguous "standard of value" is difficult to define. The world oil price is by no means a free market price, but at least it has the merit of representing the "going price" for oil in world trade. A well-developed world trade in thermal coal or natural gas does not yet exist and, as these trades develop, there is no necessary reason why the prices that emerge, if the markets are competitive, should bear a fixed relationship to oil prices rather than being determined by supply/demand conditions for coal and gas. Indeed, for natural gas, because of the importance of transportation costs one might see the emergence of different prices in different regions. Similarly, for electricity, tariffs are now regulated in all IEA countries, the principles of tariff setting vary, and there is no unambiguous definition of what "economic" tariffs would be.

A second difficulty has to do with the structure of existing government interventions, particularly through subsidies. For example, coal production is subsidized in Japan, West Germany, and the United Kingdom, and in West Germany the subsidy has been coupled with import restrictions, although these have recently been modified to the point where they do not represent a practical constraint on coal imports. In Canada and the United States, natural gas prices are controlled at relatively low levels. Although such practices may not be efficient from a strictly economic point of view, they do reflect the expression of strong social objectives and, as important, they are in many instances consistent with and supportive of national and international energy objectives.

The source of the difficulties is therefore both analytical and political, with the political problems stemming ultimately from the fact that energy objectives themselves include not only encouraging economically efficient use but also promoting the substitution of other fuels for oil and increasing indigenous production of all energy sources. There may be cases where economic pricing of energy supports the efficiency objective but conflicts with other energy objectives. The focus must therefore be on assessing, in specific circumstances, the degree to which departures from economically efficient pricing practices are consistent with the overall thrust of energy policies.

The Role for Governments

No government of an industrialized country adopts a completely laissez-faire attitude to energy, nor can any government afford to do so as long as a substantial proportion of total energy requirements remains controlled by other governments. Markets often function imperfectly, and even where they do function well, in the sense of allocating resources efficiently, the outcome can conflict with other social objectives such as income and wealth distribution. For these reasons, governments have a legitimate role and a necessary responsibility to deal with the energy problem. The precise balance between reliance on market forces and an active government role will vary from country to country. But government actions are necessary, both nationally and internationally, if the overall situation is to be managed effectively.

National actions. A number of institutional constraints on achieving energy objectives have already been referred to, some more intractable than others. In general terms, there are four key roles for governments:

- A leadership role, to provide an overall framework in which the energy problem can be intelligently discussed and to focus discussion on the consequences of alternative courses of action. This is critical in the process of shaping consensus on the appropriate tradeoffs among energy objectives and competing goals.
- An information role, to disseminate practical advice to consumers and investors about the options available to increase the efficiency of energy use and switch away from oil. In their own operations as well, governments should be, and be seen to be, implementing energy options consistent with their overall energy objectives.
- A housekeeping role, to insure that the array of governmental and quasi-governmental regulatory procedures is operated in as streamlined a manner as possible to reduce the burden on applicants and provide greater certainty about the nature and length of regulatory processes.
- An interventionist role through taxation, incentives, regulations, or even direct government involvement, to overcome the effects of market imperfections or market power. Examples would include the provision of incentives to encourage the insulation of rental dwellings (recently done in the Netherlands); establishment of guidelines governing the terms on which industrial electricity generation can be linked to existing grids, as has been done in Austria; or the development of an overall "heat plan," for municipalities as has been implemented in Denmark.*

*Other examples of areas where intervention is desirable to remove institutional constraints and actions that have been taken by IEA countries are presented in *Energy Conservation: The Role of Demand Management in the 1980s* (Paris: OECD, 1981).

In general terms, it is important that governments accept that at least in current circumstances the demand for oil is a legitimate concern of public policy.

International actions. Industralized countries cooperate in energy matters primarily through the IEA, as well as through other fora such as the European Community and the annual economic summits. The objectives of the participating countries reflect their common view of two main areas of international energy policy concern:

- the need to diversify energy systems over the medium to longer term to reduce dependence on oil and
- the need to protect against serious disruptions in oil supply and excessive oil price increases caused by temporary and relatively minor imbalances between demand and supply.

In support of the first area, IEA member governments have committed themselves to agreed Principles for Energy Policy (1977) and Principles for IEA Action on Coal (1979). Within these general frameworks, performance is monitored on a regular basis with a view to identifying areas where results can be improved. Recommendations for strengthening energy policies are made annually, and IEA ministers have agreed to accept the criticism of their colleagues and to consider the recommendations in developing their national policies.

The second area requires more operational forms of international cooperation, at least as a contingency. In times of serious disruption, it is important that all countries contribute to a solution so that none of them is disproportionately burdened. The Agreement on an International Energy Program, which provides for the sharing of oil in an emergency, provides a practical and operational response to such situations. Similarly, when imbalances arise, as was the case following the Iranian revolution and the Iran/Iraq war, the potential for sudden and massive price increases exists. Such price increases, as were experienced in 1979, represent a market response to such situations but can be extremely harmful to the collective interests of industrialized countries and, indeed, to the world economy. It needs to be borne in mind that in a market where there is a high degree of political control over supply, market forces can really act only on the demand side, particularly in the short term, and the result can be extraordinarily large and costly increases in oil prices. It is therefore only prudent for governments to prepare contingency plans, including the possibility of collective government action, which could be brought into effect to supplement market forces.

SOCIETAL CONSTRAINTS

Societal constraints derive from basic concerns about the impact of energy options on the quality of life. Often, the source of these conflicts comes from the need to reconcile competing views about consequences and risks of actions whose effects can in general not be measured and, even where measurable, are subject to different and sometimes conflicting interpretations. There is, however, an important distinction that needs to be drawn between objectives and options and between short-term and longer term implications. It is a generally accepted proposition that the three elements of the energy problem as set out earlier—the need to move to sustainable energy systems, the need to reduce dependence on oil, and the need to protect against oil supply disruptions—are inherently consistent with sustaining and increasing the quality of human life. Meeting these challenges will, in the short term, reduce international tensions and eliminate a serious potential constraint to satisfactory economic growth and development. Over the longer run, it will result in cleaner and more benign energy systems. The problem is essentially how to get from here to there. It is essentially over questions of how fast it is necessary to move and what roads to take that conflicts arise between energy objectives and quality of life considerations.

The rapidity with which the oil situation changed fundamentally in the early 1970s, and the magnitude of the shocks that have been imposed on the world economy, have seriously increased the problems of adjustment. The fastest depleting available resource is time, and the options are correspondingly reduced.

Rather than a smooth transition from oil-based systems to sustainable and benign energy systems, we are now faced with the need for an "intermediate transition," a need to find nonoil "bridges" to the future. The options are essentially conservation, coal, and nuclear, but no one option can do the job alone. And although conservation raises few difficulties, coal and nuclear are generally regarded as less environmentally attractive and more dangerous than oil. It is in the environmental and safety areas that conflicts are most pronounced and most immediate.

Environmental Issues

Environmental concerns are both general and specific, long term and short term. Specific, short-term concerns differ greatly from country to country but typically find their expression at local levels in relation to particular production or use decisions. It is difficult to generalize about such situations since they depend on local situations and attitudes and are governed by standards and procedures that vary widely from place to place. For example, in Japan population densities and land scarcity make it much more difficult to site coal-fired power plants than in many European countries, and this is also true of the northeastern United States compared with other parts of that country. In Italy and Spain, the problems are no less severe but derive more from the strength of local opposition than from geographical difficulties in finding sites. In Denmark, the situation is quite the opposite, and the share of coal in electricity generation has increased from about 20 percent in the early 1970s to almost 85 percent today, in part because the environmental consequences of using coal were judged to be less severe than the costs of continuing to rely on oil or the risks involved in introducing nuclear power. It is precisely this kind of overall calculation that is appropriate, and necessary, if environmental concerns are to be placed in a proper perspective, but the problem is that the costs and benefits are, understandably, viewed quite differently by those directly involved and others less directly affected. It is for this reason that some countries (particularly Japan, Italy, and Spain), in addition to having stringent environmental controls, are now experimenting with mechanisms to increase the benefits to local communities of new coal or nuclear power plants, through lower power rates or direct cash payments.

At a more general level, there is growing concern about the impact of acid rain, particularly in Scandinavia and Canada, and about the longer term climatic effects of increasing the concentration of carbon dioxide. Neither of these issues is related solely to coal combustion, although the projected increases in coal use have served to focus them. Both are very serious concerns and require increasing analysis and attention. The acid rain issue is much better understood at this point than is the carbon dioxide issue, and the major questions that require attention are the availability and cost of prevention or abatement technologies to reduce emissions of sulfur dioxide and nitrogen oxide, as well as the demonstration and commercialization of newer and cleaner combustion technologies, such as fluidized beds and coal/oil or coal/water mixtures. With respect to carbon dioxide, there is still a great deal of uncertainty about the ultimate impact of greater carbon dioxide concentrations on climate and, before more definitive conclusions can be reached, more scientific work appears to be necessary —particularly in the areas of global climatic modeling, the role of the oceans as carbon dioxide "sinks," and the effects of carbon dioxide concentrations on cloud formation.

Health and Safety Issues

All energy sources have health and safety risks attached to their development and use, as do virtually all activities in modern industrialized societies. Even though the health and safety record of the nuclear power industry has been relatively better than that in other energy supply industries, nuclear power is the main focus of concerns about health and safety. In part this is because of the very different nature of the risks attached to nuclear power: they are risks that have potentially very grave consequences but with an extremely low probability of occurrence; and they are primarily involuntary rather than voluntary risks—unlike other energy supply industries, the risks of nuclear power are not primarily confined to those who choose to work in the industry; they are generalized to the population at large.

As well as the more immediate safety issues related to reactor operation, there are concerns about ultimate disposal of radioactive waste (where options have been defined and tested at the laboratory stage but not yet implemented in any country) and broader issues of proliferation.

As a consequence, nuclear power has been and will likely continue to be a difficult political issue in virtually every industrialized country. The degree of difficulty, however, varies among countries, with some countries apparently experiencing little or no opposition to nuclear power; and the degree of public acceptance also seems to vary with perceptions about the severity of the overall energy situation. Member countries of IEA have agreed that nuclear power is an indispensable element in addressing the energy problem and that stronger and more effective actions are necessary to increase pubic understanding of reactor safety, implement waste management and disposal pograms, streamline licensing procedures, insure that regulatory practices do not unnecessarily constrain investment, and reinforce the reliability and predicitability of trade in nuclear fuels and technology under appropriate safeguards. The main task is to contribute to an environment in which discussion can proceed in an objective and dispassionate way.

Resource Conflicts

Serious potential resource conflicts have only recently emerged as major considerations and tend to be associated with the synthetic production of liquid fuels—either alcohol from biomass, which raises the question of competing energy and food requirements; or fuel from oil shales, which require large amounts of water that may result in inadequate supplies for other uses. These competing uses will become important constraints on the rate of deployment of technologies in these areas.

Although it is important to begin now to develop and test the technologies that might be necessary to provide liquid fuels in the future, the time frame in which alternative liquid fuels will be required on a commercial scale is still uncertain and will depend essentially on the success achieved in substituting other sources of energy for liquid fuel now used essentially for heating purposes. Certainly for coal, as an example, both the current state of technology and the economics suggest that it is more sensible to use coal-generated electricity to free up oil wherever possible, before generating liquids from coal on a commercial scale.

Distribution of Income and Wealth

A final area that deserves consideration is the very complex relationship between energy options and income and wealth distribution. The increases in oil prices through the 1970s have led to very abrupt and large shifts in income distribution, both among countries and, in those industrialized countries having domestic energy resources, within countries as well. After the initial oil shock in 1973–74, the response was to resist such changes where possible or, at least, to introduce policies that would make their impacts easier to absorb. Thus, many countries eased their fiscal and monetary policies to try to compensate for the shift of wealth from oil-consuming to OPEC countries, and others, which had domestic oil production available, imposed price controls to constrain the transfer of income from consumers to producers within their countries. It has generally been recognized that such responses have only delayed the necessary overall economic adjustments and have worked against increasing the efficiency of energy use and the achievement of the required structural changes in energy systems. As a result, the response to income distribution concerns is, with two notable exceptions, no longer through direct control of prices but through government measures to capture the economic rent and, depending upon the country concerned, use the additional revenues to facilitate the overall economic and energy adjustment process and provide targeted relief to those most adversely affected by the price increases. The two exceptions are natural gas price controls, which continue to exist in the United States but are being phased out over the next four years, and the structure of oil and gas price controls in Canada. In the latter case, it is the constitutional quarrel over how the economic rents on existing resources would be distributed between the federal and provincial governments that has been the biggest stumbling block to moving to world price levels for oil.

The underlying concerns that give rise to the societal constraints described above are real and legitimate. In our democratic and pluralistic societies they find their ultimate expression at the political level, and it is there that they must be reconciled. But the process of reconciliation must take place within a context that effectively balances the concerns of potential gainers and losers against the broader public interest.

EDUCATION AND INFORMATION

The energy problem is so complex and affects our lives in so many ways, both direct and indirect, that it may well be difficult to agree on precise formulations of the nature of the problem, and it would certainly be unrealistic to look for agreement about the relative importance of the options. Nevertheless, a broad measure of public consensus is necessary if options are to be pursued effectively, and the shaping of this consensus requires, if not agreement, at least informed discussion.

As we indicated above, governments have a critical role to play in providing information and leadership in this area, but they cannot do the job alone. Others with responsibility—industry, labor organizations, academic and research institutions, communities—must participate in broadening the base for and bringing about a sustained and informed discussion on energy problems and options at local, national, and international levels. The issues that need to be addressed are the issues dealt with in this paper and, more broadly, these Symposia.

But education and information, in substantive terms, must also go beyond the need to shape consensus. There is a practical component required. Consumers need better and more readily available information about the comparative efficiency and costs of options available to them, whether they involve buying a new appliance, investing in an automobile, or making a decision about insulation or alternative heating systems. Industrial users need practical information about the potential for efficiency improvements in their own operations and how to achieve these improvements; about the availability and cost of alternative fuels; about the technological options available or under development; about the environmental standards and procedures they would have to comply with if they were to switch from oil to other fuels; and, in some cases, about government programs that could help them economize on energy or oil use. Governments, as well, need much better specific information than many of them have about the detailed use of energy in our economies and, in particular, about its efficiency in specific uses; about the relative costs and effectiveness of various options, including measures to promote efficiency; and indeed, in some cases, about the effectiveness of their own energy programs.

Many IEA governments are active in these areas, through measures, such as appliance labeling, insulation advice and assistance programs, and auditing and advisory services for small-and medium-sized businesses; and one of the main objectives within IEA is to facilitate a continuing exchange of experience among countries about the effectiveness of such information and education measures, so that countries can learn from one another. The agency is also concerned with the sharing of experience and information in technology development, and there are currently about fifty collaborative R&D projects going on under IEA auspices.[7]

In the final analysis, the shape of future energy systems and the pace of transition will not be determined by the construction of analytical scenarios but by the choices of individuals acting in what they perceive to be their own interests. Better information, and effective networks for disseminating it, are essential both to help define that interest and to identify the options available to meet it.

FINAL EVALUATION OF OPTIONS

Over the next twenty years, which is the focus of this paper, the options are relatively limited. Reliance on oil, and particularly oil imported from potentially insecure sources, is too high. The available options are to reduce use through efficiency gains and through substitution—the latter requiring an increased use of coal, nuclear power, and natural gas. It does not appear that any of these options alone will be able to make a sufficient contribution rapidly enough to ease the overall problem; all will be necessary. Nor does it appear that new forms of energy, either synthetic or renewable, will make a major contribution to overall energy balances within the next twenty years, even though they may play a substantial role in some areas and applications. It is, however, important that basic development work on these new technologies proceed, because they will be called on to make an increasing contribution in the next century.

Criteria

The most important criterion in dealing with the energy problem over the next twenty years is to reduce dependence on oil to more acceptable levels. Over the longer term a transition away from fossil fuels will be required. Over the shorter term, contingency plans need to be developed to protect economic performances from large and sudden increases in oil prices.

Relative Importance of Alternative Considerations

Different countries approach the basic problem in different ways, depending upon their own particular resource situations, philosophies, and institutional and social structures. These differences, however, tend to result in differences of emphasis on the importance given to various options in the light of particular societal and institutional constraints and, consequently, to differences in the choice of energy policy instruments. There is broad-based agreement among the industrialized countries on the overall nature of the problem and on the range of options available to deal with it.

Recommendations

1. The primary focus of energy policies should be to achieve the greatest overall efficiency of energy use and to promote diversification of energy supplies in order to reduce dependence on oil to more acceptable levels.

2. All options that are technologically and economically viable will have to contribute to this task. In particular, greater reliance will have to be placed on coal and nuclear power.

3. Governments have an important responsibility to implement comprehensive and effective energy policies. Dependence on oil should be recognized and accepted as a legitimate area of public policy concern.

4. Market forces should be relied on primarily to bring about greater efficiency in energy use and the development of alternative supplies, but there are, nevertheless, important areas where governments must act positively to (a) create an overall climate conducive to investment in energy-saving techniques and new energy production; (b) streamline and provide greater certainty about institutional procedures governing energy use and production decisions; (c) insure that market signals are recognized and acted upon; (d) overcome market imperfections; and (e) where appropriate, speed up the response to market forces through incentives or disincentives.

5. Better information should be made available about energy problems, their economic, social, and geopolitical implications, and the practical choices individuals have in dealing with them. Governments have an important role to play in this process, but other actors in the social community must share this responsibility.

6. It is important that research into and development of technologies for renewable energy sources and synthetic fuels continue at a healthy pace, since these technologies will be required to bring about an eventual transition to sustainable energy systems.

7. International cooperation in energy matters should be continued and strengthened, not only at the government level but at other levels as well.

NOTES

1. Currently, the membership of the Organisation for Economic Co-operation and Development (OECD) includes the following countries:

Australia, Austria, Belgium, Canada, Denmark, Finland, France, West Germany, Greece, Iceland, Ireland, Italy, Japan, Luxembourg, the Netherlands, New Zealand, Norway, Portugal, Spain, Sweden, Switzerland, Turkey, the United Kingdom, and the United States. [eds.]

2. *Energy in a Finite World*, Executive Summary (Laxenburg, Austria: International Institute for Applied Systems Analysis, May 1981), p. 39.

3. See, for example, the excellent analysis by J. Jimison, "Energy—Is There a Policy to Fit the Crisis?" (Washington, DC: US Government Printing Office, 1980), Committee Print 96-IFC-57.

4. A. Tait and D. R. Morgan, "Gasoline Taxation in Selected OECD Countries, 1970–79," *International Monetary Fund*, Staff Papers, January 1980.

5. See, for example, *Steam Coal Prospects to 2000* (Paris: OECD, 1978); *Coal—Bridge to the Future*, Report of the World Coal Study (Cambridge, MA: Ballinger, 1980); and "Report of the IEA Coal Industry Advisory Board" (Paris: OECD, December 1980).

6. See, for example, *Energy Policies and Programmes of IEA Countries, 1980 Review* (Paris: OECD, 1981).

7. For more details, see the *Annual Report on Energy Research, Development, and Demonstration Activities of the IEA, 1979–80* (Paris: OECD, 1980).

CHAPTER FOURTEEN

Energy Supply, Demand, and Policy in the United Kingdom

*David le B. Jones**
Deputy Secretary
United Kingdom Department of Energy

INTRODUCTION

The United Kingdom is fortunate. We alone among major western industrialized countries are effectively self-sufficient in energy. We have very large resources, indeed, of coal (estimated at about 45 billion tons, which should last at least three hundred years at present rates of extraction) and significant resources of oil and gas (those remaining are estimated at 1,900–4,075 million tons and 600–1,800 million tons of oil equivalent respectively). We have the technology and skilled men and women for a major nuclear program. We lack commerical resources of uranium, but our stocks are large.

But for the United Kingdom self-sufficiency in energy is temporary. Unless oil and gas reserves prove markedly greater than we now estimate, we shall sometime towards the end of this century again become significant net importers of energy. Our task in the next fifteen or twenty years is to make good use of our energy wealth not only in the short term but also to prepare for the day when we again become net importers. This involves developing wider economic, industrial, and social policies to insure that wealth is used to strengthen the basis of our economy. And in the energy sector with which this paper is concerned, it means using our resources prudently with an eye to the future and preparing well in advance the replacement of indigenous production of oil and gas when it starts to decline.

*This paper is submitted in a personal capacity. It does not necessarily represent the views of the UK Department of Energy.

The United Kingdom cannot isolate itself from the wider world energy scene. We are a major trading country with an open economy. As was dramatically illustrated in 1978–79 the United Kingdom suffers along with the rest of the OECD area from scarcity of oil or sharp rises in its price. In any case, the fact that net self-sufficiency in energy is temporary by itself means that we share the common western objective of reducing dependence on imported oil and, as the Venice Summit put it, breaking the link between economic growth and the consumption of oil. And our industries have to compete with those of other industrialized countries and of the developing world. They will not be able to do this if they have to pay continuously and significantly more for their energy than competitive industries because we follow policies that disregard what is happening in the rest of the world. For all these reasons, domestic energy policy needs to be seen in its international context. The United Kingdom plays a full part in cooperative international efforts on energy, particularly in the European Community and the International Energy Agency but also more widely.

UK ENERGY STRATEGY—OBJECTIVES

The basic energy strategy objectives of the United Kingdom have for many years been to insure adequate and secure supplies of energy used efficiently and provided at the lowest practicable cost to the nation as a whole. This statement of objectives implies that a wide range of factors will have to be taken into account in formulating policy. The insuring of adequate and secure supplies means making good use of our indigenous resources of energy. Using those supplies effectively raises the question of energy conservation. But there are other less obvious considerations implicit in the statement of objectives. Secure supplies means taking some precautions against unpleasant surprises. Providing supplies at lowest practicable cost means that the cost of precautions has to be weighed against the risks being run. And the reference to cost to the nation as a whole introduces environmental, social, and regional as well as financial costs.

So, the objectives are easy to write down but far from easy to apply in practice. Their achievement involves a series of trade-offs—for example, the tradeoff between the financial costs of keeping open uneconomic pits and the social and supply security costs of closing them; the advantage in terms of supply security of reducing demand for oil and the advantage of providing an additional market for coal weighed against the financial and industrial costs of doing so; the immediate benefits to the economy brought by maximum North Sea oil production against the benefits of maintaining net self-sufficiency over a longer period.

The balance of advantage in these tradeoffs will be liable to change at any time as a result of events which can be neither predicted nor controlled. Uncertainties that currently affect the United Kingdom include the availability of oil from the oil-producing countries, the price at which it will be traded internationally, the level and pattern of future energy consumption, and the speed and cost at which new energy technology will be developed. Policy must be sufficiently flexible and robust to cope with rapidly changing events. As a UK White Paper on Fuel Policy in 1965 said, it "must maintain all possible room for manoeuvre by refraining from making, earlier than is demonstrably necessary, major changes which it may be impracticable to reverse." (White Paper on Fuel Policy, October 1965, Cmd. 2798, par. 36.) It is with the problems of devising such a policy that the remainder of this paper is concerned. There are three main strands to this task: first, to make the best possible use of United Kingdom Continental Shelf (UKCS) hydrocarbon reserves and to prolong high levels of production of these fuels; second, to develop those sources of energy necessary to meet the United Kingdom's longer term requirements, most notably coal and nuclear power but also combined heat and power and the less assured sources of supply such as the renewables; and third, to insure that all fuels are used appropriately, efficiently, and cost effectively.

UK ENERGY SUPPLY AND DEMAND

Table 14–1 shows UK inland consumption of energy in 1950 at 228 million tons of coal equivalent (MTCE) and in 1980 at 328 MTCE. In 1950 coal provided 90 percent of primary energy. Nearly all the remainder came from imported oil. Thereafter the role of oil increased substantially. By 1960 it provided 25 percent of total inland energy consumption, while coal's share was reduced to 74 percent. The introduction of natural gas and nuclear power and continued growth in the role of oil further reduced coal's share during the 1960s. By 1972, oil accounted for 48 percent of inland energy consumption (its largest share), coal for 36 percent, and natural gas and nuclear for 12 percent and 3 percent respectively.

Last year coal provided 37 percent of primary energy; oil, 37 percent; gas, 21 percent; nuclear, 4 percent; oil and hydro, 1 percent. Energy imports amounted to some 70 million tons of oil equivalent (mtoe)—about 36 percent of inland energy consumption—of which 56 million tons were oil. But this was partly offset by energy exports of 55 mtoe, all except 2 mtoe of which were oil. Though the volume of oil imports exceeded that of oil exports, the value of the latter marginally exceeded the former because North Sea crude oil is of higher quality than imported crude.

Table 14–1. **United Kingdom Energy Consumption**[a]

	Tons of coal equivalent ($x\ 10^6$)				
	1950	1960	1970	1973	1980
Coal	204	199	157	133	121
Oil	23	68	150	164	121
Natural gas	--	--	18	44	70
Nuclear	--	1	10	10	13
Hydro	1	2	2	2	2
Total	228	269	337	353	328

[a]Primary fuel input basis, excluding nonenergy use and marine bunkers.

Forecasting the energy future is a hazardous business, but projections properly used are an essential instrument in formulating energy policy. Investment projects in the energy sector have very long lead times. Decisions can be taken sensibly only if the decisionmakers have some view of possible future outlooks. However, it will be a sure recipe for wrong decisions and a wasteful use of resources if the forecasts are treated as rigid blueprints or master plans for the future. The best projections are those which turn out to be least wrong. Projections need to be subject to continuing revision in the light of changes both in the energy scene and in the wider economy, and particular care has to be taken with the assumptions on which the projections are based. There will be many pressures on the forecasters to bend the assumptions to suit particular interests. Discussion will sometimes be fraught with dissension and even heated. Some concessions to the demands of special interests may be politically unavoidable, but the usefulness of the forecasts will depend on the extent to which those pressures are resisted.

The UK Department of Energy uses for its projections an econometric model of the energy sector of the UK economy. The most recent projections were published in 1979. A variant based on lower assumptions about economic growth was introduced in 1980. A new set of projections using improved modeling techniques and taking account of developments since 1979–80 is now being prepared.

Table 14–2 summarizes the 1979 projections and the 1980 variant. A further variant has been introduced for a lower nuclear contribution than in 1979. The table shows that the United Kingdom is likely again to be a net importer of energy in 2000, although in the low-growth case only on a relatively small scale. Oil and gas will still make a major contribution to supply and demand but in the case of oil a somewhat smaller one than now. The gap will be made good mainly by coal and nuclear power. More efficient use of energy is expected to have held demand to about 20 percent below what it would otherwise have been. It would be imprudent to forecast in any detail the outcome of the work now in hand on new projections, but I will hazard one guess: that a combination of (1) low economic growth and (2) improved energy use efficiency resulting from higher prices and other measures may reduce still further our total demand for energy, so that we can still be close to energy self-sufficiency in the year 2000.

Table 14–2. **Energy Projections to 2000**

	Tons of coal equivalent (x 10⁶)	
	1980ᵃ	2000ᵇ
Total UK demandᶜ	344	445-510
UK coal demand	121	128-165
Total UK indigenous supply	337	390-410
Coal	130	137-155
Gas	55	62-65
Oil	137	100
Nuclear and hydroᵈ	15	88-95
Net fuel imports	20	35-120

ᵃThis column represents the last year for which actual data are available.
ᵇThis column represents UK energy supply and demand projections to the year 2000 that were published in 1979. The projected demand estimates assumed that, due to conservation, there would be a 20% reduction in demand for useful energy by 2000. Furthermore, in the instance of demand, the lower and upper units of the projections were based on assumed yearly economic growth rates of 2.0% and 2.7% respectively. (In 1980, these demand projections were used as the basis for a lower growth sensitivity variant—not depicted here—that assumed a yearly economic growth rate of about 1% to 2000, resulting in total UK demand of around 400 Mtce, with coal contributing some 115 Mtce and nuclear and hydro together contributing 65 Mtce.) In the instance of supply and net fuel imports, projections were based on a variety of factors and not necessarily on the aforementioned economic growth cases.
ᶜIncluding nonenergy uses and marine bunkers.
ᵈIf the nuclear contribution in 2000 were not to exceed 22 GW (the total which would follow from the electricity industry's plan to order a further 15 GW of nuclear stations over the ten years beginning in the early 1980s), the nuclear and hydro figure would be reduced to some 55 Mtce, with most of the balance resulting from this reduction probably falling on the demand for coal. For instance, in the case of the 1% economic growth rate variant mentioned above, coal demand would be likely to increase from 115 Mtce to 125 Mtce.

THE FRAMEWORK OF ENERGY STRATEGY— INSTRUMENTS, INSTITUTIONS, AND CONSTRAINTS

The United Kingdom's energy economy, like its economy as a whole, is a mixed one, and it operates in an international energy economy which is also mixed. The level of energy consumption, the way in which it is divided among the different forms of energy, and the efficiency with which energy is used depend in the main on the disaggregated decisions of millions of consumers—companies, public authorities, and private individuals. Refining and distribution of solid fuels is partly in their hands. The private sector predominates in oil production, and oil is traded in the international market—a market, however, where price is determined partly by supply and demand but also partly by the decisions of the governments of a few producing nations, particularly Saudi Arabia.

But the public sector in the United Kingdom's energy economy is also large. Electricity production (with some exceptions for private generation) and distribution, gas distribution, and virtually all coal production and some coal distribution are publicly owned. So is the UK Atomic Energy Authority. The British National Oil Corporation (BNOC) is the largest trader in UKCS crude oil.* (A fuller statement on the organization of the UK energy industries is in the Addendum at the conclusion of this paper.) And perhaps more important, decisions on such questions as the scale of the nuclear contribution to electricity supply, the pattern of development and rate of depletion of North Sea oil and gas, and the size of the coal industry raise such important wider issues—macroeconomic, environmental, social, regional—that the UK government is inevitably and deeply involved in them. That would be the case even if the energy industries were predominantly in the private sector.

In this complex situation, successive British governments have sought a middle course between two extremes of energy policy. One extreme would be an attempt at central direction based on central planning. This might in theory permit optimum allocation of resources. But it presupposes that those in the center—in other words, the government and its officials—are all wise

*The government has announced its intention to introduce legislation in the 1981–82 parliamentary session to (1) deprive the British Gas Corporation of its monopoly rights to supply gas to industrial consumers and to purchase gas produced on the UKCS and (2) enable the introduction of private capital into BNOC's oil-producing business. The oil-trading activities of BNOC will remain as a wholly government-owned operation.

and all seeing. Even the less extreme variant of setting indicative but firm targets for the contribution of the different forms of energy would, experience suggests, be likely, at least under British conditions, to lead to rigidities and the wasteful use of resources. At the other extreme, it might in theory be possible to treat the publicly owned industries as if they were in the private sector and leave everything to the market. However, this would ignore the fact that the interests of the various industries impinge closely upon one another. The electricity industry, for example, is the main customer of the coal industry, and it takes some 70 percent of the latter's total sales. It is the sole outlet for nuclear power. It competes for its sales with the gas industries, the oil industries, and to some extent the coal industries. The price relativities between gas and electricity largely determine their respective shares of the market, and that in turn reflects back on coal demand. Policies which may be good for one industry may not necessarily be in the overall interests of the nation or make for a rational use of energy.

The course adopted by the British governments since the war has been to allow the market to operate but to set a framework that influences the way in which the market operates so as to achieve wider national objectives. The key to this approach is pricing policy, but the government also influences the market through its direct influence on the nationalized industries, through its control over North Sea developments, and through the use of taxes and incentives.

Pricing Policy

The basic approach to pricing policy was set out clearly as long ago as 1952 in the report of the Committee on National Policy for the Use of Fuel and Power Resources:

> The best pattern of fuel and power use will be promoted not by the direct intervention of the Government but the exercise of the consumer's free choice of his fuel and power services—provided that competition between the fuel industries is based on prices, tariffs and terms of supply which closely correspond to the relevant costs of supply, and that the consumer is enabled to make an informed choice. (Cmd. 8647, par. 232)

Economic pricing is recognized by the Summit nations, by the International Energy Agency, and by the European Community as a key factor in solving longer term energy problems. Prices need to give both the producer and the consumer reasonably accurate signals about the cost of energy supplies. This will insure both that consumers give due weight to the efficient use of energy compared with other demands on resources and that investment in energy production is deployed most effectively.

What constitutes economic pricing? In the United Kingdom it is generally taken to mean that energy prices should reflect the continuing cost of supply of each fuel and take account of market forces both within the country and internationally. In practice, British overnments have consistently allowed oil prices to be determined by world market prices and have set their faces against any system of dual pricing under which UKCS oil would be provided to the UK market at a price lower than the international market price. As world trade in other fuels increases, so their international price levels will have a growing influence on UK prices for those fuels. United Kingdom gas prices to industrial consumers are broadly related to the competing oil product in order to balance demand and supply: in the case of firm gas, to the price of gas oil; and for interruptible supplies, to the price of heavy fuel oil. The gas industry, as a result of a decision in 1980, is raising the price of domestic gas over and above the rate of inflation by 10 percent per year for three years, but this price is still significantly below the economic level. United Kingdom coal prices, which are based on average costs, are broadly competitive with imports except at some coastal sites close to import facilities. Electricity is not internationally traded to any significant extent, nor is it readily replaced by an alternative fuel for many users. So, long-run marginal costs of production provide the basis for determining prices.

Progress thus has been made in getting UK energy prices onto an economic basis. However, this has created a new problem of industrial competitiveness. While the large majority of industrial energy consumers in the UK are paying prices comparable to those paid by their competitors, large and highly intensive consumers of gas and electricity are paying more than their competitors not only in North America but also on the continent of Europe (see Table 14–3). This is due partly to genuine differences in costs of supply —for example, electricity costs are lower in France because of the scale of that country's nuclear program and hydropower resources —and also to the strengthening of sterling against major European currencies in the second half of 1980. But this price difference also seems in some cases to be due to the fact that others are not following economic pricing principles to the same extent as the United Kingdom. Unless there can be some international understanding about the principles of energy pricing and arrangements

Table 14–3. Fuel Prices in Selected Countries

	Gas (p/therm)[a]		Electricity (p/kwh)[b] 25 MW; load factor of		Coal (£/ton)[c] high volatile (pithead price)
	Firm	Interruptible	58%	82%	
United Kingdom	22.4-28.3	19-24.5	2.7	2.5	41.1
Belgium	19-22	19	2.5	2.2	46.9
France	18-23	18	1.8	1.6	41.7
West Germany	18-31	17	2.3	1.9	53.2
Italy	20		2.4	2.2	
Netherlands	19-21	19	2.8	2.6	
United States	12.4		1.8		15.3

[a]Gas prices for European countries are November 1980 British Gas Corporation estimates from the *National Economic Development Council Energy Tast Force Report* NEDC (81)15. The US gas price is from the American Gas Association, November 1980, and represents the average 1980 industrial price, using a mid-1980 exchange rate.

[b]Electricity prices for Eurpoean countries are provisional January 1981 estimates by the Electricity Council from NEDC (81)15. The US electricity price is from the US *Department of Energy Monthly Energy Review*, July 1981, and represents the average US industrial electricity price for March 1981 at the exchange rate then current.

[c]Pithead prices for European countries are October 1, 1980, estimates from the EEC Bulletin "Carnet de Prix." The US pithead price is from *Coal Week* and is directly comparable.

to see that those principles are put into effect, countries increasingly will be driven to underprice energy in order to protect their own industries. The results will be inefficiencies in the use of energy and delays in the development of new forms of energy, thus prolonging the dependence of the western world on oil.

Other Instruments of Energy Policy

The other instruments available to the British government to influence the framework within which energy policy operates are numerous and varied. Perhaps the most important are the following:

- The corporate plans of the nationalized energy industries are discussed with the government, and their investment programs (totaling £2.5 billion a year) are settled by the boards with the approval of the Secretary of State for Energy. This enables the government to make sure that the total energy investment is not excessive in relation to the likely future level of demand but at the same time is adequate to meet that likely level, perhaps with a little to spare. Examination of the corporate plans also enables the government to satisfy itself that a reasonable return is expected—achievement is another thing—on the investment of the nationalized industries and to see that the strategies of the industries fit into wider government strategies.
- Financial targets for the nationalized energy industries are agreed between the government and the industries, and financing limits are set by the government. This is an important means of influencing the pricing policies of the industries. The aim is to set targets and limits at a level that can be achieved by economic pricing combined with rigorous efficiency and cost saving.
- The pattern of primary fuel use—both short and long term—in the electricity supply industry can be influenced by the government. The choice of fuel for the new power stations can be determined by the exercise of the government's statutory authorization powers in relation to proposals put forward by the boards. More immediately, the government can influence the choice of fuel used to meet current demand within the flexibility of the existing system; for example, by altering price relativities through taxation policies or in appropriate cases by more direct intervention in the merit order under which power stations are operated. This reflects the fact that the electricity industry's use of fuel is decisive for the future of both the coal and the nuclear industries.

- The government has powers to control development and production in UKCS oil and gas fields. This enables it, if it wishes, to influence rates of depletion in the wider national interest away from those which the interests of the producing companies would dictate.
- The government can influence the disposal of UKCS oil through BNOC, which is entitled in most cases to 51 percent of the oil produced in the North Sea. This is of particular importance for supply security.
- The level of taxation on fuels is decided by the government. Considerations other than those of energy policy are important to decisions here—particularly the need to raise revenues. But taxation can promote the efficient use of energy, as with the tax on gasoline now standing at 84p per imperial gallon, and it can influence the pattern of use between fuels, as with the tax of £8 per ton on heavy fuel oil which encourages the use of coal in place of oil.

- The government has certain powers to make grants or loans to promote specific developments that are in the longer term national interest but would not be undertaken for commercial reasons alone. A current example is the new scheme announced in the 1981 budget—a scheme under which £50 million will be available over the next four years to pay grants of 25 percent towards the costs incurred by firms replacing oil-fired

Table 14–4. **Breakdown of Estimated UK RD&D Support to the Energy Sector, 1981–82**

	1981–82 estimated expenditures[a] (£ x 10^6)
Offshore	20.5
Domestic (renewable)	15.5
IEA	4.0
Conservation R&D	0.2
ETSU[b]	3.5
MATSU[b]	1.3
Subtotal (UK Dept. of Energy's R&D)	45.0
Conservation demonstration	3.0
UKAEA net R&D	219.0
Grand total	267

[a]Estimated expenditures are at outturn prices (i.e., the prices in effect at the time of expenditure).
[b]Respectively, the Energy Technology Support Unit and the Marine Technology Support Unit at the Atomic Energy Research Establishment at Harwell.

boilers with coal-fired ones. And, like all major governments, the United Kingdom gives substantial financial support to research, development and demonstration in the energy sector. In 1981–82, this support is expected to total £267 million, of which £219 million will be on nuclear research and development (see Table 14–4).

- The government with the approval of Parliament can legislate, for example, to set standards for energy conservation and efficiency.

Some Constraints on Energy Strategy

Energy strategy in a democratic society—and indeed in any society—is subject to constraints arising from the other concerns and interests of that society. Like the instruments of policy, these constraints are many and varied. The most important is the need for the government to carry with them large numbers of individuals and powerful organizations. Energy conservation policies can be carried through only if they command the confidence of many millions of consumers. The major oil companies and major consumers have their own interests and policies which do not always coincide with those of the government. But the public corporations in the energy sector also have their own interests and policies, and they often have the backing of powerful trade unions in urging these interests and policies on the government. There is a natural concern about nuclear power, but not on a scale which has seriously inhibited nuclear development in the United Kingdom. In the area of pricing policy governments for political reasons have from time to time found it impossible to accept the level of energy pricing that economic principles would require. A particularly clear example of this occurred between 1971 and 1974, when prices in all the nationalized industries were held down in the interests of a prices and incomes policy. Recovery of prices to economic levels is still not complete in all cases. Government policy cannot disregard constraints like these; policy has to be developed in a way which as far as possible reconciles these constraints and commands as wide a measure of support as possible. The process of reconciliation will be far from easy.

Important specific constraints are as follows:

- **Financial.** The borrowings of that part of the UK energy sector which is publicly owned count against public expenditure and the public sector borrowing requirement. So does direct government support for the coal industry, for research and development, and for specific objectives. When public expenditure and the public sector borrowing requirement are strictly limited, the energy sector has to compete with other claimants for limited resources, and projects which stand up on their merits may be squeezed out.

• **Environmental.** Energy development has an effect on the environment—not always harmful but often controversial. In the United Kingdom any clash between energy and environmental interests is normally resolved on a case-by-case basis within the framework of well-established planning legislation and policy and if necessary after a public inquiry. There are opportunities for planned energy developments to be included, at different stages of definition, in regional planning strategies and the development plans of local authorities. In practice, as a result of close cooperation between the energy industries and those concerned with the environment (particularly the local authorities), a satisfactory reconciliation has generally been achieved that has allowed necessary energy development in a way which has minimized any damage to the environment. A different type of environmental problem can arise when developments on the UKCS are close to the shore or other interests such as shipping or fishing. However, it has generally been possible to reconcile these differences by consultation. For example, local fishermen in the Moray Firth (off the northeast coast of Scotland) had fears about the impact of oil activity on their livelihoods. Those fears were allayed by the drawing up by the oil and local fishing industries of a code of practice designed to minimize the interference of oil operations with fishing activities.

• **International.** Our energy policy in the United Kingdom has to be developed in ways consistent with our international obligations. Even if it were otherwise desirable, our commitments under the Treaty of Rome and the General Agreement on Tariffs and Trade would not allow us to prevent the export of oil and, indeed, other fuels produced in the United Kingdom. This has important implications for the pricing of oil, and through that, for the United Kingdom's whole approach to energy policy. The United Kingdom is committed in the European Community, the International Energy Agency, and by agreement among the Summit countries to broad guidelines of policy designed to reduce dependence on oil through the development of alternative fuels and through energy conservation. We are also committed to participate in the IEA sharing scheme in the event of a sharp reduction in oil supplies.

THE MAIN LINES OF UK ENERGY STRATEGY

This paper has so far concentrated on the framework of energy policy and the factors which influence it. It is now time to see how these considerations are currently resolved in developing the main lines of energy strategy enumerated above (UK Energy Strategy —Objectives, par. 3), on the basis of the pricing policies that were also described above (Pricing Policy, pars. 1–3).

Exploitation of North Sea Hydrocarbon Resources

United Kingdom oil production in 1980 was just over 80 million tons, and we nearly achieved net self-sufficiency. Production in 1981 is expected to be up to 95 million tons with net exports of up to 15 million tons. Production is expected to rise to a peak in the mid 1980s and thereafter decline gradually until some time in the 1990s when we become net importers again. Further exploration is being encouraged. The Seventh Round of offshore production licensing, launched in August 1980, attracted considerable interest from the international oil industry, and seventy-nine licenses have been awarded to date. Territory on offer lay in various parts of the UKCS—some in well-developed acreage, particularly the mature oil province of the northern North Sea; some in less developed or new areas, including deep waters north of Shetlands where exploration techniques will be required at the frontiers of oil technology.

A limited search for petroleum has also been taking place onshore in Great Britain for the past sixty years, and a number of small fields, mainly in the Midlands (total 1980 production— 525,000 barrels) have been producing for upwards of twenty-five years. The success rate on landward operations has been lower than offshore, and individual discoveries have been much smaller. Nevertheless, with the increases in oil prices and the relatively low cost of landward operations, there has been an upsurge in interest onshore over the past two years. This has been stimulated further by discoveries in southern England where, for example, Wytch Farm (currently 5,000 bpd) is considered to be potentially as large as a small offshore field.

In 1980, consumption of gas in the United Kingdom was over 17 billion therms. Over 75 percent of the gas supplied came from the UKCS, and most of the balance was imported from the Norwegian sector. Consumption of gas in the United Kingdom is expected to peak sometime in the late 1980s or early 1990s at up to 30 percent above current levels and may thereafter start to decline. Gas from the UKCS and imports is likely to be sufficient for our needs until around the turn of the century. It may then be worthwhile to supplement natural gas supplies with substitute natural gas (SNG), probably manufactured from coal. The exact timing of this will depend on the evolution of demand, discoveries of new supplies, and import contracts over the next twenty years.

With natural gas, as with oil, the aim is to maintain adequate levels of supply for as long as possible. Exploration for further gas supplies continues in the Southern Basin of the North Sea and elsewhere. The government has decided against an integrated pipeline system financed or guaranteed by the government to collect gas from fields in the northern sector of the North Sea; they expect, however, that the bulk of this gas will in practice be collected as a result of decisions of the individual operating companies.

The rapid development of natural gas and oil production during the last two decades has been of immense benefit to the United Kingdom economy. Table 14–5 shows the contribution to government revenues and GNP expected from North Sea oil and gas. As the table shows, during this financial year some 5.5 percent of general government revenue is expected to come from North Sea oil and gas. Some 4 percent of United Kingdom GNP is currently being generated from production of oil and gas.

Table 14–5. **Expected Contribution of North Sea Oil and Gas to the UK Economy**

Tax revenue	81-82	Fiscal years 82-83	83-84
At FY 79-80 prices (£ x 10⁶)	4 ½	4 ¾	5 ¼
As % of gen'l gov't rec'ts in that year	5.5	5.7	6.2

Direct contribution to GNP	Calendar years			
	1981	1982	1983	1984
At FY 79-80 prices (£ x 10⁶)	7.2	7.8	8.7	10.2
As % of GNP in that year	3 ¾	4	4 ½	5

Coal

As stated at the outset, UK coal reserves are very large — sufficient to last for at least three hundred years at present rates of consumption. Current production and consumption are both about 120 million tons per year, of which about 70 percent is consumed in power stations. Most UK coal is deep mined, with just over 10 percent coming from opencast sites. And the bulk of UK production is used at home, with imports and exports currently running at 4 million tons and 9 million tons respectively.

Coal is now regaining its competitive position against oil. With such large and secure UK supplies, coal can play a vital role in meeting longer term energy requirements. But if coal is to play this role, a financially sound and competitive coal industry is essential. That will require a massive investment program designed to transfer production from older, less efficient collieries mainly in the Northeast, South Wales, and Scotland, to modern efficient pits mainly in Yorkshire and the Midlands. The scale of this investment effort is shown by the tabulation that follows:

National Coal Board Capital Investment (in millions of September 1981 £):

1976–77:	510
1977–78:	560
1978–79:	690
1979–80:	820
1980–81:	820
1981–82:	~800

This reconstruction of the UK coal industry is subject to important constraints. The older uneconomic pits are often in isolated communities that will wither and die if their pits are closed. Substantial sums (£70 million in 1981–82) are being spent to ease the social problems to which pit closures give rise. For example, miners receive substantial redundancy benefits and fairly generous assistance with the cost of moving to new jobs at other collieries. But even with help on this scale, the problems are substantial. There is a natural resistance to pit closures among those concerned. So, as a matter of deliberate policy, many pits are being kept open longer than economic factors alone would dictate. At the same time, some of the new collieries will be in areas where coal mining has not hitherto taken place. There will be opposition to such development on amenity and social grounds, and this opposition will have a chance to express its views through full-scale public inquiries before decisions are taken. The results of such inquiries are by no means a foregone conclusion.

Nuclear Power

Like coal, nuclear power is a vital element in the United Kingdom's long-term energy strategy. In 1980 about 12.5 percent of the United Kingdom's public sector electricity generation came from the eleven nuclear power stations operated by the electricity utilities. This proportion may increase to about 20 percent when the three stations currently nearing completion are fully operating. Construction of two more stations has recently started, and it is planned to complete these before 1990. In December 1979 the Secretary of State for Energy accepted that the electricity supply industry's foreseen need to order some 15 GW of nuclear power capacity over the ten-year period beginning in the early 1980s represented a reasonable prospect for planning purposes. The timing of each order will depend on how demand evolves over the period. The stations will be an important step towards providing cheaper electricity in the United Kingdom, so long as they can be built to cost and on time. The Central Electricity Generating Board (CEGB) regard nuclear plants as the economic choice, and benefits will start to accrue to the consumer as more nuclear plants are commissioned.

The stations so far built or under construction use either Magnox or advanced gas-cooled reactors. The CEGB plan that their next station should use a pressurized water reactor. Thereafter the choice of reactor type for new stations will be taken as the need arises, but the orders envisaged by the CEGB will consist entirely of thermal reactors. No decision has yet been taken by the government on the development of a commercial fast reactor.

The development of nuclear power, like the development of the coal industry, is subject to constraints—mainly the natural concern of public opinion about safety, proliferation, and waste disposal. In the United Kingdom, successive governments have sought to meet this concern by allowing the fullest public debate and inquiry before decisions are taken. This has not eliminated controversy, but it has served to keep the controversy within the democratic process.

Energy Conservation

The starting point for the British government's energy conservation policies is economic pricing (see Pricing Policy, pars. 1–3 above). But experience suggests that the price message needs to be reinforced. This is done by:

- A wide range of publicity, information, and advisory services to influence domestic and industrial consumers. An example is a scheme under which, since 1977, 45 thousand one-day consultancy surveys have been carried out in industry and commerce at the expense of the government. These surveys have identified potential savings, often obtainable with little or no investment, of over £10 per annum for each £1 of government money spent. Over 50 percent of the consultants' recommendations have been implemented.
- Government financial support for research, development, and demonstration in conservation. This is expected to total almost £10 million in 1981–82 (see Table 14–6). Three million pounds of this will be spent on the Energy Conservation Demonstration Projects Scheme aimed at promoting the introduction of new applications of established conservation techniques, processes, and products, as well as the more rapid uptake of novel or improved techniques. Over eighty projects have now been approved or are under way.

Table 14–6. UK **Departmental Expenditures on** RD&D **for Energy Conservation,** FY **81–82**

	$£ x 10^6$ at 1980 survey prices
Research and development	
Depts. of Industry and Energy	4.0
Dept. of Transportation	0.4
Dept. of Environment	1.96
Demonstration projects	
Dept. of Energy	3.0
Department of Industry	0.48
Total	9.84

- Giving a lead in the public sector. The Property Services Agency (PSA), responsible for central government buildings in the United Kingdom has reduced energy consumption in its buildings by 36 percent since 1972, when it started a conservation program involving detailed surveys and target setting. Its approach has involved identifying and implementing a range of measures, both in plant management and capital improvements, which are of very general applicability, highly cost effective in terms of capital invested, and economical in terms of staff effort. Significant energy savings have also been achieved in other parts of the public sector, notably education and health service buildings, and local authority housing.

- Setting of energy standards and targets by statute or by agreement. Regulation makes a subsidiary but important contribution to energy conservation. Building regulations set high energy-efficiency standards for new buildings. It has recently been announced that these standards will be uprated for domestic buildings, to take account of the increased thickness of roof and wall insulation which have become cost effective as energy prices have risen. Roofs will have to be insulated to achieve a u value of 0.35, which corresponds to 100 mm of loft insulation, and walls a u value of 0.6. Also, in 1980 the government reduced to 19°C the statutory maximum temperature for the heating of nondomestic buildings, a measure which was estimated to save 1 million tons of coal equivalent each year. The Energy Conservation Act, which passed through Parliament this year, gives the government powers to set mandatory energy efficiency and safety standards for all types of space and water heaters, including commercial, domestic, and industrial central heating boilers. It will be implemented by a series of Orders over the next few years laying down the precise standards to be applied. In other circumstances, development of voluntary targets may be more appropriate. For example, the government has agreed with the UK motor manufacturers' association upon a target of 10 percent improvement in the fuel efficiency of new cars by 1985 from a 1978 baseline. The National Economic Development Council is encouraging the use of target-setting and monitoring techniques in a range of industries.
- Encouragement of local community interest in the more efficient use of energy. The Department of Energy and the Manpower Services Commission have recently given financial assistance to the National Council of Voluntary Organisations to launch an initiative to increase the number of such community projects which, sponsored by voluntary organizations, are designed to insulate the homes of the elderly and disabled, using teams of the younger unemployed.

At a time of financial stringency, organizations in both the public and private sectors tend to concentrate limited resources on their own central objectives. Many energy conservation projects have short pay-back periods and are in the best interests of the consumer. But these projects are often not central to the objectives of the organization concerned, and it may not always be possible for an organization to recoup the benefits of investment in conservation—for example, a local authority that invests in improved insulation of the public housing estate may not easily be able to recover that investment through increased rents. So, government subsidies or other incentives, although not in general justified, may have a role in helping to overcome those imperfections in the market. One example is the boiler conversation scheme referred to above (see Other Instruments of Energy Policy). Another is the government's Home Insulation Scheme, which provides assistance to houseowners and tenants to insulate their attics and to lag pipes and hot water tanks. Grants of 65 percent of the cost—90 percent in the case of old age pensioners—are available up to a total of £65 and £90 respectively.

Longer Term Prospects

Oil, gas, coal, nuclear power and conservation will provide the main lines of UK energy policy for the rest of this century. But as oil and gas production run down, it will be necessary to produce new and alternative powers of energy. Those currently under study or development include:

- The manufacture of substitute natural gas and synthetic oil from coal. The British Gas Corporation has a substantial development program for SNG which is now moving to a commercial-scale demonstration of the British Gas/Lurgi Slagging Gasifier. The work is ahead of UK needs but gained an early impetus from the keen US interest and support for the process. Coal liquefaction is less advanced, and discussions are now taking place between the government, the National Coal Board, and private companies on the construction of a pilot plant of 25 ton per day capacity for further development of oil from coal technology. There must be a doubt whether it will ever be economic to produce oil from UK coal. The pace of this program may therefore depend on the prospects for exploiting the technology in countries where coal can be produced cheaply.
- Combined heat and power (CHP). CHP systems have the potential significantly to reduce our total energy requirements provided that the demands for heat and electricity placed upon them can be broadly matched, and they may be an important source of heat for the domestic sector when supplies of natural

gas decline. The government has announced a program to test the feasibility of combined heat and power and district heating in UK conditions. Following the preliminary stage, nine cities have been selected for the second stage. This will comprise a detailed examination of the potential and feasibility for CHP/district heating in these nine areas. Industrial CHP is already well established in this country. The government believes that the main impetus for its further development will come from industry, but the development of some worthwhile schemes is being encouraged through part funding.

- Prospects for the renewable sources of energy (solar, wind, wave, geothermal, and tidal). These prospects are far less certain than for conventional fuels. In many cases technical problems still have to be overcome, and capital costs will have to be reduced substantially for the renewables to compete with conventional fuels in most areas in the United Kingdom. The government is financing a research program at a cost of over £15 million in the current financial year (see Table 14–7). Research is now approaching the stage at which decisions will be needed on whether to progress to much more expensive schemes such as large-scale demonstration projects, or whether to cease work on some renewable sources which are unlikely to be exploitable commercially. A large aerogenerator is being built on Orkney to generate electricity in an area where wind generation is considered to be approaching economic viability in comparison with the cost of laying power cables from the mainland. The UK wave energy program and the report of a committee on the possibility of a tidal barrage in the Severn Estuary are under review.

Table 14–7. UK **Department of Energy Expenditures on Renewable Sources of Energy**

| | *Actual or estimated expenditures,*[a] *by fiscal year (£ x 10^3)* | | | | | | |
	78-79	79-80	80-81	81-82	82-83	83-84	84-85
Wind	145	503	722				
Wave	2,240	3,496	3,321				
Geothermal	205	1,507	2,352				
Solar	167	1,457	1,281				
Tidal	108	722	1,408				
Total	2,865	7,685	9,134	15,622	13,638	14,490	15,214

[a]Expenditures are at March 1981 prices except for those estimated for fiscal years 81-82 through 84-85, where the estimates are outturn prices (i.e., the prices in effect at the time of expenditure). These estimates cannot be disaggregated until decisions are made on individual projects.

SOME QUESTIONS ON UK ENERGY POLICY

A description of what is being done may give the impression that all is going smoothly. So it is proper to add a final section to this paper which tries to expose *some* of the questions which will face those dealing with UK energy policy in the coming years. Of course, in the event of unforeseeable circumstances, the questions with which those concerned have to deal may prove entirely different!

The Balance between Investment in Supply and Investment in Conservation

Investment in energy supply is currently running at about £6 billion a year. Although it is difficult to produce accurate estimates of total investment in improved energy efficiency, we know that in the domestic sector it amounts to around £450 million per year, and we can safely assume that total investment will therefore be considerably lower than investment in supply. On the other hand, government surveys have established a large potential for cost-effective investment in conservation measures. For example, the UK Department of Industry has estimated that in industry, savings of about 6 percent are available from measures that have pay-back periods of one year or less (for an expenditure of about £400 million), and a further 13 percent savings is available from measures with pay-back periods of between one and four years (costing £2,000 million). Examples of these projects are given in Table 14–8. It has been estimated that, in the domestic sector at the end of 1980, there were about 5 million dwellings without any attic insulation, some 10 million without cavity wall insulation, and about 2.5 million without hot water tank insulation. In most cases these residential insulation measures could be expected to be generally cost effective for the individual consumer (see Table 14–9). In addition, the net present values, suitably discounted, of the conservation measures in Table 14–9 and the low cost of certain measures in Table 14–8 suggest that some investment in energy conservation is more cost effective from a national point of view than marginal investment in energy supply.

But it is by no means simple to switch from investment in supply to investment in conservation by administrative fiat. Investment in supply is handled in discrete blocks by large organizations. The government can, if it chooses, know fairly precisely what is going on with supply and can influence it. Equally, those energy supply organizations—both management and unions—can

Table 14–8. **Examples of Energy Conservation Projects in Industry**

	Capital cost of projects (£ x 10⁶)	Potential oil saving (MTCE)	Value of saving (£ x 10⁶)	Payback period (years)
Waste heat recovery	450	2.5	250	1.8
Insulation of plant & services	35	0.3	35	1
Replacement and enhancement of furnaces, ovens, and kilns	100	0.37	37	2.7
Dryers	27	0.06	6	4.5

Table 14–9. **Costs and Savings from Insulating and Draftproofing Untreated Dwellings**[a]

Measure	Number of untreated dwellings	Total cost (£ x 10⁶)	Savings per year (GJ x 10⁶)	Savings per year (£ x 10⁶)	NPV[b] at 5% (£ x 10⁶)	Payback period (years)
Attic insulation (100 mm)	5[c]	500	56	160	2,500	3
Cavity wall insulation	10[c]	2,500	100	290	3,000	8
Hot water tank insulation	2.5[c]	60	33	125	1,250	0.5
Draftproofing	14	700	52	150	850	4.5
Total		3,760	241	725	7,600	7

[a]Assuming no change in internal temperatures.
[b]Net present value.
[c]Figures from Audits of Great Britain (Home Audit), indicating status as of the end of 1979.

be expected to protect their investment program with all the pressures at their disposal. On the other hand, investment in conservation is highly disaggregated. The government has no complete picture of what is happening, and its ability to influence circumstances in favor of conservation is limited. Moreover, with investment in supply, those concerned can be more or less sure that extra production capacity will in due course result. With investment in conservation, however, the result may be higher standards of comfort in the home or a more careless use of energy in the home or in industry. These are real problems. But the issue is so important that it deserves and is getting much attention.

The Organization of Work on Energy Conservation

In the United Kingdom the task of promoting energy conservation is decentralized. Each government department is responsible for encouraging conservation within its area of responsibility and for allocating funds for this work. This has the advantage that the question of conservation can be addressed within the standard framework of administrative and financial relations between government and outside organizations. But as explained above (Energy Conservation, par. 2), it may also mean that in times of stringency energy conservation receives less attention than the more immediate concerns of the particular department. It has been suggested that one way of overcoming this might be to set up a centralized coordinating organization charged with the promotion of energy conservation. Such a body might also provide a more effective counter than now exists to the powerful interests of the energy supply industries. Such considerations explain why the question of an energy conservation agency is coming increasingly on the agenda of debate in the United Kingdom.

Rate of Depletion of North Sea Oil

Figure 14–1 shows recent estimates, submitted by the Department of Energy to the House of Commons Select Committee on Energy, of North Sea Oil production and of UK oil consumption until 1990. In the middle of the present decade there is expected to be a period when production exceeds UK consumption by a total of 200 to 300 million tons. Should the government take steps —effectively the only means open to them is to require the producing companies to cut production—in order to reduce this surplus? This is an issue on which the conflict of relevant considerations is particularly sharp. The need for maximum government revenue at a time of economic difficulty points to maximum production from the North Sea. On the other hand, UK supply security will be enhanced by maintaining net self-sufficiency in oil for as long as possible. The answer to the question of whether the United Kingdom's oil is likely to be of greater value if it is (1) sold quickly or (2) left in the ground in the expectation that the price will rise, depends on what are little more than guesses about likely trends in

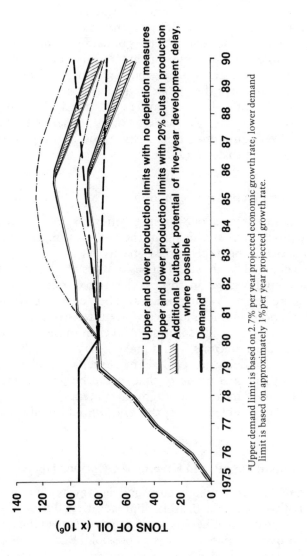

Figure 14–1. Possible Oil Production Ranges, 1980–90, with Demand Projections

[a]Upper demand limit is based on 2.7% per year projected economic growth rate; lower demand limit is based on approximately 1% per year projected growth rate.

the world crude oil price for the rest of the century, or even bigger guesses about future movements in the exchange rate of the £, and on the discount rate used in comparing future with present benefits. Not surprisingly, the outcome of these guesses is inconclusive! What is more, these conflicting factors cannot be assessed in academic isolation. The government is under heavy pressure from the oil companies—whose interests are not necessarily the same as those of the nation and who have powerful means of pressure at their disposal—to leave the choice of depletion rates up to them. But the government will soon have to reach a decision on this question, for otherwise, the passage of time will remove the possibility of reducing production. And to do nothing is just as much a decision as to do something.

The Cost of UK Electricity Production

It was pointed out above (Pricing Policy, par. 4) that costs of electricity production are higher in the United Kingdom than in countries with cheaper fuel sources—partly because of the natural advantages enjoyed by those countries, but also because much of the primary energy used in the UK electricity industry is coal, which is more expensive than nuclear. High-cost electricity is a potentially serious problem for electricity-intensive sectors of UK industry. There is no quick solution. And even in the long run, to build more nuclear stations would make heavy demands on limited resources available for capital investment, could require a contraction of the coal industry that would be unacceptable in social and regional terms, and might promote a degree of public controversy about nuclear power that the United Kingdom has so far avoided.

Diversification of Supply and Keeping the Options Open

The uncertainties which surround energy policy make it desirable both to adopt a flexible approach that does not close options too soon and to diversify supply so as to spread the risks of interruption over as many sources as possible. But such policies have their costs. Should we, for example, order nuclear power stations earlier than may be strictly necessary in order to keep the nuclear plant industry in being and the nuclear option open? Should we proceed with the Severn Barrage in the interests of diversification of supply, when the analysis to date suggests that it will be a more costly source of electricity than nuclear power? Should we allow for some surplus supply capacity against the risks of interruption? And if so, at what level and at what cost?

CONCLUSIONS

This paper has not described an energy strategy based on rigid plans or even on indicative targets. Nor has it suggested that everything can be left to the market. It has sought to describe an energy strategy which is essentially market based but which recognizes that in the UK energy sector, there are many imperfections in the market, and that there are major issues of wider national interest which cannot be overlooked and which require government intervention in the working of the market. This approach to energy strategy may lack the apparent attractions of the more simple and extreme approaches. Its implementation amid many uncertainties and the pressure of conflicting interests is fraught with difficulty. But despite all the difficulties, it is a strategy that is well suited both to a mixed economy and to the uncertain and unpredictable world in which policy has to be formulated.

Acknowledgments

The preparation of this paper has been much assisted by helpful and constructive comments and detailed contributions from many colleagues. I am grateful to them all. Without their help the paper could not have been completed in its present form. However, the opinions expressed and any errors of omission or commission are entirely my responsibility.

I am grateful also to Audits of Great Britain (Home Audit) for allowing reproduction of some of the material in Table 14–9.

ADDENDUM: THE ORGANIZATION OF UK ENERGY INDUSTRIES

Coal: The National Coal Board

In 1947, the coal mines passed into public ownership by means of the Coal Industry Nationalisation Act 1946, which set up the

National Coal Board (NCB) as a statutory corporation to manage the industry. Its members are appointed by the Secretary of State for Energy.

The NCB has, with limited exceptions, exclusive rights over the extraction of coal in Great Britain, but it is empowered to license private operators to work small mines and opencast sites. It also has powers to work other minerals, when discoveries of such are made in the course of searching for or working coal, and it can engage in certain petrochemical activities beneficial to the future of the coal industry. Retail sales remain largely in private hands, although the NCB makes bulk sales to large industrial consumers.

At the end of March 1981 there were 211 NCB collieries in operation grouped into twelve areas, each controlled by a director responsible to the NCB.

In 1980–81 the NCB's income was £4,211 million, including sales of coal totaling £3,608 million. Although the NCB made a trading profit of £69.5 million, it recorded a net loss of £52.8 million after taking account of interest payments and after receiving government grants totaling £253.9 million. The NCB's borrowing limit is £4,200 million.

Retail sales are largely in private hands; there are currently around six thousand coal merchants in Britain who deliver mainly into the domestic market. The NCB makes bulk sales to large industrial consumers and public sector users. Some 70 percent of coal sales in 1980 were to the electricity supply industry.

Gas: The British Gas Corporation

For many years gas was produced from coal, but during the 1960s, when growing supplies of oil were being imported, there was a switch to producing town gas from oil-based feedstocks. However, a more significant change began in the late 1960s following the first commercial natural gas discovery in the UKCS in 1965 and the start of offshore gas production in 1967. Supplies of offshore natural gas grew rapidly, and natural gas has now virtually replaced town gas.

The Gas Act 1948 brought the gas industry in Great Britain under public ownership and control in 1949. As a result of the change to natural gas necessitating more centralized control of production and transmission, the British Gas Corporation (BGC) was set up in 1973 under the Gas Act 1972 to replace the Gas Council and Area Gas Boards. Its members are appointed by the Secretary of State for Energy. The BGC's powers in connection with its main duty of developing and maintaining an efficient, coordinated, and economical system of gas supply and of satisfying reasonable demands for gas are to search for and extract natural gas and any oil discovered in the course of searching for gas; to manufacture or acquire, transmit, and distribute gas; to manufacture, supply, or sell gas by-products; and to manufacture, install, maintain, or remove gas plants and fittings. The BGC has about 106,000 employees. The government has announced its intention to introduce legislation in the 1981–82 parliamentary session to deprive the BGC of its monopoly rights to supply gas to industrial consumers and to purchase gas produced on the UKCS.

Natural gas is not available to Northern Ireland, and the industry there, which is controlled by nine municipal undertakings and four private sector companies, uses town gas produced from oil feedstocks. In 1970 the government decided that the large-scale expenditure required for a natural gas pipeline from Great Britain to Northern Ireland could not be justified and that it could not subsidize the industry's operations. Discussions have taken place between the Department of Energy in Dublin and the Northern Ireland Department of Commerce about the terms on which a natural gas supply might be made available to the public from the Republic of Ireland.

In 1980–81 the turnover of the BGC and its subsidiary companies amounted to £4,300 million, of which sales of gas accounted for £3,700 million. Post-tax profit declared for the year was £146 million. Recently, the BGC has been wholly self-financing. It has repaid all its long-term debt to the government and is depositing its cash surpluses with the National Loans Fund. Under the Gas Levy Act 1981, the BGC is required to pay a levy to the government in 1980–81 and subsequent years; the amount of the 1980–81 payment was £129 million.

Electricity

The first public supply of electricity in Great Britain was in 1881. In 1948 all municipal and private undertakings in Great Britain were vested in the British Electricity Authority and fourteen regional boards. The present structure of public corporations was established in 1957. The Electricity Council is the central coordinating body for the statutorily independent elements of the electricity supply industry in England and Wales. It has a general responsibility for promoting the development and maintenance of an efficient, coordinated, and economical system of electricity supply. Electricity is generated and transmitted nationally by the Central Electricity Generating Board (CEGB), which is responsible for the operation and maintenance of power stations and the main transmission system (national grid). Twelve area electricity boards are responsible for regional distribution and for the retail sale of electricity. The members of all these boards are appointed by the Secretary of State for Energy. The chairman and two designated members of the CEGB, together with the chairmen of the area electricity boards, are among the members of the board of the Electricity Council.

In Scotland, two boards—the North of Scotland Hydro-Electric Board (NSHEB) and the South of Scotland Electricity Board (SSEB)—generate, distribute, and sell electricity. Their members are appointed by the Secretary of State for Scotland.

In Northern Ireland these activities are carried out by the publicly owned Northern Ireland Electricity Service.

The electricity supply industry in England and Wales had a turnover in 1980–81 of £7,112 million. On a current-cost-accounting basis, its operating profit was £303 million, which after interest payments became a net loss of £272 million. The current borrowing limit of the industry is £6,500 million, and in 1980–81 outstanding borrowings reached £5,286 million. Gross borrowing in the year amounted to £571 million. The industry's capital requirement was £1,284 million, of which 93 percent was financed internally.

Nuclear Energy

Civil nuclear research and development is mainly in the hands of the UK Atomic Energy Authority (UKAEA), a public corporation set up in 1954 and financed mainly by the government. Its members are appointed by the Secretary of State for Energy. It advises the government on all matters relating to the civil application of nuclear energy. The electricity generating boards are responsible for operating nuclear power stations. The design and construction of nuclear stations is in the hands of the National Nuclear Corporation (NNC), which is a private sector company but with a minority government shareholding of 35 percent held through the UKAEA. Thirty percent of the shares are held by the General Electric Company Limited, and the remaining 35 percent by a consortium of manufacturing interests. An independent Nuclear Installations Inspectorate under the Health and Safety Executive is responsible for licensing the construction and operation of commercial nuclear installations in the United Kingdom.

The Oil Industry

Private companies. The UK petroleum product market is supplied by private sector oil companies including the major US and European companies, most of which operate refineries in the United Kingdom. The nineteen refineries have a nominal crude oil distillation capacity of about 130 million tons per year, which although underutilized provided net product exports from the United Kingdom of about 5 million tons in 1980.

Two of the world's major international oil companies, British Petroleum (BP) and Shell Transport and Trading, have their head-quarters in the United Kingdom. In terms of turnover they are the largest industrial companies in Britain and are the largest and third largest respectively in Europe. Petroleum sales by BP, in which the British government has a 40 percent shareholding, averaged over 3 million bpd in 1980, and its gross turnover was more than £25 billion. Of its 118,000 employees, nearly 40,000 are in the United Kingdom.

The major international oil companies are also engaged in exploration and production work on the UKCS, as are over fifty independent oil exploration and production companies. About half of the production from the UKCS in 1980 was refined in the United Kingdom.

British National Oil Corporation. The British National Oil Corporation (BNOC), a public corporation set up in 1976, is engaged in two main activities: it is an oil trader on a large scale, mainly by virtue of its right, through participation agreements with other oil companies, to purchase 51 percent of most of the oil produced on the UKCS; and it is a substantial enterprise engaged in North Sea oil exploration, development, and production. The government has announced its intention to proceed with legislation enabling the introduction of private capital into BNOC's oil-producing business. The oil trading activities will remain as a wholly government-owned operation. There is no present intention for BNOC to go into refining.

CHAPTER FIFTEEN

Summary

Session Chairman/Integrator:
Keichi Oshima
Professor Emeritus
University of Tokyo

This session had a good discussion and fruitful outcome, thanks to the excellent quality of its position paper and case study, as well as its experienced and articulate participants.

The first part of the work session's discussion centered on an identification of energy's present status and the issues surrounding it. The position paper described the reduction of dependence on imported oil since 1973 in the OECD or IEA countries as a group.

From 1973 to 1980, oil decreased from 53 percent of all primary energy to 50 percent, and imported oil decreased from 68 percent of all oil consumption to 62 percent. From 1973 to 1980, energy demand and oil demand increased more slowly than did economic growth; the ratio of total energy to GDP decreased by 12 percent, while the ratio of oil to GDP decreased by 20 percent.

Although the participants recognized the importance of the above changes, they agreed that the primary problem was the transition to expensive energy which began in 1973 and will continue for the rest of the century. Thus, the focus of the discussion was on balancing energy supply and demand in the next twenty years.

All participants agreed that the future of energy supply and demand is highly uncertain. Some participants argued that cooperative action by the IEA countries might prevent any increase in real

In addition to the chairman/integrator and those presenting papers, the work session included the following assigned participants: P. D. Boettcher, Umberto Colombo, M.S. Farrell, Måns Lönnroth, Margaret Maxey, and Macauley Whiting. The chairman/integrator was assisted by David Reister, who served as rapporteur.

oil prices; other participants argued that the real oil price would probably double by 2000. There is now an oversupply of oil on the world market; however, given the prevailing uncertainty about the future, it was emphasized that industrialized countries should continue their efforts to achieve their energy goals.

One participant presented a case for increased reliance on electricity and called for government intervention to build more power plants, while other participants cautioned about the dangers of overcapacity and argued that no new power plants may be required in the next twenty years. In any case, in order to cope with the expected uncertainty, future strategies have to be more flexible and incorporate broader considerations.

Ten years ago, most energy planners expected energy demand to grow exponentially, but such rapid growth stopped in 1973. Since 1973 and especially since the second oil crisis in 1979, the IEA countries have witnessed substantial energy conservation and fuel switching. The causes of these changes are a reduction in economic growth and an increase in energy prices. The participants noted that it is possible to recognize qualitatively that millions of decisionmakers in different places are responding to price changes causing energy conservation and fuel switching. However, the quantitative energy demand models are inadequate, and even if they were adequate, they would require forecasts of economic growth and energy prices—forecasts which are uncertain.

While the participants agreed that coal, nuclear power, and natural gas will play important roles, they felt that it would be a mistake to undertake a massive crash program to develop and switch to new energy supply sources without taking due consideration of the adverse effects on energy-related industries of too rapid a transition. The participants agree that investment capital allocations to supply and demand options should be done primarily by market forces, with limited government intervention.

The agenda for the Symposium contained a list of social constraints to be considered. The participants agreed that societies have multiple goals; these goals may conflict with each other and may inhibit the expansion of the energy system. General social goals include (1) full employment and maximum economic growth; (2) national security; (3) considerations of equity, justice, and income distributions; and (4) environmental quality. On the other hand, energy goals include considerations of (1) lowest possible cost; (2) efficient use; (3) the reduction of oil imports, and (4) adequate and secure supplies. Too much emphasis on any one goal may have negative impacts on the attainment of other goals. Balancing one goal against the other requires a political process.

The participants agreed that the allocation of energy resources is best accomplished by the market. The market balances supply and demand for various forms of energy; it determines prices and allocates investments among various options that supply or conserve energy.

The participants agreed that government should not attempt to allocate energy resources or control energy prices. The proper role for government is to correct market imperfections efficiently and quickly. Both the position paper and the case study make a case for appropriate governmental action. The arguments are briefly summarized here:

1. Government should offer leadership. Energy policy is determined by the collective actions of millions of households and corporations. Government should analyze options, encourage debate, and help the formation of a consensus on energy policy.
2. Government should provide information and should encourage the dissemination of practical advice on options available to conserve energy and switch from one form of energy to another. (The actual provision of information could be performed by government or by the private sector.)
3. Government should regulate where appropriate; for example, government should regulate natural monopolies such as electric and gas utilities. Regulations are needed to protect the environment, public health, and public safety. However, regulations should attempt to be as efficient as the marketplace by being flexible and taking account of their costs to the economy, and they should be operated in as streamlined a manner as possible to reduce the burdens they impose.
4. Government can intervene directly in the economy where appropriate by measures such as taxation, direct investment, and selection of sites for energy projects. Again, these measures should be taken with full consideration of avoiding distortion in market forces.
5. Government should participate in international agencies such as the International Energy Agency that encourage cooperation among major oil-importing countries.
6. Government should support long-run R&D on technology that produces or conserves energy.

An important issue discussed during the work session was the allocation of petroleum between the developed and the developing countries. One participant argued that since the developed countries have more options, they should make an extra effort to conserve petroleum and make it available to the developing countries.

Electricity-related issues were discussed extensively, because electricity is playing the most important role in switching from oil to other forms of energy such as nuclear power and coal, and also because electricity demand, especially for household use, is expected to increase in the future in many countries. For example, both Japan and Italy generate a large fraction of their electricity by using imported oil, and both wish to build new power plants that will not consume petroleum. A small country may not have the luxury of developing both coal- and nuclear-fired plants. Some countries have chosen coal (e.g., Denmark) while others have chosen nuclear power, (e.g., France). All countries have, in their energy planning, been surprised by the price elasticity of electricity demand. Higher electricity prices diminish demand and slow the transition to coal and nuclear power. Nevertheless, the participants concluded that for the IEA countries as a whole, both coal and nuclear power have important roles to play.

An electric utility is a natural monopoly and is under governmental control by regulation or public ownership. In this regard, electricity pricing policies were discussed. Since high prices are unpopular with publicly elected regulatory commissions, the regulated price of electricity tends to be less than what the price would be on an average cost basis. However, to encourage conservation, electricity price should be based on the replacement costs rather than average costs, and furthermore, if the price of electricity is too low, there is not enouh money to invest in new generating plants. On the other hand, the effects of international competition must be recognized. In the European common market, the United Kingdom bases its electricity prices on replacement costs while other countries base theirs on average costs. Consequently, electricity-intensive industries in the United Kingdom have a trade disadvantage. In order to encourage proper new investments, all of the IEA countries need to agree on replacement cost pricing for electricity and other forms of energy.

In many countries, government regulations to protect the environment and public health are cumbersome and cause long power plant construction times. If the regulations could be made more efficient, the shorter construction times would both reduce costs and lessen demand forecast uncertainties.

As part of this, government should set general targets for environmental and health standards and should then let industry determine the most cost-effective means of reaching those targets. For example, government air pollution regulations usually have point source emission limits for each point of combustion, whereas a sulfur tax would be a more efficient means of regulation in such a case.

In conclusion, the participants agreed to endorse the position paper's recommendations on specific issues, and they also agreed that an uncertain future requires flexible energy strategies and institutions.

[There were no comments made at the Symposium's summarizing plenary session or submitted following the Symposium which pertained specifically to this work session. However, for comments relevant to the work session's topics, see especially chaps. 6 and 18.]

Work Session on Industrialized Nonmarket-Economy Nations

CHAPTER SIXTEEN

Energy in Hungary: Problems and a Possible Strategy

László Kapolyi
State Secretary, Ministry of Industry
Hungarian People's Republic

ENERGY PROBLEMS OF THE SOCIALIST COUNTRIES AS REFLECTED IN HUNGARIAN ENERGY POLICY

Long-Term Energy Forecast

There is a close relationship between the considerable increases in the costs of raw materials and energy resources, as seen in the current epoch-making changes in world economy, and the uneven geographic distribution of these raw materials and energy resources as well as the impracticably high use of the different raw materials compared with their available reserves. According to a long-term forecast, by 2020 the utilization figures of the unevenly distributed geological resources will amount to 83 percent for petroleum, 52 percent for natural gas, and only 2.5 percent for coal.

Responsibility for the considerable divergence between energy demand and supply lies with the uneven geographic distribution and unreasonably high utilization rate of the world's geological resources. Because of high demand, the share of energy resources amounts to 10 percent in the world's commercial turnover today. The increased demand of the East European socialist countries can be met only by energy resources transported over long distances to the consumer.

There are different scenarios for projected global energy use, with the long-term scenarios to the years 2020 or 2030 differing by as much as 40 percent in their projections. However, these forecasts agree that the growth in energy demand can no longer be met by the fossil fuel reserves known at present. New sources must be developed to meet the energy demand on a different basis, with the level of development dependent on whether a higher or a lower rate of economic development is chosen by the world. Nonrenewable fossil resources will cover 40 to 63 percent of the total demand, depending on the rate of growth, while 16 to 46 percent of the demand will be met by new sources, in addition to nuclear power plants now in use. This means that after the turn of the century a considerable part of the total energy demand can be met only by thus far unknown techniques, the majority of which are now in a research phase.

Hungarian Energy Forecast

Like other countries of the world, Hungary is increasingly adapting itself to the new situation of the world energy market. The long-term forecasts prepared after the dramatic rise of world market prices in 1973 predict a continuously decreasing rate of total energy consumption and as part of this a similarly decreasing rate of electricity consumption. The reduction is conspicuous. The very modest demand predicted in the long run indicates that energy policy as a subsystem of economic policy has adopted the only viable strategy throughout the world—namely, that there is no reasonable alternative to energy conservation.

In the early 1970s, the energy consumption in Hungary was higher than justified by the state of development. In general, 1 percent of economic growth required an additional energy expenditure of 1 percent. In the present phase of development, the government strategy dictates a slower economic growth rate. Simultaneously, the structure of production and products is being modernized by means of intensive technical development with a view towards further reducing specific energy and raw material demand. The target is to keep the growth rate of specific energy demand at, or below, 0.5 percent while maintaining a growth rate of 1 percent in per capita GNP. Hence, there will be a flexibility of less than 0.5 in the development of energy as compared with the earlier figures of 1 to 1.1.

This pattern has been confirmed by the experiences over the period following the modification of the trend in 1978. While the annual growth rate of primary energy consumption had been 5.4 percent between 1975 and 1978, the same growth rate reduced to 0.7 percent between 1978 and 1980. The growth rate of electricity consumption during these two periods also differed considerably, amounting to 7.5 percent in the earlier period and to 1.9 percent in the recent one.

Hungary's energy and raw material strategy is capable of meeting the demands of the national economy by increasing the supply assortment appropriately. This strategy suggests increasing geological research and distributing energy among producers and consumers by means of highly developed energy-technology system combinations while improving the utilization processes.

Features of Hungarian Energy Policy Strategy

For the future, the combined energy and raw material strategy suggests simultaneously using coal, petroleum, and natural gas in an effort to meet electric power demand while optimally combining the development of coal- and nuclear-based electricity production. Today, the share of hydrocarbons in Hungarian electric power production is decisive, but by the end of the century, coal and nuclear energy will predominate.

The share of fuel oil and goudron in petroleum will be gradually reduced by increasing the degree of processing gradually, improving the yield of white products in two steps: first to 75 percent and then to 85 percent. The higher degree of processing will be appreciable first of all in case of valuable aromatics and engine fuels. Simultaneously, new coal utilization techniques will be increasingly introduced to meet consumer demand, and efforts will be made to use up-to-date techniques for more than 25 percent of total coal production by the end of the century.

The above indicates why a sophisticated utilization of resources is so important. For example, when bauxite is exploited, there are many other components produced in addition to the useful one. Waste is inevitably produced during processing, and such waste contributes considerably to environmental pollution. Therefore, such extraneous components of the processing system should also be reduced because of environmental considerations.

Common Characteristics in the Energy Strategies of Various Countries

When comparing our long-term energy strategy to the energy and raw material policies of other countries, either socialist or capitalist, several common characteristics can be seen. Everywhere, priority is given to achieving in energy cost savings. There is no alternative to energy conservation. Processing petroleum to as high a degree as possible and disengaging this product from the area of energy supply are problems of nearly similar importance. In different industrial sectors of the national economy, the energy resources themselves—both domestic and imported sources—are utilized as processed to various extents according to well-defined functional goals. In every country, efforts are made to decrease the degree of dependence on imports. Programs of stepping up electric energy production are based almost everywhere on coal and nuclear energy. Simultaneously, petroleum, processed to a higher extent than previously, can be taken out of the area of energy supply. Heat production methods, along with those of electric power generation, are being extended everywhere. The utilization of industrial, agricultural, communication, construction, and communal technologies that save both materials and energy is progressing at rapid pace. The use of agricultural waste materials represents a real possibility for saving energy.

Everywhere, extensive programs of energy conservation have been worked out in order to hold specific energy requirements to as low a level as possible. A common characteristic of such energy conservation programs is the fact that their implementation is bound to technical innovation.

Within the European COMECON countries, the problems associated with the supply of raw materials and energy have been considered to be of prime importance for central economic policies during the past decade. Mutually exploiting the benefits derived from a system of an international division of labor is of great importance in every stage of the process. When designing drafts for long-term cooperation programs, special attention should be given to the following considerations:

• Domestic reserves of energy and raw material should be drawn into the economy to a maximum degree and utilized rationally.
• The use of scientific and technical achievements that cut down production as well as transportation costs should be accelerated.
• In general, the COMECON countries should cooperate in the field of joint extraction.

Subsequent to World War II, the European COMECON countries developed their economies by using materials and energy on an intensive basis. This resulted in specific utilization costs that were disproportionately high when compared with the level of these countries' development. From one point of view, it meant that production efficiency diminished, since the extent of ultimate utilization increased less than that of overall production. Some of the reasons for these specific indices of utilization are:

- in various production sectors, the share of raw materials and energy is especially important;
- the technologies for converting and utilizing raw materials and energy are relatively out-of-date; and
- the level of reusing secondary raw materials and waste energy is not appropriate.

Among COMECON countries, reserves of raw material and energy are distributed rather unequally. Except in the Soviet Union, the reserves available to the COMECON countries that are considered to be important at a national economic level are limited to a few raw materials and energy sources. In almost all of the European COMECON countries except the Soviet Union, mineral production has not been able to keep pace with growing needs. Consequently, all the East European countries have had to import energy and raw materials.

In order to achieve savings in investment costs, some countries imported various products even in cases where their own natural conditions didn't necessitate it. The ratio of the number of raw material and energy production projects to the total number of industrial projects diminished substantially, even during the first half of the 1970s. As a result, the increased gap between domestic demand and domestic production helped increase pressure on the import sector. The principal suppliers of raw materials and energy are more and more reluctant to trade their "investment-intensive" raw materials for finished products. Since these raw materials and feedstocks, as well as energy sources, are easier to market than finished products, they are used more and more as exports to compensate for imports from capitalist trading partners.

In addition, the political energy strategy of some European planned economy countries contains other specific elements. These are detailed below on a country-by-country basis.

The *Soviet Union,* besides fulfilling its domestic needs, exports a considerable proportion of energy resources to the European planned economy countries and also to some market economy countries. It meets a considerable amount of the petroleum, petroleum product, and natural gas import needs of the East European socialist countries, as well as providing large quantities of coke, coking coal, and electric power. A considerable portion of the energy material reserves are, however, located in the eastern parts of the Soviet Union, while their exportation and consumption are directed towards European users. The increasing lack of balance in the structure of production and consumption requires on the one hand a major increase in the trans-

Ural production and on the other hand an ever increasing distance of transportation from these areas.

Petroleum and coal transportation to the far-eastern ports of the Pacific Ocean considerably increases the role that the Soviet Union's Asian territories play in the fuel traffic. During the period 1976–80, the Asian territories of the Soviet Union accounted for 90 percent of the increase in coal demand, 100 percent of the increase in gas requirements, and more than 100 percent of the increase in oil demand.

Under the Soviet Union's present energy structure, conventional fossil fuels are predominant and, according to forecasts, are the sources on which the primary energy supply will be based at the turn of the century. In spite of considerable efforts made to construct hydro power stations, that source's share of total electric energy output has been decreasing since the early 1960s. The cornerstone of Soviet energy policy is to develop nuclear energy considerably. On that basis, nuclear energy will play an increasingly important role in fulfilling the future energy needs of the European countries.

As in most other countries, there is a tendency in the Soviet Union to decrease the proportion of electric energy based on hydrocarbons. In the future, a considerable part of the increased electric energy output will result from the utilization of surface-mined coal. With regard to geological conditions, the majority of surface mining operations will be located in the Asian regions, which entails serious allocation problems. In the European areas, natural gas and oil play a dominant role in electrical development, and they meet two-thirds of the fuel demands of industrial boilers with their technological and air pollution control advantages.

In the Soviet Union, energy models have been worked out that allow the application of different strategies, with regard to different discrete and aggregated needs, by responding with a diverse combination of resources. The target function is to minimize the costs involved in the fulfillment of a given need—a need which might refer to the gross needs of either total energy demand or of some sectors to which priority is given.

The strategies described in the studies are based primarily on the low-cost Siberian oil and easily developed natural gas deposits, as well as on the brown coal and lignite reserves of the trans-Ural regions that are surface mineable and on the nuclear power plants to be located west of the Volga River.

In the *German Democratic Republic,* the necessary energy

sources are provided primarily by the development of the country's own raw material and energy base, the majority of which relies on domestic brown coal reserves. The level of energy materials imported—which imposes an ever growing load on the national economy—is strictly controlled, and, with improved coal enrichment techniques, directly usable energy materials and chemical feedstocks of high quality will be available from their own sources on a continuously increasing scale. Parallel to this, there is a continuous development in the productivity of nuclear power plants. To insure a continuous and stable development of the national economy over the long range, efforts are being made towards low and further decreasing growth rates in energy demand. A major proportion of increased energy demand should be met by reasonable and economic methods of energy consumption.

To insure an increase in the production of raw materials, fuel, and energy based on domestic sources, efforts are being concentrated on the following tasks:

• a major increase in brown coal production, accelerated development of high-quality coals, more intensive utilization of brown coal, and further building up of nuclear power plants;
• the provision of heat energy with a considerably higher degree of thermal conversion efficiency;
• the establishment of new heat-generating installations built for brown coal firing, with district heat supply provided from brown coal-firing plants and nuclear power plants;
• the conversion of some of the oil-fired installations to use with solid fuels; and
• the increased utilization of wastes and secondary raw materials.

In the present five-year plan, an annual 4.5 to 5 percent decrease in specific energy consumption is scheduled, with the savings in raw brown coal projected to reach 70 million tons by 1985.

Realization of 33 percent of the intended conservation is planned to come from the fields of rational energy consumption, energy conversion, and energy transfer; 39 percent from the technological processes applied in the low- and high-temperature areas; and 14 percent from the utilization of secondary energy sources.

In the *Czechoslovakian Socialist Republic*, the target plan is to decrease labor-intensive features of the national economy by means of energy conservation measures. In the forthcoming period, the maximum increase in energy consumption is planned to be 0.5 to 0.6 percent for a 1 percent increase in the national income. On the basis of the examination performed, the main reasons for the present unfavorable energy utilization are as follows:

- solid fuels—mainly brown coal of low calorific value—make up a high proportion of total energy consumption;
- energy-intensive products represent a high proportion of total industrial production;
- in the engineering industry, the manufacturing of material-and energy-intensive products plays a key role; and
- certain firing installations in the different branches of industry operate with low efficiency.

Therefore, the target of the energy economy program is to eliminate all hidden forms of wasting energy in the whole national economy. This is to be accomplished by means of different technical organization measures to reduce energy losses and by the application of semiconductors and illumination techniques, by the use of energy-effective technology, and by improving the structure of the national economy through introducing less energy-intensive products.

In the forthcoming period, nuclear power plants are expected to handle a major proportion of the increase in energy demand. Simultaneously, the capacity of hydro power stations with water storage will be increased, using as a back-up the constant capacity of nuclear power plants. The increase in brown coal production will be moderate. There will be only a slight increase in natural gas utilization by increasing imports. In the following years, the current level of oil imports will be maintained, but at the same time, the use of petroleum derivatives for energy purposes will be decreased and the proportion of the white products from oil will be increased.

In *Bulgaria*, several years' experience has been gained in the field of operating nuclear power plants equipped with 440-mw--capacity Voronez-type reactors. Simultaneously, construction of the 1,000-mw high-unit-performance nuclear power plants has started.

Bulgaria has an important background in the use of low calorific lignites for power plants. In addition to the application of lignite to power plants, the technology of household briquette production from lignite has been developed and is currently being used.

With the purchase of manufacturing licenses, the development of state-of-the-art diesel engines has started within the framework of international cooperation. These diesel engines will replace the gasoline truck engines, with their high rates of fuel consumption.

Relationships between Economic Growth and Raw Materials in the Hungarian Energy Policy

The most challenging feature of economic development is to meet ever increasing national demands for material and nonmaterial goods. National production has to satisfy end-use consumption and accumulation; it must also fill the consumption needs of production; and—to smoothly facilitate all this—it must take part in the international division of labor.

These demands determine how to divide the GNP dynamically according to branches and goods. Using past or projected data, the balance of branch relationships tells us what the outputs of the different branches constitute in the national product and how these outputs are used.

From a holistic view of the national economy, this division is rather rigid, since dynamic development and major modifications are attained only during rather long periods of considerable structural change. Thus, the global proportion of mineral resource use within national production does not change considerably in the short run; major modifications can only be attained in the long run by means of selective economic development.

As for the proportion of domestic to imported resources, the possibilities for modification are more flexible, but a modification of resource composition does not affect the global division mentioned already. For example, examining data for the past ten years, we find that the value of domestically produced mineral resources constituted a somewhat decreasing part of the national product, but this minor decrease was offset by the fact that a 1 percent growth in the national product necessitated a 2.0-2.6 percent growth in imported minerals as raw materials.

The unchanged proportion of consumption results from the fact that, with some exceptions, the major part of the demand for raw minerals does not match changes in the production pattern closely, but instead constitutes a rigid demand to be satisfied. This rigidity of demand can be seen most strikingly in the case of energy resources where the consumption level, once attained, must with some exceptions be inevitably satisfied, and further development can only be influenced through major changes in the production pattern. Throughout the world there is a growing tendency to determine future economic growth through careful consideration of up-to-date economic and energy factors, taking into account environmental constraints, investment restrictions, and the risks of dependence on imports.

The statement that only a minor part of demand is shaped by production pattern changes also holds true for other raw materials. This is due to the fact that specialization of the production pattern initially exerts influence mainly on the products and only rarely —and after a longer period—affects whole branches of production which in turn change the patterns of raw material consumption.

The close relationship between economic development and the consumption of mineral resources is of great importance in studies examining the efficiency of managing raw minerals. When considering the different ways of purchasing raw minerals, the primary consideration in weighing demands is to secure satisfaction of those demands under the conditions most favorable for the national economy. The efficiency studies generally must answer the question of whether the demand for vital raw minerals should be satisfied through imports under the most favorable conditions possible or through domestic production. These studies must also define the most favorable ratio of domestically produced resources to imports.

In the long range, the production pattern—and the consumption of raw minerals—naturally is not so rigid; it is shaped within the framework of optimization as decided by the central management of the socialist planned economy.

During this management process, decisions are made and economic policy goals are set on the basis of information reflecting the economic environment. Production decisions are reached through the optimization and weighing of demand. These decisions use as their basis the technical coefficients derived from efficiency studies on the use of production factors and from studies of final consumption. The production decisions determine the detailed demand side: the distribution of production. A detailed comparison of the supply and demand sides yields the import demands and the export possibilities, first in a division according to branches and then in a detailing of products and their assortment.

During the decisionmaking process, the sectoral directing organizations must represent the production offerings and conditions while the functional organizations represent the requirements of the target intentions. Dependent upon the level of decisionmaking and the restrictions on the different sources, the target intentions may address a total investigation of the efficient use of national resources (e.g., the highest possible growth rate of the national product) or simply the utilization efficiency of different limited resources (e.g., productivity, energy intensiveness, material intensiveness, etc.).

This systems-oriented programming of production patterns also projects changes in the long-range demands for raw mineral resources. In the case of those few minerals where domestic production is greater than the domestic demand, the efficiency studies should not weigh alternate solutions but should determine the optimum efficiency of the resource's domestic production and processing within the entire production structure.

In the framework of economic development, the management of raw mineral resources should meet growing demands due to economic development and should extend the domestic production of raw minerals to areas which—when taken together with their vertical structures—can complement other possible developments of the national economy. In the vertical processing structures and management of these raw minerals, imports are considered if domestic production, even under favorable conditions, cannot satisfy demand or if, through specialization of the production pattern, efficient and modern vertical processing structures can be established which gain a place in the production pattern despite the import of raw mineral resources.

In both cases, the demands to be met can be considered within the framework of economic development planning. The ways to meet demand are determined by means of the optimized result of the efficiency studies in interaction with the shaping of demand.

Planning Economic Development

The conscious shaping of economic development is based on planning the economy. In this framework, action programs evolve for the development of socioeconomic activity on a firm conceptual basis. Thus, planning the economy is a means to realize economic policy and consists mainly of planning the supply of energy and raw materials and the vertical production structures. At the same time, economic policy is a system of measures taken for the sake of the goals and the execution of economic development; it classifies goals and sets priorities to reach those goals.

Thus, on the basis of incoming information, economic policy must determine the sources and activities that yield the necessary use values within the production pattern, insuring appropriate conditions as well as the flow of value and material for these activities. In these economic policy decisions, the optimized variants of each domain as well as the balanced system should be considered.

Economic policy decisions shape the major processes and proportions of economic life while considering social preferences. At the same time, they pertain to the ways of reaching desired development and determine the means of economic control. The different decisions are naturally not independent of each other, for together they constitute the economic policy.

In the short run, the decisions and aspirations concerning basic modification of the production pattern are largely limited by existing circumstances, the structure of production and end consumption, and the actual state of the different resources. These limiting factors are the strongest in the extractive branches of production, whereas in the processing branches there are greater possibilities for short-term changes. Primarily in the case of production patterns within the different branches, the degree of freedom in planning and the possibility for structural changes is greater in the long run in every branch. Thus, it is much easier to work out different but consistent variants, and the choices among these variants are also greater.

A conceptual economic policy is actually a development plan giving perspective and direction and containing the desired pace of economic development, the structural changes, the main data on the development of living standards, the development priorities, and the changes in international economic relations and regional problems. Thus, a very important and sometimes determinant part of the economic policy is the strategy for energy and raw materials management—a strategy which constitutes a basic part of the central control of socioeconomic processes.

The economic plan, by means of projected complex long-range programs, synthesizes the economic goals for a given period and the resource allocations necessary to reach those goals. As part of this, optimum distribution of resources is sought in order to optimally realize the economic development goals.

Economic planning takes into account the needs of society and the actual limits and potential interchangeability of resources, and it seeks to insure harmony between goals and resources in order to realize important economic development plans. A maximum growth in the efficiency of national production is attained through an optimum allocation of resources.

In shaping plans and alternatives for development, the capabilities of the national economy must be borne in mind. A strong sense of the factors involved in optimum resource allocation must be kept in mind if the economic goals are to be achieved.

In planning for economic growth, the role of raw materials, especially raw mineral resources, is particularly important, because of the high capital investment and long time needed to set up new mineral production facilities and the risk involved if the geological circumstances are not exceptionally favorable. In case of imports, these aspects are reflected in purchase prices and present a particular problem since the demand for raw mineral resources grows at approximately the same pace as the economic development itself.

A decrease in raw material use intensity can be attained—without hurting efficiency demands—by increasing the product's level of processing; by developing an assortment and quality of product which is modern, competitive, and favorably salable on the world market; and by increasing production yields. In planning for economic development, the basis for considering raw mineral resources—imported or domestic—is the extent to which they meet the above criteria and—supplemented by the processing phases— become part of a production pattern that insures maximum efficiency. In conjunction with this, complex efficiency studies extending over the entire national economy assess and help determine the development of the different branches (among them, that of mineral resource management) which constitute the nation's economic growth under the given limiting circumstances.

To meet general requirements, some innovative bases of production must be formed and, with the use of state-of-the-art production and processing technologies, internationally competitive factories must be established. In this regard, the Hungarian vertical processing structures based on home mineral resources usually prove to be favorable. This means—if the right solutions are found and applied—a concentration of production and products, thereby automatically satisfying the requirement of reducing industrial assortment and increasing the production volume.

Economic development planning for a long-term domestic strategy is vital because the realization of development generally takes six to eight years. Furthermore, established extracting facilities are in place for even longer and cannot be converted into other branches.

The Role of Technical Development in the Power and Raw Material Supply of the Hungarian Economy

In the power industry, achieving a dynamic balance between

available sources and potential demands depends on the complementary interaction of the following conditions:

- the provision of energy resources with a minimum expenditure and the modification of their structural pattern towards a more favorable direction;
- the development of efficient power production and transformation processes;
- the development in all fields of equipment with low power consumption levels; and
- the establishment of optimum operating conditions for power systems internal to the country as well as those interconnected with foreign countries.

By satisfying these conditions, the growth of raw material and power demand related to actual economic growth can be kept as low as possible.

With respect to power consumption, the main aims of technical development within the different branches of the Hungarian economy are listed below by sector.

Industry:

- in metallurgy: to develop ore dressing, introduce high-field-strength magnetic separation, and increase the air temperature and furnace pressure of blast furnaces as well as their evaporation cooling by the injection of decomposed natural gas, and by all these means to reduce coke consumption;
- in steel production: to attain an optimum ratio of electro-furnace and converter technologies and a broader application of continuous casting and heat recuperation equipment into process cycles;
- in cement production: to restore coal firing and introduce precalcination;
- in lime production: to install new coal-fired plants with up-to-date technology and to utilize lime-fly-ash mixtures of lower heat demand instead of cement;
- in brick production: to increase the gas-silicate ratio and the conversion of brick-making to coal; and
- in general: to develop the means for heat exchange, concentration, and drying processes as well as for heat recuperation equipment; and to develop modern measuring and control systems for power cycles by means of a broader application of computer techniques and micro processors.

Agriculture:

- to expand the utilization of wet feed storage and the development of low-power-demand drying technologies in order to minimize the power demand of drying processes;

- to develop low-power-demand soil cultivation methods;
- to utilize bio-heat for the establishment of an optimum indoor barn temperature in animal husbandry;
- to establish optimum combinations of prime movers and machine tools and to propagate technologies featuring several operations within one run in the field of agricultural machinery;
- to utilize biomass, initially for firing purposes and subsequently for biogas and methanol production; and
- to develop the equipment and process sequences required for power production from agricultural and forest by-products and other wastes.

Communication and transport:

- to realize optimum communication patterns and develop modern means of mass transit, influencing industrial and settlement development policies in order to reduce the demand for transport;
- to develop production cooperation and good distribution and deposit networks, with special attention to transport demands; and
- to design construction and equipment that has lower specific fuel consumption requirements and to further develop electric- and diesel-power transport.

Communal power consumption, including households:

- to improve thermal insulation and develop insulated doors and windows so as to minimize the energy consumption of buildings for heating purposes;
- to implement programmed day and night temperature control;
- to redesign powered household equipment and tools in order to reduce their power consumption;
- to utilize renewable power sources (solar, geothermal, wind) to satisfy household energy demand;
- to develop efficient coal-firing equipment that can be easily automated and has environmentally benign properties; and
- to develop methods and means for minimizing heat supply losses and methods for the selection of an optimum supply system in the field of district heating.

A POSSIBLE STRATEGY FOR HUNGARIAN ENERGY AND RAW MATERIALS POLICY

In General

From a strategic point of view, it is fundamental to achieve the target system set by the socioeconomic policy. The specific situation of raw mineral resources must be taken into consideration. The process of exploiting primary mineral resources is and must be gradually upgraded, partly because of the nonrenewable nature of these resources and partly because of the inflexibility of the economic structure in utilizing them. In the past ten to fifteen years, demand reduction, replaceability, reprocessibility, and complex utilization have come to the front.

The strategy is not based on a choice between the processing industries and the extractive industries. The point of decision is whether domestic raw material or imported raw material is provided for as the basis for the development of the different processing industries. In the first instance, mines are constructed and operated in the country, while in the second, exporting industries are developed to compensate for the imports. Typically, the investment costs required to develop the coal mining industry in Hungary lie on the order of magnitude of one-half to two-thirds when compared with the investment required to compensate for petroleum imports by exporting the product of some processing industry.

Due to the intrinsically interwoven nature of an economy, the national economy has no predetermined energy demands. A flexible strategy must be elaborated in energy policy, even for one possible trend of global economic development. Here, the structure of resource distribution within the economy and the dynamism of this structure are of fundamental importance. Included in this set of resources are both social and natural resources.

The strategy of energy economies is determined by the range of demands resulting from the variable development of the macroeconomy and by the system of natural parameters (i.e., the given resources) as well as by the extent to which allocable live and materialized labor are used on an international scale. In elaborating the structure, it is necessary to be well prepared, both in advance and continuously thereafter, for changes and to be able to respond flexibly to such changes.

Therefore, it is necessary in considering energy systems that alternatives be elaborated and that forecasts be taken into account, because here the development requires a long time and entails relatively high costs to bring about the required facility. Within the framework of a development which has been decided, a most up-to-date vertical processing structure should be built on the basis of extractive facilities.

A trend towards technological improvement has been experienced internationally in the past six to eight years: technical innovations have been concentrated in this field. This permits complex demands to be flexibly met by the exploitation of upgraded natural resources. There is an increasing demand that the multifaceted potential of each deposit be realized. This requires that a vertical production system be the basis for a production process containing much valuable work and producing competitive products.

The increased share of extractive industries in the Hungarian national economy is justified by the effective vertical production structure which insures a high degree of processing and, by means of improvement in the production and product structure, guarantees energy rationalization and improved yields and which (indirectly, through an increased degree of processing) reduces the raw material and energy needs per unit of GNP. In addition to reducing the raw material and energy needs, the production and energy structures are consciously controlled in a way such that the import dependence of the national economy will also decrease, due to more efficient use of the nation's resources and novel techniques for using its energy supplies.

To further justify increased extraction in the country, it should be noted with respect to prospective extractions, the geological conditions as explored by geological research as well as the conditions of extraction are more favorable than the conditions under which domestic mineral resources are currently being extracted. Thus, the overall rise of production costs which has contributed to the well-known shift in the world market price relationships has little effect on overall expenditures in this instance.

A survey of the explored and expected domestic mineral resource base as well as the vertical production systems that can be built on this base show that it's not the efficiency of extraction which sets production limits for the next twenty to thirty years but rather the engineering constraints, especially when the long construction times and the available resources are taken into consideration.

Considering what has been said above, the reasonable trend of development for the next twenty to thirty years is to increase the overall national production, but doing so differently for each type of explored new mineral resource. As a result of technology development, present demand can also be reduced even if production is increased two or threefold, because a two to threefold increase in productivity can also be expected.

In accordance with what has been said when discussing the vertical structures, technological development also affects the utilization of raw mineral resources and energy supplies. As a result, the specific raw material and energy demand can be improved, and products better equipped to meet the demands of the processing industry can be developed.

Needless to say, national raw mineral production will be developed only if this development requires a lower expenditure than that attributable to imports, directly or indirectly by means of exchangeable products; in other words, this domestic development must comply with the optimum production structure of the national economy. Factors suggesting such a development are as follows:

• increased safety because of the national production;
• an expected further shift in world market price relationships;
• increasing difficulties in selling export products to compensate for the cost of imported raw materials due to competition on the world market;
• efforts to cope with unemployment in the capitalist countries; and
• the high import content in our export products together with difficulties in increasing the exports required to compensate for this import content.

Of course, the efficient development of national energy production and the importing of energy sources are alternatives which are not mutually exclusive.

Possibilities of Energy Resource Production

Within the overall development of the production of national raw minerals, the extent to which the various raw materials are developed is different. Such differences are discussed by resource in this section.

Coal. The reasonable extent of developing coal mining is determined partly by the circumstances of other energy sources.

Coal and nuclear energy are considered to be reasonable alternatives in the country in the long run.

In the different national economies, there is a close relationship between economic development and the use of energy and raw materials, and it is difficult to modify this relationship in the short run. However, as the conditions of the world economy change, this relationship and especially its proportions change, not as a result of central decisions but rather because of modifications taking place in the structure of the national economy's production and consumption—modifications that come about either spontaneously or through central plans and the regulatory system. Hence, changes result in new relationships tending toward equilibrium, and continued economic development requires that the demand for raw materials and energy be met again, although this time in compliance with a modified structure of production and consumption.

The increased, compensating role which coal is playing in the national energy balance is supported by research and development in the field of coal production—R&D that is increasing throughout the world. As a result of this R&D activity, an industrial-scale application of innovative utilization techniques can be expected within a reasonable time, and these techniques will enable the demands of a wide range of consumer energy needs to be met by coal.

Hungary's present coal mining annually exploits only 0.7 percent of its economically available coal reserves, with only 0.3 percent of its lignite exploited by surface mining. Hence, an increase in the present production rate is technically feasible and, in accordance with what has been said so far, absolutely justified economically. In view of the available coal resources and assuming a simultaneous dynamic modernization of both extraction and processing techniques, a development in coal mining corresponding to 800–900 PJ of coal-based heat per year seems reasonable. Assuming a practicable growth in the current rate of production, this can be achieved by the end of this century.

With such a growth in production and depending on the growth rate of the national economy and on the reduction of specific energy needs, more than 30 percent of Hungary's energy demand can be met by coal produced in domestic mines, and thus about 50 to 60 percent of the long-term energy demand can be met by domestic sources, provided uranium ore and hydrocarbons are also available under reasonable conditions in the long run.

Wide and thorough efficiency investigations showed no significant difference from an economic point of view between the electricity production by coal-based power plants and that by nuclear power plants, and it was also shown that neither was superior to any other practice. Hence, *it is justified to take coal and nuclear energy in approximately identical ratios as the bases for new power plants, as suggested by the long-term plan.* The precise share of coal-fired and nuclear power plants in additional electricity production can be determined over a longer period, in accord with the concentration of resources over space and time. Therefore, power plant construction is realized in cycles, constructing only one power plant at any one time. In this case, an average of 50 to 60 percent of Hungary's coal production would be used for electricity production in power plants, and, in accordance with increasing production, increasing amounts of coal could be used to substitute for or reduce the growth of hydrocarbon utilization, using the up-to-date processes now under development.

To meet the energy demands of commercial consumption (boilers, furnaces, etc.), agricultural consumption, and household consumption, provisions will be made for modern efficient facilities and equipment that require minimum live labor and possibly are produced in the country, and for efficient space-heating systems that do not use hydrocarbon fuel at all or economically and reasonably use coal. In planning for facilities and equipment, the fact that in commercial and utility energy consumption much of the primary energy provided by coal must, to be usable, be converted to secondary forms (such as district heating, gas, electricity, coke, etc.) will be taken into consideration.

A domestic ability to meet 60 to 70 percent of Hungary's coke demand can be brought about by developing increased metallurgical coal production and a greater proportion of metallurgical concentrate in the output. This development involves different industries, and it requires that the domestic coking capacity be increased and that the noncoking part of black coal be utilized in power plants in a vertical production system.

As an additional development, the production of lignite-based breeze and briquetting coke for utility consumption and for consumption in less critical industrial processes may be introduced. The reducing gas produced during the production of breeze which can be used in the iron ore mixture may be utilized directly in the hearth to substitute for coke.

Petroleum and natural gas. On the basis of the known resources and expected results of geological research, the present rate of petroleum and natural gas production can be maintained for several more decades. In terms of heat value, this production rate matches the present rate of coal production, meeting about 25 percent of Hungary's energy demand at present. It is expected to meet about 10 to 15 percent of total energy demand until the end of the century.

Petroleum. In processing about 2 million tons of petroleum, both domestic and imported, efforts are being made to increase the proportion of high-grade fuel in order to meet inelastic consumer demand for products such as engine fuel and white products (through such processes as cracking and visbreaking).

An important requirement of our long-term energy and raw material policy is that hydrocarbons used as fuel be employed for energy purposes only—in other words, for firing only as dictated by high-grade chemical engineering processes and as guided by technical and economic considerations. In order to increase Hungary's competitive ability on the world market, the same requirement is imposed on using hydrocarbons as feedstock in the chemical industry: they are to be used to produce only up-to-date, high-return goods where a significant degree of value can be added in their processing and where this processing can be efficiently executed.

Natural gas. Based on the country's existing natural gas resources, a gross production capacity of 6 to 6.6 мrd m^3 has been developed. Assuming geological research of the required intensity and fruitfulness, the same production rate can be expected over the next few decades, and this rate can be slightly increased by exploiting low-grade inert gas deposits.

Because natural gas currently is economically available and has technical assets, it can be applied to many uses in chemical engineering and the production of energy. To increase its use efficiency, natural gas will be distributed to consumers that offer the highest economic use efficiency ratios. Therefore, the limited existing investment resources should be used for increasing the range of utility and commercial consumers that can efficiently use this product for constructing plants to store natural gas for energy-intensive purposes, and for manufacturing value-intensive products in the chemical industry.

Uranium. With the technical and financial assistance of the Soviet Union, uranium ore has been explored for and mined in Hungary since 1950. Due to the current highly developed mining techniques, the uranium mining efficiency has trebled since 1970. However, considering the magnitude of available mineral resources, the production can not be expected to increase quantitatively as it has done in the past.

The question of whether the uranium needs of Hungary's nuclear power plants can be met by domestic production is answerable only with a knowledge of the fuel processing techniques to be used. Advanced technological processes could multiply the effective degree of fuel productivity, and, by reprocessing the spent fuel leaving the reactor, a closed cycle could be brought about.

A Combination Strategy for Energy Policy

Within a reasonable strategy for energy policy, the main domestic energy resources—coal, petroleum, natural gas, and uranium—will be processed to the highest possible degree before consumption. In the overall energy picture, the vertical systems of coal, petroleum, natural gas, and uranium production interact at many points.

In the Hungarian energy picture, different conclusions can be drawn about the appropriate proportions of resource extraction and allocation to meet domestic energy demand. As has been said earlier, according to findings to date, neither uranium production nor petroleum and natural gas production can be increased quantitatively, and even their present rates of production can be maintained in the decades to come only if geological research meets with at least the same success that it has so far. On the other hand, known coal reserves will permit the production rate of coal to be increased considerably over the long run (eighty to one hundred years), and innovative techniques of using coal may constitute an important contribution to future energy development.

From the point of view of Hungary's entire energy system, the contribution of other energy sources—water resources, geothermal energy, agricultural waste, and so forth—is insignificant, but the on-site use of these alternative energy sources is usually economic and thus justified. Such subsystems of alternative energy use within the entire energy system include using water resources for powering agricultural irrigation, using geothermal energy for space heating, using agricultural waste for fueling crop-drying facilities, and so on. In addition, different energy resources may interconnect in a special way, as, for example, in the case of a pump-storage water power plant which is filled by means of coal- or nuclear-based electrical power.

In general, the Hungarian energy picture is very complex, and its resources should be considered both singly and in combination. This complex energy picture is dynamic in space and time, the spatial situation being fundamentally determined by the location of the raw materials, including, of course, the expected mineral resources, with the parameters of the situation changing continuously as geological research advances. Accordingly, a "life diagram" of energy resource deposits—from the extraction process up through the waste phase—is taking shape from geological research, both retrospectively and in planning for the future on the basis of present knowledge. There are many different factors affecting the destiny of a resource deposit, with technical development, which involves each phase of the vertical production system, as the most important factor. This technical development is realized by innovations to insure that the receptor system be adaptive enough—a development which shows the dynamism of the energy picture through time. As part of this, it is important to carefully prepare the connection between the innovative and adaptive systems, taking into account that both systems are in some ways rather inflexible.

Because of the variable geographic distribution of energy resource deposits, international relationships are also dynamic through time, depending as they do on both the specific development of the national economies concerned, and on their political relationships. Within a multifaceted energy-technology system, the characteristics of the system are subject to dynamic changes over time: for example, the coupling of electricity and heat production can change daily, weekly, or monthly.

Finally, nuclear energy appears in the picture as the basis for electricity production either in addition to coal or, considering the percentage increase in efficiency offered by fast breeder reactors (an increase which lies an order of magnitude higher than the present level), as a singular alternative to coal. This rate of development in resource efficiency dynamically prolongs the supply of domestic energy resources considerably.

CHAPTER SEVENTEEN

The Nature of the Energy Problem in Yugoslavia

Naim Afgan
Scientific Adviser
Boris Kidrič Institute of Nuclear Sciences

INTRODUCTION

The energy policy of any country strongly depends on the country's geopolitical characteristics, which describe the essential parameters necessary to understand its political and economic structure. Yugoslavia is a southern European country with a population of about 22 million people living in an area of 255,800 square kilometers. It is a socialistic republic with a federal structure of six sovereign states. Within the federation, each of the six federal states has its independent but commonly agreed-upon energy policy. Yugoslavia is a strongly motivated developing country with one of the highest economic growth ratios in Europe over the last twenty years. In its development it has gone through three typical periods. The first period was characterized by the post-World War II reconstruction of its economy, the primary aim being to reestablish its mainly agricultural economy; the second period was characterized by a central planning economy model designed to assist in the beginning of the country's industrial development; the third and current period has been characterized by the introduction and development of a self-management political system which has as its aim the establishment of a free market system within a socialistic economy.

ENERGY SYSTEM DEVELOPMENT

Energy Resources

Yugoslavia does not belong to the group of countries that

enjoy a wealth of energy resources. With its primary geological energy resources totaling 8.64 billion tons of coal equivalent (tce), which corresponds to 363.83 tce per capita, such resources are only about 75 percent of the average level for Europe as a whole, with six European countries having at their disposal per capita energy resources that exceed Yugoslavia's. In addition, it should be noted that Yugoslavia's population constitutes around 0.6 percent of the world's population, but its energy consumption constitutes only 0.4 percent of the world energy consumption and its total energy resources constitute only 0.1 percent of the world's energy resources. The structure of primary energy resources is given in Table 17-1.

Table 17–1. **Energy Resources in Yugoslavia**

	Units		*tce (x 10^6)*	%
Coal	10^6 t	204	91	1.1
Brown coal	10^6 t	1,936	1,097	12.7
Lignite	10^6 t	19,232	5,706	66.0
Oil	10^6 t	400	390	4.5
Oil shale	10^6 t	200	300	3.5
Gas	10^9 Nm^3	172	140	1.6
Uranium[a]	10^3 t U_3O_8	36.5	910	10.5
Hydro energy	10^9 kwh/year	67.7	8	
Total			8,642	100.0

Source: D. Savić, and D. Veličković, "Thermoenergetic Resources in Yugoslavia," Symposium I on Energetics in Yugoslavia (Belgrade: Serbian Academy of Sciences and Arts, 1968). In Serbo-Croatian.
[a]Used only in thermal reactors.

The major part of Yugoslavia's primary energy resources is low-caloric lignite coal.[1] Most of these lignite resources are available through open pit mines.

Yugoslavia's coal resources are situated at three places: Kosovo, Kolubara, and Tuzla. This has enabled the development of three large open pit mines with highly mechanized excavation processes, thereby making possible very economically favorable coal production.

Domestic oil production currently meets only one-third of the present consumption.[2] Yugoslavia's two main oil basins are the Panonian basin and the Dinarian basin, with oil reserves together totalling 400 million tons. It should be mentioned that only 30 percent of these basins have been adequately explored. Recent offshore drilling has produced promising results, and there are now high expectations of more oil reserves.

Gas reserves in Yugoslavia are estimated to be 172 billion

normal cubic meters (Nm^3). However, this estimate does not include the most recent discoveries, which are still to be evaluated but which look promising. Most of Yugoslavia's gas resources are in the same basins as its oil resources.

Yugoslavia's oil shale resources contribute 3.5 percent to its total energy resources.[3] Oil shale resources are currently estimated at approximately 1.1 billion tce, but due to insufficient investigation, only 0.3 billion tce are proven reserves. There are two main basins, Aleksinac and Sinj. The former is considered to be the major location, and most of the prospecting has been performed there.

Yugoslavia's uranium resources are still uninvestigated for the most part. The present estimated level ranges between 30,000 and 40,000 tons of uranium oxide (U_3O_8). [4] There are around 10,000 tons of proven uranium resources whose extraction is economically justified. It should also be mentioned that these figures include results obtained from 50 percent of the territory covered by uranium prospecting.

A high hydro power potential is one of the geographic characteristics of Yugoslavia.[5] With an average potential of about 65 Twh/year, Yugoslavia's hydro power is derived from three main watersheds: the Black, the Adriatic, and the Aegean seas. The Black Sea watershed, the Adriatic Sea watershed, and the Aegean Sea watershed encompass 70, 21, and 9 percent of Yugoslavia's territory respectively, and they respectively provide 65, 29, and 6 percent of its average hydro energy potential.

Yugoslavia's solar energy potential is estimated to be roughly 363,000 Twh/year.[6] The average insolation in Yugoslavia is 320 w/m². Demand for low-temperature heat could be partly satisfied from the available solar energy, but present high investment costs deter the extensive use of solar energy. The Adriatic coast is the most favorable place for the prospective use of solar energy in Yugoslavia.

Hot water geothermal sources of energy currently meet only a rather small portion of Yugoslavia's total energy demand.[7] There are quite a number of potential geothermal heat sources in Yugoslavia, but their capacity is very limited and their water temperatures very seldom exceed 100°C. However, as some of these geothermal sources are in rural areas, they well could serve as additional energy sources in improving agricultural production.

Energy Consumption

Yugoslavia's total primary energy consumption was 11.8 million tce in 1957, and thereafter it increased at a rate of 6.7 percent, so that it reached 49.1 million tce in 1979.[8] Figure 17-1 shows primary energy consumption in the period 1957–76. It can be seen that over this period, total annual consumption of solid fuel increased 1.6 times, but for the same period, consumption of liquid fuel, natural gas, and hydro energy increased 12.7, 41.5, and 6.5 times respectively. During the same period, the average rate of increase for solid fuel was 2.63 percent; for liquid fuel, 14.9 percent; for natural gas, 21.5 percent; and for hydro energy, 9.7 percent.

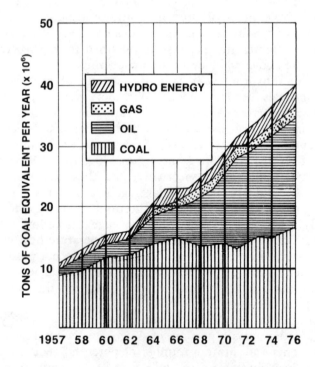

Figure 17-1. **Primary Energy Consumption in the Period 1957–76**

These uneven rates of increasing consumption of the different energy sources have affected the primary energy consumption structure. Figure 17-2 shows the structure of primary energy consumption for the period 1957–75. It can be seen that over this period, solid fuel's proportion of total primary energy consumption was halved, but during the same period the proportional consumption of liquid fuel, natural gas, and hydro energy increased by multiples of 4, 12, and 2 respectively. Solid fuel's proportion of total primary energy consumption was 83.7 percent in 1957, dropping to 41.2 percent in 1976, whereas during the same time, liquid fuel increased from 12.3 percent to 46.6 percent. There are two main reasons for these energy structure changes. The first is the low price of liquid fuel during the 1960s, and the second is the necessarily higher proportion of liquid fuel concomitant with rapid industrial development.

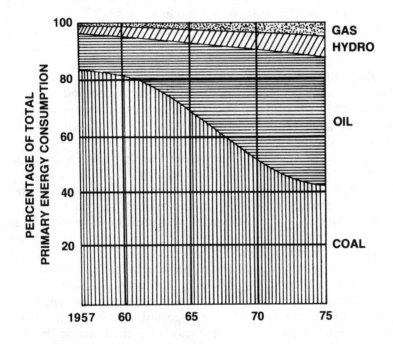

Figure 17-2. **Structure of Primary Energy Consumption in the Period 1957–75**

The structure of final energy consumption is closely related to the level of industrial development. In its early stages of development, the Yugoslav industrial sector's proportion of total final energy consumption was about 48 percent, reaching a level of 52.3 percent in 1975. During the same period, the agricultural sector's proportion of total final energy consumption increased from 0.9

percent in 1957 to 7.4 percent in 1975. For a complete picture of the effect of industrial development on the structure of energy consumption, it should be noted that Yugoslavia's total final energy consumption has increased from 9.1 million tce in 1957 to 27.6 million tce in 1975. With regard to changes in the structure of final energy consumption, it also should be mentioned that an increased standard of living has strongly affected domestic energy consumption.

Electric Energy System

The current installed capacity for the production of electric energy is 13,890 MW, with 59 TWh produced in 1980.[9] This production capacity is divided between hydro and thermal power plants. The production capacity of hydro power stations recently reached 6,120 MW. Until a few years ago, hydro power stations played the leading role in satisfying Yugoslavia's electric energy demand. However, in 1977 Yugoslavia's thermal power electricity generation capacity attained and began to exceed its hydro power capacity (see Figure 17-3), and this thermal power capacity can be expected to grow in the future.

Yugoslavia is generally self-sufficient in its electric energy production, with sometimes a small export of electricity and sometimes a small import. In 1979, the largest portion of total electricity production—39.1 percent—was used by the industrial sector. The second largest portion was used for household purposes. As shown in Figure 17-3, electric energy production increased nearly sixfold from 1960 to 1980. Since a nation's standard of living usually has a high correlation with its electric energy consumption, increases in the national average income per capita generally correspond to increases in electric energy consumption. In this respect, however, data for Yugoslavia are unlike comparable data for other countries: in Yugoslavia, increases in average income have not fully kept pace with increases in electricity consumption. The main reason for this discrepancy is that in Yugoslavia the development of a rather extensive electric grid has enabled an unusually large portion of the population to have access to electricity. In addition, Yugoslavia's electric energy pricing policy has stimulated the extensive use of electric energy for household purposes.

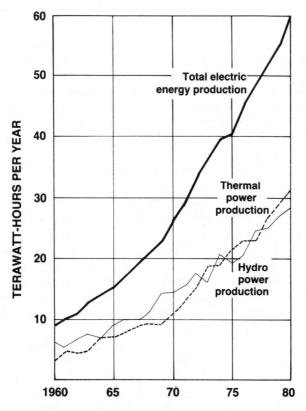

Source: adapted from *Yugoslav Power Industries, 1980*—preliminary data (Belgrade: Union of the Yugoslav Power Industries, January 1981). In Serbo-Croatian.

Figure 17-3. **Electric Energy Production in the Period 1960–80**

LONG-TERM ENERGY POLICY

The main aim of Yugoslavia's long-term energy policy is to attain the conditions required for a constant rate of increase in total national income. In order to achieve this goal within an international energy market which lately has not been very stable, there will be an effort to achieve the most favorable conditions to stimulate the long-range development of local energy resources. Yugoslavia, with large resources of low-caloric coal, has decided to stimulate investment in the lignite open pit mines and use of this resource for electric energy production. At the same time, research and development programs to investigate and adapt coal gasification and liquefaction processes that will be adequate for the lignite resources at large lignite mines are being launched. Through an analysis of a scenario which assumed maximum use of the lignite for electric energy production[10] it was realized that although available lignite has a low sulfur content, one limitation on the development of a large energy production complex would be the environmental capacity of the area in which the complex is located. It thus was recognized that this capacity should not be exceeded under any condition.

One of the milestones of Yugoslavia's energy policy is the diversification of its supply of energy sources. Yugoslavia currently imports about 45 percent of its supplies to meet total primary energy consumption. Even with the intended maximized use of domestic energy resources, dependence on imported liquid and gas fuel will continue. Yugoslavia, as one of the nonaligned countries, shares its political views with many of the oil-producing countries and is strongly motivated to be a customer of those who do not impose any political conditions on the export of their oil. Implementation of this position means that plans for future liquid and gas imports should be directed toward those countries that are willing to meet this criterion. It is also felt that the diversification of energy imports will offer more flexibility for short-range policies that might be affected by local disturbances.

Regarding future domestic sources of liquid and gas fuel, intensified prospecting for new oil and gas resources is planned. Since most previous drillings have been to a maximum depth of less than 3,000 meters, where the oil reservoirs are located in sand sediments in the Tertiary complex, it is expected that given the geological and tectonic structure of the basins, good results could be achieved by increasing the drilling depth. An increase in the drilling depth to 6,000 meters therefore is planned.[11] Due to specific natural conditions and to a lack of financial means, the exploration of possible offshore oil and gas resources in the Adriatic Sea has been rather limited. However, recent marine and seismic investigations there have been promising. Based on their structural and geological characteristics, the Adriatic offshore regions appear to be likely locations for the discovery of new oil and gas reserves.

In the long run, nuclear energy is planned as one of the base-load suppliers of electric energy in Yugoslavia. When the maturity of this energy source has been proven, it is intended that nuclear power will be introduced into the country's electric energy system and that support will be given to the general industrial development necessary for nuclear power's infrastructure.[12] Nuclear power, with its exacting demands, has stimulated the development of special industries which are able to satisfy its safety requirements. Their development will enable the increased productivity of these vital industries. Another objective of Yugoslavia's planned nuclear energy program is the rational use of the limited nuclear fuel resources available. Political constraints on the availability of nuclear fuel services has become one of the obstacles to the normal introduction and development of nuclear energy. Those countries which plan to join the group of the countries now enjoying the peaceful use of nuclear energy must take into consideration the present and future political conditions imposed by those with developed nuclear fuel cycle service technologies and corresponding industries. This consideration is one of the essential reasons why nuclear energy programs in most countries, including Yugoslavia, will be planned to include development of the corresponding infrastructure industries necessary to satisfy the demand for required nuclear fuel cycle services.

Situated in southern Europe, Yugoslavia benefits by having higher than average insolation. For this reason, the country's long-term energy planning also takes into consideration the potential use of solar energy. With its advantages, solar energy is planned as one of the major prospective contributors to the energy supply of the resort regions, where low-temperature heat is needed for heating and cooling requirements. If a breakthrough in solar technology facilitates the use of solar energy for other purposes, a larger role for solar energy is also envisaged.

The use of energy in rural areas is becoming more important, due to its increasingly large proportion of the country's total energy demand. In view of the availability and amount of biomass wastes in rural areas, measures will be taken to stimulate the greater use of these wastes.

Hydro power was for a long time one of the most important and cheapest energy sources in Yugoslavia. The country's total hydro energy potential amounts to about 60 TWh/year. Fifty percent of this potential is currently used. As will be further explained in the forecast section of this paper, much of this unrealized potential is planned for use in the near future. Because the development of some of this hydro potential would entail a large investment, it would have to be designed as a complex irrigation and agriculture project with energy production as one of its ancillary benefits. Water is becoming one of the national resources that should be used in a way that optimizes its total benefit to the national economy, and its most important use is expected to be for purposes other than energy.

RESEARCH AND DEVELOPMENT PROGRAM

Along with the process of planning a long-term energy policy, it was felt necessary to devote particular attention to the corresponding effort towards the organization and management of a long-term energy research and development program. Only a long-term energy program enables the practical application of research and development achievements to the utilization of potential energy resources. In order to correlate the long-term energy policy with energy research and development, the following objectives are taken into the consideration:

(i) A research and development program related to the better utilization of coal should be aimed at increasing the efficiency of converting coal into secondary energy. This should be achieved by investigating the processes needed for coal liquefaction and gasification and also by exploring the improvements needed to enable coal's more efficient use in today's modern thermal power plants.

(ii) The development of depth drilling methods and their use in geological and lithological surveys for potential oil and gas resources should be the aim of a research program related to liquid fuel prospecting.

(iii) A research and development program in the field of nuclear energy should have four main focuses:

• Development of the nuclear fuel cycle technologies needed for the economic and rational use of nuclear fuel. In this respect, emphasis should be placed on access to the new technologies required for uranium ore processing and enrichment, nuclear fuel element production and reprocessing, and radioactive waste treatment and disposal—the last requiring both the technology for and the development of safe and reliable, temporary and permanent radioactive waste disposal methods. It is expected that other countries may share these objectives, which will offer the possibility of establishing a joint program in this field. In connection with this, attention should be given to the proposal made by the Yugoslav delegation at the 1978 General Conference of the International Atomic Energy Agency in Rio de Janeiro. This proposal suggests that an international pool be established to stimulate the joint efforts of all countries interested in creating the financial, material, and technological conditions needed for their own development of all fuel cycle phases.[13] This pool should be a free access organization for all countries that are willing to share with other members of the pool their financial, material, and technological means of realizing all phases of the nuclear fuel cycle, in accordance with the pool's agreed conditions. In proportion to its own contribution, each member country will have the right to use the available joint resources of the pool for its own use. It is expected that the joint long-term planning of nuclear fuel and nuclear fuel conversion systems will enable all interested countries embarking on nuclear energy programs to have access to all phases of the nuclear fuel cycle without taking on an unduly great burden in the beginning phases of their nuclear energy programs.

- Development of the safety and safeguard methods and methodology needed to guarantee the peaceful use of nuclear energy with minimum risk to the environment. Special attention should be paid to the safe and reliable disposal of radioactive waste, and to the development and implementation of adequate quality control systems to insure the optimal participation of domestic industries in the production of nuclear steam supply system components.

- As a component of a nuclear energy research and development program, measures to insure the development of the techniques and methods needed for active participation in the international fast breeder reactor program. It is expected that countries interested in long-term cooperation in the field of nuclear energy will join efforts to take full economic, social, and technical advantage of the use of fast breeder reactors.

- The organization by interested partners of coordinated research in the field of fusion energy development. Participation in this energy source's development should be geared to the expectation that fusion energy will be the main source of energy by the beginning of the next century.

(iv) A research and development program in the field of renewable energy sources should be focused on the development of new techniques and processes to increase these sources' degree of efficiency. In this sense, solar energy and its storage systems should be given priority.

(v) The modeling of energy systems and their adjustment to changes in the energy structure that are necessitated by the national economy's development should be one objective of an energy research and development program. Research activity to pursue the development of energy planning methodology and its adaptation to local political and social conditions is necessary, and with continuous evaluation and assessment of energy forecast structures, it is expected that it will be possible to obtain the relevant information needed by decisionmaking bodies.

(vi) International cooperation is seen as a necessary condition for the positive achievement of a long-term energy research and development program. In this respect, it is expected that the present cooperative energy program among different countries will be enlarged. Also, cooperation with other developing countries is envisaged as one possible means to foster and promote mutual connections. The organization of a joint energy program which would include several developed and developing countries might prove to be one of the most beneficial approaches to planning long-term energy research and development. Since financing might be a limiting factor, it would be desirable if some of the surplus capital available in the oil-producing countries was oriented in this direction.

INVESTMENT CAPITAL REQUIREMENT

One of the immanent problems of the developing countries is their shortage of investment capital.[14] The strong correlation between the availability of investment capital and the rate of increase of total national income has become one of the limiting factors in the development of many developing countries. This problem is even more pronounced in the energy sector, due to the recognized correlation between gross national production per capita and respective energy consumption.

The World Energy Conference Group's forecast[15] for total primary energy consumption in the developing countries projects an increase from 1.8 billion tce in 1980 to 3.9 billion tce in 2000. This implies that the amount of capital needed in these countries for investment in the exploration, development, and conversion of primary energy sources for the twenty-year period under consideration will exceed that which has been invested for energy throughout these countries' history. It is obvious that the requisite amount cannot be obtained solely from capital available in the developing countries, although it is expected that the developed countries will contribute to this investment.[16] Since the scarcity of some primary energy resources is becoming pronounced in the developed part of the world, it seems reasonable to expect that, for prospecting efforts in regions of the developing countries where the discovery of new energy resources is highly probable, the capital invested will also be beneficial to the developed countries. However, it must be borne in mind that a new approach must be developed to establish fair relationships which guarantee that the host country will gain favorably by this type of joint venture. Furthermore, these ventures should be based on the long-term interest of both parties.

Because of electric energy's high investment requirements, financing may prove to be a limiting factor in the growth of this mode of power. According to studies by the International Bank for Reconstruction and Development,[17] power expansion investment requirements remained at about 7–8 percent of the gross fixed capital. It is estimated, however, that a shift to a higher capital cost plant would force developing countries to raise this proportion to about 10–12 percent. This, in turn, would correspond to a range of about 2–4 percent of the Gross Domestic Product forecasted for developing countries. While these figures do not represent an insuperable burden when considered as a long-term average, they tend to conceal the critical difficulties which will be encountered by many developing countries over the transitional period characterizing the introduction of several capital-intensive technologies into their power-producing sector.

In order to prevent financing terms from becoming the essential criterion in the selection of bids submitted by vendors, it will be necessary to establish a system of international financing which offers some prospects for stability in the conditions and terms of loans designated to help developing countries bridge a particularly difficult financing gap in their energy development planning. This system would prove to be of great assistance in meeting the potential energy needs of most of the developing countries.

Besides direct investment in the energy production sector, which is very important for short-term development plans, it is also necessary to devote a portion of the total financial resources to research and development programs that are oriented to the medium and long term.

Yugoslavia is one of the developing countries which has invested a substantial part of its total available investment capital in its energy sector. In the last twenty years, approximately 15 percent of the total available capital invested in industrial development has been directed towards its energy sector.[18] However, in order to achieve the forecasted domestic production of primary energy, it will be essential to insure that this percentage is raised to about 20 percent. Since investments in the exploration of new primary energy sources must be committed well in advance of their expected return, the heavy financial burden which must be borne in the initial stage of an energy development program is particularly apparent. In addition, the need for plants which convert primary energy sources such as oil and coal into power and into other forms of usable fuels will make this financial burden even more onerous.

ENVIRONMENTAL PROBLEMS

An environmental quality suitable for human life has been recognized in Yugoslavia as a constitutional right of the people. The energy sector is becoming one of the largest pollution-producing sectors, and it is self-evident that attention must be paid to pollution control in the vicinity of large plants that consume primary energy to produce power.[19] Since 1974, Yugoslavia has had regulations in the form of a public law which imposes controls governing the quality of the surroundings of large power plants. This public law imposes regulations which limit river water temperatures to a maximum of 28°C. It also prohibits the sulfur concentration in the air around power stations from exceeding 0.15 mg/m^3. The radiation dose in the area around nuclear stations is limited to 5×10^2 Seiverts per year (1 Sv = 100 rems). Besides these maximum limits, there are regulations which impose the corresponding methods of pollution control and the methodologies to be used for evaluating and assessing the design criteria used in the construction of power plants.

During the development of the energy sector, it was realized that there are two types of locations which produce pollution of a major and potentially hazardous nature. The first is the thermal and nuclear electric power station—one of the most intensive consumers of primary energy at a single location. At present, there are three relatively large thermal power complexes, each having a capacity of around 1,500 mwe. A system of monitoring the surroundings of those energy complexes has shown that the limitations imposed by the public law have been followed. The second type is the urban area—also a major consumer of primary energy. Oil has mostly been used for heating in the large urban areas, and adverse climate conditions in some of the large cities have caused violations of the limitations imposed by environmental controls. Increasing gasoline consumption by the transport sector in urban areas has contributed to the problem, and as a result, most of the large cities have already exceeded environmental control limitations. For this main reason, many countries must investigate the possibility of reorganizing their major energy-consuming sectors by imposing new regulations on primary energy use.

In order to prevent further exacerbation of pollution problems in locations close to energy complexes, it has been decided in Yugoslavia to extend the present monitoring system to insure full environmental control. Also, it is estimated that in view of the present level of technology and the quality of lignite at open pit mines, a limit should be imposed on the total extent of lignite production as well as on that of electric energy production. According to the energy demand forecast, the three major open pit mine locations will have their electric energy production limited to around 13,250 mwe at Kolubara, 13,200 mwe at Kosovo, and 6,830 mwe at Tuzla.[20] At the Kolubara location, which is 20 miles southwest from Belgrade, a limit of 8,150 mwe is planned,[21] thereby allowing 500 tons of sulfur dioxide per day, 50 tons of ash per day, and 16,500 mw thermal per day in the year 2000. But even if there is a significant effort devoted to the analysis and study of these environmental problems, there is still a lack of full understanding about the problems we are going to face.

EDUCATION PROBLEMS

It is widely recognized that the availability of qualified manpower is one of the essential conditions for the success of any energy development program. This statement is especially relevant for those developing countries whose energy sectors are attempting to contribute to their future economic progress.[22] The Yugoslav educational system is traditionally organized, with its main emphasis on those sectors which have manpower requirements sufficient to offer employment opportunities. In this respect, there are essentially three levels of education, with specialization in accordance with the economic sector. The energy sector is recruiting its manpower from schools oriented to the electric, mechanical, and process industries. The traditionally high standards of some of these schools have insured a high quality of education which has served from the very beginning as the foundation of Yugoslavia's energy program. It should be mentioned that some of these schools are now offering their services to facilitate the development of the education systems in some of the developing countries that are less far along.

It has been realized that the introduction of new energy sources will require a corresponding development of manpower. Education in the field of nuclear energy is one of the areas which is in a long-term planning stage, in anticipation of permanent manpower recruitments for Yugoslavia's prospective nuclear energy program. In this regard, nuclear energy research centers have had and will continue to play an important role.[23] The estimated nuclear energy manpower requirements for the years to come are as follows: in 1980, 240 specialists; in 1990, 700 specialists; and in 2000, 2,000 specialists.

Besides the direct recruitment of the manpower needed for the planning, construction, and operation of power plants, the manpower needs of a research and development program must also be emphasized. Since this program will require some specialists who are not currently available in Yugoslavia, it is expected that through programs of international organizations and through bilateral exchange programs with the developed countries, there also will be opportunity for education in these specialized fields.

Since, for any of the developing countries that are less developed, the education of manpower will be an especially heavy burden in promoting their energy programs, the availability of Yugoslavia's education capacity might be of great assistance. Yugoslavia has a number of bilateral agreements with countries that feel the need for this kind of assistance. I strongly believe that extension of this education program to other countries will be possible as well.

POWER INDUSTRIES DEVELOPMENT

Yugoslavia is planning to have power plants with a total capability of about 40,000 mwe in operation by the year 2000. It is estimated that, in present terms, this will cost a total of about 50 billion dollars, or about 2.5 billion dollars per year. This constitutes a significant challenge to develop local industries that should take part in this power program development. Based on the experience of existing industries, it is expected that industries producing heavy components will in all probability obtain support for their further expansion. Some of the present domestic industrial output has gone into the design and construction of Yugoslavia's power plants. For example, it should be noted that almost 95 percent of

Yugoslavia's hydro power plant components are designed and produced domestically, and that work by local industries has constituted roughly 50 percent of the execution of Yugoslavia's thermal power plants. Furthermore, although the first nuclear power plant was a turnkey-based contract with Westinghouse, the participation of Yugoslav industries in its construction was substantial, amounting to about 35 percent of the total invested capital.

Since the power production sector also includes mining industries and relevant machine production industries, the chance for these industries' further development could also be envisaged. The present total mining output capacity is 46.8 million tons per year, with 125 million tons per year expected in 1990 and 200 million tons per year expected in 2000. This expansion will require increasing the mining capacity every year. Since most of Yugoslavia's mining is surface mining, this will require a corresponding increase in the surface mining machine capability, which will give production opportunities to the corresponding industries.

ENERGY DEMAND FORECAST

Primary Energy Demand

In the next five-year plan, it is predicted that the Gross National Product will increase by an average annual rate of 4.5 percent. For the following fifteen years, up to the year 2000, the Gross National Product is expected to increase by average annual rates of 5 percent, 4.5 percent, and 4.0 percent for the periods 1985–90, 1990–95, and 1995–2000 respectively. Accordingly, primary energy demand is expected to be as shown in Table 17–2 and Figure 17–4. The structure of primary energy demand is given in Table 17–3 and Figure 17–5. It should be noted that the main primary energy sources for the rest of this century are expected to be lignite, oil, and gas, with an optimal economic use of hydro energy. Lignite production is expected to increase approximately fivefold by the turn of the century. For the same period, it is expected that oil consumption will double and that gas consumption will increase more than fourfold, while domestic gas production remains on the level achieved in 1980. Electric energy production is expected to be based mainly on thermal power stations, which are scheduled to produce 50 percent of Yugoslavia's electric energy in the year 2000.

Table 17-2. Primary Energy Production and Consumption Forecast

Energy source		Units	1980	1985	1990	1995	2000
Hard coal	Production	10^6 t	0.6	0.6	0.6	0.6	0.6
	Consumption		4	5	6.5	8	10
Brown coal	Production	10^6 t	11	12	12	12	12
	Consumption		11	12	12	12	12
Lignite	Production	10^6 t	32-36	49-58	68-84	116-142	162-203
	Consumption		32-36	49-58	68-84	116-142	162-203
Oil	Production	10^6 t	5.1	6.6	7.3	6.7	5.3
	Consumption		16	21	27	34	38
Natural gas	Production	10^9 Nm^3	2.7	4.2	4.0	3.4	2.4
	Consumption		3.8	6.7	10.3	14.4	18.4
Electric energy	Production	10^9 kwh	60	88-102	121-156	165-209	213-280
	Consumption		60	88-102	121-156	165-209	213-280

Source: S. Vrhovac, L. Ljubiša, and M. Simonović, "Development and Use of Nuclear Energy in the Energetics of Yugoslavia," Conference on the Energetics Development in Yugoslavia, Opatija, 1980, Yugoslav Committee for World Conferences. In Serbo-Croatian.

Table 17–3. **Energy Structure Forecast**

	Source as percentage of total				
	1980	1985	1990	1992	2000
Hydroelectric energy	6.5	5.8	5.5	5.0	4.0
Natural gas	8.6	11.1	12.9	13.0	13.9
Liquid fuel	46.0	44.6	42.6	40.0	36.4
Coal and nuclear	38.9	38.1	39.0	42.0	45.7
Total	100.0	100.0	100.0	100.0	100.0

Source: D. Spasojević, I. Mihajlović, et al., "Prospective Application of Nuclear Power Plants in the Technology of Complex Use of Coal in REIK Kolubara," Mining and Energetic Industries Consortium "Kolubara," Belgrade, 1980. In Serbo-Croatian.

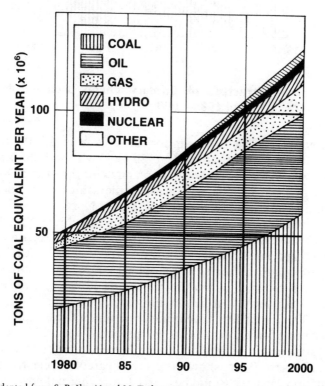

Source: adapted from S. Bošković and M. Todorović, "Long-term forecast of the energy consumption in Yugoslavia," *Energy and Environment* (Belgrade: Yugoslav Union for Protection and Development of the Human Environment, June 1980). In Serbo-Croatian.

Figure 17–4. **Primary Energy Demand Forecast for the Period 1980–2000**

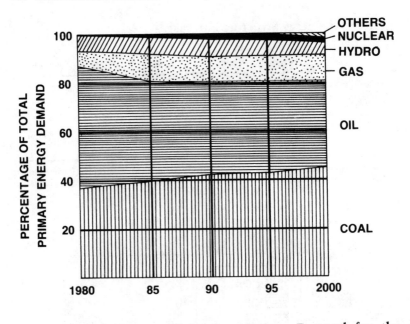

Figure 17–5. **Structure of Primary Energy Demand for the Period 1980–2000**

Electric energy consumption in 1980 was 2,210 kwh per capita. It is planned that the level in 1990 will be 5,455 kwh per capita, with 8,739 kwh per capita planned for the year 2000. This means that Yugoslavia will achieve the same consumption of electricity per capita in the year 2000 that Norway, Sweden, and the United States had in 1974. In order to achieve the electric energy production forecasted, it will be necessary to construct a greater capacity of electric power production. In the period 1985–2000, a total capacity of around 22,000–25,000 mwe should be built. It is expected that in the period 1980–90, a capacity of around 7,000 mwe will be put into operation, and in the period 1990–2000, the additional capacity will amount to 15,000 mwe.

Figure 17–6 shows one of the potential options for a power plant structure that could satisfy the forecasted electric energy demand. As the figure shows, the largest part of the electric energy production capacity would be derived from thermal power plants. The total hydroelectric power plant capacity in the year 2000 is expected to be about 13,000 mwe. Only recently, Yugoslavia put in operation its first nuclear power station, but according to the scenario in Figure 17–6, an additional capacity of 6,000 mwe from nuclear power plants would be introduced to Yugoslavia's electric grid by the year 2000.

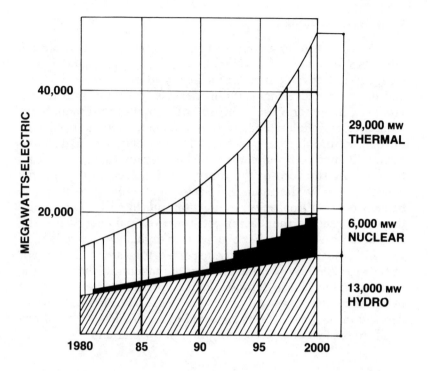

Figure 17–6. **Electric Power Plant Structure for the Period 1980–2000**

It is planned that the energy structure will be changed in accordance with the main objectives of Yugoslavia's energy policy. Although a substantial increase in the total consumption of liquid fuel resources is expected, it is also expected that this resource's percentage portion of Yugoslavia's primary energy structure will decrease over the course of the forecasted period.

It should be noted that the most recent primary energy demand forecast for Yugoslavia has lowered the former predictions of oil and gas consumption in the years to come. This lower prediction is based upon the present effort to achieve substantial substitutions for oil and gas in the future. But it must be said that apart from this underestimated oil and gas consumption forecast, there were no actions taken to stimulate the substitution process. It was also realized that the substitution process would entail capital-intensive investments—a fact which has to be considered within the framework of total investment capital availability, since the highest oil and gas savings could be achieved in the industrial and domestic sectors.

Final Energy Demand

The structure of a country's final energy demand is closely related to its industrial development. Yugoslavia embarked upon its industrial development immediately following World War II and has achieved a high rate of increase in its Gross National Product. Since energy use is very intensive in the high productivity industries, it is understandable that new energy sources need to be introduced into the energy system. The rate of industrial development will appreciably affect the final energy forecast for the years to come. By the turn of the century, it is expected that energy consumption by the industrial sector will constitute about 46 percent of total final energy consumption. In particular, metallurgy will become the leading consumer of final energy over the next decades. Also, a rather high portion of final energy consumption will be attributable to the transport sector—around 24 percent in the year 2000. When these factors are taken together with the substantial increase in domestic consumption expected as a result of an increased standard of living, Yugoslavia's final energy demand in the year 2000 is expected to reach a level that corresponds to present energy consumption in Switzerland.

CONCLUSIONS

From the above, the following points can be concluded.

(1) Intensive economic development in the last several decades in Yugoslavia has been the main force promoting rapid development of the energy sector. In Yugoslavia's change from a typical developing country with a rather low national income to its present level, an energy strategy that corresponds to the development of the individual economic sectors has proved to be indispensable.

(2) Since the energy sector is one of the most capital-intensive sectors, it is apparent that the rate of economic development has been strongly correlated with the availability of financial resources to develop Yugoslavia's energy sector. This proves to be a fundamental limitation on a developing country's ability to achieve its optimal rate of economic development. For this reason, the international community should devote more attention to the assistance of those developing countries, in order to increase the availability of their energy sources.

(3) A shortage of investment capital has been recognized as the immanent problem for most developing countries and has been one of the factors limiting the expansion of their energy sectors. As a developing country, Yugoslavia has borne witness to the fact that a shortage of investment capital was one of the factors limiting its optimal socioeconomic development. For this reason, we in Yugoslavia are in agreement with all those who have found the same obstacle to their development and are very determined to look for more justice in how the world's wealth is shared. We strongly believe that the world community in its long-term planning should see as being in its own interest the development of favorable conditions that will stimulate the availability of energy resources in the developing countries.

(4) For those countries embarking on the development of their energy sectors, a lack of appropriate manpower has become one of the limitations in taking full advantage of the discovery of domestic energy resources. In this respect, it is important to establish an educational system that will support the development of the energy sector. By having an established and traditionally good educational system, Yugoslavia was able to provide sufficient manpower at all levels of education in a relatively short period of time. This has proved to be a very important factor in the development of its local power industries and in sustaining a high availability of current and newly developed energy source units.

(5) Planning for long-term research and development is the key to economic development in any country. For this reason, the energy sector's development should be correlated with its corresponding research and development programs. Even Yugoslavia is not a very good example of planning long-term research and development programs in its energy sector, and its case could verify the fact that a long-term research and development program would be of great benefit to the development of many of its economic sectors, including its energy sector. At this point, it should be emphasized that a joint international effort would be of great importance in meeting this need, especially if it created cooperation among the developed and developing countries. Within the framework of future development, participation of the scientific communities from the developing countries in the long-term energy research and development program will represent a significant contribution towards solving future energy problems.

(6) Yugoslavia's energy forecast has shown that some energy problems will be left for solution after the turn of the century. This means that any energy policy should take into account the future

prospects offered to the energy sector by the stimulation of the options available through soft energy concepts. In addition, since it is planned that a majority of Yugoslavia's energy will be imported from the energy-producing countries, it is expected that Yugoslavia's future energy policy must be diversified in order to be able to overcome local disturbances in the exporting countries —disturbances which could happen even with our best wishes for the growth of mutual understanding and with our efforts to attain peace in the world.

NOTES

1. M. Simonović, "Solid Fossil Fuel Resources and Its Possible Development and Use," Symposium I on Energetics of Yugoslavia (Belgrade: Serbian Academy of Sciences and Arts, 1968). In Serbo-Croation.

2. Lj. Paradjanin, "Potential Possibility for Oil and Gas Production in Yugoslavia," Symposium I on Energetics of Yugoslavia (Belgrade: Serbian Academy of Sciences and Arts, 1968). In Serbo-Croation.

3. D. Jujić, M. Novaković, and V. Živanović, "Geological Potential and Possibility for Oil Shale Production in SR Siberia," Conference on the Energetics Development in Yugoslavia, Opatija, 1980 (Belgrade: Yugoslav Committee for World Energy Conferences).

4. S. Putnik, V. Omoljev, V. Jokanović, and V. Živanović, "Status and Perspective of Nuclear Fuel Resources in Yugoslavia," Opatija, 1980 (Belgrade: Yugoslav Committee for World Energy Conferences).

5. M. Pećinar, S. Olujić, and M. Melentijević, "Hydro Power in Yugoslavia," Symposium I on Energetics of Yugoslavia (Belgrade: Serbian Academy of Sciences and Arts, 1968). In Serbo-Croation.

6. V. Gburčuk and K. Jovanović, "Available Solar Potential in Yugoslavia," Energy and Environment, Sarajevo, June 1980 (Belgrade: Yugoslav Union for Protection and Development of Human Environment). In Serbo-Croatian.

7. "Demand and Possibility for Better Use of Geothermal Energy in Our Country," Conference of the Energetic Development in Yugoslavia, Opatija, 1968 (Belgrade: Yugoslav Committee for World Energy Conference). In Serbo-Croatian.

8. S. Boškovic and M. Todorović, "Long-Term Forecast of Energy Consumption in Yugoslavia," Energy and Environment, Sarajevo, June 1980 (Belgrade: Yugoslav Union for Protection and Development of Human Environment). In Serbo-Croation.

9. S. Vrhovac, L. Ljubiša, and M. Simonović, "Development and Use of Nuclear Energy in the Energetics of Yugoslavia," Conference on the Energetics Development in Yugoslavia, Opatija, 1980, Yugoslav Committee for World Energy Conferences. In Serbo-Croatian.

10. H. Požar, "A View of Future Supplies of Electric Power in Yugoslavia by Means of Conventional Forms of Energy," Journal of Yugoslav Committee of the World Petroleum Congress (Zagreb: August 1974), pp. 369–390. In Serbo-Croatian.

11. J.. Ugrinec, "Potential Basins and Regions for Oil and Gas Pros-

pection," Conference on the Energetics Development in Yugoslavia, Optaja, 1980 (Belgrade: Yugoslav Committee for World Energy Conferences).

12. L. Ljubiša, "Yugoslav Electric Power Industries," in J. C. Denton and N. H. Afgan, eds., *Future Energy Production Systems* (Washington: Academic Press and Hemisphere Publishing Corporation, 1976).

13. N. Afgan and M. Osredkar, "International Nuclear Fuel Cycle Pool," Conference on the Use of Nuclear Reactors in Yugoslavia, Belgrade, May 1978 (Belgrade: Boris Kidrič Institute of Nuclear Sciences).

14. M. Sadli, "World Energy Productivity: the Nature of the Problem for the Developing Countries," International Energy Symposium I, Knoxville, TN, October 1980 (Knoxville: The 1982 World's Fair).

15. "Report on World Energy Demand to 2020, Conservation Commission of the World Energy Conferences, London, 1977.

16. "Some Energy Problems and Issues in Developing Countries," Report of the Secretary General — Committee on National Resources, United Nations Economics and Social Council, Istanbul, 1979.

17. "Evaluation and Definition within the Scope of INFCE of the Specific Conditions in the Needs of the Developing Countries," IAEA paper to Working Group 3/21, Vienna, 1978.

18. R. Protić, "Current Energy Problems in Yugoslavia," *Nafta* 7–8, Journal of Yugoslav Committee of the World Petroleum Congress (Zagreb: August 1974), pp. 355–366. In Serbo-Croatian.

19. Z. Zarić, Lj. Ćuk, M. Golubović, A. Kneževic, P. Krasulja, and D. Podunavac, "Socio-Economic Development, Energy, Environment—Present Status and Future Development," Energy and Environment, Sarajevo, January 1980 (Belgrade: Yugoslav Union for Protection and Development of Human Environment). In Serbo-Croatian.

20. See reference cited in note 10 above.

21. D. Spasojević, I. Mihajlović, et al., "Prospective Application of Nuclear Power Plants in the Technology of Complex Use of Coal in REIK Kolubara," Mining and energetic industries consortium "Kolubara," Belgrade, 1980. In Serbo-Croatian.

22. L. P. Grayson, "The Design of Engineering Curricula," *Studies in Engineering Education* 5 (Paris: UNESCO Press, 1977).

23. N. Afgan, A. Kocić, and M. Mataušek, "Nuclear Research Centres in Developing Countries and the Manpower Development for Nuclear Power Programs," International Symposium on Manpower Requirements and Development for Nuclear Power Programs, Saclay, 1979 (Vienna: International Atomic Energy Agency).

CHAPTER EIGHTEEN

Summary with Selected Comments

Session Chairman/Integrator:
Ioan Ursu
First Vice-President
Romanian National Council
for Science and Technology

INTRODUCTION: THE NATURE OF THE PROBLEM

As a preliminary caveat, it should be noted that the category of countries addressed in this work session—"industrialized non-market-economy nations"—includes a variety of countries of different geopolitical conditions, historical backgrounds, civilization patterns, and levels of development. In the agreed language of the Symposium, among them one can definitely identify "industrialized" and "industrializing" as well as "energy-surplus" and "energy-deficit" nations. Furthermore, assessments of the levels of these countries' development depend on the assessment criterion used. Thus, some of these countries could be identified as "developed" in terms of their accumulated fixed assets, reflecting diverse industrial infrastructures with high unit outputs; as "developing" in terms of national income, reflecting their net production intensities; and as underdeveloped in terms of, among other things, their levels of labor productivity and energy intensity and productivity and their standards of living. And, as a final caveat, it should be mentioned that some participants in this work session pointed out that their countries are neither typically "industrialized" nor particularly "nonmarket."

In addition to the chairman/integrator and those presenting papers, the work session included the following assigned participants: John H. Gibbons, Zsolt Kohalmi (who presented László Kapolyi's paper), Igor Makarov, Georgio Santuz, Bogumil Staniszewski, Ferenc Szidarovszky, and Dan V. Vamanu. The chairman/integrator was assisted by Mitchell Olszewski, who served as rapporteur.

During the work session, two remarkably enlightening background papers were presented: a position paper by Dr. László Kapolyi, State Secretary of the Hungarian Ministry of Industry; and a case study by Professor Naim Afgan, senior scientist with the Boris Kidric Institute of Nuclear Sciences in Belgrade, Yugoslavia, both of which gave comprehensive accounts of the technical and socioeconomic dimensions of the energy problem in the category of countries under consideration. Based on these accounts and on the substance of the work session's debates, a trial composite pattern describing the countries under consideration may be identified. According to this pattern, it can be seen that owing to the way in which the energy and economic systems of the "industrialized nonmarket-economy nations" evolved, many of them entered the oil supply disruptions of the 1970s with the following common characteristics:

- An economic structure dominated by energy-intensive industries that generated a substantial share of the total national income. (In some cases, such industries have not yet entirely payed back the capital invested in them, a situation which makes decoupling energy consumption from GNP generation a problem, particularly in the short run.)
- Capital stocks that had been designed for relatively low energy prices rather than for the realities of escalating fuel costs.
- Centralized energy supply systems, in many instances— systems keen on insuring the accessibility of primary energy resources and using national electric power grids that were, in the case of the European socialist countries, internationally connected. (However, the primary resources of these supply systems already suffer from a certain fatigue, since rich, easily accessible, high-grade, and low-cost domestic fossil deposits are now approaching exhaustion.)
- An energy consumption pattern dominated in most cases by an industry addicted to hydrocarbons and suffering in some cases from the common deficiencies of infant industries—that is, relatively low thermodynamic efficiencies of their hardware, unsophisticated management, and insufficiently productive energy use.
- An economy under pressures stemming from raw material shortages, in several instances—shortages particulary of raw energy materials, and especially oil, due to limited domestic production, diminishing reserves, and difficulties in obtaining access to the resource markets because of severe price increases, trade imbalances, and external debts.

In short, the energy constraints of the present times affected, although to differing degrees, the economies and development policies of the industrialized nonmarket-economy countries— some developed, others developing, many well grown in their infrastructures but in some instances still immature in their economic structures, performances, and efficiencies.

Against such a background, it was found that *the problem* that the countries under consideration have to face in evaluating their energy options is inherently twofold:

- in the short term—to *sustain themselves under the deterioration of the present global energy/economic system*—that is, to insure adequate energy supplies in order to enable their economies' continued operation and reach decent, projected development goals; and
- in the medium and long term—to *acquire the capacity to formulate, implement, and manage an energy transition* to more sustainable and satisfactory energy systems.

There is no trace of a complacent "enjoying precedence" in the short-term task formulated above: the continued operation of the existing energy/economic systems of the countries under consideration is a matter of neither egocentrism nor free choice— it is a matter of survival. Hence, finding interim solutions to short-term difficulties is an unavoidable prerequisite if these countries are gradually to undertake the considerable efforts involved in a transition to more sustainable energy systems.

FINDINGS

The finding that emerges from the above perception of the problem is almost self-generating. It consists in *securing continuity and dynamic stability of both the subsistence and the development levels of the countries under consideration through resilient changes in their energy/economic systems, their modes of work, and their lifestyles.*

Basic Attitudes

In a world of instabilities that often spoil the original, positive values of the interdependence created by the scientific and technical revolution, it is believed that a sound way to reinstate these values is to *achieve or consolidate national and collective self-reliance* in ways that support the overall energy, economic, and social welfare. Energy programs in the countries under consideration (among which are included this author's country, Romania) recognize national and collective self-reliance as a crucial value, or recognized goal.

To achieve the appropriate degree of energy self-reliance in a reasonable time, *national mobilization of endogenous efforts* and adequate coordination of actions through *consistent international cooperation* are essential. From this perspective, the virtues of planning, popular participation in decisionmaking, and better economic self-management are deemed important to make the societies under consideration better prepared, at least in principle, for the quick and responsive strategies essential in times of crisis. However, it would be wrong to think that the creation, maintenance, and functioning of such mechanisms is a simple task. For example, it was noted during the work session that the centrally planned economies' apparent freedom of decision on investment policies does not preclude the objective actions of a series of economic laws that are duly and increasingly recognized in that area.

Assumed and Contemplated Options

For a number of decades, the countries under consideration have entertained coherent, formal, government-enforced energy policies. In times of globally cheap energy, at least a marginal excess of supply capacity was common wisdom. Today, however, the thrust of the energy policies has clearly shifted to one of *rationalizing energy services, tailoring demand to these services, and adjusting supply to the reevaluated demand.* No matter how sane such a focus is, it is clear that these countries' energy demand will continue to grow in absolute terms until the end of the century, along with their intended economic growth, particularly when the long lead time required to significantly decouple economic growth from energy consumption is taken into account.

Under these circumstances, there is obviously no place in the energy policies of these countries for radical, exclusive alternatives. The key word is the *mix:* centralized *and* decentralized, conventional *and* nonconventional, nonrenewables *and* renewables, "hard" *and* "soft" paths *merging.*

On the *supply* side, *diversification is the key to an acceptable mode of increasing energy production.* It involves the following, among other things:

• increasing premium nonrenewable domestic reserves through intensified geological survey;

- assuring low-grade, inaccessible, remote, and costly domestic nonrenewable resources as reserves, mainly through innovative technologies to tap and process them in forms economically convenient to end users; and
- promoting renewable energy sources, wherever economically or socially justifiable, and improving their competitiveness as a practical although still modest approach to ease the burden on nonrenewable sources and as a paradigm of future energy systems in the "great mix" era.

On the *demand* side, the emphasis should be on *conservation as a key to increasing energy productivity*. This involves different time horizons and echelons and should include the following:

- planning production levels consistent with the need to maintain adequate geological reserves;
- improving the thermodynamic efficiency of existing hardware through, among other things, secondary recovery of used heat, cogeneration, and other "total energy" concepts;
- substituting energy-conserving materials and technologies for energy-intensive ones to the greatest extent possible;
- using innovative industrial processing techniques to reduce industrial energy intensity;
- improving production management and logistics, supporting standardization (of processes, components, subassemblies, products, etc.), and reshaping building codes and other such legal and administrative methods;
- developing "energy conscious" styles of industrial and civil engineering design, architecture, work patterns, and lifestyles;
- making gradual changes in production patterns by deemphasizing energy-intensive branches and emphasizing low-energy, high-skill ones;
- reassessing demand in light of the enthalpic heat content and unit shaft or electric power effectively required by end users; and
- using system analysis methods to optimize the selection of options.

At the *interface of supply and demand*, efforts should be made to adjust supplies which are diversified but difficult to manage (in terms of cost and environmental and other side effects) to a demand picture that is revised but aggressive and unforgiving (in terms of its needs for continuity and quality of supply). Of prime consequence are *supply reallocations;* for example:

- shifting the balance of resources used for electricity generation from hydrocarbons to mainly coal, hydro power, and nuclear power;

- siting major energy-generating centers within economical thermal transport distances of major consumers, in order to facilitate "total energy" concepts;
- undertaking massive cogeneration to supply energy for both industrial processes and district heating;
- decentralizing the local utilities' base loads and operating them on renewable resources, with adequate conventional back-up; and
- saving energy through low entropy—that is, through more infusion of *information* in the energy/economic system, more use of microelectronics and so forth in the hardware, more systems analysis and operational approaches in the software, and greater use of more sophisticated control techniques throughout the supply and demand area.

A series of *supportive actions* are envisaged for the countries under consideration. Among these are the following:

- technical and managerial improvements in the performance and reliability of the national electric power grids, with improved hardware, software, and personnel training;
- improved methods of energy data gathering and analysis for energy forecasting;
- state monitoring and control of fuel and energy consumption on an end-user basis;
- state regulation of fuel and electricity prices, with an aim of making them stimulative of conservation;
- substantial state budgetary support for energy research, development, and demonstration, as well as state subsidies to consumers to facilitate penetration of new energy-saving or-supplying technologies;
- encouragement of and support for local commitments to energy self-reliance;
- the enactment of appropriate laws, regulations, and so forth;
- education to develop an "energy conscious" public and to design and implement patterns of a more energy-conserving society; and
- commitments to projects of bilateral, regional, and international cooperation in the fields of RD&D, equipment manufacturing, technology sharing and exchange, resource surveying and exploitation, information flows, education and training and so forth. (The views expressed on this point at the work session endorsed the findings of the UN Conference on New and Renewable Sources of Energy. In this context, the vital role of such international commitments for the developing countries, especially for the least developed ones, was strongly emphasized during the work session. The view was expressed that the countries under consideration, many of them already commit-

ted to programs of assistance to the developing countries, are in a position to serve, alternatively and selectively, as both donors and beneficiaries in the above fields. Also, international cooperation was noted as being particularly necessary and effective to optimally develop energy resources that are sometimes commonly shared.)

As a result, it was felt that (1) supply-oriented and demand-oriented options are highly interlinked in the frameworks of the nonmarket economies and indicate inextricable relationships between domestic production and consumption—a built-in feature of such systems; (2) likewise, long-term and interim short-term options can be difficult to distinguish, for short-term initiatives are germs or indications of longer term trends or commitments; and (3) the options are not mutually exclusive but are, in fact, complementary. The balanced mix of such options helps preserve the virtues of the centralized system—a well-established and inevitable outcome of the industrial era—while diminishing this system's inherent vulnerability; it also allows in a noncontradictory manner for the contributions of decentralized options, nonconventional technologies, and renewable resources, without prejudicing their possible roles in the future.

In summary, for the countries under consideration, mixed solutions to mixed challenges were felt to be one of the wisdoms of the current energy transition.

Constraints, Challenges, and Opportunities

A comprehensive view of these limitations and possibilities followed naturally from the work session's investigation of the nature of the problem and options for its solution, as presented above, but further elucidation can be obtained from a more detailed examination of some potential restrictive elements. One in particular was the assumption that all action in the energy field—whether technical, economic, social, or strategic—takes *changes* and long lead times.

Changes or substitutions were found necessary in the supply and demand patterns and the structures of economic production and energy generation, conversion, transportation, distribution, and end use. Almost invariably, this involves *technology substitutions*, no matter whether the technology is dealing with energy or information issues.

Making generalizations about technological change is a risky exercise. However, attempts were made to develop a pool of opinions about the feasibility of technology substitutions in the countries under consideration. Among the problems influencing decisions on technology in a broad sense, the following were identified:

- **The given energy/economic structure.** The unceasing supply of this structure with primary resources is of essence for sustaining and developing the economy.
- **Technology availability.** Although many of the countries under consideration possess important endogenous capabilities to generate technology, the rapid technological progress globally makes technology transfer, particularly to the developing countries, of exceptional importance. (In this context, the prolonged absence of the International Code of Conduct on the Transfer of Technology as well as the conspicuous accretion of restrictions and embargoes on technology trade and transfer, many of them relevant to energy, were found by the work session to be particularly regrettable and embarrassing. Another serious source of concern was that innovative efforts in the energy technology field are affected, as are all the other civil sectors, by the unmatched diversion of R&D capabilities and budgets to the military sector.)
- **Substitution costs.** A particular effect of this general constraint was emphasized with regard to countries that have limited resources (which is the case with many of the countries under consideration)—namely, that the *physical costs* of substitutions may sometimes weigh more in the decision process than their *capital costs*. This relates to the relative domestic scarcity of materials required by modern technologies such as special pure metals and alloys, polymers, glasses—materials that are durable and heat-, shock-, and corrosion-resistant, and so forth—many of which are energy intensive themselves and are usually produced in strictly controlled quantities for current uses *other than* substitutions. (In particular, crash substitution programs to recover low-enthalpy heat or to utilize renewable energy sources in some instances required massive initial supplies of such materials, and these programs were found to be potentially hindered by such a constraint, even though the applications resulting from them could pay back their incorporated energy content in a matter of a few years.
- **Capital costs.** The capital costs of substitutions (e.g., interfuel or conservation options), particularly of retrofittings, were also recognized as important moderating factors. In this context, it was noted that the past years' developments in the world economy have narrowed considerably the credit markets—markets overheated particularly by unbearable interest rates —which casts a shadow on the prospects of investments in

substitutions in all countries of limited resources, particularly in the developing countries. Therefore, options requiring minimal capital costs (e.g., control technologies using microprocessors) are of particular interest.

- **Uncertainties.** Uncertainties were found to result partly from the *long lead times* required to implement new technologies under the constraints of the existing technical and economic systems, especially in light of the uncertainty associated with future energy demands (and especially since the higher demand levels will be prolonged by the former constraints). These uncertainties can result in potentially high economic penalties if the supply capacity provided greatly exceeds or falls short of demand.

Other uncertainties were found to follow from the risks associated with substitutions. Unspecifiable but potentially grave risks attend, for example, the hasty implementation of "promising" but immature technologies that may turn out to be obsolete well before they pay back the investments made in them.

Nonmarket economies essentially are systems whose components have high long-range correlations with each other, and thus the success or failure of any one component, including the energy economy, reverberates strongly throughout the system. Therefore, although the system has ample resources for resilience, its limits of resilience must be carefully delineated, since crossing them necessitates sudden and severe adjustments in the infrastructure whose costs may be painfully internalized by society.

The energy issue, if not properly managed, may drive all the world's economies to the verge of disruption, and many of the countries under consideration in this work session see themselves as no exceptions in this regard. This again emphasizes the crucial value of *a number of options,* to be acquired in a timely fashion by these countries by strengthening their technological capability and encouraging, at all costs, rich energy and technology mixes.

Societal Response and Participation

Insuring the social acceptability of the energy and energy-related options was seen as a major task and a permanent concern of the decisionmakers in the nonmarket economies.

Several of the work session's participants emphasized that the energy programs in their countries by their very nature reflect the needs and aspirations of their societies. Some participants identified a permanent goal of their countries' systems as one of better equipping the systems to integrate popular wisdom and will when national commitments to developmental strategies and programs are made. Some speakers pointed to heavy local participation in energy decisionmaking (particularly in siting new energy facilities) as an example of this.

Criteria relating to the environment, to health and safety, to the quality of life, and to other governing factors can be clearly observed in all of the countries under consideration, through both their energy and technology policies and their prophylactic and corrective measures, including laws, regulations, and institutions. Like everywhere else in the world, purely economic and noneconomic (e.g., environmental) arguments are weighed in these countries on a case-by-case basis, in the recognition that both categories of argument must be judged in terms of their immediate social effects *and* their long-term effects. As an example, it was remarked that the advent of nuclear power is seen in some of the countries under consideration not only as an imperative need to fill foreseen energy gaps that cannot be met in time by either conservation or other alternatives, but also as a way to deemphasize in the long run the role that coal unavoidably will have to play over the next two decades as an interim energy solution for many countries, but one which entails many economic, social, and environmental hazards.

Insuring an option's social acceptability seems to depend heavily on the degree of societal cohesion around definite ideals and values—values which can best be assessed in times of hardship. In this regard, the opinion was expressed that an *increased awareness* and ever improving *knowledge* of energy helps to achieve *participative societies* that are less inclined to exacerbate constraints and more prepared to assume the inherent risks and, eventually, hardships of the energy transition, confident of this transition's long-run potential to normalize, in rational and equitable terms, their energy situation.

With regard to the global debate on the criteria, such as environment factors, affecting energy policies, it was noted that the efficacy of this debate so far has been limited by some selective inattention, as David Rose would put it. However, it is also a fact that an all-encompassing multidimensional treatment of a society's problems is not yet always and everywhere available in real life. Whether one likes it or not, time is still vital, thus inducing sequential approaches. By way of example, it was noted that when a country is suffering from a fuelwood shortage ("the second energy crisis"), one should not confine oneself to warning of the deadly environmental perils of advanced deforestation; one would do better to *help* that country acquire adequate knowledge and skill on improved wood plantations and rational harvesting, land reclamation, more efficient and economic burning and other conversion techniques, better road and rail infrastructures, appropriate industries to yield better profits from the wood, and alternative energy sources—thereby eliminating the dangers of deforestation and the environmental offense by the very process of meeting the first priority—the fuel and energy supply. Orienting debates and, particularly, international cooperation and assistance to coherent and substantive *action* was seen by the work session's participants to be essential, since action is always closer to people and easier to appreciate than the philosophy behind it.

A clear determination of the merits, limits, and shortcomings of the nonmarket *and* market systems was found to be a profitable analytic exercise. However, the belief was expressed that there is room in this world for a pluralistic global approach to energy and other critical issues. The important thing is that each nation be free to pursue its own path, the one most responsive to its needs and value system, without imposing burdens on others.

RECOMMENDATIONS

In the 1970s, faced with numerous endeavors to assess the global energy situation and its prospects, we gradually learned to discriminate between conjectural and fundamental energy difficulties and, hence, to perceive the so-called energy crisis as a long-term *challenge* and an exciting *opportunity.*

Today, in 1981, the emerging concept of *"the energy transition"* seems to accommodate our perception of the energy issue quite well. It is apparent that no blitzkrieg can solve the problem. There are no quick or easy fixes for this problem except, perhaps, for some of its short-term ramifications. Various options must be continually scanned, assessed, and pursued by government, the private sector, and international organizations on the assumption that they are *trial solutions* for a *trial time period.*

Basically, then, one is left with (1) the valuable awareness that each individual, community, nation, and international association is on board the same ark en route to another energy order and (2) the underlying conviction that it is within mankind's grasp to gradually and peacefully achieve that better energy order through scientific knowledge, technological capability, economic wisdom, and, most of all, political will. This philosophy is in no way defeatist, nor does it avoid the task of continually working on the present energy systems, economies, domestic and foreign policies, and international relationships, in order to mark steps towards their betterment. Committed attitudes, projects, plans, and programs *are* needed; physical, capital, and human resources *should* be better managed; and additional sources *should* be mobilized. Options must be talked in all candor, with awareness of their limits but confidence in their vital potential to keep the energy transition moving. And all these have to be done on the understanding that, as Hans Landsberg said during Symposium I, "...policy decisions in both the public and the private sectors [must] face the stubborn reality that whatever steps are taken in the energy arena have repercussions for other societal goals."*

In order for such talks and encounters to be meaningful, it is important that those primarily concerned be included. Pragmatic as well as moral considerations lead to the conclusion that a larger number of participants from the developing countries, known to be those most frustrated by the present status and future prospects of the energy problem, should be invited to take part in the next phase of the International Energy Symposia Series. Their presence would draw attention to the basic yet often ignored truth that *there cannot be peace in a hungry world.*

On the other hand, no matter how wide the participation, the wisdoms arising from the Symposia Series must be *exposed*, not *taught*, to the world. This is in line with the main commitment of these Symposia, which is to enhance international communication on a crucial global issue—energy—and thus to help build further mutual understanding, tolerance, and confidence, all of which are unavoidable prerequisites of viable, sincere, effective, and far-reaching international cooperation.

*H. H. Landsberg in *World Energy Production and Productivity*, Proceedings of the International Energy Symposium I, ed. R. A. Bohm, L. A. Clinard, and M. R. English (Cambridge, MA: Ballinger, 1981), p. 183.

SELECTED COMMENTS

DAVID LE B. JONES

I was in the group on industrialized market-economy countries. Listening to Professor Ursu's statement on the nonmarket economies, I was struck by the many similarities between the market and the nonmarket industrialized economies in terms of both their problems and their policies. In the work of our group, in Professor Oshima's summary of its discussions [see chap. 15], and in Professor Ursu's remarks, there was throughout a recognition of the need for a mixed approach; a recognition that there were not sharp choices to be made but rather a need for nuclear power, renewable energy sources, coal, oil, and more efficient energy usage; a recognition of the need for both centralized and decentralized solutions; and a recognition of the need for a pluralistic and global approach. There were, of course, some differences of emphasis, but I doubt if they were greater than the differences of emphasis which arise at various times in the policies of individual countries with dissimilar market economies—for example, the United States and France.

The question I would like to raise is the framework for decentralized decisionmaking—in other words, how to get the right balance between reliance on the market (that wasn't the phrase Professor Ursu used, but I think it was the same thought) and government intervention in a market which is very imperfect. In the market-economy group, we gave a lot of attention to pricing policy and to how prices, particularly for commodities such as gas and electricity that are not traded internationally, ought to be fixed. Professor Ursu also mentioned pricing policy—I think he said there was state regulation of the prices of electricity and coal. One question I would like to put to him is the basis on which these regulated prices are fixed—is an effort made to relate these prices to supply costs, and if so, how is this done? And I would like to ask if he could describe in a little more detail how the state monitoring of the end uses of energy demand to which he referred is carried out—how deeply does this process of state monitoring go, and how is it reconciled with a decentralized approach? It seemed to me, listening to Professors Oshima and Ursu, that there well could be a case for more comparative study and discussion of energy policies in the market and nonmarket industrialized economies—not at a political level but at the technical and expert level which has been going on in this Symposium.

KEICHI OSHIMA

I had a very similar impression to what Mr. Jones said. The only thing I want to say is that as for recommendations for the next Symposium, I think, as some of the other people have emphasized, that the interfaces between the four national groups—the nonmarket and market industrialized (although "market" may not be a good basis for definition); the energy-surplus and energy-deficient developing—will be the really important concern, particularly when considering future problems to be solved and actions to be taken on future uncertainties. Thus, what I propose is that the next Symposium should focus on these interface issues.

AMORY B. LOVINS

I should like to fuse some earlier contributions by Dr. Usmani and others into a question for Guy Pauker. In the spirit both of the proposal in our book *Energy/War** for the Fund for Renewable Energy Enterprise and of similar proposals made at the UN conference in Nairobi, I wonder why one should limit concessionary exports to coal. For example, in the United States, where our biggest energy supply expansions in the past years have come from renewable energy sources, we have at least an equally attractive opportunity to export photovoltaics, for example, and the means of making them indigenously. This would bring genuine energy independence rather than leading to dependence on a different kind of fuel export, concessionary or not; it would also bring —particularly to rural areas in developing countries—the unquestioned benefits of electrification to which Dr. Usmani correctly referred.

*A. B. and L. H. Lovins, *Energy/War: Breaking the Nuclear Link* (San Francisco: Friends of the Earth, 1980; and New York: Harper & Row, 1981). This book expands on the authors' article (with L. Ross), "Nuclear Power and Nuclear Bombs," *Foreign Affairs* 58 (Summer 1980).

JOHN H. GIBBONS

I had the pleasure of spending my time this week in the work session on industrialized nonmarket-economy nations, and I believe that, other than the rapporteur, I was the only person involved in those discussions who wasn't from Eastern Europe. I learned many things and congratulate Professor Ursu for his careful preparation and fine leadership in that session.

To add a postscript to Amory Lovins' remarks, there is another form of energy export which we discussed in our session. It is a two-way flow, and it doesn't involve a depletable resource such as coal but rather involves the flow of information on technology related to, for example, the more efficient conversion and use of energy in various economies. It seems to me that while we tend to think of solutions to energy problems in terms of material resources, some of the most important solutions are, instead, intellectual—that is, they entail the ingenious application of human intellect through technology to provide energy-related amenities and necessities with less energy per se.

I would also like to underscore for Professor Ursu some of the interesting observations made in our session. We spoke frequently of the role of information and education in making wise energy decisions. Time and time again, the notion of energy price surfaced as one of the most effective conveyors of information for consumers. It seems to me that there is a great commonality not only of energy problems (such as the need for more efficient use) but also of solutions such as price signals that transcend the diverse forms of governance under which we live.

It also seems that we witness rich opportunities around the world for both bilateral and multilateral cooperative ventures in matching the self-interests of various countries in research and in widespread attempts to use energy more efficiently and to take better advantage of local resources.

Our needs are common, and the extent to which our nations can make progress will directly ease the burdens of the others. I hope that such a common cause will, in the 1980s, engender some uncommonly sensible acts of international cooperation.

ZSOLT KOHALMI

I took part in the industrialized nonmarket-economy work session, and with regard to several of the comments about the commonalities between it and the market economy session, I am going to try to give a very short answer on why these commonalities should occur. The reason is this: although we have antagonistic political ideologies, we don't have an antagonistic approach to Mother Nature, and the latter is what we are facing here. I hope that we also do not have antagonistic sets of basic human needs, for that is also what we are facing.

I want to make a few comments on the role of international cooperation. It has been emphasized, but I think very few specifics have been given, either of good examples or of bad ones. I would like to mention two.

First, in our work session's group of countries, genuine efforts to construct hydroelectric stations in common on bordering rivers—a natural way to go—have really paid off. These decisions were made on "nonmarket" bases, but in the marketplace, too, everything looks very good. The same benefits can be achieved from several countries making common investments and working together to extract major—for example, Siberian—gas resources.

Second, and more negatively, obstacles to cooperation can occur, particularly in the marketplace. When we talk about free trade in the context of this Symposium, we may well be thinking about the flow of oil, gas, or whatever. But this flow is impeded if any government elects to qualify the *equipment* used to extract oil as strategic and to place it on an embargo list. This is an action which can only aggravate the energy situation for all of us.

IOAN URSU

First, I am impressed by the perceptiveness of many speakers here about the commonalities in energy and energy-related issues among countries with different economic systems. Surprise at such commonalities only indicates some basic lack of understanding about how close we really are, or could be, to true international cooperation. It also measures the extent of the communication gaps between various regions coexisting in this world which we share. Here we are, with one more reason to salute events like this Symposium.

In regard to other comments made here, I certainly agree with and understand the remarks by Dr. Kohalmi and Professor Afgan [see chap. 24]. I also appreciate the pertinent remarks by both Dr. Gibbons and Mr. Jones, some of them of an interrogative nature, which challenges our spirit. Full answers would require a much too ambitious endeavor for the short time allotted.

Take, for instance, the issue of *domestic pricing* in the so-called nonmarket economies. In this respect, we are now living in an interesting time of transition from a traditional policy of ample subsidies to the production sectors (subsidies which acted as an umbrella for the embryonic infrastructure of industries built up from scratch) to a policy of economic self-sustainment, self-financing, and self-management, for all of which "efficiency" is the key word and more and more the passport to life for these enterprises. It takes a profound knowledge of the situation to realize how deep and, indeed, shocking the impact of such a shift can be for the economies, work methods, and even lifestyles of our countries. While gradually lifting the protective umbrella from the production sectors (which *is* a sort of "deregulation" of a peculiar nature), the state control of prices—a well-established function in our economies—tries, to the extent possible, to protect the consumers, the people. But everywhere, there are unavoidable constraints and limits, and the *international pricing* of energy, raw materials, technology and credits—the behavior of the world market in general—is one of the most severe.

About *decentralizing the energy supply:* I must have been in a hurry during my oral summary, since it was my intention to point out clearly that for the particular energy patterns—and histories —of our countries, only a *mix* of approaches—that is, centralized *and* decentralized, conventional *and* nonconventional, nonrenewable *and* renewable, hard *and* soft—could possibly suffice as an energy policy. In our growing economies, where the need for adjustments is as heavy as the difficulty of assuming and paying for them, there is no place for radical, exclusive alternatives. Al though our energy supply systems are by nature highly centralized, we take seriously complementary decentralized solutions, in terms of locally based energy "islands" that are *potentially* autonomous in critical situations, local sources such as micro hydro power plans and community biogas plants, coal-fired district heating units, and so forth.

I cannot make a long story short. However, by inviting all the participants who were concerned with the outcome of our work session to reach me in writing or by any other way they might deem appropriate, I hope that we may continue our dialogue, to the benefit of improved communication and shared insights.

Work Session on Energy-Surplus Industrializing Nations

OPEC and the Oil-Importing States in the 1980s (with a Note on Mexico): Economic, Political, and Security Relations

Theodore H. Moran
Director, Landegger Program in
International Business Diplomacy
Georgetown University School of Foreign Service

INTRODUCTION

The history of attempts to analyze the future prospects for OPEC provides more than sufficient grounds for humility. In the first decade after the formation of OPEC in 1960, the major oil companies predicted that the real price of oil would not rise for technical or economic or political reasons above $2 per barrel until well into the twenty-first century. Shortly after the first oil crisis of 1973–74, Milton Friedman predicted that the fourfold price rise would stimulate a sufficiently rapid turnaround in the market to cause the breakup of OPEC.[1]

The mid 1970s saw the onset of a new perspective, led by the US Central Intelligence Agency, that the world faced energy "gaps," absolute shortages, and rapidly escalating prices. The consensus forecast was that the world had to complete, in short order, the transition to the end of the oil era.

Now, at the beginning of the 1980s, a new orthodoxy is emerging. With oil prices 700 percent higher in real terms after the "second oil crisis" of 1979 than they were in 1973, conventional wisdom is taking the form that the world can look forward to glut for at least a decade; that demand for OPEC exports will never again be as high as it was; and that the cartel has lost most of its power to influence events. In this future of market weakness, the behavior of OPEC can largely be taken for granted, according to the new orthodoxy, as the members scramble to maintain producer cohesion and the ability to act in the cartel's common economic interest. The new orthodoxy of complacency toward OPEC and the energy crisis rests on the assumption, of course, that the security context for OPEC can be maintained in relatively stable fashion.

This paper will attempt to provide a framework for analyzing three questions: (1) How much of OPEC's power has been lost? (2) How is OPEC likely to determine its price and production decisions in the 1980s? and (3) What is the connection between oil supplies, on the one hand, and the political and security relationships between the key producer and consumer states, on the other?

THE SETTING FOR OPEC BEHAVIOR: SUPPLY AND DEMAND FOR OIL

In the aftermath of the second oil crisis following the fall of the Shah of Iran, OPEC exports dropped from 28.8 Mbd in 1979 to 24.7 Mbd in 1980 with a further fall in the second half of 1981 to 20.4 Mbd (see Table 19–1).

Table 19–1. OPEC **Exports and Noncommunist World Oil Demand, 1973–81** (millions of barrels per day)

	OPEC exports	Oil demand	OPEC share of demand
1981 (8 months)	20.4	39.2	52.0%
1980	24.7	49.0	50.3%
1979	28.8	51.6	55.9%
1978	27.9	50.9	54.8%
1977	29.6	49.5	59.8%
1976	29.3	48.0	61.0%
1975	25.6	45.2	56.7%
1974	29.1	46.3	62.8%
1973	29.5	47.9	61.6%

Source: J. H. Lichtblau, "The Limitation to OPEC's Pricing Policy," Petroleum Industry Research Foundation, August 21, 1981, p. 4; and *Petroleum Intelligence Weekly,* October 12, 1981.

The decline in energy consumption has been due in part to the sluggishness of economic growth rates in the large industrial energy-consuming states. But there has also been a marked drop in the ratio between the increase in energy usage and the growth in GNP in the OECD countries, from a historical relationship of one-to-one (one-percentage-point growth in GNP producing a one-percentage-point increase in energy usage) to a low of 0.6 to 1 for 1980. Studies by the Harvard Business School Energy Group, the Ford Foundation, and Resources for the Future have suggested that the possibilities for additional conservation are greater and less painful than had previously been assumed.[2] At the least some momentum of increased efficiency should continue as the capital stock of the industrial states is turned over, with less wasteful cars, airplanes, homes, and industrial buildings replacing older models.

These trends have led to what *Petroleum Intelligence Weekly* calls "the increasingly popular view" that the demand for oil has peaked and the demand for OPEC is already past its zenith.[3]

On the other hand, even small changes in highly uncertain assumptions can dramatically alter the market projection by 1985. Consider, for example, what happens to the US government's "midrange" base case* if the hypothesized OECD economic growth rate is raised by one-half a percentage point, from 2.6 percent per year to 3.1 percent (even though the ratio of energy consumption to GNP growth is held constant at 0.5), and the aggregate noncommunist economic growth rate is increased from 3.0 percent to 3.5 percent. The result is a jump in the demand for OPEC oil by 1.9 mbd, from 26.0 mbd to 27.9 mbd (see Table 19–2).

Table 19–2. **Noncommunist Oil Consumption and Supply, 1985** (million barrels per day of oil equivalent)

	Actual 1980	"Midrange" base case in 1985	Higher economic growth rate case for 1985
Oil consumption			
OECD	38.8	35.6 (2.6% growth)	36.1 (3.1% growth)
OPEC	2.7	3.5 (5.8% growth)	3.8 (6.3% growth)
Other	8.2	8.9 (4.1% growth)	10.0 (4.6% growth)
Total	49.7	48.0 (3.0% growth)	49.9 (3.5% growth)
Oil production			
OECD	15.5	14.6	
Other	5.7	7.9	
Communist	0.9	(-0.5)	
Total	22.1	22.0	
Required OPEC output	27.8	26.0	27.9

(Totals may not add due to rounding)

Source: US Department of Energy, *Energy Projections to the Year 2000*, July 1981 (DOE/PE-0029), pp. 2–9 and 2–13.

*The US government projections are taken for illustration not because they are considered particularly prescient but because the model is publicly explicated in some detail. In fact, however, the US government's "midrange" forecast for OPEC production in 1985 (26.0 mbd) is not far different from others that are available such as Lictblau's (26.9 mbd) or Ait-Laoussine's (26.6 mbd).[4]

Table 19–3. **Maximum Likely OPEC Production in 1985** (million barrels per day)

	Actual 1980	US government low case 1985	Ait-Laoussine 1985	US government high case 1985
Algeria	1.0	0.7	0.9	0.8
Ecuador	0.2	0.1	0.1	0.2
Gabon	0.2	0.1	0.1	0.2
Indonesia	1.6	1.4	1.3	1.5
Iran	1.6	2.0	2.5	2.5
Iraq	2.6	2.2	3.0	3.5
Kuwait	1.7	1.2	1.7	1.5
Libya	1.8	1.2	1.5	1.5
Nigeria	2.1	1.4	1.9	1.8
Qatar	0.5	0.4	0.3	0.5
Saudi Arabia	9.9	6.7	9.5	8.7
United Arab Emirates	1.7	1.2	1.8	1.4
Venezuela	2.2	1.9	2.0	2.1
Natural gas liquids	0.8	1.2	(included above)	1.6
Total OPEC production	27.9	21.5	26.5	27.8

Source: US Department of Energy, Energy Projections to the Year 2000, July 1981 (DOE/PE-0029); N. Ait-Laoussine, in Petroleum Intelligence Weekly, September 28, 1981.

This pushes demand for OPEC oil against or above the high range considered likely for OPEC output by energy forecasters (see Table 19–3), even if Iran and Iraq are able to end hostilities and restore production to sizable proportions (a combined total of 6 MBD) by 1985.

It is instructive to look at what happens to market tightness under more optimistic scenarios (from the consumers' point of view): suppose slightly higher world economic growth occurs, while budgetary constraints in the high absorbers Nigeria, Venezuela, and Indonesia induce them to cut prices until their productions average at least 90 percent of full system capacity. This adds 0.5 MBD to the US government "high maximum" figure for OPEC production, postponing the moment when oil markets become tight only until mid 1986.

Finally, suppose world economic growth rates remain sluggish throughout the 1980s. This postpones the moment of market tightness a little further until sometime in 1987. In addition, it should be remembered that expectations play a large role in the oil industry. The anticipation that demand might begin to overtake supply will begin to be registered in market behavior more than a year before the event. Moreover, the oil market is extraordinarily fragile; the 1973–74 crisis witnessed an actual shortfall of less than 7 percent, yet it released a quadrupling of oil prices; the 1979 crisis involved an actual reduction of 4 percent for one quarter, yet it produced a price jump of 150 percent.[5]

Thus, even with optimistic assumptions about the evaluation of international oil markets, the world is likely to face growing oil tightness by the mid 1980s.

In the other direction, pessimistic scenarios are also plausible. The extraordinary degree of conservation evident in consumer behavior in 1980–81, for example, has been based on a reaction to sharply escalating oil prices and the perception that such dramatic climbs would inexorably continue. A period of stable or declining real prices could slow or even reverse the drive toward conservation. Iran and Iraq might continue in an uneasy state of confrontation that did not permit them to sustain production above a combined total of 3.5 MBD. These or other pessimistic scenarios would produce tight oil markets as early as 1983–84.

In short, there are great uncertainties about the actual evolution of oil markets in the 1980s, and small changes in the optimistic assumptions prevailing at the present will markedly alter the relation of supply and demand. For planning purposes, it seems reasonable to conclude that by the mid 1980s the condition of international oil markets will again be left to the discretion of the handful of Persian Gulf exporters whose felt need to produce at the level required is highly uncertain.[6]

More ominously, over the longer term some analysts predict that OPEC must become a more gulf-centered organization: rising internal consumption, according to Nordine Ait-Laoussine, could force Indonesia, Ecuador, and Gabon out of OPEC and into the role of net oil importers by 1985, with Algeria, Nigeria, and Qatar suffering a similar fate after 1990.[7]

By the year 2000 the proportion of OPEC exports produced in the gulf might grow from 67 percent in 1980 to almost 90 percent. If Mexico were to join OPEC, the percentage of gulf-origin oil would drop but not, of course, the absolute amount.

Thus, the world may enjoy a period of relief from tight oil markets in the early years of the 1980s, but it is by no means assured over the medium to longer term of escaping the fate of living dependent upon a small number of fragile and problematic gulf regimes.

UNDERSTANDING THE DETERMINANTS OF OPEC BEHAVIOR

The preceding section has argued that the world oil market

remains fragile, with small changes in expectations about the evolution of supply and demand producing large changes in behavior, and that the medium-term forecast about the evolution of supply and demand is highly uncertain. Nevertheless, the short-term forecast is for continuing glut, barring a major interruption in the security of the exporting states (see section on "Security and Stability" below).

If consumers were faced with a competitive industry on the supply side, the simplest information about production capacity in relation to demand would be sufficient to predict falling real prices into the mid 1980s. But how will a core group of the OPEC members—composed of Saudi Arabia, Kuwait, the United Arab Emirates, Qatar, and perhaps Libya—whose surplus of oil revenues and financial reserves above budgetary needs offers the governments discretion in how much to produce, exercise that discretion?

The difficulty in understanding how to answer this question comes not from the fact that economists have too little to say about cartel behavior but that they have too much.[8]

The predominant framework for predicting how OPEC will behave is derived from constructing an optimal price trajectory for the group as if it were a perfect monopoly trying to maximize the economic benefits from its oil resources. The contention of this approach is that economic self-interest provides the best predictor of the cartel's prices and production decisions over time. In short, the *most likely* price for OPEC will be the best price for OPEC oil; and the best price, in turn, is the price that maximizes the economic returns received by the cartel. Certainly, a price that is too low deprives the members of revenue, but so does a price too high, by inducing conservation, stimulating substitution, and damaging the economic health of the consuming countries. A price policy that is too timid will prove self-denying; a price policy that is too aggressive will prove counterproductive.

This reduces the forecasting problem for OPEC behavior to the task of tracing the price path that would be chosen by a rational monopolist. On the one hand, the revenue stream must be adjusted to reflect the impact of the price chosen at each moment on the structure of subsequent supply and demand. On the other hand, the revenue stream must be discounted at a rate that reflects income lost by postponing earnings rather than taking and investing them in the present. By estimating elasticities of supply and demand, however, and assigning an appropriate discount rate, the monopoly pricing problem is readily solved.

But the situation is more complex in the case of an exhaustible resource like oil.[9] A commodity with a fixed stock can generate an implicit return for its owner not only by being extracted and sold but also by being left in the ground and appreciating in value. For this reason, the rational monopolist in the oil industry must add considerations about the rate of exploitation that balance oil in the ground and money in the bank. To achieve this balance, the value of a resource in the ground must be growing fast enough to equal the value of future sales for a producer to be willing to leave it there. Hence, the choice between exploiting or conserving the resource will depend on whether the monopolist expects its net price to increase exponentially at a rate equal to the discount rate on future earnings. If a higher rate is anticipated, then self-interest dictates postponing production and benefiting from the more rapid appreciation of the hydrocarbon assets in the ground. If a lower rate is anticipated, production in the present should be hastened.

The gain to producers from the successful cartelization of a nonrenewable resource like oil comes from the manipulation of the pace of exploitation: the price path of the monopolist will be initially higher and subsequently lower than a competitive price path, as shown in Figure 19–1.

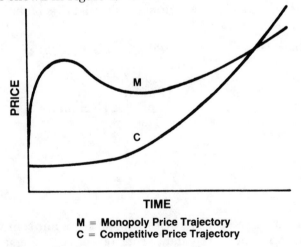

M = Monopoly Price Trajectory
C = Competitive Price Trajectory

Source: T. H. Moran, "Modeling OPEC Behavior: Economic and Political Alternatives," *International Organization,* Spring 1981.

Figure 19–1. **Monopoly and Competitive Pricing**

This simple scheme for forecasting how OPEC will behave assumes that the exogenous variables of supply and demand for energy determine both the opportunities and the constraints of the would-be cartel. Among other problems to be analyzed later, this approach immediately suffers from the representation of the OPEC governments as a single rational actor.

Even within the economic framework of OPEC modelers, the individual OPEC countries have differing "rational" economic interests depending upon domestic social pressures, revenue needs, alternative sources of export earnings and fiscal income, hard currency financial assets, and geological reserves. Hence, they have different discount rates for present versus future earnings and different strains or pains associated with holding spare capacity or not developing additional capacity. Ultimately the members of OPEC have different preferred price and production paths for the exploitation of their petroleum reserves.

The principal simplifying technique to accommodate this diversity of economic preferences has been to break OPEC into subgroups according to the extent of oil reserves and the need for current reserves. Those governments without pressing cash needs who must worry about the long-term outlook for oil will, by hypothesis, be more dovish or moderate on oil prices than those who want to press the market for all it can bear in the short term by pushing prices up. The two groups thus might be broken out as follows:

Price pushers	*Price "moderates" (by assumption)*
Algeria	Saudi Arabia
Ecuador	Kuwait
Gabon	United Arab Emirates
Indonesia	Qatar
Iran	
Iraq	
Libya	
Nigeria	
Venezuela	

The detailed analysis of this distinction in economic self-interest between "price pushers" and a core of "price moderates," or between "hawks" and "doves," reveals two things: first, that there will be a substantial difference of perhaps 30 to 60 percent between the preferred price path of the two groups (see Figure 19–2); second, that the economic costs to whichever group is unsuccessful in prevailing will be great (perhaps $600 billion over the next forty years).[10]

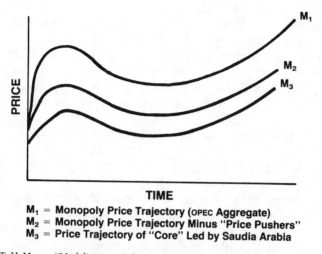

M₁ = **Monopoly Price Trajectory (OPEC Aggregate)**
M₂ = **Monopoly Price Trajectory Minus "Price Pushers"**
M₃ = **Price Trajectory of "Core" Led by Saudia Arabia**

Source: T. H. Moran, "Modeling OPEC Behavior: Economic and Political Alternatives," *International Organizations*, Spring 1981.

Figure 19–2. **Alternative Optimal Cartel Prices**

One would anticipate therefore that the struggle between the two groups, if they follow their economic self-interests, will be great. Within the framework of those who model OPEC behavior according to economic self-interest, however, the Saudis should play a key leadership role in keeping the cartel on a moderate course. With the largest reserves and the lowest discount rate, the Saudis have the greatest interest in insuring against long-term conservation and substitution. They should use their vast productive capacity to resist the price pushing of the hawks, and discipline other members of the cartel if the latter do not go along with the Saudi price preferences by taking away their market.

There is a growing school of energy analysts who argue that this economic self-interest model correctly explains current Saudi actions. Throughout 1981, the Saudis resisted the call of the price hawks for a quotation above $34 per barrel. At the key OPEC meeting in Geneva in May and afterwards, the Saudis held firm, refusing to lower their production below approximately 10 Mbd unless the others agreed to a unified price at $34 per barrel.

Clearly, there was a preoccupation within the Saudi leadership that oil prices had risen too far from the point of view of the long-term health of the cartel. As Sheik Yamani argued: "We do not want to shorten the life-span of oil as a source of energy before we complete the elements of our industrial and economic development, and before we build our country to be able to depend on sources of income other than oil. In this respect, the Kingdom's interest might differ from those of its OPEC colleagues. In OPEC, there are countries that will stop exporting oil before the end of the eighties; for such countries the life-span of oil should not extend beyond that time. But if the life-span as a source of energy ends at the close of the present decade, this will spell disaster for Saudi Arabia."[11]

In short, he explained, "We think [the rise in oil prices] went too far."[12]

On this basis, the economic self-interest school of OPEC analysts tend to argue that the behavior of the cartel, led by the Saudis, can be taken for granted. Faced with the increasingly evident reaction of consumers to high oil prices, OPEC will be held to a line of moderation. And if some of the more aggressive members deviate from the moderate line, the Saudis will force them back. Moreover, they will do this on their own without the need for political inducements.

Douglas J. Feith argues, for example, "Whether [Saudi] rulers are 'pro-Western,' 'friendly' or pleased with any or all of the elements of US foreign policy is an issue 'linked' to Saudi oil policy only in diplomatic rhetoric and in the minds of those who do not actually bear responsibility for turning Saudi oil into money."[13]

Eliyahu Kanovsky makes the case most straightforwardly, "Why make political concessions to persuade the Saudis to do what their economic interests dictate?"[14]

Thus, with a weak market and increasing conservation forecast to the mid 1980s, economic self-interest should ensure moderation well into the future. After a decade of trying to cope with OPEC's power, an era of complacency can replace an era of crisis.

OPEC BEHAVIOR OBSERVED, 1973–81:
ECONOMIC VERSUS POLITICAL MOTIVATIONS

Is such complacency justified?* Is economic self-interest a powerful enough force to constrain OPEC's options through the 1980s? Can the Saudis be depended upon to keep the cartel's price and production policy "moderate" (with prices most likely falling in real terms) through the 1980s?

*Again, the assumption of this section is that the Saudi regime remains stable and intact through the 1980s. For a treatment of possible instability and upheaval in Saudi Arabia, see the section below on "Security and Stability."

The historical record suggests that such complacency is not justified: an analysis of the key decision points for intra-OPEC decisionmaking between 1974 and 1981 demonstrates that while market parameters may provide the outer bounds for price and production behavior, *political* and *security* considerations are crucial in shaping how the core of OPEC (led by Saudi Arabia) act.

The detailed examination of the six crucial episodes in the determinations of OPEC policy for 1974–81 has been done elsewhere.[15] The results are summarized here. They show that no economic model accounts for, or is even consistent with, the outcomes of OPEC policy choice. Instead, political and security concerns, especially those related to the broader issues in Arab–Israeli relations, constitute the "high politics" of OPEC behavior as led by Saudi Arabia.

Key OPEC Decision Points, 1974–81

There have been six occasions since the end of the 1973 embargo when the question of determined Saudi leadership within OPEC on behalf of moderation have come to a head.

• The first crisis in intra-OPEC bargaining, in 1974, is similar to that in 1981. In 1974, the Saudis feared that they had pushed oil prices too high. The world was experiencing the most severe recession of the postwar period. Under pressure from the United States, they agreed to a special auction of 1.5 mbd that would lead to a unilateral lowering of the OPEC marker. In response to strong language from their fellow OPEC members, especially the Shah of Iran, however, they canceled the auction. As a compromise, they carried out a formal price reduction that was more than offset by royalty and tax increases. To make the price rise stick, they absorbed a disproportionate share of the production cutback. "Their actions," the *Oil and Gas Journal* editorialized, "destroyed any illusions the world may have had about them as seekers of lower crude prices."[16] The desire not to offend threatening neighbors outweighed economic or market considerations.

• At the end of 1975, with the world economy struggling back from near-depression, Sheik Yamani again indicated concern about the health of both the oil market and the international economic system. Leaving the OPEC price meeting dramatically in the middle, he asked his country's leadership if he could use the kingdom's 4 mbd in excess of capacity to enforce a freeze or "small increase" in the price of oil. In the face of an OPEC majority against him when he returned, however, he assented to keep his spare export capacity in reserve and went along with a 10 percent rise.

• The fall of 1976 showed the first instance of determined Saudi leadership in the face of opposition within OPEC. In fact, the world economy was reviving, inflation was down, and oil markets were stronger than they had been since the embargo of 1973. But to send a signal to the new US President, Jimmy Carter, the Saudis decided to fashion an oil policy "that will serve political purposes similar to those for which the kingdom led the oil battle in October 1973."[17] When the other OPEC members balked at accepting a small real decline in the price of oil, Yamani left the meeting and ordered his country's full installed capacity put to use to dictate price unilaterally. The resulting two-tier price split was the most serious challenge OPEC had to cope with until 1981.

• In 1977, global economic activity was again relatively strong. Yet the Saudis pledged to President Carter that oil prices would remain frozen through 1978. Despite some wavering, they stuck to their word after the American leader neutralized any opposition from the Shah. This era of "price freeze" coincided with the first debate in the US Senate about the Saudi request for F-15 fighters.

• Contrary to conventional wisdom, the "second oil crisis" of 1979 was not produced, directly, by the loss of Iranian production that accompanied the fall of the Shah. Rather it came with the abrupt Saudi decision after his overthrow to mandate a 2.4 MBd reduction in production that had helped make up for the Iranian shortfall in the midst of a weakening global economy. This caused the Aramco partners to cancel third party contracts, which drove those buyers into the spot market. The spot market skyrocketed. Distressed by the Camp David treaty between Egypt and Israel and by Israeli behavior on settlements and autonomy, internal divisions in the Saudi leadership prevented a consensus on producing more than the normal ceiling of 8.5 MBd despite the near doubling of oil prices until six months after the onset of the oil crisis, when the Saudi Crown Prince Fahd publicly denounced the Camp David approach, called on the United States to enter into a direct dialogue with the Palestine Liberation Organization (PLO), and pledged a new output level of 9.5 MBd in a gesture that Ambassador Robert Strauss, the Arab-Israeli negotiator, was asked to carry to President Carter to demonstrate the political nature of the oil production decision. Fahd achieved a new consensus on oil production policy within the Royal Family only by using the occasion to send Washington the message that the Saudi perspective on the Arab-Israeli peace process and on the ultimate status of Jerusalem could not be ignored or taken for granted.

• Throughout 1981, the Saudis have again been preoccupied with the weakness of the oil market. They have simultaneously been eager to establish a firm relationship with the new administration in Washington. This latter concern has been reinforced by Soviet activities in Afghanistan, the Yemens, and the Horn of Africa. In this context, the large arms sale package including AWACS and enhanced equipment for the F–15s has become both a symbol of US-Saudi friendship and a concrete manifestation of the Saudi determination to protect its oil facilities from an attack. Thus, economic and political-security considerations have reinforced the Saudi willingness to stand firm against the more hawkish members of the cartel. It is not clear that the economic motive alone would be compelling.

The examination of these six key historical episodes suggests that the Saudis in fact take a leadership role in OPEC very cautiously, acting most forcefully when there is political or security advantage to be gained in their relations with the consumer states, principally the United States. Two times they went against the strong protestations of their neighbors to use price and production policy to send signals about Middle East and defense issues (1976 and 1977). Two times they stepped away from a position of price leadership under pressure from their neighbors (1974 and 1975). When they acted resolutely in the 1977 case, they stuck to a personal promise to the president of the United States, but only after he had eliminated any hostile reaction from Iran. When they exercised an OPEC leadership role in 1981, political and security issues in the Saudi relationship to Washington, as well as economic concerns, assumed importance simultaneously.

The simple model economic self-interest is not consistent with the diversity of Saudi actions from raising prices and accepting a smaller market share (1974, 1975), to freezing prices and taking a larger market share (1976, 1977, 1981), to precipitating a large price jump but offering a discount to shift the market losses onto others (1979). Nor can Saudi behavior be explained by looking at how tight oil markets are prior to the pricing decision. Considering the ratio of demand for OPEC oil to OPEC capacity, prices were raised twice when the market was relatively slack (75-percent-capacity utilization preceding the 1974 decision; 72-percent-capacity utilization preceding the 1975 decision). Prices were restrained twice when the market was tighter (84-percent-capacity utilization preceding the 1976 decision; 78 percent preceding the

1977 decision). Only once, in 1979–80, was there a clear market push involved in a price rise; but the market push was not inevitable had the Saudis willed it otherwise. Only once, in 1981, was there a major preoccupation with the long-term health of the oil market reflected in a price freeze.

If the calculation of economic self-interest cannot be relied upon to determine Saudi willingness to exercise forceful cartel leadership, still less dependable is the state of the world economy. Three times the Saudis pushed for lower oil prices when the global economy was relatively strong and inflation comparatively low —the two-tier split at Doha in 1976–77; the price freeze of 1977– 78; the stand against the OPEC hawks in the first half of 1981. They abandoned a leadership role on behalf of moderation three times when the global economy was weak and inflation comparatively high—the 1974 auction with the world heading into recession; the 1975 "stand off" increase with the world in bad straits; the second oil crisis of 1979 with the international economy again fragile. This behavior pattern undermines even the most generous formulation of the economic optimization idea, which suggests that Saudi oil reserves are so large and the period of exploiting them and caring for the financial assets they generate so long that the Saudis should identify their self-interest with the long-term health of the world economy and the strength of its financial institutions. It also undermines the hypothesis of "interdependence" as the motivator of Saudi policy; namely, that a growing web of financial, commercial, and investor ties leads to a mutual appreciation of the common good on the part of energy-importing and energy-exporting states.

More forcefully put, Saudi behavior during the "second oil crisis" of 1979 demonstrates, as did the oil embargo of 1973–74, that the kingdom's leadership is willing to run large risks to the international economic and financial system to press its political objectives on Middle East issues.

On the basis of the past record, the industrial democracies will have to conclude that Saudi Arabia will continue to condition the kingdom's oil policy on the state of its security and political relationship with the United States rather than being ruled by economic self-interest and compelled to act forcefully on behalf of moderation. In the medium term, to the mid 1980s, this means that the Saudis could continue a policy of high production, forcing the price of oil down in real terms, or steadily cut back production to allow oil prices to firm. It means that they could continue to expand production capacity for use in emergencies or when supply and demand come into balance again, or they could let new capacity expansion slip. Over the longer term, when oil markets do begin to tighten, the world could find the kingdom's production capacity standing between 12 and 14 mbd with special "surge capacity" above that, or reduced below 10 mbd with no surge capacity built in. And, between the short term and the long term, it would be imprudent to assume that the kingdom will not from time to time make a dramatic gesture on oil policy, as in 1973 and 1979, to send the signal to Washington that its security position in the gulf and its political interests in the Middle East cannot be overlooked.

But, to make matters even more complicated and less satisfactory, the vulnerability of the kingdom conditions what it can realistically promise as the reward for success in advancing national political goals. Thus, Prince Fahd has promised "good things" in oil policy if the Arab-Israeli controversy is settled on terms acceptable to the Saudis. Yamani has referred to "a new era" in the same context. But the impact on other OPEC members of a large production jump and the hostile security threats in the region that would be inevitable if prices declined precipitously effectively foreclose the option of large production increases and deprive the promise of credibility. As a consequence, the linkage between Saudi oil policy and the Arab-Israeli controversy is asymmetrical: lack of progress toward a long-term solution, including East Jerusalem, that the Saudis consider acceptable or are able to live with, can have costly negative consequences for oil policy; success in achieving a long-term solution is likely to offer, at best, only modest direct positive benefits.

THE SECURITY AND STABILITY OF THE PERSIAN GULF

The preceding section argues that Middle East policies on the part of the United States and other industrial democracies that attract wide segments of the Arab community as well as insure the security of Israel can help protect against the downside risk of "oil shocks" and "unfriendly signals" from the gulf. There is a second reason, as well, to link the policies of the industrial oil-consuming states toward the Middle East to the behavior of oil markets in the mid 1980s: such an approach toward Middle East issues can also help to eliminate a major source of instability in the oil-producing states of the gulf.[18]

A primary threat to the integrity of the regimes in the gulf comes, of course, from the Soviet Union. Explanations for the

Russian move into Afghanistan split into (1) those who see the invasion as the reflex of a Great Power to the existence of instability or a power vacuum on its border; (2) those who see the invasion as part of a premeditated plan to tighten the grip on the jugular of the West; and (3) those who see the invasion as the outcome of habitual fishing in troubled waters. All three explanations converge, however, in suggesting that internal disintegration in Iran, or Saudi Arabia (or Kuwait, Bahrein, the United Arab Emirates, Iraq) could involve Soviet intervention.

Clearly, the oil fields of the gulf have become, figuratively speaking, the Berlin of the 1980s. In this context, American (and British and French) efforts to strengthen their military forces in a way that can deter Soviet adventurism is vital to the defense of the West.

But the prospect of Soviet troops invading over the horizon of their own accord, so to speak, has never seemed as probable to the governments in the gulf as domestic revolution (perhaps aided by the Soviets) stirred up with the rallying cry that the regimes have given up on Jerusalem and the Palestinians while identifying themselves with an America that is unresponsive to Arab needs. This potential for anti-American instability provides a linkage between the security of the gulf states and broader us policies toward Arab-Israeli problems.

This kind of linkage may be transmitted directly through Palestinian minorities in Kuwait, the United Arab Emirates, Bahrein, and to a lesser extent, Saudi Arabia. But it is also important for Arab societies having to absorb a younger indigenous generation of technocrats and intellectuals returning from abroad with more enthusiasm for pan-Arab and pro-Palestinian ideas than their elders and greater willingness to reject the idea of dependence on the United States. The first peak of students sent abroad for education after the oil revenue jump of 1973 is just now returning to the gulf states. In Saudi Arabia alone, for example, the society must absorb 8,000 foreign-trained students in 1981 in comparison with 2,500 the preceding year. The more negotiations about the West Bank and Jerusalem appear stalemated, the more likely that a dissident religious leader or military officer or fundamentalist figure might be able to rouse a mass movement against a Royal Family that derives its legitimacy from the role of guardian of the Muslim Holy Places. It is within this setting that the Soviet propensity to fish in troubled waters becomes a real threat. A Soviet presence in the oil fields is more likely to come not from naked aggression but from insertion in a domestic conflict that

links frustration at an Arab-Israeli stalemate with anti American-
ism and a charge that the incumbent rulers have sold out their
birthright.

This illuminates a possibly fatal flaw in the idea of building a
"strategic consensus" in the region against the Soviet Union in the
absence of (or to substitute for) a policy toward the West Bank and
Jerusalem that can mobilize broad Arab support. The intensification of the security relationship with the United States, in isolation, could end up precipitating the kind of instability that the
relationship is ostensibly designed to prevent.

The proposition that there is some degree of linkage between
us-Middle East policies and the security of oil facilities in the gulf
must be limited, however, in two important ways.

First, even a (hypothetical) overnight settlement of the Arab-
Israeli dispute would not insure the stability of the regimes in the
gulf. The process of rapid modernization is inherently destabliz-
ing. It exacerbates income disparities, religious and ethnic cleav-
ages, and culture shock. In addition, there are separate and
independent national hatreds and rivalries as exemplified by the
Iran-Iraq war.[19] Progress on Arab-Israeli issues would not eliminate
these internal and external threats to stability.

Second, the prospect of a comprehensive peace in the Middle
East might itself increase the threat from splinter groups in the
Arab world dissatisfied with the outcome. The latter might try to
undermine the peace process by stirring up trouble in the oil-
producing states.

Despite these qualifications, however, the contribution of an
Arab-Israeli peace to the long-term prospects for stability in the
gulf remains real and significant.

Progress on Arab-Israeli issues will bring the added dividend of
facilitating closer military (and intelligence) cooperation with the
Saudis, the Emirates, and Oman. Even without bases, the ability of
American forces to operate effectively in a crisis would be greatly
enhanced if the Saudis, for example, could be persuaded to over-
build air facilities, naval facilities, and communications networks
that would be required to accommodate the rapid deployment of
us troops if the latter were called upon. But, to the exasperation of
American security officials obsessed with the nightmare of confronting the Soviets in an emergency, the governments of the gulf
refuse. The ability of the weak (on whose existence the powerful
depend) to play chicken with their own destiny may inspire a
whole new generation of academic game theorists. The gambit is
to cry wolf, and refuse bricks for the house. But if the beast ever

does appear, there may be no way to save the situation in time. We know it. They know it. But in the absence of progress toward a Middle East settlement, the resistance persists.

In short, the challenge of providing for the security of the oil facilities of the gulf in the 1980s cannot be seen through the prism of us-Soviet relations alone. An effort to move toward resolution of the Arab-Israeli controversy must be an integral part of the strategy to ensure stability in the gulf.

A NOTE ON MEXICO: PRODUCTION POLICY AND INTERNAL STABILITY

Mexico enjoys all of the benefits of OPEC pricing, without any of the pressures on a cartel member to share in the common apportionment of spare capacity. In fact it may acquire a "preferred supplier" status in relation to the United States due to its possession of secure routes in North America and its nonparticipation in oil embargo activities.

The analysis of Mexican oil production behavior[20] divides into opposing camps: (1) those who argue that buyers (principally the United States, but also Japan and Europe) should grant non-energy-related concessions to induce the Mexicans to produce more than they otherwise would; and (2) those who argue that Mexican domestic revenue needs will drive hydrocarbon exports to expand as rapidly as is technically and politically feasible (subjecting the country to strains from rapid modernization as severe as those encountered in the Persian/Arabian Gulf).

The comparison of Mexico with the high population-high absorption states in OPEC supports the latter point of view. There are, still, two areas in which American policies must be reinforced to insure the continuity of Mexican oil policy: (1) the preservation of the emigration "escape-valve" (albeit via a guest-worker permit) to offer relief to pent-up pressures in Mexico; and (2) the continuation, and expansion, of liberal trade principles for us imports from Mexico to assist the flourishing of labor-intensive industries in Mexico. (For additional arguments on behalf of the second point, see the other papers prepared for this Symposium dealing with the problems of Third World development, LDC debt, petrodollar recycling, and the need for the preservation of access by developing nations to the markets of developed nations.)

Both of these areas will be under close scrutiny and intensive debate in the early 1980s. Backsliding on the part of the United

States in either will impact adversely on the prospects for domestic instability in Mexico and jeopardize access to oil exports.

SUMMARY CONCLUSIONS

The first two sections of this paper argued that while the short-term prognosis is for oil price weakness, the evolution of oil markets through the 1980s remains highly uncertain with the possibility of supply and demand closing together even before 1985. Moreover, to maintain a balance between supply and demand in the mid 1980s will require the cooperation of that core of gulf states, led by Saudi Arabia, who have discretion about continuing production above their basic revenue needs. Thus, these sections concluded by rejecting the emerging conventional wisdom in the oil-consuming societies that they can become complacent about the energy crisis. Rather, the large industrial states must continue the movement toward replacement cost pricing for energy, reinforce their efforts to build sizable stockpiles to protect against supply interruptions, reiterate their commitment to International Energy Agency emergency allocation plans, and consider adopting oil import tariffs to reflect the national security premium of relying on uncertain foreign sources of supply.

The third and fourth sections of this paper focused on the determinants of OPEC behavior and found, on the basis of the empirical examination of the crucial OPEC decisions since 1974, that economic motivations are not sufficient to explain the actions of the core OPEC states, in particular Saudi Arabia. The kingdom's actions on price and production are not consistent with simply trying to maximize its own economic returns from hydrocarbon resources or with trying to protect the international financial system in which it has a growing stake. Rather, Saudi oil policy choices have been more closely keyed to the state of the political and security relationship with the United States. Indeed, the kingdom has been willing to expose the international economic system to large risks to insure that its views on Arab-Israeli issues are not overlooked or ignored. As part of the strategy to assure oil supplies, the major consuming states, led by the United States, must pursue policies that seek to combine the security needs and political perspectives of both the Israelis and the Saudis. It should be noted, however, that for reasons elaborated previously here, the principal gain from linking oil and Middle East issues will come in helping to lessen downside risk, not in putting an end to the energy crisis. The oil that can be found in the West Bank, so to speak, cannot be counted on to break OPEC.

The theme of an important (albeit limited) linkage between the Arab-Israeli struggle and oil supplies was also present in the fifth section, which analyzed the prospects for the security and stability of the gulf. Soviet aggression remains a central threat to the oil-consuming nations' interests in the region, and the strengthening of military capabilities to deter aggression must be encouraged. But more plausible than naked invasion is the prospect of indigenous upheaval (perhaps abetted by the Soviets) brought to coalescence with the charge that the regimes in power have lost their legitimacy by selling out to a United States that is unsympathetic on Arab issues. In assessing the stability of Saudi Arabia, for example, the status of East Jerusalem is particularly salient due to the guardianship role exercised by the Royal Family. Once again, however, the link between Middle East questions and stability in the oil-exporting states must be carefully circumscribed: much of the destabilization inherent in rapid modernization is unrelated to Arab-Israeli tensions. But it is clear, nevertheless, that an effort to build in the area an anti-Soviet consensus that does not include active American-sponsored movement toward a Middle East peace or that tries to substitute for such movement, may end up exacerbating the potential for the instability it is intended to prevent.

The sixth section shifted the focus to the largest oil-exporting country outside OPEC, which is Mexico. It suggested that internal needs for revenues are likely to drive the country to expand oil and gas exports as fast as is technically and politically desirable into the mid 1980s. This may aggravate the strains within Mexican society similar to the experience of the rapidly modernizing states within OPEC. To help keep such strains manageable, it is important for the United States to reinforce two policies: first, to keep open the option for Mexican workers to seek jobs in the United States through a "guest worker" or other similar program; and second, to maintain and expand the system of relatively unimpeded importation of labor-intensive goods from Mexico.

In conclusion, the present period of glut in world oil markets marks not an end to the "energy crisis" but a period of respite in which to strengthen relations between producer and consumer states and work to make the international energy system more resilient in the face of ever possible but unexpected disruption.

NOTES

1. Milton Friedman, *Newsweek*, March 4, 1974.
2. R. Stobaugh and D. Yergin, eds., *Energy Future: Report of the Energy Project at the Harvard Business School* (New York: Random House, 1979); H. H. Landsberg, Chairman, *Energy: The Next Twenty Years* (Cambridge, MA: Ballinger, 1979), A Ford Foundation Study; S. H. Schurr, Project Director, *Energy in America's Future: The Choices Before Us* (Baltimore: The Johns Hopkins University Press for Resources for the Future, 1979).
3. *Petroleum Intelligence Weekly*, September 28, 1981.
4. J. H. Lichtblau, "The Limitations to OPEC's Pricing Policy," and N. Ait-Laoussine in *Petroleum Intelligence Weekly*, September 28, 1981.
5. R. J. Lieber, *Energy and the Troubled Partnership: Alliance Consequences of the Energy Crisis* (Washington, DC: The Brookings Institution, draft, p. 1–7, 1981).
6. Under the US government's "Maximum Likely OPEC Production" in 1985—which, at 27.8 MBD, is still lower than production in the 1970s—the Saudis must produce 8.7 MBD; the Kuwaitis, 1.5 MBD; the United Arab Emirates 1.4 MBD; and Qatar, 0.5 MBD.
7. N. Ait-Laoussine, *Petroleum Intelligence Weekly*, op. cit.
8. For a detailed analysis of attempts to model OPEC behavior using economic optimization and simulation techniques, see T. H. Moran, "Modeling OPEC Behavior: Economic and Political Alternatives," *International Organization*, Spring 1981, from which parts of this paper are derived with permission of the Board of Regents of the University of Wisconsin System. A revised version of that article will appear in J. Griffin, ed., *The Future of OPEC*, forthcoming.
9. H. Hotelling, "The Economics of Exhaustible Resources," *Journal of Political Economy*, April 1931. See also R. M. Solow, "The Economics of Resources or the Resources of Economics," *American Economic Review*, May 1974; F. Peterson and A. Fisher, "The Exploitation of Extractive Resources: A Survey," *Economic Journal*, December 1977; R. Pindyck, "Gains to Producers from the Cartelization of Exhaustible Resources," *Review of Economics and Statistics*, May 1978; J. Sweeney, "Economics of Depletable Resources: Market Forces and Intertemporal Bias," *Review of Economic Studies*, February 1977; and J. Stiglitz, "Monopoly and the Rate of Extraction of Exhaustible Resources," *American Economic Review*, September 1976.
10. E. Hynilicza and R. S. Pindyck, "Pricing Policies for a Two-Part Exhaustible Resource Cartel: The Case of OPEC," *European Economic Review* 8, 1976. See also R. S. Pindyck, "The Economics of Oil Pricing," *Wall Street Journal*, December 20, 1977, and "OPEC's Threat to the West," *Foreign Policy*, Spring 1978; T. D. Willett, "Conflict and Cooperation in OPEC: Some Additional Economic Considerations," *International Organization*, Autumn 1979; S. F. Singer, "Limits to Arab Oil Power," *Foreign Policy*, Spring 1978; H. Ben-Shahar, *Oil: Prices and Capital* (Lexington,

MA: D. C. Heath, 1976); B. A. Kalymon, "Economic Incentives in OPEC Oil Pricing Policy," *Journal of Development Economics* 2, no. 4 (1975); and P. L. Eckbo, *The Future of World Oil* (Cambridge, MA: Ballinger, 1976).

11. "Yamani Takes A Look At The Future For Oil," lecture at the University of Petroleum and Minerals, Darnman, January 31, 1981, reprinted in *Petroleum Intelligence Weekly*, March 9, 1981.

12. *New York Times*, April 20, 1981.

13. D. J. Feith, "Saudi Production Cutback An Empty Threat?" in the *Wall Street Journal*, March 30, 1981.

14. E. Kanovsky, "On Saudi Oil Policy," *New York Times*, December 19, 1980. For a similar point of view, see S. F. Singer, "A Letter to the Secretary of State," *Washington Post*, March 22, 1981, and "Oil Importers Wonder: Can Glut Continue?," *Washington Post*, June 28, 1981.

15. T. H. Moran, "Modeling OPEC Behavior: Economic and Political Alternatives," *International Organization*, op. cit.

16. *The Oil and Gas Journal*, November 18, 1974.

17. This was Yamani's explanation of the Saudi decision on television in Jiddah, December 22, 1976, reprinted in *Middle East Economic Survey*, January 10, 1977.

18. For a detailed examination of problems of security and stability in the gulf, see T. H. Moran, "The Middle East and the Gulf: What is the Linkage for US Policy?" forthcoming.

19. More than a dozen wars have been fought among the regional states in the past decade and a half. W. J. Levy, "Oil and the Decline of the West," *Foreign Affairs*, Summer 1980.

20. E. Wonder, "US-Mexico Energy Relations," copy, July 1981; S. K. Purcell, ed., *Mexico-United States Relations* (New York: Proceedings of the Academy of Political Science, Volume 34, No. 1, 1981); F. Hampson and K. J. Middlebrook, "Energy Security in North America," in D. A. Deese and J. S. Nye, eds., *Energy and Security* (Cambridge, MA: Ballinger, 1981); D. Ronfeidt, R. Nehring, and A. Gándara, *Mexico's Petroleum and US Policy: Implications for the 1980s* (Washington, DC: Rand for the Department of Energy, 1980).

CHAPTER TWENTY

Energy and Economic Development in Arab OPEC Countries

Hassein Askari
President
Askari, Jalal and Sheshunoff International

INTRODUCTION

A common characteristic of OPEC countries is that they are all, at the present time, net exporters of oil. Beyond this common element, the OPEC countries are about as diverse as any group of countries. Member countries differ dramatically as to their endowments of energy and other factors of production, resulting in a wide spectrum of economic development prospects. In some countries, the nation's energy surplus is limited, while in others it can be expected to last on a substantial scale well into the next century. Some members have large populations, such as Indonesia (139 million) and Nigeria (82.5 million),[1] while other members— Gabon, Qatar, and the United Arab Emirates—each have populations of less than 1 million. In terms of standard measures of economic output, three countries within OPEC are at the top of the international ranking, with per capita incomes above $15,000— Kuwait ($17,270), Qatar ($16,590), and the United Arab Emirates ($15,590)—while Indonesia ($380) and Nigeria ($670) are at the low end of the international scale.[2]

In this paper, the task of analyzing the energy and development prospects of this diverse group of countries has been made somewhat easier by focusing on the seven OPEC members that are Arab countries: Algeria, Iraq, Libya, Kuwait, Saudi Arabia, Qatar, and the United Arab Emirates. The following aspects of these Arab oil-exporting countries are considered here: their economic characteristics, their energy resources and balance of energy supply and demand, their economic achievements, and their economic prospects. The paper then concludes by analyzing the countries in light of past historical parallels and future possibilities and constraints.

ECONOMIC CHARACTERISTICS

The basic economic facts on these seven Arab OPEC countries are set out in Table 20–1. Divergencies are substantial. Algeria, with 2.4 million square kilometers and more than 18 million inhabitants, is physically the largest and most populous country, while Qatar, with a little over 10 thousand square kilometers and only 226 thousand inhabitants, is the smallest. In 1978, the economic sizes (GNP) of the nations ranged from $3.7 billion in Qatar to $626 billion in Saudi Arabia, with per capita incomes ranging from a low of $1,580 in Algeria to a high of $17,270 in Kuwait. These latter divergencies, in GNP and in GNP per capita, would be even more pronounced today since the full impact of the 1979–80 oil price increases is not reflected in the 1979 figures. Although these and other basic economic differences are clearly evident, these economies share some distinct underlying characteristics.

Table 20-1. **Basic Economic Statistics, 1979**

Country	Size (km²)	Midyear population (x 10³)	GNP at market prices (1979 US $ x 10⁶)	GNP/capita (1979 US $)
Algeria	2,400,000	18,235	28,940	1,580
Iraq	483,317	12,643	30,430	2,410
Kuwait	24,280	1,266	21,870	17,270
Libya	1,760,000	2,850	23,390	8,210
Qatar	10,365	226	3,750	16,590
Saudi Arabia	2,240,000	8,495	62,640	7,370
United Arab Emirates	77,000	833	12,990	15,590

Source: 1980 World Bank Atlas, Washington, DC, World Bank, 1981.

Oil, to differing degrees, dominates these seven economies. Oil is both the major component of their GNPs and the dominant source of their export revenues (see Table 20–2), and with the possible exceptions of Algeria and Iraq, the economies are one-resource economies. Oil is not only dominant; it dwarfs the rest of the economies of all but Algeria, contributing roughly 70 percent of the other six countries' GDPs. (In most industrialized economies, on the other hand, the share of all extractive industries in GDP averages around 5 percent.) This heavy dependence on a single export product exposes these countries to unusual risk in a world of volatile prices. Moreover, the exhaustible characteristic of oil has direct bearings on the future economic activity of these oil-exporting countries and necessitates judicious diversification of their economies if they are to maintain their levels of economic prosperity over time.

Table 20–2. **The Role of Oil, 1979ᵃ**

Country	Ratio of oil exports to GDP (%)	Share of oil exports in total exports (%)
Algeria	30	92
Iraq	65	99
Kuwait	77	95
Libya	64	99
Qatar	77	95
Saudi Arabia	66	99
United Arab Emirates	70	95

ᵃEstimated from data in *International Financial Statistics* (International Monetary Fund) and various national sources.

To different degrees, all the oil-exporting countries face one important question—the optimal transformation of their limited and nonrenewable oil wealth. In most of these countries, oil and associated gas are the only significant natural resource. Agricultural potential is restricted by rugged terrain, harsh climate, the absence of adequate water resources, and rudimentary infrastructure. Industrial development is limited by the dearth of complementary factors and especially by the limited supply of professional and skilled labor.

Oil is the major natural wealth of these countries. For the present, the depletion of oil will, in terms of traditional accounting methods, result in oil revenues and thus in high levels of national output. However, if the depletion of oil does not result in the development of a productive nonoil sector, then current output will be derived from a depletion of oil wealth—a process which is unsustainable in the long run. In other words, the national outputs of the major oil exporters are not comparable to those of economies where output is a flow derived from a sustainable productive base. In the case of these oil exporters, however, national output is derived from an exchange, or transformation, of a nonrenewable asset. Thus, potential output is maintained in these countries only if oil is not extracted or if extracted oil is efficiently transformed into other productive capital.

A simple comparison of a diversified economy to an economy based totally on oil may be useful. In a diversified economy, economic output from agriculture and industry can be expected at a constant level for all future generations, as long as the soil does not deteriorate, capital is replaced, and natural disasters do not occur. More likely, if the capital base is increased through positive net investment, technological change, and other measures to enhance productivity, output will in fact increase over time.

In an economy totally based on extraction of exhaustible resources, this constancy or growth of national product is not assured. At the extreme, if a country had x units of oil reserves and it produced them all in the first year, then its national product, given standard national income accounting practices, would be the price of oil multiplied by x. But in the second year, its national product would be zero unless it had invested some of the first year's proceeds to establish a productive base. In any event, its national product in the second year would be only a fraction of its national product in the first year, because even if it invested all its revenues from the first year, the return on the investments would in all likelihood be less than 100 percent.

From the above example, it can be seen that in essence, the national product of an extractive economy is not comparable to the national product of a diversified, nonextractive economy. Conventionally calculated national product in these countries embodies a great proportion of asset transformation as opposed to economic production.

We can, therefore, deduce that current national output figures for extractive-based economies are inappropriate measures of these countries' sustainable national products. Conventional national output figures are overestimates of the sustainable long-run outputs of these countries. Given that oil resources are finite, it is intuitively evident that the depletion of oil for consumption alone, without savings and investment, will eventually result in a zero level of national output for an economy that is totally based on oil. These economies are currently involved in a process of partial asset transformation as opposed to production from a sustainable economic base. This partial asset transformation is the exchange of their oil wealth or assets for other assets and consumption. Because some of the oil assets are consumed, the asset transformation process can only be considered as partial.

A direct and obvious implication of such a view of an oil-based economy—or of any economy heavily dependent on extractive (exhaustible) industries—is that these economies should have high national savings rates. More precisely, given two economies that are essentially similar except that one has a heavy dependence on extractive industries, the latter should have the higher national savings rate. This higher national savings rate is necessary if the economy is to maintain its level of national output after its extractive resource is depleted. In turn, current savings should be used to build a nonoil productive economic base and to acquire foreign assets. Also, in turn, the available opportunities in these areas affect the rate of oil depletion.

The domestic economic development potentials of the oil-exporting Arab countries are diverse, to say the least. In Algeria and Iraq, agriculture has been a source of output in the past and has the potential of becoming even more important in the future. However, in Kuwait, Libya, Qatar, Saudi Arabia, and the United Arab Emirates, there is little potential for a large agricultural sector. In Algeria and Iraq, the availability of nonoil resources and professional and skilled labor has promoted some industrialization. In contrast, in Kuwait, Libya, Qatar, Saudi Arabia, and the United Arab Emirates, the harsh physical environment and reliance on imported materials and labor has made expansion of industry and services both costly and difficult.

In all of these countries, diversification away from oil has become an important policy goal, given the implied risks of total reliance on one commodity whose price is primarily determined in the international market. Besides these usual arguments favoring diversification, diversification and the increased role of the private sector have become important in some of these oil-exporting countries for at least another reason. Given the dominance of oil and its ownership by the government, it is difficult to provide gainful employment and purchasing power to citizens unless the economy is diversified. The oil and petrochemical sectors provide limited employment opportunities and only indirectly benefit the private sector and citizens. Thus, diversification is the mechanism to distribute income to citizens and to enhance the role of the private sector. In turn, this process will ameliorate social conditions by distributing the benefits of oil more widely.

But given large differences among the countries in their endowments of basic production factors, the implied strategies are different. For Kuwait, Qatar, and the United Arab Emirates, diversification has meant exporting more oil than necessary to finance domestic consumption, and it has meant investment in order to build substantial foreign assets. These countries have implicitly decided that the development of large domestic industrial and agricultural sectors is not viable. Instead, they feel that domestic infrastructures, social services, human capital, and financial services should be encouraged. The development of human capital and regional and international financial services is therefore one avenue of limited diversification. The other is foreign investments. In other words, these countries, given their limited domestic potentials, have decided to export more oil than is necessary for domestic financial needs in order to achieve diversification away from oil, in part through the acquisition of foreign investments.

At the other extreme, Algeria, given its more limited oil resources and larger domestic development potential, is attempting to achieve diversification away from oil through domestic development financed from oil and gas exports and from substantial borrowing on the international capital markets.

The strategies of Iraq, Libya, and Saudi Arabia, in differing degrees, fall in between the two extremes. They are achieving diversification away from oil through both domestic development and the accumulation of foreign investments, with the latter intended to finance future domestic development.

The economic opportunities and choices faced by all seven countries are in large part determined by their domestic energy resources. And, as will be seen in the next section, individual energy resource endowments vary substantially across the countries.

ENERGY RESOURCES AND THEIR BALANCE OF SUPPLY AND DEMAND

Although a heavy dependence on oil is the major common characteristic of these seven economies, their oil supplies and reserves, which are the major factors in determining their future economic prospects, differ greatly. Per capita proven oil reserves range from a low of 450 barrels in Algeria to a high of 51,264 barrels in Kuwait (see Table 20–3). This fact implies a wide variation in the oil wealth available to achieve diversification and economic development: as shown in Table 20–3, at a price of $34 per barrel Algeria's total oil wealth per capita is $15,300, compared with Kuwait's

Table 20–3. **Reserves and Production of Oil and Gas**

Country	Proven oil reserves as of 1/1/81 (bbl x 10⁹)	Proven gas reserves as of 1/1/81 (ft³ x 10²)	Proven oil reserves per capita (bbl)	Dollar value of per capita proven oil reserves (at $34 per bbl)	1980 oil production (bpd x 10⁶)	1980 gas production (ft³ x 10⁹)
Algeria	8.2	131.5	450	15,300	1.0	517.0
Iraq	30.0	27.4	2,373	80,682	2.6	18.0
Kuwait	64.9	30.8	51,264	1,742,976	1.4	291.3
Libya	23.0	23.8	8,070	274,380	1.8	132.8
Qatar	3.6	60.0	15,929	541,586	0.5	38.3
Saudi Arabia	165.0	110.0	19,423	660,382	9.6	310.2
United Arab Emirates	30.4	20.8	36,494	1,240,796	1.7	205.5

Source: for total reserves and production, International Petroleum Encyclopedia 1981 (Tulsa, Oklahoma: The Penn Well Publishing Company, 1981).

$1,742,976. These oil reserve figures can also be viewed from another perspective. At current production levels, Algeria's oil will last roughly another 22 years, whereas Kuwait's will last 127 years. But to assess the future availability of oil to finance diversification and economic development, a closer examination of energy demand and supply in these countries is necessary.

The future availability of oil for export will be largely determined by domestic demand for oil and by oil production policies. Domestic demand for oil in these seven countries has grown rapidly and can be expected to continue growing at rates above the world average. The historical growth of these countries' consumption of refined petroleum products is shown in Table 20–4. As can be seen, recent annual growth rates of petroleum product consumption have, on an average, been around 20 percent. The reasons for such rapid growth in energy consumption are in part a function of these countries' growth in their levels of income, economic development, and urbanization. Given their economic characteristics, these countries had low levels of energy consumption in the 1960s and early 1970s. But rapid economic change (through real income growth and urbanization) from a low base of energy consumption has resulted in a large percentage growth of energy consumption. This growth in consumption has been further encouraged by energy prices which are, on an average, substantially below world levels. (It is reported that "most domestic energy prices tend to be only 10–50 percent of the world market prices."[3]) Finally, this rapid growth of energy consumption has been and will continue to be influenced by the harsh climate of the region and by these countries' diversification into petrochemicals.

Several estimates of future developments in oil supply and demand exist. While, as expected, there is substantial agreement on estimated demand developments, there is much less consensus on supply. Table 20–5 gives the projection of Dr. Fesharaki, a recognized and respected expert in the field. Dr. Fesharaki projects more than a 200 percent increase in domestic oil consumption over the period from 1979 to 1990. He also predicts a substantial reduction in oil production. The impact of these demand and supply projections on exports is quite dramatic, and, depending on

Table 20–4. **Consumption of Refined Petroleum Products in Selected Arab Countries, 1967–77**

	Annual level of consumption (bpd x 10^3)											Average annual growth rate in consumption (%)	
	1967	1968	1969	1970	1971	1972	1973	1974	1975	1976	1977	1967–77	1972–77
Algeria	24.1	27.5	32.3	39.5	40.6	47.0	60.4	57.5	70.1	82.9	93.9	14.3	13.17
Iraq	62.3	64.9	66.4	61.9	70.5	76.2	85.2	116.8	183.1	174.3	184.4	13.4	20.04
Kuwait	11.2	11.8	12.5	14.4	15.6	16.4	16.9	19.3	23.9	30.0	31.1	11.0	14.67
Libya	13.8	18.5	20.9	16.4	19.9	24.4	32.7	37.3	51.2	54.3	63.0	16.4	18.80
Qatar	1.2	1.3	2.4	1.6	1.7	2.0	1.7	3.2	4.2	5.7	7.1	18.1	25.28
Saudi Arabia	23.5	27.6	39.0	42.5	47.1	54.8	68.2	87.7	117.2	158.0	160.2	22.0	23.88
Abu Dhabi	0.7	1.4	2.0	2.2	2.3	3.2	4.6	6.4	10.7	12.4	9.7	31.0	25.81
Total OPEC[a]	516.9	585.8	634.0	670.1	736.7	825.4	970.6	1,102.1	1,343.9	1,633.6	1,805.3	13.4	16.21

Source: A. Al-Janabi, "Estimating Energy Demand in OPEC Countries," Energy Economics Vol. 1, No. 2, April 1979, Table 1.
[a]Totals are for all OPEC members and do not represent a sum of the figures given above.

Table 20–5. **Production, Consumption, and Exports of Oil** (bpd 10^6)

Country	1979 (actual)			1985 (projected)			1990 (projected)		
	Domestic production	Domestic consumption	Exports	Domestic production	Domestic consumption	Exports	Domestic production	Domestic consumption	Exports
Algeria	1.14	0.14	1.0	0.9	0.27	0.63	0.9	0.45	0.45
Iraq	3.38	0.20	3.18	4.0	0.34	3.66	4.5	0.53	3.97
Kuwait	2.25	0.04	2.21	1.5	0.06	1.44	1.0	0.09	0.91
Libya	2.06	0.08	1.98	1.6	0.17	1.43	1.2	0.31	0.89
Qatar	0.50	0.01	0.49	0.5	0.01	0.49	0.5	0.02	0.48
Saudi Arabia	9.20	0.25	8.95	6.3	0.46	5.84	6.3	0.84	5.46
United Arab Emirates	1.82	0.05	1.77	1.8	0.07	1.73	1.0	0.11	0.89
Neutral Zone[a]	0.56	—	0.56	0.5	—	0.50	0.5	—	0.50
Total	20.91	0.77	20.14	17.1	1.38	15.72	15.9	2.35	13.55

Source: F. Fesharaki, "Global Petroleum Supplies in the 1980s: Prospects and Problems," *OPEC Review*, Vol. IV, No. 2, Summer 1980, p. 37.
[a]Kuwait and Saudi Arabia.

developments in real oil prices, there could be a very noticeable effect on oil revenues. Even if production is assumed to be stable, there would at least be a very adverse effect on the exports of one country—that is, Algeria. In the case of Algeria, the growth in domestic consumption alone will reduce oil exports by about 50 percent.

But even for the more fortunately endowed oil exporters, the rapid increase in domestic energy consumption should be seen as undesirable insofar as it is a product of waste. Moreover, prices that are set at an artifically low level will result in a mix of factor inputs which is suboptimal, processes which are inefficient, and industries in which the oil exporters do not have a comparative advantage. Although all of the governments are aware that the availability of oil and gas for export has in large part determined the pace of their economic achievements and will continue to do so in the future, they have found it difficult, for social and political reasons, to restructure domestic energy prices. Although in time the magnitude of the energy waste will become more and more pronounced, it may then be politically even more difficult to increase energy prices, since society will have for a longer period based its decisions on cheap energy, without which many processes and industries would be unprofitable.

ECONOMIC ACHIEVEMENTS

The oil price increases of 1973–74, resulting in large current account and budget surpluses, afforded the oil-exporting countries unusual opportunities to increase the pace of their economic development. Given that oil revenues constitute, on an average, about 90 percent of budget receipts in six of the countries and about 55 percent in Algeria, government spending was increased rapidly. Although national output increased substantially, it was also marked by the emergence of severe bottlenecks, especially in shortages of labor, port capacity, distribution systems, and power. These constraints on supply in turn resulted in high rates of inflation. But the inflow of expatriate labor, growth of infrastructure, and, beginning in 1976, more restrained fiscal policies reduced the level of inflation. These rapid economic changes may be best observed by examining in more detail developments in Saudi Arabia.

Saudi Arabia's First Five-Year Development Plan (1970–71 to 1974–75), was launched prior to the oil price increase, with anticipated expenditures of 41 billion riyals. At present 1 riyal, or SRL, is equal to 30 US cents. The objectives of the plan were to develop basic infrastructure and to expand and improve government services. But, as a result of the oil price increases, actual expenditures during the period amounted to SRLS 78 billion with an 11.6 percent annual growth of real nonoil GDP of 11.6 percent. During the latter years of the first plan, it was realized that the absorptive capacity of the economy was still the limiting factor to more rapid development and growth. Thus the major objectives of the Second Five-Year Development Plan (1975–76 to 1979–80) were physical infrastructure, manpower resources, social services, and the increased participation of the private sector. Planned expenditures were set at SRLS 498 billion, with 48 percent of the total to be allocated to physical infrastructure, human and economic resources, and social development. However, in view of the high level of inflation, actual expenditures exceeded planned expenditures by 37 percent.

The achievements of the second plan have been substantial. Real nonoil GDP grew at an annual rate of 14.8 percent. There were a number of notable attainments: port capacity was expanded greatly (commercial berths increased from 24 to over 130), the extent of paved roads was increased by 76 percent, 350,000 new residential dwellings were constructed, the number of hospital beds was increased by 68 percent, school enrollment was up by 35 percent. Such progress was achieved while reducing inflation from 31.6 percent in 1975–76 to 2.4 percent in 1979–80. These achievements were made possible by large expenditures, careful planning, and a substantial inflow of expatriate labor, with the proportion of expatriate labor increasing from 28 percent in 1974–75 to around 45 percent in 1979–80. Over the entire decade of the 1970s, real GDP grew at an average annual rate of 11.4 percent, with nonoil GDP growth at 13.2 percent. During the same period, the number of hospital beds increased from 6,787 to 12,525, and school enrollment increased from 0.5 million to 1.5 million. (For more detail, see *Annual Report*, Saudi Arabian Monetary Agency, Riyadh.)

Saudi Arabia has recently changed its emphasis from infrastructure, which up to now has taken about 80 percent of the planned expenditures, to achieving a diversification of the economic base. This process, however, is subject to two important constraints. First, the extent and the specifics of the diversification process will incorporate both long-term efficiency and social requirements. Long-term efficiency is indicated, in part, by global comparative advantage and regional specialization patterns. Social requirements are seen in an Islamic framework of providing equal opportunities for education, medical attention, housing, and other social services to all in order to accommodate individual development. Second, the authorities feel that given the emphasis of structural change, increased dependence on expatriate labor would at this time be inappropriate and unnecessary. Therefore, to maintain an acceptable rate of growth in the nonoil sector without fueling inflation, the authorities foresee an increase in labor productivity from at least five sources: (1) the shift from construction to more productive sectors, (2) the replacement of unskilled expatriate labor with more skilled labor, (3) additional efforts in education and vocational areas, (4) increasing the overall skills of the indigenous employed labor force, and (5) increasing the capital/labor ratio of the economy.

The major vehicle for this orderly construction and diversification of the productive base is the Third Development Plan. Under the third plan, projected expenditures for the public sector, excluding transfer payments, foreign aid, and national defense, are set at $235.7 billion. Of this total, development spending will be $211.35 billion, with the remainder allocated for administration, local subsidies, and contingency reserve. The major thrust of the plan is in four areas: heavy industry, light industry, establishing a viable industrial and agricultural base, and establishing the required level of professional and skilled manpower to sustain the Saudi Arabian economy. At the end of the plan, the role of the private sector is expected to increase further.

Although infrastructure and transportation were stressed under the second plan, they will still require large allocations over the foreseeable future. The third plan allocates $11.4 billion to road construction, $1.35 billion to railroads, $10.75 billion to civil aviation and the international airport, and $7.16 billion to seaports. Wire and wireless communications will receive $8.71 billion under the plan.

Although the largest single industrial enterprise in history, the gas-gathering scheme, is near completion, and Jubail and Yanbu are about to embark on their petrochemical careers, Saudi Arabia still requires large expenditures in electricity and industry. Under the third plan, an allocation of $15.8 billion has been made for increased electrification, and $60.4 billion has been targeted towards industrialization. The plan calls for the development of new industries, including cement, industrial gas, intermediate petrochemical products, glass, metallurgy, spare car parts, animal feed concentrates, building materials, and farm products.

The increase in production by Petromin, Saudi Arabia's state-owned oil company, would include an increase in output by local refineries from 120,000 barrels per day to 640,000 barrels per day by the end of the plan period. Petromin is also to boost its export refineries' production to 1.33 million barrels per day from the present 580,000 barrels per day, while increasing its production of lubricating oils from 500,000 barrels per year to 2 million barrels per year. Also planned is an increase in the kingdom's oil storage capacity from 9.6 million barrels to 22.75 million barrels, while the kingdom's pipeline system for oil products is to be expanded to carry 511,000 barrels a day, up from 112,000 barrels per day. The crude oil pipeline system is to carry 2.2 million barrels per day, up from only 51,000 barrels per day.

In view of the educational requirements of the kingdom and their important role in future development, the plan calls for educational expenditures on the order of $23 billion. Direct expenditures on health will amount to $11.6 billion. The plan also attaches increasing importance to agriculture and water resources; such allocations amount to $19.2 billion.

The successful implementation of the third plan should go a long way towards achieving the desired diversification of the Saudi Arabian economy, which is still heavily dependent on oil. In 1979–80, the share of the oil sector in GDP was still around 66 percent. The value added in agriculture was only about 1 percent of nonoil GDP, while agriculture employed roughly 25 percent of the total labor force. Cultivated land has constituted less than 0.5 percent of Saudi Arabia's total area. Water has been the major constraint. To overcome this problem, the government has in the past constructed more than fifty dams and has developed irrigation networks. In addition, the government has constructed drainage facilities to reduce salinity, and the Agricultural Development Bank has given interest-free loans to encourage land improvement and cultivation. The contribution of manufacturing to GDP is also small, with a contribution (excluding refining) to GDP of around 2 percent in 1979–80. The government's activities in industry are concentrated in a few large petrochemical projects, but the government is now trying to increase the private sector's role in industry by giving substantial incentives such as interest-free loans for up to 50 percent of a project's cost for fifteen years and tariff exemptions on imported inputs.

Economic achievements and constraints in the other countries differ mainly by degree. Kuwait, Qatar, and the United Arab Emirates have placed relatively more emphasis on the service

sector; on an average, economic growth in these three countries
has not been as rapid as in Saudi Arabia, and inflation in the three
countries, which was not as high as that in Saudi Arabia in 1975,
has been reduced but is currently above that in Saudia Arabia.
These three countries have employed a larger proportion of expatriate labor in their labor forces than has Saudi Arabia and have
averaged about 80 percent expatriate. On the other hand, Iraq and
especially Algeria, having larger absorptive capacities and being
better endowed with complementary factors of production, have
shown more diversification than Saudi Arabia and thus did not
experience the same jump in inflation during 1974–75. Their
larger populations have also meant that expatriate labor has been
used to a very limited extent.

ECONOMIC DEVELOPMENT PROSPECTS

Although rapid economic progress has been achieved in these
seven countries during the 1970s, future success may not come as
easily. The growth of nonoil GDP has been largely fueled by expansion of construction, services, and trade. It is one thing to build
infrastructure and factories, but it is another to have these factories
produce a competitive product. In the longer term, nonoil GDP growth
will have to come increasingly from the development of agricultural/
industrial/financial goods and services that are competitive in the
international market. To adopt an inward policy of import substitution may be an easier course for these countries in the short run, but it
will not solve their long-term development needs, since import substitution has invariably proved to be a difficult policy after simple
products have been so substituted. The task of selecting projects and
industries in which these countries have a comparative advantage
will be difficult, given their generally limited endowment of complementary factors of production.

The process of diversification will present unique challenges
for the governments of these countries. Given that oil receipts are
such a large proportion of their domestic GDPs, the governments in
these countries are the major suppliers of investment resources.
Thus, the governments have to determine the areas of comparative
advantage and diversification. Up to now, petrochemicals have
been identified as one area of diversification. The basic reasons
underlying this choice have been two: (1) these countries are
endowed with an abundance of the basic raw material inputs and
(2) petrochemicals are capital-intensive, and these countries have

relatively high capital-labor ratios. For these reasons, it has been generally accepted that these countries should develop petrochemicals and thus increase the export value of their raw materials by adding value domestically.

However, these reasons are insufficient by themselves to justify the large-scale development of petrochemicals, as many other questions need to be answered. Comparative advantage cannot be assessed on the basis of one product and its domestic cost of production. Instead, it should be evaluated on the basis of relative costs in a multicommodity framework. In the cost calculation, the cost of plant construction and world prices for all inputs should be used. If it turns out that Middle Eastern petrochemicals can compete only with large subsidies (such as a price for feedstock below its opportunity costs) for the foreseeable future, then the value of such diversification would be dubious. But given the limited facts and figures for the necessary calculation, only in time can we determine the success of diversification into petrochemicals.

A sound policy of economic diversification is not only prudent and wise to reduce the dualistic nature of these economies; it is also necessary in order to provide citizens with diversified and gainful employment opportunities. A nation, no matter how rich, would find it difficult to survive socially by living off the returns of its financial assets.

Besides decisions regarding diversification, the governments of these countries face numerous socioeconomic problems. With large increases in government expenditures and no system of income taxes in the gulf countries, the oil wealth has not equally benefited all segments of society. Given the lack of comprehensive data, it is difficult to judge how income distribution has changed in recent years. But in any case it would be prudent for the governments to establish at least a framework and system of income taxation, to provide against the time when it may be necessary to address distributional problems and also as a source of government revenue as oil resources are exhausted. Given the long period necessary to establish a working tax system, it would be prudent to start earlier rather than later.

Rapid economic change has been accompanied by large increases in the size of these countries' expatriate labor forces. The benefits of such labor inflows to the labor-importing country are clear, but there have also been costs. The inflow of such labor in many countries has led to social difficulties. In many instances, the imported labor has been housed in restricted areas, and they have not received the normal benefits available to citizens—educational and health services, social security, land ownership privileges, and so forth. In other instances, the immigrant workers have received greater benefits and have not adapted to the local society and its customs. As a result, realities have led to social tensions. Furthermore, in countries where foreign labor is a large proportion of total labor force and total population—Kuwait, Qatar, and the United Arab Emirates—the political implications for unrest are clear. Finally, much of the immigrant labor force are still committed to their countries of origin where in many cases their families reside. Thus, when sufficient funds have been accumulated, the foreign workers will often return home, taking with them their acquired skill and experience. In turn, the labor-importing countries may have to retrain new labor from abroad.

Socioeconomic problems are not limited to the problems surrounding expatriate labor. Rapid economic change has put enormous stress on the social fabric of society. Values which have been adhered to for centuries are being eroded by the change and the increased exposure of these countries to the outside. Thus, these governments are confronted not only with economic problems but with a host of social problems which invariably accompany rapid change.

CONCLUSION

These economies show many striking similarities and dissimilarities; the most important fundamental difference contributing to observed dissimilarities among them is the relative sizes of their oil endowments. For all these countries, however, oil has dramatically affected developments in their economies. Furthermore, given that oil is finite in quantity and for several of the countries their only known physical resource, they all must transform their oil assets into productive nonoil domestic assets and attractive (real-yield) foreign assets if they are to be successful over the long run.

To assess the likelihood of such success from an historical perspective, it may be useful to compare the circumstances surrounding economic changes in the Middle Eastern oil-exporting countries to those surrounding the gold rush era,[4] as the magnitude and swiftness of changes in the Middle East recalls a few historical parallels. Such parallels can be seen in the nineteenth century gold rushes, particularly in California in the late 1840s and early 1850s and in the Yukon and Alaska in the 1890s.

In both the gold rushes and the oil boom, the consequent inflation has been clearly of a demand-pull nature, coupled with severe supply bottlenecks. On the demand side, purchasing power multiplied severalfold within a few months. Not only did the local per capita money supply increase significantly, but the number of consumers also climbed dramatically. The characteristic "rush" of population associated with gold in the nineteenth century has its counterpart in the Middle East in the mid 1970s.

The gold rushes occurred in areas quite isolated from the major suppliers of basic goods—in western North America, months by ship around Cape Horn from Atlantic coast suppliers. (The gold sites in the Yukon and Alaska were in addition a hard journey inland from entry ports.) Local production during the gold rushes was limited—for example, California then had only a minimal agricultural base—and inventories of basic goods at the onset of the gold rushes reflected the low population levels before the discoveries.

In the oil boom of the Middle East, physical isolation has not been the specific cause of supply bottlenecks, but the transportation infrastructure, often inadequate even before 1973, has simply not been able to handle the rapid expansion of imports. For example, ship demurrage charges in 1974 approached and then exceeded actual transport costs as freighters backed up in established ports like Jeddah, Khorramshahr, and Dubai for unloading waits as long as two hundred days. The often poorly graded highways of Turkey began to crumble under convoys of overloaded trucks bound for Iran and Saudi Arabia, while delays at undermanned customs stations all the way from Europe to the Middle East added weeks to truck journeys that might have taken ten days in 1972. Furthermore, while the local population of the Middle East is much more numerous than it was in California or the Yukon, local production has been limited by a paucity of manpower trained in the specific skills needed for increased output. These Middle Eastern countries were not and are not developed industrial economies; local production does not satisfy domestic demand for items such as automobiles, and most of these countries have little domestic capacity even for basic consumer goods.

The marked degree of inflation that has occurred in gold and oil boom regions has been essentially a geographically limited phenomenon, and with the easing of supply bottlenecks, the most exaggerated price increases have been slowed down and then reversed over time. Since most of the goods imported into these boom regions have been purchased at world prices and have represented minor shares of world production, the increase in local demand has had little effect on international price trends.

The transformation of an economically backward area with an exhaustible resource is critically dependent upon the quality of the investments. In the case of the California gold rush, California was put on the road to becoming a major producing region of the United States—first as a supplier of agricultural goods and other natural resources, then as a manufacturer. In the Yukon and Alaska, on the other hand, the gold rushes were short-lived phenomena with little permanent economic effect.

There are many similarities between the economic circumstances surrounding the gold rush era and the current economic position of many OPEC countries. The basic reasons for such a parallel are (1) the discovery or increase in value of an exportable natural resource and (2) fundamental similarities between the OPEC countries' economic structures and those of the gold rush areas.

In all cases, there has been a rapid expansion of the money supply. But in the face of this increase in purchasing power, the economy has been unable, in the short and intermediate time frame, to expand domestic production of goods and services. In addition, due to distance or to physical constraints of infrastructure (ports, roads, warehousing, etc.) and to some extent because of the time required for procurement and transportation of imports, imports have not increased rapidly enough to satisfy demand. As a result of this excess demand, domestic inflation has been very rapid. And in the Middle East, many-times-disastrous government economic policies, such as those permitting uncontrolled and haphazard expansion of imports and domestic production, have resulted in a great deal of economic waste, such as food rotting on board ships. More importantly, domestic expansion has occurred in industries in which the country probably does not have a comparative advantage, while others with such advantages, as in Iraq's agriculture, have been neglected.

Such prevailing economic forces have in most instances resulted in several similar economic patterns. Foreign labor or labor outside of the region have moved in to supplement domestic labor supply. In OPEC countries, as with the gold rush economies, this labor movement has been pronounced and has consisted of both skilled and unskilled labor. With the initiation of these rapid forces for change, these countries have accelerated towards being market economies, and with such a change in economic structure, a more sophisticated financial and banking system has developed. But during this transition prices increased, the increases being most pronounced in nonsubstitutable goods where imports would not supplement domestic availabilities. With higher prices, domestic supplies responded positively. However, in the short run this expansion was not enough to satisfy demand, even when supplemented with imports.

The interesting question from the Middle Eastern and OPEC perspective is whether, in view of the recent greater availability of financial resources, a sustained drive to development like that in California is probable, or whether the boom may be a temporary phenomenon, as it was in Alaska and the Yukon. To determine this, we should consider who ends up benefiting in such economies, what investments attract the beneficiaries, and where these investments are located.

Certainly much OPEC capital is reinvested regionally, just as gold capital was retained in California. But some early indications are not promising that these investments will pay off in the Middle East. There are also signs that, as in the Yukon, private capital holders are looking elsewhere for future investments.

There are, however, two major differences between the gold rush cases and the Middle Eastern countries. First, the oil boom seems likely to last somewhat longer, giving local governments the opportunity of gaining experience from the excesses of the mid 1970s. This opportunity does not extend equally to all the Middle Eastern countries. Though most people are aware that not all the Middle East is equally endowed with oil, there is a tendency to forget how uneven the distribution is. For this purpose, the figures discussed earlier (the value of per capita oil reserves) are particularly revealing. Second, the economic potential of these seven countries differs from that of California in at least one other important respect. In California, the high quality of agricultural land, the availability of water, the favorable climate, and the existence of other complementary factors of production were all favorable elements for rapid economic development of the region. In the case of the seven oil-exporting countries discussed here, however, there is a dearth of these favorable elements for development, and thus oil wealth cannot be as easily transformed into a base for domestic productivity.

The transformation of this oil wealth will thus face many pitfalls in these countries. But given the magnitude of such wealth, the richer oil states have considerable leeway to get their investments in order. However, they increasingly may have to look abroad for promising investment opportunities, given their large current account surpluses (see Table 20–6). But other producers (notably Algeria and Iraq) enjoy less leeway and must take greater care to avoid waste, lest the Yukon example become appropriate when oil production begins to decline, perhaps as soon as the mid 1980s.

Table 20–6. **Current Account Balance**[a]

Country	1974	1975	1976	1977	1978	1979	1980
Algeria	584	-1,650	-871	-2,307	-3,526		
Iraq	2,855	2,971	2,160	2,200	2,200	8,200	
Kuwait		6,684	7,172	5,645	6,967	15,440	
Libya	1,900	96	2,581	3,058	1,297		
Qatar	1,667	1,130	1,020	426	692	1,710	
Saudi Arabia	24,022	17,059	17,127	16,679	2,900	13,800	46,000
United Arab Emirates		3,900	4,600	3,700	3,200	5,500	10,500

Source: Estimated from various national sources.
[a]Goods, services, and private transfers, in millions of dollars. Blanks in table indicate no data available.

The effects and consequences of the gold rush and the oil boom on the political and social fabric have also been great. As an example, the gold rush in California resulted in an evolution in institutions such as government. These evolutions out of the mining camps and towns provided the new society with its framework for order. This framework was thought crude at first, but it developed rapidly and promoted an environment that over the next hundred years made California the nation's most populous and wealthy state.

In the Middle Eastern experience, two different cases should be briefly mentioned—Kuwait and Iran. The Kuwaiti approach has been to give foremost priority to social programs for its citizens. Industrialization has taken a back seat; instead, oil revenues have been used to develop the foremost social services in the world and to accumulate surplus revenues which have been invested abroad to provide a "social" fund for all future generations of Kuwaitis.

The Iranian experience is, on the other hand, much more complicated and much less promising. Iran's oil revenue per capita is much less than that of Kuwait; this fact coupled with Iran's potential for economic development encouraged its government to embark on a rapid industrialization program. But even more important than industrialization was Iran's commitment to military expenditures, other prestige projects, and investments. This approach resulted in a neglect of agriculture, education, health services, and other social expenditures—in fact, due to large military expenditures and foreign commitments, even industry received little attention. Specifically, even the form of the commitment to industry was ill conceived, and few long-run advantages were given to the economy. These glaring facts coupled with dismal income distribution contributed in large part to the political and social disruptions of 1978–79. That is, the neglect of society at large and the phenomenal wealth of a few ripped apart the social fabric.

In short, although the gold rushes and the oil boom have many similarities, the end result may be different. The gold rushes in Alaska and the Yukon were short-lived with little permanent impact. To California, the discovery of gold was in large part a blessing, but to Iran, the oil boom could up until now be more appropriately referred to as the "oil curse." For other oil-exporting countries, the results have been a mixed bag. But the future is uncertain, and success depends both on the extent of oil wealth and on how it is transformed into productive domestic assets and foreign holdings. Fortunately, several of the Arab oil-exporting countries have substantial oil reserves and large holdings of foreign assets to accomplish their economic development goals.

Thus, it is important to note in conclusion that although some of these countries have substantial oil reserves, they cannot be classified as prosperous until they have developed an alternative and efficient economic production base to replace their oil as it is depleted. The development of an efficient nonoil base is likely to be difficult and is by no means certain. Cooperation from the industrialized countries will be critical in the areas of technology transfer and open markets for the exports of OPEC countries.

NOTES

1. Mid-1979 population figures are from the 1980 *World Bank Atlas* (Washington, DC: World Bank, 1981).
2. Per capita income figures are for 1979 from the *1980 World Bank Atlas*, op. cit.
3. A. Shihab-Eldin, G. T. Abed, and S. S. Al-Qudsi, "Requirements for the Development of Additional Energy Sources," p. 10, from a seminar on Development Through Cooperation between OAPEC, Italy, and Southern Europe, Rome, Italy, April 7–9, 1981.

4. This comparison relies heavily on an unpublished paper—for further details, see H. Askari, J. T. Cummings, and H. Reed, "Middle East: Gold Rush or Economic Development," University of Texas at Austin, 1978.

CHAPTER TWENTY-ONE

Summary with Selected Comments

Session Chairman/Integrator:
Mohammad Sadli
Professor of Economics
The University of Indonesia

The category "energy-surplus industrializing nations" comprises a wide range of countries, from high-per-capita-income countries such as Kuwait and Saudi Arabia to poor or low-middle-income countries such as Indonesia and Nigeria; from large agricultural countries to small city-states. The options for industrialization also differ widely, from import substitution to export orientation based on the country's natural resources endowment or labor abundance. The only thing this group of countries has in common is that they are oil exporters and therefore have relatively comfortable balances of payments. Their development needs, however, are still great, implying that their dependence on the world economy for their development inputs and for their exports will still be crucial for the decades to come. Even the countries with surplus balances of payments depend on the world economy as a marketplace for their investments.

This work session focused its attention on two major aspects of these countries' problems and their role in the world economy: their role as major oil and gas exporters (specifically, in this respect, the OPEC countries) and their common as well as individual development problems.

In addition to the chairman/integrator and those presenting papers, the work session included the following assigned participants: James E. Akins, Fereidun Fesharaki, Shunsuke Kondo, Robert L. Loftness, Hans H. Landsberg, and Luis Sedgwick Baez. The chairman/integrator was assisted by Wallace Koehler, who served as rapporteur.

OPEC

It is popular to perceive the OPEC countries as a cartel; that is, a small number of suppliers controlling a dominant portion of the market and working in collusion to set prices to maximize revenues. In fact, their actual behavior cannot meet all the conditions of an effective cartel. OPEC is an intergovernmental organization, relying for its decisions on the rule of full consensus, which in practice is not often achieved. Traditionally, the OPEC countries have been unable to agree on planned production rationing among the members. But nevertheless they set, or try to set, prices, even without an agreement on production control.

Other observers cannot see that OPEC really dictates prices. In their opinion, the market—that is, the balance between supply and demand—determines the price, and OPEC only ratifies the outcome. The two major price shocks, in 1973 and in 1979, were initiated by incidents not controlled by OPEC as an organization, although some of its larger members had much to do with the events. The failure of the market to return to its original position, however, could be attributed to the role of OPEC in trying to sustain prices after these had gone up because of other reasons.

Although OPEC consists of thirteen member countries, a handful of Persian Gulf states are responsible for the bulk of exports. Among these countries Saudi Arabia stands out, at the moment, with about half of OPEC exports. Hence, an understanding of Saudi Arabia's interests, its behavior, its decisionmaking processes, and its political dynamics would be very important to gauge the prospect of the future oil market.

OPEC is ostensibly an intergovernmental organization with solely an economic purpose, that is, to set prices for the members' crude oil exports. In the members' decisionmaking process, they stay away from politics and try to rationalize their decisions solely on economic grounds. This by itself is a very difficult process of integration, because the member countries' economic interests are widely divergent—for instance, consider the difference between the so-called high absorbers and the low absorbers.

Again, the outcome will depend on the interests of a small number of gulf countries, among which Saudi Arabia is the most important. Although economic interests, especially in their long-term aspects, have been important to understand Saudi Arabia's decisions, other interests (i.e., political and security interests) have played and will continue to play an overriding role in their decisions. The combination of these factors always makes it extremely difficult to predict the outcome of OPEC decisions and the future of the international oil market. Nevertheless, one could probably observe that Saudi Arabia's role, which is a moderating one, is more prevalent in times of surplus or balance in the market and less in times of tightness. But even in the latter case, Saudi Arabia's moderate price decisions have lowered the effective average price for all OPEC crudes.

In the past decade the international crude oil market has been characterized by instability and fragility in the outcome of its prices and outputs, perhaps reflecting the basic instability in that part of the world from which the bulk of the supplies come. Even today, despite the emergence of a market weakness, the world in the 1980s remains dependent upon decisions of, and happenings in, a small group of gulf countries who have substantial latitudes in production levels. The experience of the past suggests that the world oil markets will continue to be quite fragile, with small perturbations leading to large swings in prices.

Moreover, there are other variables, outside those inherent in the gulf area, to be reckoned with. The industrialized countries have succeeded remarkably in conservation and have been able to even reduce their oil consumption, although the economic recession has also been responsible for this. On the other hand, a majority of developing countries still have high rates of commercial energy consumption which in the intermediate future will be difficult to meet from diversified sources. Moreover, the socialist countries could become larger importers of crude in the future. On the other side of the demand/supply balance, however, some non-OPEC developing countries are improving their oil and gas production, thanks to the greater price incentives the present market conveys. All in all, supply and demand are precariously balanced.

Complacency about the "end of the energy crisis" or "the moderation of OPEC" (driven by economic self-interest) is not justified. The need to use the current period of market weakness to strengthen relations between producers and consumers is paramount. Because of the dominant position of the gulf countries in the amount of total exports, progress towards a solution of the Middle East problems is of crucial importance. Greater political stability in the Middle East is not a condition for the solution of the global energy problem, but at least it may reduce the risk of abrupt supply reductions and concomitant price jumps.

Given the fragility of the oil market and the interest of both producers and consumers in its smooth functioning and relative stability, the rationale for consumers to stockpile and for producers to have excess capacity remains strong.

The Yamani proposition for an orderly movement of crude oil prices through world inflation and economic growth towards a long-term equilibrium price has its attractions and should be considered for the mutual benefit of producers and consumers alike, with the health of the world economy and the interests of the developing countries at heart. Of course, one should recognize the inherent difficulties in such a proposition, which are legion. Orderly and more predictable movements of crude prices are probably in everybody's long-term interests in principle, but several baseline numbers and quantifications need to be agreed upon. For instance: What should the ultimate equilibrium price be? As related to what alternate energy source (synthetic fuels?)? When should this equilibrium be achieved (at the end of the century?)? From what base should the orderly movement of the price start? Using what inflation indexes and escalators? and so forth. It would be extremely difficult to expect early agreement on a global basis on these values, but at least discussions, deliberations, and a regular dialogue could be started.

From the standpoint of the large industrialized developed countries, continued dependence on the uncertain outcomes of the decisions of a small group of exporter countries suggests consideration of a national security premium for imported oil, although there is a possibility that the imposition of an import tariff could have a counterproductive impact on OPEC output decisions.

To induce the large oil-exporting countries to continue their production policies, oil-importing countries, especially the large industrialized nations, should remove obstacles—institutional, attitudinal, and otherwise—to the former's investment of surplus funds in the latter, especially for investments in real assets. Investments in monetary instruments should be given greater assurances of real value maintenance.

The Iran-Iraq conflict remains an important source of instability. On the one hand, if peace could come about there is the possibility of an early upsurge of exports from this region; on the other hand, there is always the danger that the conflict may spread, invoking disaster scenarios. Peace may also have to wait for political changes in both countries.

The recycling of petrodollars, which became a serious concern after the first oil crisis and again after the second one, apparently is no longer perceived to be an unsolvable problem. The international money market is able to cope to a large extent with the new situation. Yet, the distributional impact is still important, since it affects a number of developing countries having structural problems in their system (an example is Turkey). The international system, including multilateral institutions such as the International Monetary Fund and the World Bank, should pay attention to this problem.

DEVELOPMENT PROBLEMS

As was observed in the beginning, the energy-surplus developing countries make up a divergent group, including both high absorbers which still need capital inflow from outside and others that are low absorbers. The former generally have more diversified economies, while the latter have much more narrow bases—in other words, mainly oil.

The Arab countries are rich in GNP terms, but they realize that such is the result of a process of consuming their exhaustible endowments, and hence they feel themselves not well off in the same way as the established industrialized countries. Their major problem is managing the transformation of their economies, using only their once-over stock of natural wealth, into a self-sustaining and productive economy. What they have to avoid is the fate of places such as those which, after the gold rush was over, became ghost towns.

This transformation and diversification of their economic basis has two aspects, an international one and a domestic one. The first one is related to diversifying and safeguarding the value and improving the real rate of return of their holdings abroad; that is, the investments of their current surpluses. They need a good and unobstructed investment climate outside.

Investment opportunities for the surpluses of the gulf countries should be enlarged and improved, everywhere in the world. Traditionally, investments in monetary assets have a rather low rate of return in real terms, except perhaps in periods like today of high interest rates. Investments in real assets should be made more possible or facilitated; restrictions should be removed or reduced; and diversification should be encouraged. Investment and incentive regimes in the receiving countries should give positive inducement to the utilization of the oil-rich countries' surpluses. The developing countries should make greater efforts to attract these funds by providing greater incentives or guarantees to overcome the hesitation of potential investors who are still unfamiliar, especially with respect to developing countries, with the game of overseas direct investments and portfolio investments.

The domestic aspect concerning economic transformation differs from country to country. For the oil countries that lack significant agricultural sectors, the problem is finding the proper so-called comparative advantage for their industrial development and diversification away from oil extraction. The tendency is to reinvest oil revenues in oil- and gas-related industries and services. Although gas and oil may be abundant production factors, it does not necessarily follow that petrochemicals constitute a pertinent comparative advantage. A petrochemical industry requires vast amounts of capital, technological expertise, and infrastructure, all of which could make it a high-cost industry in developing countries, and international product competitiveness may require pricing the inputs at less than their opportunity costs. However, comparative advantage is not strictly a theoretical phenomenon, nor a static one. It also has entrepreneurial and innovational dimensions, and it has to be judged in a longer term context. Joint venture engagements with multinational companies could produce viable propositions.

Investments in petrochemical industries and refineries also have a locational dimension. Traditionally, such industries are market oriented, and thus, locating the industry at the raw materials source may result in an efficiency handicap—for instance, in the shipment of the products. Oil- and gas-producing countries, however, often have a preference for the latter approach. By locating the industries within their countries they hope to have greater employment spread effects and external economic effects, thereby aiding the transformation of their economies. This socioeconomic and political preference should be recognized.

The above may have longer term implications for the international distribution of labor and specialization; for example, the world industrial and trading system has to cope with a shift towards the OPEC countries in the incidence of the petrochemical industries location. The slate of crudes and products entering the international market will also experience changes, calling for international accommodations.

Other petroleum-exporting countries have mineral resources and agricultural resources such as timber that are waiting to be exploited. These countries also wish to invest their oil revenues to develop these resources and increase domestic value added by engaging in processing. The same problems of comparative advantage and location will be encountered. Left to competitive market forces, many timber processors and smelters may go near the market—in other words, the industrialized countries. These processing industries, however, are seen as important instruments to transform the oil-exporting countries' economies by providing a basis for industrialization. The oil-exporting countries are prepared to subsidize the growth of such industries in their infancy. The problem is to see to it that such infant industries eventually mature.

Although the energy-surplus developing countries are "richer" or have greater resources than the other developing countries, their development needs require the same international assistance and cooperation. They may not need foreign aid, but foreign direct investments and technical or technological cooperation with multinational companies will strengthen their industrialization. The products of their domestic industries need market outlets in the world, and hence, the trading system should not have too many barriers for these new entrants.

The large OPEC countries want to build up and enlarge their capacities in transportation, refining, processing, and other downstream activities. For these they need technical cooperation, equity participation, and access to markets for their products and services. The establishment of broader based multinational enterprises may be a very appropriate thing in the near future.

Oil-producing countries with large population bases and relatively underdeveloped agricultural sectors have other development problems. Their efforts to raise food production and develop other agricultural potentials are often frustrated by the impact of their oil economies. High wages in the oil industry and related activities draw labor away from agriculture for food and other smallholders' activities. The exchange rate is distorted by the oil revenues in the sense that it cannot serve both the oil and modern sectors and the traditional sectors at the same time. The result is often an overvaluation vis-à-vis the needs of the traditional sectors. Food becomes cheaper to import than to produce locally. Export-oriented industrialization is made equally more difficult.

Because of abundant or ample oil revenues in the oil-producing countries, domestic resource mobilization is often lax. Apart from tax revenues from oil, the growth of domestic tax revenues frequently is lagging. Income taxation cannot counteract the rapid growth in income differentials, and income distributions worsen. Countries with these problems should make an honest effort to tighten up income taxation in order to improve domestic resource mobilization and prevent social distortions which act as destabilizing factors.

Usually, domestic fuel prices are set far under international levels and involve disguised or open subsidies. The subsidy system progressively permeates the economy, distorting prices and resource allocations.

For petroleum-exporting countries that still are in the lower income brackets and have yet to go the long road of development and industrialization, these distortions in domestic savings and resource allocations will jeopardize their long-term growth prospects. They can ill afford to tamper with the orderly functioning of the market mechanism for resource allocation. Hence, their public policies should underscore the reality that economic resources are significantly scarce for them as well and that the price system should reflect realities.

CONCLUDING REMARK

The energy-surplus countries have a very important interface with the industrialized countries and also with the oil-importing industrializing countries, and this interaction and interdependence can only grow in significance. The degree and extent of international cooperation will increase, and the management of this interdependence calls for our special attention. The developing countries' party line, including that of the OPEC countries, is that energy problems should be the subject of international dialogue but that these problems should not be disassociated from overall development problems and needs and their relation to the world economic system. The process of global deliberations and negotiations, endorsed by the Cancún Summit, should be duly started after expeditious preparatory talks.

SUMMARY OF RECOMMENDATIONS AND OPTIONS

1. The current complacency about the "end of the energy crisis" is not justified. The need to use the present period of market weakness to strengthen relations between producers and consumers is paramount. The dominant position of gulf countries in the amount of total exports should be recognized. Progress towards a solution of the Middle East problems is of crucial importance.

2. Given the fragility of the oil market and the interest of both producers and consumers in its smooth functioning and relative stability, the rationale for consumers to stockpile and for producers to have excess capacity is strong.

3. From the standpoint of the large industrialized countries, continued dependence on the uncertain outcome of the decisions of a small group of exporter countries suggests consideration of a national security premium for imported oil.

4. To induce the large oil-exporting and oil-surplus countries to continue their production policies, oil-importing countries, especially the large industrialized nations, should remove obstacles—institutional, attitudinal, and otherwise—to the former's investment of surplus funds in the latter, especially for investments in real assets.

5. The recycling of petrodollars is now no longer perceived to be an unsolvable problem. Yet, distributional impacts are still important. The international system, including institutions such as the International Monetary Fund and World Bank, should be made more capable of rendering assistance.

6. The developing countries should make greater efforts to attract the funds of the petroleum-exporting countries by providing greater incentives and guarantees.

7. The major OPEC countries are inclined to diversify into petrochemical industries, refineries, and other oil-related activities and aim at export markets. The international trading and investment regimes should take this into account, make the necessary accommodations, and take part in this process of international changes of location and specialization.

8. Developing countries, including oil-producing countries, should not fail to improve their domestic resources mobilization. They would be well advised to improve their income tax collections, for this might improve social equity and contribute to greater stability. Their public policies should underscore the reality that economic resources are significantly scarce for them as well and that the price system should reflect these realities.

9. The orderly movement of crude prices through world inflation and economic growth towards a long-term equilibrium price related to a relevant alternative energy source should be considered for the mutual benefit of producers and consumers

alike. Although it would be extremely difficult to expect early agreement on the crucial variables, at least discussions, deliberations, and a regular dialogue could be started.

10. The global economic system will witness an increase in the importance of interdependence and necessary cooperation. The management of such interdependence calls for special attention. The process of global deliberations and negotiations should be duly started. Later this process should enable regular policy consultations on a global basis.

SELECTED COMMENTS

SHUNSUKE KONDO

One of the key lessons we should learn from history is the importance of long-term perspectives in our policymaking processes. In this context, more stress should be put on the projection that some of the present oil-exporting countries will lose their exporting capabilities in a few decades, and it should be recommended that these countries develop strategies of structuring their energy supply sectors through proper mixtures of conventional and alternative resources, including solar and nuclear power.

Work Session on Energy-Deficit Industrializing Nations

CHAPTER TWENTY-TWO

Introduction and
Part A: The Energy Problem
in the Oil-Importing
Developing Countries

*Joy Dunkerley**
Senior Fellow
Resources for the Future

Part B: Case Studies of
Burundi and
the Cameroon

*Emmanuel Mbi**
Economist
The World Bank

INTRODUCTION

The title of this chapter's Part A defines the major common characteristic of this group of countries—these countries are all to a lesser or greater extent net oil importers. In almost every other respect they differ widely. At one end of the spectrum are countries like Korea and Brazil with annual per capita incomes of $1,500 (US $) per year. These countries have such large and dynamic manufacturing sectors that they are close to graduating into the class of industrialized nations. At the other end of the spectrum are countries like Ethiopia and Upper Volta whose sole activity is primitive

---

*The authors note that the views expressed herein are their own and should not be interpreted as necessarily representing those of Resources for the Future or the World Bank.

agriculture, yielding per capita incomes of under $200 per year. This wide variation in standards of living is reflected in the amounts of energy consumed. Per capita consumption of commercial energy in Korea, for example, is sixty to eighty times higher than in Ethiopia and Upper Volta, more than compensating the much higher consumption of traditional fuels in these two countries.

Consequently, it is difficult to insure comprehensive treatment of the energy-related problems of such a diverse group. In this work session of the Symposium, the first task is to analyze the problems of the group as a whole, with reference where appropriate to subgroups based on income, economic structure, degree of energy dependence, and potential energy sources. To illustrate further the diversity of countries included, case studies of countries within this group that differ dramatically—South Korea, the Cameroon and Burundi—are presented in the work session [see chap. 23 and Part B of this chap.]. In this way, attention should be drawn to the fact that within the broad context of a common set of energy problems, the oil-importing developing countries are differently affected, have different possibilities for solving these problems, and will require different types of assistance.

PART A: THE ENERGY PROBLEM IN THE OIL-
IMPORTING DEVELOPING COUNTRIES*

THE NATURE OF THE PROBLEM

The basic energy problem for these countries is how to secure adequate and reliable quantities of energy to enable minimal needs to be met and economic development to proceed, without serious effects on the environment and on equity. This is all the more difficult because the linkages between these areas are not well understood. But to take one area—the relationship between energy consumption and economic growth—one central fact stands out. Apart from exceptional circumstances, energy consumption rises as economic activity expands. A growth in output is associated with increases in energy consumption as both cause and effect. The rise in output requires increased inputs of energy as well as other factors of production, and higher incomes stemming from this increase in output are used to buy more energy services as well as other goods. This holds true in both industrial and developing economies.

But there is little reason to suppose that energy consumption and economic growth move together in exactly the same proportions. In the industrial countries, in the fifteen or so years prior to 1973, both energy consumption and economic growth did indeed show a similar increase. Since 1973 their energy consumption has risen on average by only one-half of their increase in Gross Domestic Product.[1]

The experience of the oil-importing developing countries is more difficult to assess because of their dependence on poorly documented traditional fuels [fuelwood, dung, etc.], which account for an average of 30 percent of their total energy consumption.[2] On the one hand, there is strong evidence that commercial energy consumption expands more rapidly than economic growth. Thus, from 1960 to 1970—a period of rapid development for the

*This analysis and particularly the section dealing with supply options draws heavily on *Energy Strategies for Developing Nations* by Joy Dunkerley, William Ramsay, Lincoln Gordon, and Elizabeth Cecelski published by Resources for the Future, Inc. (Baltimore: Johns Hopkins University Press, 1981). The author wishes here to acknowledge her debt to her coauthors in preparation of this paper.

oil-importing developing countries—commercial energy consumption in these countries rose by 7.2 percent per year, well in excess of the 5.6 percent average annual increase in Gross National Product.[3] In other words, for each 10 percent increase in Gross National Product, commercial energy consumption in these countries rose by 13 percent.

On the other hand, it appears that consumption of traditional fuels in these countries, though rising, is rising less rapidly than Gross National Product. Whether the overall energy intensity of these economies increases or decreases as development proceeds will depend on which trend dominates —the increasing intensity of commercial energy use or the decreasing intensity of traditional energy use. In any event, given current population and economic growth rate assumptions, consumption of both traditional and commercial fuels is expected to increase substantially in the future. If traditional fuel availabilities are to keep up with rural population growth, they must expand by about 2 to 3 percent per year. If biomass fuels are to be used as feedstock for new liquid or gaseous fuels or for expanded industrial use, production will need to increase much faster. For commercial fuels, the World Bank estimates that oil consumption in developing countries will increase by over 6 percent per year up to 1990—a high figure, even though it is a sharp reduction from projections made not long before. This implies an increase in oil imports from the current 4.5 mbd to 7.6 mbd and a more than doubling of oil import bills (in constant dollar terms).[4]

Apart from the difficulties imports of this size will impose on the economies of the oil-importing developing nations, these imports will also contribute in a major way to defining the balance of the international market for oil. A recent International Energy Agency report estimates that with present plans, OECD oil imports will rise by 2 mbd between 1979 and 1990. Imports to centrally planned economies will rise by an additional 2 mbd and to developing countries by a further 3 mbd, requiring in total an extra 7 mbd from OPEC by 1990.[5] As estimated OPEC exports are assumed to remain unchanged, this could result in much higher prices for available supplies. Alternative projections based on substantially lower OECD consumption levels would bring the market theoretically into balance, but only just. Either way, the level of oil imports into the developing countries will be critical.

Increasing the availability of both traditional and commercial fuels to the levels estimated by the World Bank and others could raise major though rather different sets of difficulties in the two sectors.

Traditional Fuels

Prospects for increasing supplies of traditional fuels are complicated by the fact that the existing situation—particularly for forest products, which provide about 70 percent of total traditional fuel supplies—is already difficult.

While fuelwood energy resources are in principle renewable, overconsumption erodes the resource base. If more wood is taken from the forest (due to the fuelwood and timber needs of a growing population or to land clearing for agriculture) than the annual increase in forest growth, eventually the forest resource base will disappear. Data on this important topic are somewhat patchy, but there seems adequate evidence to indicate that the resource base of tropical forest is presently under considerable stress, losing 16 million hectares annually.[6] Moreover, this overall figure does not reveal the situation in certain regions and countries where the losses have been much more severe. Even if the task of reconstituting the forests started immediately, the situation could get worse before it gets better; like many other forms of energy, there is a long lead time in developing new fuelwood supplies.

The results of a shrinking forest base are important and far-reaching. First of all, a special hardship is imposed on the rural poor, who are the largest users of traditional fuels. Firewood is the major, perhaps the only, form of fuel available to 2 billion people —about one half of the world's population. There are a limited number of responses, all of them unsatisfactory, to fuelwood shortages. Users can, for example, reduce their energy consumption from already low levels by going without hot meals, or they can devote more money or, in practice, more time to collection. Buying wood fuels in some cities of the Sahel has been estimated to require one-fourth to one-third of the average laborer's income, illustrating in concrete terms the African maxim that "it takes more time to heat the pot than to fill it."[7] In other areas—for example, Nepal and central Tanzania—it takes an estimated 200 to 300 person-days of work per year to gather a family's fuelwood.[8] Another response is to cut fruit trees and other economically valuable species for wood, thus reducing both future food and income yields.

In all these cases, the quality of life of a major part of the world's population deteriorates. Nutritional standards fall, and valuable time which could be spent in more productive or enjoyable pursuits must be devoted to the arduous and unpopular task of firewood collection.

The destruction of forest and ground cover also has serious environmental consequences. Tree removal—particularly in upland areas—can eliminate topsoil, destroy reservoirs by siltation, and thereby reduce food production, perhaps requiring increased food imports. Shortages of fuelwood will also lead to increased use of crop and animal residues for fuel, with severe effects on soil fertility and agricultural productivity. It has been estimated, for example, that 1 ton of cow dung burned as fuel means forgoing 50 kilograms of food grain production.[9]

The severity of these problems differs from country to country. Some countries—Colombia, Chile, Brazil, and several Caribbean countries—are so well-endowed with forest resources that they are considered unlikely to experience national shortages, though local shortages can still develop even in well-wooded regions. Many more countries are likely to develop problems of widespread severity. These include India, Pakistan, Bangladesh, and numerous African countries—in other words, the poor countries.

In brief, there already exist major problems of sustainability, equity, and environmental quality associated with current patterns of traditional fuel usage. In considering how to increase traditional fuel supplies in the future, we must realize that we are starting from a very weak position. Indeed, future developments could worsen the existing problems which already weigh heavily on those—the rural poor in the poorest countries—least able to defend themselves.

Commercial Fuels

Increasing the availability of commercial fuels also raises problems. As we have seen, consumption of commercial fuels in developing countries typically increases faster than total national output. This reflects the increasing weight, typical in a developing economy, of activities such as modern industry, transportation, and urban living that can only be carried out using commercial fuels under current practice, and it also reflects the increasing inaccessibility to traditional fuels of rapidly growing urban areas. Since the mid 1950s, the sharp rise in commercial fuels consumption has meant in practice a very rapid rise in oil consumption.[10] Other forms of commercial energy are also used—gas and coal and hydroelectricity in countries with local resources—but oil has the predominant share: currently 67 percent of total consumption, up from 60 percent in 1955.[11] Oil is therefore the mainstay of commercial energy supplies in oil-importing developing countries. Consequently, the sharp increases in oil prices which took place in 1973–74 and 1979–80 created major problems for these countries. Since oil prices are expected to keep rising in the future, these problems are likely to recur.

In brief, since oil imports tend to be a significant part of total imports and since it is difficult to reduce oil consumption or nonoil imports substantially in the short run, the increase in prices of oil and other imports affected by higher oil prices leads to a sizable increase in the total import bill. In the absence of an equivalent increase in exports—an increase made all the more difficult by oil-price-induced recession in many export markets — trade balances deteriorate, and the resulting deficits have to be financed by borrowing or running down foreign exchange reserves.

Internally, rising oil prices lead to increases in the prices of other goods. At the same time, the need for consumers to spend more for their energy services means that there is less money to buy other things, and in the absence of countervailing measures, aggregate demand falls. The rise in imported oil prices consequently has simultaneous inflationary and deflationary effects on the domestic economy.

Therefore, following an increase in oil prices the governments of oil-importing countries are faced with a variety of pressing problems—large trade gaps, inflation, and lower growth rates. In the longer term, the process of adjustment to the higher resource costs of imported oil involves a reduction in oil imports, either through increasing indigenous energy supplies or by conservation. This means that major efforts are needed to mobilize both domestic and foreign resources in order to finance higher oil bills and to make the necessary energy sector investments.

These problems are experienced by all oil-importing countries, whether industrial or developing. But a similar change in energy costs might be expected to have a more severe impact on the economies of oil-importing developing countries. For most developing countries, the availability of capital and of the foreign exchange resources to pay for essential imports is a major constraint on growth rates. If the only means of financing additional costs of oil imports is to reduce nonoil imports, the adverse effect on growth rates may be very severe. Moreover, any given reduction in overall growth rates falls more heavily on developing nations because in these countries higher population growth absorbs a larger fraction of the total Gross National Product than in industrial countries, leaving less for improvements in per capita incomes and living standards.

This is a long agenda of problems, and indeed, at the time of the first increase in oil prices, it was widely believed that the economic prospects of the oil-importing developing countries were seriously jeopardized. With the benefit of hindsight, however, it seems that for the oil-importing countries as a whole, the adjustment to the oil price increases was surprisingly manageable and certainly much easier than anticipated at the time. Current account deficits, due to larger oil bills but also to higher prices of other imports, rose by an unprecedented extent from over $11 billion in 1973 to over $50 billion in 1979 and to $70 billion in 1980 (see Table 22A-1 covering, unless otherwise specified, nonoil developing countries). But as this table also shows, these large current deficits were more than covered by borrowing from abroad. Indeed, foreign exchange reserves increased. On the other hand, outstanding debts rose and debt service ratios climbed steadily upwards.

In the developing countries, domestic economic growth rates, which had increased by an average of 6 percent up to 1972, fell by one percentage point after 1973, to 5 percent. Though this economic performance was better than that achieved by the industrial countries, even a modest reduction in growth rates is a serious matter for countries that must depend on high economic growth levels to reduce widespread poverty. A one percentage reduction, for example, could mean a one-third reduction in the annual improvement in living standards for the citizens of the developing countries.

But as we have seen, the group of countries contained in this category is both large and heterogeneous, and thus the performance of the group as a whole can mask important variations among individual countries or groups of countries. In particular, an important distinction must be made between the medium-income and the low-income countries. As mentioned above, the economies of the net oil importers as a group grew at an average of 6 percent per year before 1973. But this average covers widely different growth performances. The economies of the middle-income

Table 22A-1. **Oil-Importing Developing Countries: Balance of Payments and Economic Growth[a,b]**

	1973	1974	1975	1976	1977	1978	1979
Current account deficits of nonoil developing countries[c]	11.5	36.9	45.9	32.9	28.6	35.8	52.9
Net external borrowing by nonoil developing countries	11.0	24.9	32.2	33.5	26.1	37.8	44.5
Change in reserve assets of nonoil developing countries	9.3	1.2	-2.0	12.7	11.9	18.2	11.0
Changes in real GDP in nonoil developing countries (% change from previous year)	6.7	5.5	4.4	5.5	5.1	5.0	4.7
Of which net oil importers (% change from previous year)	6.5	5.3	4.3	5.6	5.3	4.9	4.2

Source: International Monetary Fund, *Annual Report 1980* (Washington, DC: IMF, 1980).

[a]Following IMF definitions, nonoil developing countries include net oil-importing development countries and those net oil exporters (Bahrain, Bolivia, Congo, Ecuador, Egypt, Gabon, Malaysia, Mexico, Peru, the Syrian Arab Republic, Trinidad and Tobago, and Tunisia) whose oil exports do not account for at least two-thirds of total exports nor amount to 100 million barrels per year. Algeria, Indonesia, Iran, Iraq, Kuwait, Libyan Arab Jamahiriya, Oman, Qatar, Saudi Arabia, the United Arab Emirates, and Venezuela are not included.

[b]Unless specified as a percentage, all figures are in billions of US dollars.

[c]Estimated at $70 billion for 1980.

countries that were also major exporters of manufactures grew by almost 8 percent per year, while the low-income countries' economies grew by only 4 percent, barely achieving an increase in living standards. After 1974, the economic growth rates of the former fell to 5.6 percent—quite a sharp fall, but still sufficient to continue increases in per capita incomes. During the same period, the economic growth rates of the low-income countries declined very little, going from 4.0 to 3.6 percent, but this decrease meant a halt to even the modest improvements in living standards which had taken place before.[12] Many African countries—Ethiopia, Zaire, Uganda, Ghana, Mozambique, Angola, and Zambia to name a few—actually experienced a decline in living standards, often by a substantial amount. Projections of economic activity for this group of countries indicate that the most that can be hoped for is to arrest this decline so that the citizens of these countries, if not becoming better off every year, at least do not become even poorer.[13]

The same distinction between middle- and low-income countries holds in international trade and payments. In the low-income countries, exports have remained chronically depressed, which has meant, as these countries in general do not have access to private long-term capital, that it has not been possible to increase imports. Nonetheless, their debt services and ratios of debt to exports of goods and services are very much higher than the average for all oil-importing developing countries.

In contrast, the stronger economic position of the middle-income countries, manifested in their ability to increase exports and attract long-term private capital, has permitted them to finance higher imports. But this has been achieved at a cost, and the cost is sharply increasing debt service ratios. While the average level of debt service ratio (guaranteed debt only) for the middle-income countries is still within historically manageable limits (to 1979, the last year for available data), the trend is strongly upwards. In addition, there are several countries within this group whose large borrowings seriously compromise economic management. It is estimated, for example, that Brazil's total export earnings are taken up by imports of oil and debt servicing, which includes in this case a substantial amount of nonguaranteed debt contracted by the private sector.

This question of debt—how to get it and how to service it—is one of the major questions facing the oil-importing developing countries. Shortening of maturities and higher interest charges will continue to push debt service ratios upwards.

We have concentrated here on foreign exchange costs, as these were the most immediate and dramatic manifestations of the increases in oil prices. In addition, major efforts will be needed to mobilize domestic resources to meet the investment needs of expanding domestic energy production and promoting conservation. Investment in the energy sector will expand sharply. The World Bank projects energy investment by the oil-importing developing countries at $180 billion (in constant 1980 $) over the five-year period 1981–85, and $270 billion during the period 1985–90, up fourfold from the pre-1973 period.[14] It is within this context that the World Bank announced plans for a new energy affiliate, though the prospects for the affiliate in its original form appear unpromising at the moment.

In brief, the oil-importing developing countries face major problems in their efforts to secure adequate supplies of energy, both traditional and commercial. These problems are superimposed on already existing difficulties which are especially acute in the case of the African low-income oil-importing developing countries. We are told, however, that the Chinese ideogram for crisis is composed of two subsidiary characters, one signifying difficulty and the other opportunity. We turn now to an investigation of these opportunities.

EVALUATION OF OPTIONS: TECHNICAL AND ECONOMIC CONSIDERATIONS

Much of the discussion of ways to improve energy availabilities centers on the energy sector—in particular, on ways to increase domestic production and to increase the efficiency with which energy is used. Though this is an important topic which will occupy most of the rest of the paper, it is not the whole story. It must not be forgotten that energy is valued not for itself but for the contribution it makes to economic growth. The goal of all of these countries is first and foremost to promote economic growth, not to minimize energy imports *at all costs*. When resources of several kinds, not just energy, are in short supply, it is all the more important to allocate these resources as efficiently as possible.

Economic Policy

The need to allocate resources efficiently has implications for both general economic policy and development strategy. For some countries, especially those with major manufacturing export sectors, the most efficient policy may be reliance on export promotion to earn the foreign exchange for higher priced oil. This certainly seems for many countries to have been the most successful approach in the years after 1973. A recent series of articles traces the effects on the balance of payments and on the economic growth rates in twenty-four countries, including twelve newly industrializing countries (NICs) and twelve other less developed countries, of higher oil prices and other external shocks—effects such as a slowdown in world demand for these countries' exports and subsequent economic policies. The poorest countries of Africa and Asia are not included in the group of countries studied.[15]

Among these twenty-four countries, an important distinction is drawn between outward-looking economies (such as Singapore, Uruguay, and Korea), whose development strategies are geared to export-oriented industrialization, and inward-looking economies (such as Brazil and India), whose strategy promotes import-substituting industrialization. The results of this study indicate that within both the NICs and the other twelve less developed countries, outward-oriented economies, although suffering greater external shocks in 1973–74, made more successful domestic adjustments. Among the NICs, for example, economic growth rates in the outward-oriented countries decreased from 7.4 percent pre-1973 to 5.9 percent in 1973–76 but rose again to 9.7 percent in 1976–79. In NICs following inward-oriented policies, on the other hand, economic growth rates fell from 6.9 percent pre-1973 to 5.0 percent in 1973–76 and did not recover in subsequent years. Debt service ratios are also much lower in the outward-oriented economies than in the others.

There are several lessons to be drawn here. The first is the importance for the oil-importing countries of establishing correct policies, such as setting realistic exchange rates and providing adequate incentives for savings and investments. Among the countries examined here, the outward-looking economies were quicker to adopt these policies in response to external shocks such as increases in oil prices. The second is that the outward-looking strategy is a feasible option for a good many countries. It is not restricted to the NICs but has wider application to several of the less developed countries. It does not, however, seem to be a realistic option for the poorer countries of Africa and Asia, where overall development problems predominate and lack of resilience or economic infrastructure hampers adjustment to shocks such as higher oil prices. For these countries, other solutions must be found. Finally, the efficacy of the outward-oriented strategy will depend on the industrial countries' keeping their domestic markets open to exports from the developing countries. This may be the single most effective way of helping oil-importing developing countries cope with higher import bills.

Energy and Development

Concern for the correct allocation of resources following higher oil prices also has implications for the relationship between energy and development. Energy is an increasingly important part of development strategy. While it is important to avoid an excessive concentration on the energy sector to the detriment of other aspects of economic activity, the energy implications of development strategies must be taken into account to reflect higher energy costs and possibilities of uncertain supplies. This is often more complicated than it sounds, since the necessary processes of consultation cut across traditional ways of conducting business in most governments. Ministries or departments of energy, if they exist, are usually charged with planning and overseeing the energy sector. But equally important decisions affecting energy use and availabilities are frequently made outside energy departments, in, for example, departments of transportation, agriculture, or industry. It therefore is important to insure that the energy implications of proposed development strategies are carefully thought out and that the effects on energy of decisions made in other sectors are integrated into the planning process.

Similarly, development plans need reorientation to take account of higher energy prices. Energy considerations are of particular relevance in making certain types of choices among development tactics. One example of such a choice would be the type of industrial development to encourage. Some industries (iron and steel, cement, chemicals, plastics) are highly energy intensive. When oil supplies were cheap, it was possible for countries without major domestic energy reserves to develop such industries. Now this option is less attractive. Planning for industrial development must of course consider all factors of production, but there are a number of important industries in which high-cost energy is a major constraint.

Another example could be regional development policies. In many countries regional disparities in incomes, employment, and access to opportunities are very large, leading to plans for regional development to narrow these differences and decentralize production facilities. A further advantage of this policy could be a less rapid rise in demand on the transport sector—a critical sector for developing countries since it typically accounts for 40 percent or more of total oil use.

Increasing Domestic Energy Production

Even recognizing the necessarily changed role of energy in development, more energy will still be needed. The rise in oil prices and the possible insecurity of imported suppliers has turned attention to possibilities of increasing energy production at home. What is the energy resource endowment of the developing countries? Since fossil fuels and conventional hydroelectricity will continue to be the most important sources of energy for developing countries for at least the next two decades, we start with them. Because of the availability of low-priced Middle East oil, there was previously little incentive to develop domestic resources of conventional fuels in these countries. Now, with high oil prices and the expectation of more increases to come, these incentives are very great.

There are grounds for both pessimism and optimism in finding new *oil and gas reserves* in developing countries. On the optimistic side, it is evident that many countries have not been adequately investigated for oil and gas. On the negative side, many experts believe that only relatively small amounts of oil and gas remain to be discovered from entirely new areas, such as the nonoil developing countries. This judgment, however, applies particularly to large reservoirs. Thus, substantial reserves may still exist in small- and medium-sized fields, adequate to provide part if not all of the commercial energy consumption of many of these countries. Among the oil-importing developing countries, oil has been discovered in five countries (Benin, Chad, Niger, Ivory Coast, and Sudan) that are not producers at present, and others (e.g., Brazil, Chile, and Pakistan) are actively trying to increase exploration efforts.

The recent rises in oil prices have also renewed interest in *coal*—at one time the mainstay of the world's energy economy. Developing countries are estimated to have about 10 percent of the world's economically recoverable coal reserves, though only 2 percent of its geological resources.[16] This discrepancy may be due to a lack of recent exploration for coal. Significant deposits may exist in many African countries (Nigeria, Zimbabwe, Swaziland, and Botswana). Resources may be much higher than current estimates in Colombia and Brazil, and new sources of coal in Asia may include Bangladesh, Taiwan, Korea, the Philippines, and Indonesia. Even if countries do not have domestic coal, imported coal could be an interesting alternative to imported oil.

Renewable energy resources are thought to be a particularly attractive option for the oil-importing developing countries, since they are assumed to be more amply endowed with renewables than with conventional resources. In general terms this is true. The amount of *hydroelectricity* available in some developing countries is very large compared with foreseeable future needs. Thus, in Africa only 1.5 percent of the hydro potential is now developed, and for Asia the total is 8 percent and for Latin America, 6 percent.[17] And these may be underestimates, not fully taking into account the recent higher costs of fossil fueled thermal power generation nor the potential for small-scale hydro projects. Both *solar and geothermal resources* tend to be relatively high in quality in much of the developing world.

Biomass resources include forests and crop and animal residues, and more recently the production of special field crops for energy purposes. In general, in most nonarid countries, biomass is a potentially very abundant resource—for many countries, perhaps the only significant energy resource for practical near-term use. For all three developing continents, potential annual supplies of energy from forest growth and manure and crop residues exceed current levels of commercial energy consumption by a considerable margin. Even at the lower limits, potential supplies of biomass in Latin America and Asia are about double current levels of commercial fuel consumption. In Africa the margin is much greater.[18] The emphasis here must be on "potential." As we have seen, tropical forest lands are already under pressure. To turn this situation around and fulfill their potential, the forests of the developing world must be more efficiently managed to produce a sustained yield of wood for energy and other uses. This means that the strongest effort must be made to produce forest products in both plantations and village woodlots on a sustained-yield basis, and to improve yields through better harvesting practices, fertilization, and genetic research.

Crop and animal residues are also potentially available for energy use but in a rather different way than forest products, since these residues are often already fully utilized for fertilizer, animal feed, or, in many countries, as a replacement for firewood. In either case, it is likely that potential wastes are already being put to some economic use, and their diversion to energy uses may create additional problems.

This summary of energy resources in the oil-importing developing countries as a group gives grounds for modest optimism. But these resources are not equally distributed among this large group of countries. Some are more richly endowed than others. As Table 22A-2 shows, few countries have major fossil fuel reserves, al-

Table 22A-2. **Energy Resources of Developing Countries[a]**

		Low	Medium	High
			Hypothetical potential[b]	
		Low	*Medium*	*High*
Fossil fuel reserves[c]	*Low*	Burundi (3) Ghana (3) Malawi (3) Rwanda (3) Sri Lanka (4) El Salvador (4) Ethiopia (4) Haiti (4) Laos (4) Lesotho (4) Morocco (4) Philippines (4) Somalia (4) Togo (4) Yemen Arab Republic (4) Lebanon (3) Dominican Republic (4) Guatemala (4) Ivory Coast (4) Jamaica (4) Jordan (4)	Zambia (2) Afghanistan (3) Chad (4) Benin (4) Guinea (4) Honduras (4) Kenya (4) Mali (4) Mauritania (4) Senegal (4) · Sierra Leone (4) Sudan (4) Tanzania (4) Uganda (4) Nicaragua (4) Panama (4) Portugal (4) Uruguay (4)	Central African Republic (4) Liberia (4) Madagascar (4) Nepal (4) Niger (4) Upper Volta (4) Costa Rica (4) Papua New Guinea (4) Paraguay (4)
	Medium	Pakistan (2) Thailand (4)		Cameroon (4)
	High	Korea (1) · Turkey (3)	Argentina (1) Chile (2) Brazil (3)	Colombia (1)

Source: J. Dunkerley and G. Knapp with S. Clapp, "Factors Affecting the Composition of Energy Supplies in Developing Countries" (Washington, DC: Resources for the Future, 1980).

[a]Does not include centrally planned economies, capital-surplus oil exporters, or countries with populations of less than 1 million. Within each section, countries in the left-hand column are those considered by the World Bank to have actual or potential fuelwood problems. The figures in parentheses indicate net oil imports as a percentage of commercial fuel demand, as follows: 1 = 0-25%; 2 = 25-50%; 3 = 51-75%; 4 = 76-100%.

[b]Categories are defined as follows: Low—annual energy potential per capita under conditions of average flow from installed and installable capacity utilized twelve hours per day, less than 0.2 tons of coal equivalent (tce); Medium—annual energy potential per capita, between 0.2 and 1 tce; High—annual energy potential per capita, more than 1 tce.

[c]Categories are defined as follows: Low—less than 1 tce per capita; Medium—between 1 and 10 tce per capita; High—more than 10 tce per capita.

though several countries low in fossil fuels do have significant hydroelectric potential and an adequate fuelwood supply. However, there are a disturbing number of countries, largely poor countries in Asia and Africa, which are short of all resources and which also have currently a high rate of dependence on energy imports. This characteristic again sets these countries aside from the other group of nonoil developing countries.

The timely development of energy resources depends on the possibility of converting them into a form suitable for use by final consumers, and this in turn depends on costs and the availability of technology. The extraction costs of these different resources are difficult to assess. Very few conventional energy resources have been resurveyed in the light of current prices, so we have little firm data to go on. Many of the renewable forms of energy have been produced and consumed outside of commercial markets, so here again it is difficult to assign costs to such fuels in the context of extensive commercial development. While cost data on conversion of the conventional forms of energy are well established, some of the new conversion technologies such as methanol from wood have not been sufficiently tested on a commercial basis to yield adequate cost data. And of course, costs of many of these fuels will vary widely between locations. Despite these caveats, two conclusions can be drawn. First, the fourfold rise in real petroleum prices from 1972 to 1980 has generally improved the competitive position of other forms of energy. And second, on the basis of rather patchy data, the conventional fuels—domestically produced oil, coal, and gas—still appear to be cheaper than the others, particularly the more exotic forms.[19]

This does not mean that the oil-importing developing countries should abandon interest in the newer forms of energy. After all, they will be needed when finite supplies of fossil fuels are exhausted or unmanageably expensive. But it does mean that these newer forms of energy face quite severe developmental problems —they are more expensive for the moment, and they face greater technical problems than the conventional fuels. Thus, there are still unresolved problems in operating automobiles on the various combinations of liquid fuels, storage for solar and wind energy, biomass digestion, and more efficient charcoal kilns.

Conservation and Fuel Switching

In addition to increasing supplies of domestically produced

energy, the oil-importing developing countries can also ease their balance of payments problems by using energy more efficiently —that is, by using less energy to perform a given service. It is often thought that there is little room for conservation in poor countries where energy consumption is already low. But there is developing evidence in all sectors that important opportunities exist to save energy even at low consumption levels.[20] In the transport sector —which typically uses 40 percent of the total oil supplies in developing countries—major variations in passenger car ownership suggest that savings in gasoline could be achieved without adverse effects on economic growth or welfare. In Brazil, for example, there are forty passenger cars per million 1975 dollars of GDP, compared with only four per million in Korea. Improvements are also possible in the energy efficiency of both trucks and passenger cars, and greater attention could be paid to the many forms of public and private transport other than the private automobile —jitneys, taxis, motorized bicycles, scooters. There are limited possibilities for fuel switching in this sector, although several countries, notably Brazil, have major programs for the production of liquid fuels from sugar and other crops.

In industry, wide variations in the amounts of energy used in similar industrial processes suggest that here, too, energy could be saved—either through "housekeeping" changes or through some more substantial changes in procedures. The industrial (and thermal electric) sector also offers more opportunity for substituting other fuels for oil—gas, if possible, but also coal.

In developing countries there are two broad types of households: urban households mainly using kerosene, gas, or electricity supplemented by traditional but commercialized wood and charcoal; and rural households largely dependent on traditional fuels produced and consumed outside the cash economy. These traditional fuels, mainly used in cooking, are used very inefficiently. Technically simple improvements in cooking stoves could double efficiencies and save an estimated 100 MTOE or more of biomass annually as well as reduce the environmental costs of fuel use.[21]

As in the case of supply development, the attainment of conservation potential depends on costs, and here again it is possible only to generalize. Higher oil prices clearly have increased profitable possibilities for conservation. Indeed, at current energy prices it may be cheaper (and quicker) to save an additional unit of energy than to provide one.[22] And in conservation, a good deal of the technology—energy-efficient cars and industrial processes; even improved stoves—already exists.

INSTITUTIONAL CONSTRAINTS

Even when profitable and technically feasible possibilities of increasing energy production and promoting conservation exist, institutional constraints can still impede their fulfillment.

A major obstacle to increasing domestic energy supply and improving the efficiency with which energy is used often lies in pricing policies. In principle—for economists at least—a correct pricing policy should reflect the long-run costs of acquiring energy by incorporating a scarcity value if the energy is depletable and a security premium if the energy source is imported. This applies to the pricing of individual fuels as well as to energy as a whole.

Pricing policies in oil-importing countries—both rich and poor—can diverge widely from such goals. In many countries, subsidies have been introduced to promote selected activities such as agricultural development and forestry protection and to insure that the poor have access to energy. These aims are unexceptionable, but the costs of achieving them have risen sharply in recent years, and it may be necessary to examine whether they can be achieved by other means. Serious equity problems are raised by the prospect of higher energy prices, especially for the household and transport needs of the poor. It is highly desirable, however, for budgetary as well as efficiency reasons, that increased emphasis be given to energy pricing that reflects the replacement costs of energy used.

When discussing a politically sensitive issue such as subsidies for kerosene, it is important to realize that these subsidies benefit all users, not just the poor. For example, subsidies for kerosene combined with high taxes on other petroleum products such as gasoline can lead to major use distortions, with kerosene being used as an automotive fuel and even as boiler fuel.

While correct pricing policy is a necessary condition for increasing supplies and promoting conservation, it may not always be in itself sufficient and fully achieve these aims. A variety of other factors could hold back development. On the supply side, potential producers might hesitate to undertake energy projects despite high prices, due to the producers' uncertainty about their future freedom of action and to possibilities of expropriation. Host country agreements with foreign companies impinge on delicate issues such as depletion rates and on wider problems of national sovereignty. Another problem in petroleum development is how to attract foreign oil companies to exploit small- or medium-size wells which may produce enough oil only for the host country. These types of problems may help to explain the rather slow exploration activity in the oil-importing developing countries over the past ten years. An interesting recent development has been the willingness of the World Bank and other regional banks to act as intermediaries between producing company and host country. They can also assist in petroleum survey and exploratory drilling, which cuts down the area of uncertainty facing both companies and countries in concluding petroleum development agreements.

On the side of energy consumption, the high initial cost of some energy-efficient devices may have discouraged their introduction despite highly favorable life-cycle costs. This may account in part for reluctance to adopt new wood-burning stoves although, as we shall see below, social acceptability also seems to have been important. In commercial fuels, lack of knowledge or managerial expertise may inhibit measures to improve energy use efficiency.

SOCIETAL CONSTRAINTS

Energy policy and development does not and should not take place in a vacuum. Other societal values relating to the environment, to the distribution of income and wealth, and to full employment will compete with the claims energy makes on the limited supplies of resources, and these other societal values can divert activity from what would otherwise be a rational energy development. It is important to realize, however, that societal "constraints" can also operate in an advantageous way. They can, for example, deter excessive efforts at energy independence that might prove costly in terms of economic development.

The environmental impacts of commercial energy development in the oil-importing developing countries are similar to those in industrial countries—air pollution, oil spills, loss of life and limb in coal mining, nuclear plants safety issues, and so forth. Major hydroelectric systems can affect siltation patterns and downstream agriculture, flood large areas of productive land, and lead to an increase in parasitic diseases. The environmental impact of biomass development—when it involves widespread excessive cutting of trees—can be equally serious, endangering species and changing watershed and even local and regional climates.

It is important to realize, however, that well-managed forestry development based on improved forest management and higher yields could reverse some of these unfavorable impacts. The same is true for biogas digesters—their gas is cleaner than traditional fuels; they serve a sanitation function; and they relieve pressures on the forests.

In countries where a large part of the population lives in poverty, questions of income distribution and equity are of major concern. In rich countries it can be argued that equity considerations should be addressed by methods of supplementing income outside the energy sector, but in poorer countries practical and effective methods may not be available. Some energy technologies may be too expensive for the poor people in these countries, and other technologies, while economical over the long run, may require high initial costs. Improved stoves may fall into this latter category. A second problem is that the poor may not benefit from new energy technologies, because methods of dissemination, patterns of ownership, and local power structure favor the relatively wealthy. For example, individual biogas plants, besides having a high capital cost, need from three to five cattle to produce the dung needed for their operation, effectively restricting this technology to the well-to-do in rural areas.

Another facet of income distribution and equity is the aspect of the resource conflict that is known as "food versus fuels." This conflict could occur in a major alcohol fuels program if high-value energy crops displaced food production and thereby drove food prices up with serious consequences for the poor. Such resource conflicts could also occur with other factor inputs, but if labor was the input at issue, income distribution conceivably could be improved.

EDUCATION AND INFORMATION

Education and information covers both the transfer of technology from industrial to developing countries and the dissemination of new technologies within developing countries.

With regard to technology transfer, the basic question is to what extent the technology developed in industrial countries is suitable for transfer to and adoption in the developing nations. Much of the industrial countries' commercial energy technology has been based on saving labor, but in recent years there has been a move toward energy-efficient technologies, particularly in those industries where energy represents a substantial part of total costs. As both oil-importing industrial countries and oil-importing developing countries face higher energy costs, the development of these energy-efficient technologies is appropriate for both, although there remain other problems of saving rather than fully utilizing labor.

On the supply side, too, energy technologies developed in the industrial countries can be more or less appropriate to conditions in developing countries. Current nuclear and coal mining technology may not fit all developing countries, especially those whose electricity grids are too small to accommodate the most efficient nuclear power plants or those where abundant labor can be used in less capital-intensive mining operations.

Much biomass technology has been evolved in the developing countries, or at least within the developing country context. In the developing areas, there is already a considerable body of experience with traditional energy, including biogas, rural development with important energy components, solar cookers, "improved" wood burning stoves, social forestry projects, alcohol fuels, and so forth. The existence of such experience offers a good opportunity for technology transfer from one developing country to another.

Thus far, the industrial countries have done little in a systematic way towards developing new technologies for traditional fuels. There could be a greater role in the future, as the industrial countries also take more interest in this energy. Examples include help with sustained-yield forestry management and with alcohol fuel and synfuels research as well as with the more exotic renewables.

Acceptance of new devices and dissemination of new techniques can also raise problems. Improved stoves, which seemingly offer users relief from the burden of firewood gathering, have been notoriously slow to be adopted. This may be partly due to user unfamiliarity with the new stoves—a problem which could be overcome by careful education in the stoves' use. But reluctance to adopt the new stoves is also related to costs and can be caused by the new device's inappropriateness to serve traditional functions. For example, an improved stove can cook as well as a campfire and use less wood, but it does not keep animals away or provide a focus for family and community life the way a campfire does. Solar cookers mean both cooking outdoors, which is unusual or inconvenient in some situations, and cooking when the sun is out, which again goes against some ingrained customs. In cases such as these, a more careful attention to design appropriateness is indicated.

FINAL EVALUATION OF OPTIONS

The major problem facing the oil-importing developing countries is how to increase energy supplies of both commercial and traditional fuels without compromising economic growth, imposing excessive damage on the environment, or adding further to the burdens of the poor. We have discussed here a series of possibilities for providing this energy together with the constraints or difficulties faced in carrying them out.

This discussion of problems and options has contained a good deal of pessimism. This pessimism, typical of the outsider observing the developing countries, almost certainly underestimates the vitality and resilience of the developing world. But this pessimism, although perhaps not totally warranted, is a good fault. The alternative, a belittling of the problems being faced by oil-importing developing countries, would not be fitting in the light of their widespread poverty. Undue pessimism can, however, lead to a misdiagnosis of the problems and thus to the espousal of inappropriate forms of assistance.

Several options for overcoming energy problems in the oil-importing developing countries have been presented here. These options have been presented and discussed separately, but in practice each country would use a mix appropriate to its own resource base and level of development. The possibility of increasing export earnings may be an option more limited in application than the others, but it still could be an important policy for several countries, including particularly those who already have a significant export trade in manufactures. The other three options apply more generally to all countries, although their relative emphasis and success will depend on the resources of the country concerned. Those countries with fossil fuel resources should press ahead with their development. Similarly, many countries have important hydroelectric resources to develop. Caution must be exercised, however, to make sure that development does not outstrip demand. Almost all countries should pay more attention to biomass, not only to correct existing difficulties but also to insure that biomass—for most countries their most plentiful resource—can play an increasing and more efficient role in their energy economies.

Attention should also be paid to conservation, including both the efficiency with which energy is used and the reorientation of development strategy in the light of higher energy prices. Again, the relevance of conservation will vary from country to country. As the following case study of Burundi shows, conservation will not be a feasible option at very low levels of commercial energy consumption.

The balance between increased production and conservation would have to be struck on a case by case basis. As a general order of magnitude, the World Bank estimated that oil import savings by 1990 from an aggressive conservation and fuel-switching policy could be as great (about 1.2 mbd) as a maximum effort to raise oil production. Together, these actions could reduce oil bills by as much as $30 billion in 1990.

The main burden of resource mobilization for implementing energy strategies will fall on the public authorities and the private sectors of the developing countries. However, their actions can be supported by international financial and technical assistance motivated not only by humanitarian concerns but also by the realization that all countries stand to benefit from a successful resolution of energy problems, no matter where in the world they occur.

This assistance could take the form of:

1. Facilitating the efficient recycling of surplus OPEC funds and helping to solve the problem of developing country debt.

2. Keeping markets open to permit the developing countries to pay higher oil bills through increased exports.

3. Helping to finance domestic energy program costs. The costs involved are very large, estimated by the World Bank to be about $50 billion (1980 $) annually—about double current levels. The burden of meeting these costs could be eased by the expansion of multilateral and bilateral aid and by the encouragement of private investment. Given high petroleum prices, such investments could bring high rates of return to domestic and foreign investors alike.

4. Providing mechanisms for the transfer of energy technology (in forestry practices, alcohol fuels, and solar and conservation technology).

5. Providing technical assistance (in energy planning and energy resource exploration and appraisal).

Such a combination of domestic and foreign initiatives could be highly effective for the middle-income countries among the oil-importing developing countries. But as we have seen, the problems of the poorer countries in this group may prove to be more intractable. These countries, mainly the poorer countries of Asia and Africa, are unfortunately placed. Living standards are abysmally low and in many cases declining. Their debt burden is already high. They have few energy resources, or indeed any other resources. These countries have been the most heavily hit by higher oil prices. Having little flexibility in their economies, they are not able to adapt to the new conditions in the way the other countries can. These countries will require different forms of assistance relying much more heavily on concessional aid. Although these countries are numerous—as seen in Table 22A-2, about thirty countries would be included—their economies are small in terms of world oil imports. Highly concessional aid in the form of either financial assistance or cheap oil could be easily managed. This was a suggestion first mooted by Professor Sadli at Symposium I last year. I hope here to have expanded his theme a little in order to pave the way for the policies which are to be the theme of Symposium III.

NOTES

1. *Energy Policies and Programmes of IEA Countries, 1977 Review* (Paris: OECD, 1978).
2. J. Dunkerley, W. Ramsay, L. Gordon, and E. Cecelski, *Energy Strategies for Developing Nations,* published for Resources for the Future by the Johns Hopkins University Press (Baltimore: 1981), p. 5. There is a wide variation among countries. Traditional fuels account for under 10 percent of total energy consumption in Korea but for over 90 percent in many of the African countries.
3. J. Dunkerley and S. Matsuba, "Energy Consumption in the Developing Countries," in *US Energy Outlook: A Demand Perspective for the Eighties,* a report prepared by the Congressional Research Service for the use of the Committee on Energy and Commerce of the US House of Representatives, July 1981, Committee Print 97, 97th Congress, 1st Session (Washington, DC: US Government Printing Office, 1981), p. 421.
4. World Bank, *Energy in the Developing Countries* (Washington, DC: World Bank, 1980), p. 50.
5. *Energy Policies and Programmes of IEA Countries, 1980 Review* (Paris: OECD, 1981), p. 18.
6. K. Oenshaw, "Energy Requirements for Household Cooking Stoves" in *Proceedings of Bio-Energy 1980,* Bio-Energy Council, Washington, DC, 1980, p. 256. Quoted in Dunkerley, Ramsay, et al., op. cit.
7. W. Knowland and C. Ulinski, "Traditional Fuels: Present Data, Past Experience and Possible Strategies" prepared for the US Agency for International Development (Washington, DC: USAID, September 1979), p. 15. Quoted in Dunkerley, Ramsay, et al., op. cit.

8. J. S. Spears, "Wood as an Energy Source: The Situation in the Developing World," paper presented at the 103rd Annual Meeting of the American Forestry Association (Washington, DC: World Bank, 1978), p. 11. Quoted in Dunkerley, Ramsay, et al., op. cit.

9. J. S. Spears, op. cit., p. 5. Quoted in Dunkerley, Ramsay, et al., op. cit.

10. By an average of 7.2 percent per year.

11. Dunkerley and Matsuba, op. cit., p. 420.

12. International Monetary Fund, *Annual Report 1980* (Washington, DC: IMF, 1980), p. 12.

13. World Bank, *World Development Report 1980* (Washington, DC: World Bank, 1980), p. 6.

14. World Bank, *Energy in the Developing Countries*, pp. 6–7.

15. B. Balassa, all of the following: "The Newly Industrialized Developing Countries after the Oil Crisis," in *Weltwirtschaftliches Archiv*, Journal of the Kiel Institute of World Economics, Band 117, Heft 1, 1981, Tübingen: J. C. B. Mohr (Paul Siebeck); "The Policy Experience of Twelve Less Developed Countries, 1973–78" (Washington, DC: World Bank, 1981); and "Adjustment to External Shocks in Developing Economies," mimeo (Washington, DC: World Bank, June 1981). The newly industrialized countries referred to are Argentina, Brazil, Colombia, Mexico, Chile, Uruguay, India, Korea, Singapore, Taiwan, Israel, and Yugoslavia. The twelve less developed economies are Egypt, Tunisia, Morocco, Kenya, Philippines, Thailand, Jamaica, Peru, Tanzania, Indonesia, Ivory Coast, and Nigeria.

16. Dunkerley, Ramsay, et al., op. cit., p. 142.

17. Ibid., p. 157.

18. Statement of J. Dunkerley and W. Ramsay before the Subcommittee on International Economic Policy of the US Senate Foreign Relations Committee (The Global Economy, February 25, 1981) on the role of biomass in developing countries (Washington, DC: Resources for the Future), p. 6.

19. World Bank, *Energy in the Developing Countries*, p. 90.

20. This is illustrated in two current Resources for the Future studies on energy conservation potential in Kenya and in Ecuador's manufacturing sector.

21. Dunkerley, Ramsay, et al., op. cit., p. 124.

22. See, for example, Shell Briefing Service, *Improved Energy Efficiency*, June 1970.

PART B: CASE STUDIES OF BURUNDI
AND THE CAMEROON

As has been pointed out, it is difficult to insure comprehensive treatment of the energy-related problems in the nonoil developing countries since they are a very diverse group of countries with different consumption levels, patterns, and requirements. Income levels differ; industrial development differs; social systems differ. The rising cost of energy over the last few years has spurred interest in the development of indigenous energy resources in many of these countries. Levels of resource endowments vary, and, in the past, the development of such resources was considered uneconomic for many of these countries.

The case studies considered here concern two African countries that are both affected by the energy problem, but each in different ways. They were selected as case studies since they clearly illustrate differences even within the same category of countries, thus showing the need for caution when discussing this problem. The resource endowments of both countries differ, and so do their levels of economic development. Cameroon, unlike Burundi, has recently become an oil producer, but definitely not at OPEC levels although the potential may be there in the long term. Cameroon is also endowed with several other major energy resources, such as hydro and biomass. Burundi, on the other hand, is a country without significant exploitable indigenous resources, with the exception of peat and hydro. These case studies briefly review the energy and related sectors, isolate the kinds of energy problems, identify avenues for possible solutions based on potential indigenous resources and other planning efforts, and conclude with an assessment of obstacles to reaching those possible solutions.

BURUNDI

Burundi (see Table 22B–1) is a small country of 25,000 square kilometers located in Central Africa. It is a landlocked country and is bounded to the north by Rwanda, to the east and south by Tanzania, and to the west by Zaire. With a population of about 4.5 million, it is, after Rwanda, the most densely populated country in Africa. Burundi is classified by the United Nations as one of the least developed countries in the world, with a 1979 per capita GNP of US $180.

Table 22B-1. **Reference Figures: Burundi**

Population	4.5 million (1980)
GNP	US $180/capita (1979)
Value of petroleum product imports	US $12 million (1980)
Consumption of petroleum products	243,768 barrels/year (1980)
Production of crude oil	0

Burundi, like most other oil-importing developing countries, has been severely affected by the spiraling cost of petroleum in recent years, a factor which has had far-reaching adverse effects on the country. In addition to the rising cost of oil, the energy problem in Burundi can be attributed to several other factors, some of them beyond the country's control. These other factors include:

- severely depleted forest resources as a result of population pressure and belated efforts at reforestation;
- the geography of the country—as a landlocked country, oil imports have to be brought over very long distances and are subject to other hazards such as poor transportation facilities, the politics of neighboring countries, and so forth;
- inadequate development of other indigenous resources such as peat and hydro, owing to a lack of capital; and
- the frequent subjection of imported power from neighboring countries to lengthy and unexplained interruptions.

Currently Burundi imports all of its petroleum products.[1] Chances that petroleum will be produced in the near future are remote since there have been no significant exploration activities carried out. Some experts believe that the geological formation in certain parts of the country might reveal oil if drilling took place. However, some experts feel that only small oil deposits might be found, and therefore large oil companies might be reluctant to carry out exploration. In any event, the fact remains that all of the petroleum products are imported and for the foreseeable future will remain so. Thus, Burundi will be subject to the shocks in the world oil market. The value of petroleum imports rose fivefold between 1975 and 1980. The price of gasoline, for example, rose from US $0.75 to about $3.00 per gallon during the same period.[2] Obviously, this does not take into account the shortages and other hardships resulting from both the cost and the scarcity of fossil fuels.

As stated earlier, petroleum imports are also affected by transport problems since Burundi is a landlocked country. Imports of petroleum, like other goods, come through the ports of either Mobasa, Kenya, or Dar-es-Salaam, Tanzania. Imports that are handled through Mobasa have to be transported by road through Kenya, Uganda, and Rwanda, a distance of over 2,000 kilometers. Several problems constrain efficient transport and give rise to additional costs. These problems include the volatile political situation in Uganda as well as poor security in that country; the poor condition of existing roads; cumbersome custom formalities and vehicle licensing required by the other countries; and weight restrictions on the use of the most direct routes. About $1 is added to the price of a gallon of diesel owing to these factors.[3]

Imports handled through Dar-es-Salaam do not fare any better. Transport has to be handled by rail from Dar-es-Salaam to the port of Kigoma on Lake Tanganyika, a distance of about 1,300 kilometers. From Kigoma, imports have to be transported by barge to Bujumbura. Several bottlenecks affect the efficient handling of these imports: the congested situation and limited capacity of the port of Dar-es-Salaam, the inadequate capacity of the rail link between Dar-es-Salaam and Kigoma, the lack of equipment at the lake port of Kigoma, and an obsolete barge fleet.

Burundi relies on coffee exports for more than 70 percent of its income. Although the country's petroleum requirements are not extremely large, it faces problems that could, as happened before, result in a cutoff of oil supplies. In this event, a cutoff at coffee harvest time could be disastrous.

A substantial proportion of electricity consumed in Burundi is imported from Zaire. At times there have been unexplained interruptions in supply lasting many days, and this has obviously had the effect of halting activities that rely on electricity for their energy requirements. Total electricity requirements in Burundi at present would not exceed 80 MW.[4] Nevertheless, the country possesses a hydroelectric potential far in excess of that figure. Realization of this potential is, however, hampered by a lack of capital, a situation which could have far-reaching effects on economic development, including exploitation of Burundi's large nickel deposits estimated at 3 percent of the world's total.

Forest resources in Burundi cover only about 3 percent of the country's total area and are being severely depleted due to a large population increase and hitherto inadequate reforestation programs. Nearly 85 percent of Burundi's population relies on wood and other wood-derived products to meet most of its energy needs. At the present rate of consumption and reforestation, the wood resources of the country could be entirely depleted in about four years. Most rural industries as well as some urban ones rely on fuelwood for their energy requirements. The severe depletion of forests has also resulted in many households, particularly those in the rural areas, spending at least two days per week gathering fuel. In the urban areas most of the fuel used by households is charcoal obtained through a commercial medium. The depletion of forests has resulted in charcoal being brought from longer distances, and as a result it has become even more expensive. Between 1973 and 1980, the price of charcoal more than quintupled, and it now costs close to $0.30 per kilogram. A household of six in Bujumbura consumes about 135 kilograms per month and thus has to spend about $40 per month on charcoal,[5] while the average income in Burundi is only about $90 per month per household. Reforestation programs have been inadequate, in part because of a weak forest service and a shortage of trained staff. The ecological consequences of the depleted forests could be severe. Burundi is a hilly country with most of its farms on hillsides; thus the lack of tree cover results in severe erosion. Furthermore, as wood becomes scarce, people turn to using agricultural and animal residues, thereby depriving the soil of valuable nutrients.

Burundi has large untapped resources of peat. Reserves are presently estimated at 500 million tons (dry weight).[6] This, in theory, could supply most of the country's energy needs for several decades. Peat is found in small deposits in valley bottoms in the mountains of the Zaire/Nile divide and in larger deposits in the Akagera River Valley. The government, with the cooperation of outside donors such as USAID, is seriously exploring the possibilities of large-scale development and use of peat resources. Production is currently in the range of 9,000 tons per year.[7] These efforts are, however, being hampered by factors such as ecological and technological ones. Peat has a high water retention capacity. Extracting it in large quantities may result in serious flooding during the rainy season or excessive drying out of the soil during the dry season, making the land where peat has been extracted useless for agriculture. Land for raising crops and livestock is in short supply. Extraction of peat, which renders the land unsuitable for those purposes, thus may not be a very good idea.

Peat has a high ash and water content and consequently gives off a lot of smoke when burnt. However, if converted into coke, the quality of the peat can then be improved. This coking process may be expensive, although this expense must be weighed against the fact that the country possesses few other indigenous resources that could be inexpensively exploited. In spite of the potential contribution peat could make in meeting the energy requirements of Burundi, it perhaps should not be thought of as the long-term solution to the country's energy problem. It could and should, however, fill the energy gap in the interim, while other longer term alternatives are being explored.

The picture that emerges from all of the above is a rather complex one. We have a small developing country whose development efforts could be jeopardized by the magnitude of the energy problems it faces. There is also the question of resilience. Every sector of the society is affected by these energy problems— industry, government, commerce, household. Industries have in the past slowed down for a lack of fuel or electricity. Working hours have at one time or another had to be shifted for lack of fuel. Household outlays for the purchase of fuels continue to rise. Government expenditures on energy imports rise every year. Because of already low levels of consumption, the likelihood that conservation, particularly of commercial energy, can make a significant contribution is remote.

The prospects will not be bleak if potential solutions are embarked upon soon. The country is not devoid of resources. Exploitable hydro resources are available. Peat is available. There is evidence of geothermal resources. What is needed at present is the following:

- more devotion of efforts and capital to the development of hydro resources to replace imported electricity;
- increased reforestation and the establishment of training programs for foresters;
- improved technology for more efficient production and utilization of fuels such as wood and charcoal;
- stepped-up efforts for peat production and appropriate technology for its utilization by various sectors of the economy and population; and
- stepped-up efforts to assess, explore, and exploit other resources such as solar and geothermal.

While these potential solutions are embarked upon, equity should not be neglected. The obstacles to these potential solutions have been mentioned. These problems are, however, not insurmountable. The will and desire to tackle these problems are fortunately in abundance. Indigenous energy institutions in Burundi have undergone positive changes during the past two years and continue to be strengthened. An energy ministry under dynamic leadership was created nearly two years ago. The first year was spent bringing the hitherto widely dispersed responsibilities in the energy sector under the control of the energy ministry. This past year has been spent identifying options.

In order for these efforts to bear fruit, however, appropriate assistance is necessary. For example, while it may be difficult to get large oil companies interested in investing capital for oil exploration in Burundi, international assistance agencies could become involved. It is no secret that many oil companies have become interested in certain countries only after international agencies took the initiative to spur exploration. International institutions could also provide technical assistance in the areas of planning, training, and institution-building. While Burundi's energy ministry is under able and dynamic leadership, well-trained middle-level energy specialists are necessary to support the efforts at the top. It should be noted that one factor that has contributed to the difficulties in the energy sector in Burundi is the lack of such personnel to provide a basis for ongoing assessment of the energy sector and clearly defined and developed areas for assistance. In the forestry sector, for example, there are few if any Burundian nationals who are trained as foresters.

In summary, while the situation is a difficult one, it should prove to be an excellent challenge for the international community, both public and private. The will and desire on Burundi's part to tackle these problems are there. There is eagerness on the part of the government to work with external donors to handle the problem in the short run and develop long-term solutions. No doubt, this is a clear case where concentrated and appropriate international assistance could play a pivotal and important role. The conditions are excellent for such assistance.

CAMEROON

In comparison with Burundi, Cameroon (see Table 22B-2) is a large country with an area of about 183,000 square miles. In 1979 it had a population of 8.2 million and has an average annual population growth rate of 2.3 percent. The urban population currently makes up about 30 percent of the total population, and outside the urban areas, the population density is quite low. As in many developing countries, agriculture is the mainstay of its economy, employing about 70 percent of its work force and accounting for over one-third of its GDP. Agriculture also provides the underpinning of the country's industrial sector, for the country is one of Africa's net food exporters. The industrial sector, which consists mainly of small-scale manufacturing and a large aluminum manufacturing complex, accounts for about 24 percent of the country's GDP.

Table 22B-2. **Reference Figures: Cameroon**

Population	8.2 million (1979)
GNP	US $675/capita (1980)
Value of petroleum product imports	US $285 million (1979)
Consumption of petroleum products	7.3 million barrels/year (1981)
Production of crude oil	85,000 barrels/day (1981 est.)

Cameroon is endowed with fairly substantial forestry, hydro power, oil, and gas resources.[8] There is also evidence of lignite, radioactive, and geothermal resources. Forests cover nearly 20 million hectares of the nation's land area, and most of this forest area is as yet unexploited. Proven recoverable crude oil reserves are about 500 million barrels and discoveries continue to be made. The country's hydro power capability is equivalent to about 100,000 GWh per year, and known recoverable gas reserves are estimated to be about 6 trillion cubic feet, 95 percent of which is nonassociated. New finds continue to be made.

Although the country is potentially an energy-surplus nation, it was not until 1979 that petroleum production actually began. Until May of 1981, the 20,000 barrels of petroleum products consumed per day were all imported, in spite of the fact that the country had become a crude oil exporter in 1979. In May of 1981, a 40,000 barrels per day refinery was put into operation.

Until the production of oil started, Cameroon was severely affected by the rising costs of world oil—costs which resulted in steep price increases for gasoline, kerosene, jet fuel, diesel fuel, and so forth. For example, the costs of transportation rose steeply between 1974 and 1979 as a result of these price increases. This trend has since leveled off, with price increases now being attributable more to inflation. The price for regular gasoline is about $2.30 per gallon at present.[9]

Although potentially an energy-surplus nation, Cameroon could face problems in the energy sector. At the moment, research, planning, and decisionmaking in the energy sector are fragmented. Evolving and implementing an integrated energy strategy is necessary in order to properly develop and use the available energy resources, to avoid local shortages, and to insure maximum revenues to the government for developmental purposes. In this way, the growing demand for and increasing prices of oil and gas resources would be of considerable benefit to the country. It is evident that the optimum development and management of indigenous energy resources is the key problem facing Cameroon, given its relatively new entry in the area of energy production and its inadequate technical and organizational infrastructure.

Development of an integrated plan would no doubt require considerable training and reorganization. There is a need to train managerial and technical energy specialists. There is also a need at present to bring all energy-related activities under the coordination of a central agency such as the Ministry of Mines and Energy. This should be coupled with developing the ministry's capability to collect and analyze energy-related information on a regular and continuing basis, so that a national energy plan consistent with the economic development plan of the country could be effected.

Cameroon and Burundi are two developing countries that face energy problems, although their problems are of a different kind. On the one hand, Cameroon is endowed with abundant exploitable energy resources; on the other, Burundi faces severe problems in meeting its energy needs. For both countries, however, available resources should be directed so that they meet developmental and social goals. While Cameroon may possess the resources, the absence of integrated and set policies for the development and use of its resources could pose significant problems for it. In both countries, technical assistance for institutional development and training in the energy sector should be a major consideration.

NOTES

1. Ministry of Public Works, Mines, and Energy; Burundi; July 1981.
2. Ibid.
3. Ibid., December 1980.
4. National Water and Electricity Corporation, Burundi, December 1980.
5. World Bank, *Burundi Urban Energy Study*, January 1980.
6. National Peat Corporation, Burundi, July 1979, December 1979, and July 1981.
7. Ibid. and USAID (Burundi), July 1981.
8. Ministry of Mines and Energy, Cameroon, July 1981.
9. Personal observation, August 1981.

CHAPTER TWENTY-THREE

A Choice of Energy Policies for a Rapidly Industrializing Country: A Case Study of the Republic of Korea

Hiwhoa Moon
Coordinator of Economic Policies
Office of the Prime Minister
Republic of Korea

THE NATURE OF THE PROBLEM

The Republic of Korea [or South Korea; hereafter referred to as Korea in this chapter for brevity's sake] is known to have exhibited one of the best economic performances among the developing countries during the 1960s and 1970s. Since its ambitious First Five-Year Economic Development Plan was launched in 1962, its real GNP has grown at an annual average rate of 8.5 percent, raising the GNP per capita from $80 in 1961 to $1,600 in 1980.

This rapid economic growth was largely supported by the drastic expansion in Korea's manufacturing sector, whose output grew at 17 percent per year in constant prices, increasing its share in GNP from 8 percent to 34 percent. Exports, comprising principally manufactured goods, rose from $41 million in 1961 to $17.5 billion in 1980, representing a 38 percent per year increase.

To support the sharp expansion of GNP, manufacturing production, and exports, the demand for energy rose drastically. Thus, during 1961–80 primary energy consumption in Korea grew at 7.8 percent per year from about 9.8 million tons of oil equivalent (toe) to about 41 million toe (see Table 23–1).

One important factor that caused the sharp increase in demand and consumption during the past twenty years was the high income elasticity of energy demanded. During 1970–73 the elasticity for Korea stood at 1.04, a similar level to those of the United States, West Germany, and Japan for the same period.

Table 23-1. **Consumption of Primary Energy**[a]

	Anthracite	Bituminous	Petroleum	Hydro	Nuclear	Firewood	Total
1961	3.2 (32.7)	-	0.8 (8.2)	0.2 (2.0)	-	5.6 (57.1)	9.8 (100.0)
1966	6.1 (46.9)	-	2.1 (16.2)	0.2 (1.5)	-	4.6 (35.4)	13.0 (100.0)
1971	6.0 (28.3)	-	10.8 (50.9)	0.3 (1.4)	-	4.1 (19.4)	21.2 (100.0)
1973	7.2 (28.6)	0.4 (1.6)	13.6 (54.0)	0.3 (1.2)	-	3.7 (14.6)	25.2 (100.0)
1976	7.8 (26.3)	1.0 (3.4)	17.3 (58.2)	0.4 (1.3)	-	3.2 (10.8)	29.7 (100.0)
1980	9.9 (24.1)	3.3 (8.0)	24.0 (58.4)	0.5 (1.2)	0.9 (2.2)	2.5 (6.1)	41.1 (100.0)

Source: Data for 1961–76 from Economic Planning Board, *Handbook of the Korean Economy* (Seoul: 1980), p. 313; data for 1980 obtained from the Planning Bureau of the Korean Ministry of Energy and Resources in July 1981.
[a]In million metric tons of oil equivalent. Figures in parentheses are percentages.

During 1975–79, however, the elasticity for Korea was far greater than those of the three other countries: 0.96 (Korea), 0.42 (United States), 0.79 (West Germany) and 0.30 (Japan). This unfavorable comparison for Korea appears to have been caused mainly by two factors. In the first place, little effort was exerted by Korea for the conservation of energy after the first oil crisis; and second, the pattern of Korea's industrial development was steadily moving towards heavy energy-consuming industries.

It is interesting to note that consumption pattern by energy type has gone through substantial changes. The share of firewood, which accounted for 57 percent of total primary energy consumption in 1961, declined to about 6 percent in 1980. Declining importance was also visible in the case of anthracite, whose share dropped from 33 percent to 24 percent over the same period. This decline was the result of three major factors: the relative stagnation of domestic anthracite production, the comparative technical and economic advantages of petroleum for power generation and industrial uses, and the availability of low-cost petroleum supplies until the end of 1973. The share of oil, however, grew steadily from 8 percent in 1961 to 58 percent in 1980.

Also, the pattern of industrial demand for energy showed substantial changes as time passed. Mainly due to the rapid growth of the industrial sector, the share of industry in total energy consumption increased from 19.6 percent in 1966 to about 49 percent in 1980 (see Table 23–2).

Table 23-2. **Energy Demand by Sector, as Percentage of Total**

	1966	1974	1977	1981[a]	1986[a]
Industry	19.6	35.7	43.7	49.0	48.9
Transportation	11.6	13.5	8.7	11.3	12.3
Residential & commercial	58.0	46.0	39.5	29.1	26.7
Others	10.8	4.8	8.1	10.6	12.1
Total	100.0	100.0	100.0	100.0	100.0

Source: Korea Development Institute, *The Comprehensive Plan for Energy Supply-Demand to the Year 2000,* 1978.
[a]Projections.

Given its narrow domestic energy resource base, which is limited to only a small quantity of anthracite and hydroelectric production, Korea's reliance on imported energy has been growing steadily. Its dependency ratio on imported energy increased from 10.5 percent in 1961 to 61.6 percent in 1976 and to 69.4 percent in

1980. In particular, since the share of petroleum in total energy consumption rose from 8.2 percent to 62.1 percent over the same period and since no oil is produced domestically, the dependency on imported energy has increased sharply.

This rising dependency on foreign energy has serious implications. It puts heavy pressures on the nation's balance of payments position. Korea is expected to spend as much as $6.5 billion on imported oil in 1981. This is equivalent to 11.5 percent of its GNP and 32 percent of its export earnings, contrasting sharply with the fact that only ten years ago, in 1970, merely $100 million—which accounted for 1.3 percent of GNP and 11.1 percent of export earnings—was spent on imported oil.

AN INITIAL EVALUATION OF OPTIONS: TECHNICAL AND ECONOMIC CONSIDERATIONS

Meeting the Rising Demand

As in the case of other energy-deficit countries, several better alternatives are opened to Korea for meeting the rising energy demand. They include developing domestic energy resources; securing adequate foreign energy supplies including oil, coal, and perhaps LNG; and stockpiling.

Development of domestic energy resources.

Coal. Anthracite is Korea's main domestic energy resource. The Korea Mining Promotion Corporation (KMPC) estimates its total reserves at about 1,500 million tons. Of these, some 600 million tons, the equivalent of 32 years of production at the 1980 production level (19 million tons), are known to be exploitable. Anthracite deposits are located in the mountain areas, which require drift-type and shaft-type underground mines and labor-intensive techniques. Korea's mining productivity, therefore, is low: only about 1.1 tons per man-shift, compared with a maximum of 3 tons per man-shift in the United States. The quality of coal produced is also low, generating only 3,500 to 5,000 kilocalories per kilogram. Most of it is not suitable for coking and is briquetted and used as household fuel.

The production of anthracite has been increasing steadily from 5.9 million tons in 1961 to 18.6 million tons in 1980, although the share in total energy supply has been declining.

Nevertheless, domestic production meets only about 87 percent of total demand. The rest of the demand has depended upon imports. In 1980, a total of 20.8 million tons was consumed, of which only 18 million tons were produced domestically. The demand for 1986, the final year of the Fifth Five-Year Development Plan (FFYP), is projected at 26.5 million tons, implying a 3.1 percent per year increase, and domestic production is projected at 21.5 million tons, requiring a 3.0 percent per year increase.

To secure the long-term supply that will meet the shortage between total requirements and domestic production, overseas investment in the exploitation of foreign coal mines has been actively encouraged by the government. Accordingly, several private firms are seriously considering overseas investments seeking long-term contracts.

Power. The power sector in Korea is fairly well developed. About 99 percent of all households have access to electricity; almost 100 percent in urban areas and 98 percent in rural areas. The average household consumed 1,830 kilowatt-hours in 1980, and per capita consumption was 857 kilowatt-hours. Of the total consumption in 1980, 85 percent was attributable to industry, while residential use consumed about 18 percent.

Between 1971 and 1980, the total installed generating capacity increased from 2,628 megawatts to 9,391 megawatts, recording a 15 percent increase per year (see Table 23–3). In 1980 thermal power generation accounted for about 81 percent of the total installed generating capacity; hydroelectric, 13 percent; and nuclear, 6 percent. The reserve margin was about 33 percent in 1980.

Table 23-3. **Summary Data for Electric Utilities**

	Installed generation capacity (MW)	Gross gener- ation (GWH)	Sales (GWH)	Peak load (MW)	Reserve margin (%)
1961	367	1,773	1,189	306	-9.9
1966	769	3,886	3,008	696	2.0
1971	2,628	10,540	8,883	1,777	34.6
1976	4,810	23,116	19,620	3,807	3.9
1980	9,391	37,239	32,734	5,457	33.2
1986[a]	17,591	67,639	59,130	11,190	35.5
1991[a]	27,404		99,936	12,773	18.7

Source: Planning Bureau, Korean Ministry of Energy and Resources, July, 1981.
[a]Long-term plan targets.

The generation, distribution, and sales of electric power have been handled by the Korea Electric Company (KECO), which is jointly owned by the government and private interests. The absolute majority of KECO's shares, however, are owned by the government.

Security of crude oil supply. Since about 60 percent of Korea's total energy requirement is met by petroleum while not one drop of oil is produced domestically, it is natural that securing an adequate supply of overseas crude oil should become one of the nation's top priorities. An uninterrupted supply of oil is essential to sustain rapid industrial and economic growth. Particularly for Korea, a country with its capital city of Seoul only 17 miles away from the world's most hostile regime, a smooth supply of oil cannot be overemphasized in view of its national defense.

The demand for crude oil increased from 103.2 million barrels (bbl) in 1973 to 194.4 million bbl in 1980, indicating an annual average increase of 8.6 percent. During the same period, the combined capacity of the five privately owned refineries increased from 400 thousand barrels per day (bpd) to 790 thousand bpd. Part of the newly expanded capacity of the Honam refinery is still underutilized.

Some diversification of its supply source was necessary for Korea, because, first, no single oil-producing country was willing to supply the entire quantity Korea needed and, second, it was necessary for Korea to lessen the risk of supply disruption which may be caused by placing too much reliance on one or two supply sources only. As Table 23–4 shows, more than 95 percent of Korea's crude oil imports are from three Middle Eastern countries (Saudi Arabia, Kuwait, and Iran), a fact which lends even greater desirability to Korea's attaining more geographical diversification of its oil sources, provided that an adequate supply at reasonable cost is available.

Recently, Korea has been making progress in its efforts to achieve geographical diversification of its supply sources, mainly in the areas of Latin America and Indonesia, but the magnitude of supplies from these areas is known to be very small. For Korea, the importance of securing an adequate amount of crude will continue to outweigh the need for diversification. Considering that the FFYP, which covers 1982–86, calls for an increase of crude oil supply by as much as 35 percent, securing the necessary amount of oil on a sustainable basis deserves highest priority.

Table 23-4. **Source of Crude Oil Imports**

Country	Source of supply Share(%)	Deal	Pattern of deals Ratio(%)
Saudi Arabia	59.9	Major	51.4
Kuwait	26.9	G - G[a]	29.2
Iran	8.5	D - D[b]	16.3
Others	4.7	Spot	3.1
Total	100.0	Total	100.0

Source: Korean Ministry of Energy and Resources, *The Strategy to Overcome Energy Constraints in the 1980s,* May 1981.
[a]Government-to-government deal.
[b]Direct deal between private parties.

Stockpiling. In recent years, for almost all energy-deficit countries, stockpiling of energy refers mainly to storing crude oil. This reflects the vital importance of oil in keeping the nation's economy and industry running and in maintaining, in the case of Korea, a twenty-four-hour operating defense system. Stockpiling bears special significance for Korea since its source of crude depends geographically on a very limited area.

The government has established the FFYP target for stockpiling at sixty to ninety days' requirement of oil, and it plans to achieve this target between 1982 and 1986. When completed, the total amount stored will exceed 100 million bbl, without counting private firms' voluntary storage. Korea's principal method of storage chosen is to utilize oil tanks on the ground and some underground facilities. At present, tanks capable of storing 10 million bbl each are under construction.

Compared with other countries, Korea's target is very moderate. Japan and France, for instance, are expected to store in the near future amounts equalling up to one hundred twenty days' requirement. The main reasons for Korea's poor performance in stockpiling lay, first, in its difficulty of obtaining a volume of crude that sufficiently exceeded consumption, and second, in the vast expenses that stockpiling requires. The government estimates the combined cost of the storage facilities' construction, their operating costs, and the crude itself, will reach about $9 per barrel by 1985, with the operating costs alone at about $2.50 per barrel.

To induce domestic oil companies to store crude, the government is likely to allow them to transfer part of the storage costs to the consumer. It will also offer the oil companies some subsidies and will give preferential interest rates on loans extended to them.

Reducing the Total Cost of Energy

In this section, two options for reducing the total cost of energy for Korea will be considered. One is through the reduction of demand itself; the other is through lowering the unit cost of energy supplied.

Reduction of energy demand. There are two conceivable ways of achieving this demand reduction: sectoral restructuring of the economy and conservation.

Restructuring the economy. Like those of other countries, Korea's economy underwent structural changes in favor of energy saving following the first energy crisis in 1973. Thus, Korea's elasticity of energy consumption with regard to GNP was reduced from 1.04 over 1970–73 to 0.96 over 1975–79. This performance, however, is quite unsatisfactory compared with the performances of other countries: over the same periods, the comparable elasticities of the United States fell from 0.85 to 0.42, and those of Japan fell from 0.97 to 0.30. One of the main reasons for Korea's relatively poor performance was probably that unlike other countries, Korea was not forced to restructure its economy since its rate of exports and growth were far better than those of other countries, even without the restructuring.

Since the second oil crisis in 1979–80, however, Korea has been striving hard to shift the industrial sector of the economy to a less energy-consuming one. Various measures are being considered towards achieving this objective. One is for the government to identify certain subindustries that are thought to use excessive energy and try to keep these subindustries' growth to a minimum. Primary metallic industries producing such materials as aluminum and steel as well as chemical industries whose products include fertilizer, cement, and many of the petrochemicals belong to this "designated group."

New investments in the designated group are allowed only in exceptional cases. Various subsidies and preferential tax and interest rates that may otherwise have been accorded cease to be available for any new investments in this group. The government plans to make up with imports any production shortages of this group caused by the curb on investment. On the other hand, with regard to those establishments in the group that are already existing, the government is trying to maximize their energy efficiency through various methods such as compulsory replacement of boilers that are less than 70 percent efficient.

Conservation. In a country like Korea, with its small endowment of energy resources, conservation deserves greater attention than in the United States and other countries with abundant energy resources. How to devise the most effective tools to achieve greater conservation therefore is extremely important.

The overall planning, implementation, and monitoring of Korea's conservation drive is under the responsibility of its Ministry of Energy and Resources (MER). However, the drive in the public sector is primarily under the responsibility of the Ministry of Government Administration (MGA).

The conservation measures currently in effect in Korea have been, for the most part, directed toward residential and commercial uses of energy—uses whose portion of energy consumption is below 35 percent. These measures include closing gas stations on weekends, a campaign to turn off a light bulb in each household, the compulsory use of insulation materials in new housing and building construction, and restricting the use of air conditioners and elevators.

These measures so far seem to have produced only a limited result. For instance, gasoline consumption in Korea amounts to only 3 percent of its total petroleum consumption, and therefore, the gasoline-saving drive is more symbolic than substantial. The same verdict can be applied to the other conservation drives. In 1980, crude oil imports decreased by only 1.4 percent, while real GNP fell by 5.7 percent.

To improve substantially the yield of conservation, efforts should be directed toward conservation in the industrial and transportation sectors, whose combined portion of energy consumption is currently 55 percent and is expected to reach over 70 percent by 1986.

Table 23–5 shows the result of an energy survey conducted in 1978 by the Korea Energy Management Corporation, a semigovernment body, on 806 industrial establishments. The survey reveals that, on the average, 22 percent of energy generated was wasted; in the case of the metal and paper industries, the percentage of energy loss exceeded 30 percent.

Table 23-5. **Waste of Energy by Industry**

	Ceramics	Textiles	Chemicals	Metallic	Paper	Foods	Misc.	Total
Establishments[a]	95	225	138	112	82	112	42	806
Waste ratio (%)	8.8	23.6	19.5	31.6	30.5	23.0	10.0	21.5

Source: Energy Survey (1978) conducted by the Korean Energy Management Corporation.
[a]Indicates number of establishments surveyed.

It appears that Korea has much room to achieve greater energy conservation. Towards the end of 1980, MER and MGA formed an intensive energy conservation plan in which a broad range of incentive systems as well as mandatory regulations were utilized. However, except for the first few months, the efficiency of the new plan seems to have become much lessened as the world oil situation became more relaxed in the first half of this year.

Lowering the unit cost of energy. As the price of oil increased sharply following the energy crisis in 1973, cheaper energy was sought by almost all non-OPEC countries. In the case of Korea, mainly three types of energy were pursued to cut the unit cost: nuclear power generation, coal, and natural gas.

Korea's first nuclear power plant, located at Kori, Kyung-Nam province, began operation in 1977 with an installed capacity of 600 megawatts. The government intends to have installed a total of thirteen nuclear plants by 1991, aiming at having nuclear constitute more than 40 percent of Korea's power generation by 1991 (as opposed to its present 6 percent). When completed, the total installed capacity will be over 11 thousand megawatts. Although it may be the most economical way, nuclear power generation entails problems. Besides the probable safety hazards which may accompany almost all nuclear plants in the world, the problems faced by Korea include the difficulty of financing the extravagant cost of installation as well as the problem of recruiting and training adequate personnel to operate the thirteen plants.

Since Korea's poor endowment of domestic coal makes its domestic supply more expensive, foreign supplies of coal will have an increasing significance. Therefore, with regard to coal, emphasis has been placed on developing overseas supply sources through private and government initiatives. To facilitate coal imports, the domestic receiving facilities such as port or storage sites have to be strengthened. For the strengthening of these facilities, the government hopes to rely on as many private investments as possible, by providing such incentives as long-term loans at preferential interest rates.

Except for the moderate use of LPG available from domestic refineries, gas so far has not played a significant role in Korea, mainly because domestic gas production is slight. The use of gas, however, is expected to increase in the future. In particular, an increase in home use of gas is anticipated because of its convenience for cooking. The FFYP projects that the number of gas-reliant homes will increase from 140 thousand in 1981 to 800 thousand in 1986. Also, the government is encouraging the replacement of naphtha by gas as raw material for certain industries.

Beginning in 1983, 1 million tons of LPG will be imported annually, and beginning in 1984, 3 million tons of LNG per year will be imported. It has been announced recently that, beginning in 1984, 150 million tons of LNG will be supplied by Indonesia annually. Additional facilities for receiving, transporting, and storing gas are under construction, mostly at government cost. However, the problem of financing the massive cost of the additional facilities as well as the imported gas has not yet been resolved.

INSTITUTIONAL CONTRAINTS

Pricing Policy

The main thrust of Korea's energy pricing policy has been to promote industrial development at low cost and to subsidize the poor segment of the population.

This policy is evident in the schedule of taxes on petroleum products: 13 percent taxes on kerosene, bunker C, and heavy fuel oil; 20 percent on diesel fuel; and 150 percent on gasoline. Productive sectors and the poor get the most favorable tax rates and hence lower consumer prices, whereas consumer use of gasoline is taxed the most heavily.

In the case of coal, the domestic market price has long been held by the government at a level well below the domestic production cost and international prices. This pricing policy was designed to subsidize the poorest segment of the population, since coal is predominantly used by the poor as their main residential fuel. Because of the divergence between the production or import costs and the domestic market price, a sizable deficit in the government's coal account is incurred each year in the form of subsidies to domestic coal producers and importers.

The structure of electricity prices also favors industrial uses and penalizes nonproductive residential uses if they exceed certain limits. In extreme instances, some residential uses are charged as much as 95 cents per kilowatt-hour, when the total cost is only 5 cents per kilowatt-hour.

It is feared that the above price structure may have an unfavorable impact on resource allocation, since a long-term divergence between marginal supply costs and market prices is always likely to result when there is a misallocation of resources.

Investment in the Energy Sector

The energy sector's share of total investments has long been one of the largest in Korea. Investments in power generation, refineries, tankers, and storage facilities accounted for about 15 percent of all investments made during the Fourth Five-Year Development Plan period (1977–81). In 1980 prices, this amounts to roughly 6,775 billion South Korean won or, equivalently, approximately 12 billion US dollars. Investments during the FFYP period will reach about $24 billion, also in 1980 constant prices. Of this amount, about 55 percent will be allocated to the power generating sector, 18 percent to the petroleum sector, and 8 percent to the coal sector. Individual items which will entail big investments include nuclear power plants, resource transportation, LNG receiving and distributing facilities, oil stockpiling, and coal import facilities.

It is anticipated that financing this huge investment requirement will be difficult. Roughly one-third of the total cost is expected to be arranged from foreign sources, and the remainder is expected to be mobilized domestically.

Institutional Arrangements

The energy sector comprises a mixture of public, semipublic, and private corporations which are under the supervision of MER. Although each corporation is equipped with its own decisionmaking body, the influence of the government in pricing, financing, and establishing long-term priorities and at times in personnel management has been significant.

In particular, the government has been heavily involved with pricing and financing. Although the pricing policy for the petroleum subsector has been generally in line with international practices, which set prices at levels high enough to induce proper investments, in the power and coal subsectors the prices have been artificially held down to subsidize industrial establishments and the poor. Due partly to inadequate levels of prices and partly to difficulties in mobilizing massive amounts for investments, the government has been the main instrument through which most of the investments have been financed to develop coal mines and power facilities. The government does this through a complex system of budget allocations and subsidies.

Other areas to which attention should be directed with respect to institutional arrangements are (1) the management of coal mines and (2) the problem of supplying manpower for nuclear power generation. Private coal mines, which account for 75 percent of all coal production, still suffer from old-fashioned management techniques and lack of mechanization. Enabling managers to obtain training in modern management techniques would certainly increase the coal mining subsector's productivity. Also, further mechanization in coal mining seems necessary for both increased productivity and increased safety.

To manage and operate the thirteen nuclear power plants to be in effect by 1991, a well-planned recruiting and training program should be designed and implemented. Although recruitment and training are being conducted currently, the scale of these efforts needs to be much greater.

SOCIETAL CONSTRAINTS

Environment and Quality of Life

It is generally true that as the amount of energy available for people's disposition increases over the course of time, their life becomes easier, and hence the quality of life improves. In Korea, as more energy became available for automobiles, trains, lighting, heating, cooling, and industrial production, the average Korean's quality of life improved substantially. However, it is also true that this greater use of energy has created new problems: damage to the environment and the subsequent deterioration of the quality of life. Three types of energy have been the main causes of the greatest environmental problems in Korea: coal, petroleum, and nuclear energy.

The use of coal has increased both at home and in thermal power plants. Between 1961 and 1979, the ratio of homes that used anthracite as their main residential fuel increased from 37 percent to 55 percent. Over the same period, the number of thermal power generating plants that use coal, including imported bituminous, as their main fuel increased twofold, but during the next five years, this number is projected to increase fourfold. Furthermore, a significant number of industries, such as the cement industry, switched from oil to coal as the price of oil increased sharply. As a result, the emissions from thermal power plants and the dumping of burned-out coal began to pollute the air and soil quickly, raising grave concerns about environmental cleanliness.

No one can deny that the emissions and discharges from automobiles, refineries, and industrial plants using oil have severely damaged aspects of Korea's environment. In particular, the air pollution caused by oil-burning factories and automobiles became so serious that severe penalty clauses began to be introduced. Unless appropriate corrective measures are soon applied, this problem of air pollution is likely to get much more serious as the use of oil continues to increase.

Although the problem of pollution from nuclear power generators has not become a reality yet, the likelihood that it may become a serious environmental issue in the near future should not be underestimated. This can become a pressing reality when twelve more nuclear power generators become operational before 1991. All the liquified and solid waste materials from the Kori nuclear plant are currently kept underground and therefore do not cause any harm to the environment, at least for the moment.

The Office of Environmental Protection (OEP) came into being in 1980 as one of the branches of Korea's government. Since its birth, OEP has applied various protective and corrective measures to attain a clean environment. However, full-scale enforcement of the antipollution codes which OEP laid out cannot as yet be achieved because of its massive cost, the burden of which can hardly be borne in so short a time by the private and public sectors.

Health and Safety Issues

The prospect that the use of nuclear power will increase in the near future raises serious concern about the security of the power plants themselves and about potential employee health problems.

In particular, there is a constant danger that a radioactive fuel leakage incident, such as that witnessed at Three Mile Island a few years ago, may occur at any time.

Furthermore, employees working at nuclear power plants are constantly in danger of exposure to radioactive materials at emission levels above that permitted by health standards. Presently, health checkups are required of all Kori employees every three months. This practice should no doubt be applied to all employees who will work at the other nuclear power plants to be constructed. The current permissible criterion applied by the Korea Electric Company in the checkups is 5 rem per year for nuclear plant employees and 0.5 rem for regular office employees.

Another energy-related health problem in Korea is the danger of death caused by carbon-monoxide inhalation. For a long time in the country's history, Koreans relied on "Ondol" — the anthracite-burning floor-heating device—for heating rooms. This device, however, entails the possibility of inhaling carbon monoxide, which can lead to death.

Another safety problem—this one inherent in the continuous reliance on oil—relates to Korea's oil stockpiling. In particular, with Seoul only 17 miles away from the truce line, the safety of Korea's storage facilities involves not only economic and technical problems but also the political problem of safeguarding the facilities from the constant threat of hostile acts by the North Koreans.

EDUCATION AND INFORMATION

Education

Perhaps one of the most important education programs for Korea's energy sector is the training program for the present and future staff of nuclear power plants. For this, the Korea Electric Company established a rather bold training program which will extend for about five years. KECO projects that the number of employees at nuclear plants will increase from the present 4,400 to 9,800 by 1986. About 10 percent of the employees will be sent to foreign companies such as Westinghouse Electric (builder of the first nuclear plant in Korea) or Bechtel for training. The domestic and overseas training programs will cover such subjects as the construction, operation, maintenance, and safety reviews of nuclear plants.

In its early phase of instituting nuclear power, KECO relied heavily on foreign experts for the operation of the Kori nuclear plant, but now almost all foreigners there have been replaced by Koreans. Eventually KECO hopes to gradually foster an independent training system which would not require foreign assistance. Nevertheless, in such areas as health physics or instrumentation, continued reliance on foreign expertise is expected for some time to come. A related problem feared by the management of KECO is the possibility that a significant number of trained employees may be tempted to move to other domestic or foreign firms offering better pay and working conditions. In particular, the possibility of moving to an overseas job may be much greater than KECO's management currently expects. Such a move would be potentially quite attractive because the pay difference between Korea and the other industrialized countries is substantial.

Other areas where energy-related education is being conducted include mining and conservation. The Korea Mining Promotion Corporation offers training programs for the employees of private and public mining companies on safety precautions, techniques of soil analysis, and mining management. Also, the Korea Institute of Energy and Resources (KIER), which is under the supervision of MER, provides regular training programs on efficient energy management in private and public establishments.

Technology

Korea's energy-related research has been conducted primarily by KIER and the Korea Advanced Institute of Science and Technology. However, the scope and sophistication of this research on energy has been quite moderate compared with research in other fields done by the same two institutions.

The focus of the energy-related research has been on possible fuel substitutes, conservation, waste heat utilization, and environmental preservation. This research has sometimes been carried out in conjunction with foreign institutions or firms.

Although the government actively encourages energy research by the private sector, the result has so far been disappointing because of the rate of higher financial return offered by other investment opportunities.

The transfer of energy technology from abroad so far has also been insignificant. The record shows that only a limited number of operational and safety staff members of KECO have visited foreign institutions under the bilateral technology exchange programs.

Consumer Information

In Korea very few efforts so far have been made by the government or by business firms to educate private citizens about the many aspects of energy, except perhaps in the area of energy conservation.

Various news media were mobilized during the last few years to inform the general public of the significance, content, and benefit of energy conservation. Nevertheless, the overall result appears to have fallen short of expectations, mainly because, as stated before, the government's conservation drive was aimed mostly at relatively insignificant residential uses. If the focus of the conservation drive had been on industrial and transportation uses, the result would have been much more meaningful.

FINAL EVALUATION OF OPTIONS

Criteria for Evaluation

The criteria for evaluating a nation's energy policy can differ depending on the situation of that nation. However, for a country like Korea which may be characterized as energy deficit and rapidly industrializing, the following points may well be presented as the evaluatory criteria:

- An adequate supply of energy should fully support rapid industralization and economic growth, in addition to satisfying consumers' energy requirements.
- The cost of energy should be low.
- The energy supply should be steady.
- The portion of the supply coming from overseas should not be reliant on too geographically narrow a basis.
- Dependency on imported energy should not be excessive.
- The environmental pollution caused by energy use should be minimal.

Evaluation of Korea's Energy Policy

Based on the above criteria, one can evaluate Korea's energy policy as a fair one.

Korea has successfully achieved rapid industrialization, economic growth, and a major personal income boost during the last two decades. Underlying this economic performance was the steady and relatively inexpensive supply of energy throughout the period. Thus, Korea's energy policy was consistent with its long-term objective: steady supply of inexpensive energy to the production sectors.

Also, the current efforts to find alternative oil supply sources in areas other than the Middle East appear timely and well motivated. The investment in the exploration for and development of coal or LNG in foreign countries on a long-term basis is also justified, since it will help secure a steady supply of energy at lower costs.

The government's decision to meet electricity requirements primarily by expanding Korea's nuclear power capacity seems to be the natural result of the situation which the country faces at present. The other two sources of primary energy for generating electricity—in other words, water and fuel—are physically very limited in Korea, while the need for electricity is projected to increase rapidly as industrialization progresses fast. Moreover, nuclear power generation may be better for a clean environment than is thermal power generation which uses oil or coal.

Nevertheless, much improvement can be made in almost all aspects of Korea's energy policy. For instance, the energy conservation drive could achieve greater and more significant results than at present if greater stress could be put on industrial and transportation uses of energy.

The pricing policy in Korea also needs to be improved. The price of energy supplies for industrial uses and for the poor segment of the population has been sometimes below the marginal supply costs (e.g., bunker C oil). In the long run, this practice will tend to disturb the efficient allocation of resources and will eventually lead to a rise in energy supply prices. Therefore, the government's pricing policy should be reviewed on the basis of a well-defined balance between the need to supply cheap energy and the need to achieve an efficient resource allocation in the long run.

Another point which deserves attention is energy conservation's relationship to energy cost increases. In the past, the transfer of cost increases was almost automatic, and hence, producers had no motivation to reduce the energy content within a product. Therefore, the conservation drive will have a much better result if energy cost increases cannot be automatically transferred to the price of final products and eventually to consumers. This can be done either through consumer resistance (high price elasticities) or through government armtwisting of producers on final product prices.

The third area for improvement is related to the institutions involved with researching, formulating, and implementing energy policy. There seems to be little information exchange among MER, the research institutions and public corporations under the supervision of MER, and private energy firms. As a result, the nation's energy information is very much scattered, and hence, the efforts to formulate consistent energy policy are substantially impaired.

REFERENCES

1. Economic Planning Board, *Handbook of Korean Economy*, 1980.
2. ———— and Ministry of Energy and Resources, *The Long-Term Energy Plan 1979–91*, September 1979.
3. Korea Development Institute, *Long-Term Prospect for Economic and Social Development 1977–91*, 1978.
4. ————, *The Comprehensive Plan for Energy Supply-Demand to the Year 2000*, 1978.
5. Korea Energy Management Corporation, *Energy Survey 1978*.
6. Ministry of Energy and Resources, *The Plan for Energy Conservation*, October 1980.
7. ————, *The Strategy to Overcome Energy Constraints in the 1980s*, May 1981.
8. The World Bank, *Korea*, a World Bank Country Economic Report (Baltimore; The Johns Hopkins University Press, 1979).

Summary with Selected Comments

Session Chairman/Integrator:
Guy J. Pauker
Senior Staff
Rand Corporation

INTRODUCTION

In addition to the diligent efforts of its designated participants, our work session had the benefit of a joint session on the third day of the Symposium with the members of the work session on energy for rural development, including that session's chairman/integrator, Dr. Ishrat Usmani of the United Nations, and Ambassador Alejandro Melchor of the Philippines and Mr. D.R. Pendse of India. Our work session also had valuable participation from Dr. William Fulkerson and other observers from Oak Ridge National Laboratory and the University of Tennessee.

During the course of the work session, an important position paper was presented by Mrs. Joy Dunkerley of Resources for the Future, supplemented by a revealing comparison between the energy problems of Burundi and Cameroon written by Dr. Emmanuel Mbi of the World Bank. An impressive case study of South Korea was then presented by Dr. Hiwhoa Moon, Economic Policies Coordinator in the Office of the Prime Minister of the Republic of Korea.

In addition to the chairman/integrator and those presenting papers, the work session included the following assigned participants: Fernando Altmann Ortiz, Thinakorn Bhandugravi, José Goldemberg, Donald Klass, and Sanga Sabhasri. The chairman/integrator was assisted by Sheldon Reaven, who served as rapporteur.

FINDINGS

These three papers and the lively discussions that followed impressed upon all of us the severity of the situation confronting the energy-deficit industrializing nations and the urgent need for crash measures to assist some of the most deprived countries. We are all aware, of course, that more than one hundred countries are in the "energy-deficit developing" group. The presentations illustrated the fact that there is great heterogeneity in that group, which includes rapidly industrializing nations such as Brazil and South Korea, with per capita incomes of over $1,500, but also very poor stagnant countries, such as Upper Volta, Burundi, and Bangladesh, with per capita GNPs below $200.

We were reminded of the complexity of the concept "energy-deficit industrializing nations"—a concept covering countries that are potentially rich in energy resources but that have not yet been able to mobilize these as useful supplies, and also covering other countries that have a very limited endowment of both traditional (noncommercial) and conventional sources of energy. All these countries have, like the rest of the world, fallen into the trap created by the cheap cost of petroleum before 1973. They are using imported oil to fill the gap created either by their lack of energy resource endowments or by their past substitution of oil for other, indigenous energy resources.

Since the oil price increases from $2.00 per barrel in 1973 to $10.70 per barrel in the second half of 1980 (using constant 1970 prices), the energy-deficit oil-importing nations have suffered from increasingly severe current account deficits. Whereas prior International Monetary Fund estimates anticipated deficits of $72 billion for 1980, actually these deficits were $10 billion higher. For 1981 a prior projected deficit of $80 billion is actually likely to exceed $100 billion.

These large deficits have occurred despite efforts by energy-deficit developing countries to increase exports and to keep down the increase in imports. The deficits are due to factors largely beyond the control of the energy-deficit developing nations—in particular, the rise in the prices of their imports, the upward shift of interest rates on their debts, and the recent sharp fall in the prices of their export commodities (prices which, in real terms, are at the lowest level since 1950).

The vulnerability of the energy-deficit developing countries is easy to illustrate: it has been estimated that a $1 per barrel increase in the price of crude oil costs the oil-importing developing countries $1.5 to $1.8 billion annually, and a 1-percentage-point increase in the interest rate raises the annual interest costs of developing nations by $2 billion. Thus, at the end of 1980 the total outstanding debt of developing countries was $456 billion, of which $373 billion was owed by oil-importing developing countries, and the debt service of all developing countries in 1980 amounted to $92 billion.

The future consequences of this situation could be catastrophic, if a lack of adequate resources for investment in the two top-priority areas—food and energy—results in slower economic growth rates and even in a decline in per capita income in real terms. Despite these disturbing facts, our panel's discussions did not end on a note of gloom and despair, because we believe that solutions are available.

CONCLUSIONS AND RECOMMENDATIONS

Our work session was particularly interested in the four options outlined in Mrs. Dunkerley's paper for overcoming the oil-importing developing countries' energy problems; namely:

- Increase export earnings.
- Modify development strategies.
- Increase domestic energy supplies.
- Improve energy efficiency.

We concluded that various mixes of these options would provide viable energy strategies adapted to the specific circumstances of different countries.

In her oral presentation, Mrs. Dunkerley stressed that a number of measures could be highly effective in helping middle-income countries, specifically:

- facilitating the recycling of OPEC surpluses;
- keeping open the industrial countries' markets;
- encouraging private foreign investments in the middle-income countries' energy development;
- enabling technology transfer, especially in the biomass field; and
- providing energy planning assistance.

But the poorest energy-deficit industrializing countries have problems that are more intractable. They will continue to depend heavily on concessional aid and basic technology, although their oil imports are very small and their total needs easy to manage, as Professor Mohammad Sadli had suggested at Symposium I.

The Thai participants then pointed out that even in their country, which is endowed with some energy resources, including recently identified substantial deposits of offshore natural gas, and which exports large amounts of food grains, the burden of oil imports is very great, absorbing one-third of all Thailand's export earnings. Efforts to increase and diversify exports are hindered by the protectionism of the industrialized countries.

Professor José Goldemberg remarked that it is wrong to view a country like Brazil as underdeveloped. São Paulo is as developed as Chicago, but it encompasses only 10 percent of the country. The economies of developing countries are managed by their upper middle classes, who share the values of the developed world —values which have spread in a manner reminiscent of the Roman Empire, where the ruling elite dispersed from Rome carrying their civilization with them.

Today, the ruling elites everywhere speak the same language: for instance, the economists in the service of their respective governments speak the language of the World Bank and of the multinational corporations, and in Brazil, for example, they have implemented three of the four options outlined in Mrs. Dunkerley's paper. But the fourth option, which concerns modifying development strategies, does not enter their schemes. We must spell out what such alternative strategies would be.

Seeking concrete options, the work session group singled out a number of conclusions and recommendations which emerged as particularly salient from the prepared papers and from our discussions:

1. Many developing countries urgently need assistance in developing national energy strategies and energy plans. In some instances regional cooperation may be particularly useful and should be encouraged. Almost everywhere, education on energy matters is needed and would be beneficial.

2. The fuelwood crisis is a matter of extreme urgency for some countries. About half of the world's population relies on traditional energy sources, particularly wood, for cooking and other household needs. While some countries are still well endowed with timber resources, in others deforestation is progressing rapidly with very serious consequences to the environment and local daily life. In those countries, reforestation is an urgent task, especially since the alternative of shifting from fuelwood to oil and other imported energy sources increases the already excessive burden created by their growing energy demands.

3. The obvious need to speed up the substitution of other energy sources for imported oil raises a host of questions; the work session addressed these as follows:

- We concluded that the urgent issue of pricing, which concerns the governments of all oil-importing countries, cannot be reduced to simple formulas, since it sometimes has conflicting consequences with regard to economic growth, social equity, and energy conservation.
- We accepted as self-evident the need for accelerated exploration for and production of oil for domestic consumption in countries that now import oil.
- Our attention was primarily focused on oil's replacement with other energy resources, such as natural gas, coal, hydro power, biomass, and other new and renewable energy resources.
- There was agreement among us that electrification is imperative in the developing countries, because it is inconceivable that future generations will accept to live indefinitely without light, food refrigeration, and television.
- We also discussed at some length the prospects of substituting ethanol- and methanol-based fuels for gasoline in the transportation sector, and of coconut oil for diesel fuel.
- It was suggested that developing countries could pursue a dual strategy in supplying the energy needs of their rural populations: a decentralized approach for providing for cooking fuel and a centralized one for power generation.

ADDENDUM: A POSSIBLE ROLE FOR COAL

Although our work session group lacked the time to discuss it, I wish to bring to the attention of the Symposium an imaginative, novel idea presented in our work session by Ambassador Alejandro Melchor of the Philippines, who is a director of the Asian Development Bank. His proposal should be of particular interest not only to energy-deficit industrializing countries but also to coal producers, including our host, the state of Tennessee. In our joint session with the rural development group, Ambassador Melchor proposed a coal-export program that would have some similarities with the us government's PL 480 grain exports, which have benefited both food-deficit developing countries and the American farmer. We hope that Symposium III in May of 1982 will discuss in detail Ambassador Melchor's following brief exposition of this idea:

The *Global 2000* report commissioned by the us government

has called our attention to the clearly visible serious stress already upon us, which, barring revolutionary advances in technology and unless corrected, will make life for most people on earth more precarious and more vulnerable to disruption.

In the energy field, which is the focus of these Symposia, that report projects that even with stringent conservation measures and rapid increases in prices, world oil production will approach geological estimates of maximum production capacity in the 1990s. A rough assessment of the energy component of the global power equation shows that the immediate years ahead will be critical ones for the United States, due to its continued dependence on imported oil and the time lag required before development of alternative energy sources and the retooling of industries to such energy sources take place.

Coal, which now supplies more than a quarter of the world's energy, has been identified as the only energy source that can provide from one-half to two-thirds of the energy needs of the world over the next two decades.

The United States, which is one of the largest producers, consumers, and exporters of coal, has demonstrated coal reserves of some 450 billion tons, which at current production rates is estimated to be good for at least three hundred and fifty years, and more than adequate to meet any forseeable production expansion in this and the next century. It is therefore suggested that the US government, interested American state governments, and the coal industry should jointly study the possibility of making coal available to energy-deficit developing countries on concessional terms, if it is to be used as a substitute for imported oil. Counterpart funds generated by the sale of coal could also be applied to economic development programs in the recipient countries. Other major coal-exporting countries, particularly Australia, may wish to join in this important study.

The case study of Korea presented to our work session by Dr. Hiwhoa Moon provided a graphic illustration of coal's potential as a substitute not so much for oil as for firewood for household usage. From 1961 to 1980 the amount of coal consumed as primary energy increased from 3.2 million tons of oil equivalent (MTOE) to 13.2 MTOE, while the consumption of firewood decreased from 5.6 to 2.5 MTOE. While some of the coal is used for coking, much is briquetted and used as household fuel.

Evidently, a coal export program not only could alleviate the balance of payment problems created by oil imports, but also could help reduce the devastating environmental impact of excessive reliance on firewood. But the transition to coal will, of course, require the creation of a new infrastructure, including transportation and distribution modes, and will entail massive efforts to change consumer habits, an undertaking which is particularly difficult in the household sector.

SELECTED COMMENTS

MARCELO ALONSO

Referring to the work session on the energy-deficit industrializing countries, for which I think the summary was excellent, I would like to emphasize one point which in my opinion is not very clear in Dr. Moon's case study of South Korea—a case study which shows the difficulties in which the energy-deficit industrializing countries may find themselves in relation to the exigencies of the world in which we live. If my understanding is correct, by the end of the 1960s and the beginning of the 1970s, Korea had decided on a new policy of industrialization based essentially on energy-intensive industries. That policy was adopted shortly before the period of the 1973–74 oil crisis, and it had begun to be implemented. We have witnessed how Korea has progressed since that time, but during that period, as the oil crisis became more serious the performance of Korea's economy was very seriously affected and had very little leeway to shift in one direction or another. I don't know if Dr. Moon would like to elaborate on that.

However, the point I really want to raise, using Korea as an illustration, is that the developing countries that have an energy deficit and want to move ahead with industrial development are subject to tremendous variations, tremendous uncertainties against which they can have very little protection, considering the world as it is at this moment. One could suggest, for example, that these countries should increase their exports to increase their incomes, but who is going to buy their exports? That shows another serious limitation, and I believe that if we want to encourage the industrialization of the developing countries, some new form of international cooperation involving trade, financing, and energy is needed.

D. R. PENDSE

My comment relates to the summary report of the work session on energy-deficit industrializing countries. One of the solutions referred to was to modify development strategies. This is a very important point, indeed. At present, regardless of whatever little national-level planning has been done in energy-deficit countries, an increase in the consumption of fossil fuels for energy has become an inescapable condition of any growth and development that can be achieved. The Third World countries do not know of development strategies that will bypass this route to development. That is why there is an urgent need to modify the development strategies. The World Bank has also recognized this, but the question is really one of how to go about doing it. I do not know if much thought was given to this during the work session in question.

NAIM AFGAN

Being from Yugoslavia, which, as you know, is a developing country with a socialist economy and a self-management system, I have learned that our energy problems are substantially different from those of other countries under consideration in the "industrialized nonmarket-economy nations" work session. But during the discussions of that work session, I also learned that this group of countries shares problems in common with many other countries regarding long-term energy research and development. Since long-term R&D might be a problem for many of the countries discussed at this Symposium, I would like to use this opportunity to propose to the organizers of the third Symposium that more attention be devoted to those problems shared by most of the countries. Specifically, I think that the subject of a consensus could be more easily obtained on long-term R&D programs than on other subjects discussed during this Symposium. The Symposia Series thus became the springboard for launching a new scientific and technology program in the field of energy.

In the above regard, I particularly would like to emphasize the importance of a long-term international R&D program for the developing countries. As was stated in the report of the work session on energy-deficit industrializing nations, the major obstacle to the faster development of indigenous energy sources in those countries is a shortage of the capital needed for long-term development. In this respect, an international program on long-term energy R&D would greatly help the conjunction of the developing countries' individual efforts. If it is envisaged that some of the developing countries will join the developed part of the world in the not-so-distant future, it seems profitable to include the local and sometimes disorganized energy programs now going on in an internationally agreed-upon cooperative program, in order to attract available financial resources. Such a program would be of great assistance to many groups interested in international scientific cooperation on energy. The 1982 World's Fair might be a good place to initiate an organized international system in the field of energy with one of its outcomes the creation of a center of excellence to increase the availability of energy resources in developing countries.

GUY J. PAUKER

Several of our distinguished colleagues addressed questions or comments to our work session, and I will take them in the order in which they were presented.

First, as to the comment of Dr. Alonso, who raised the issue of what lessons we should derive from the fact that South Korea had initially decided on a policy of industrialization based on energy-intensive industries: I must say that this is an issue which we discussed. However, knowing in advance that I would be the last of seven chairman/integrators reporting on the work of their sessions, I assumed that my popularity would be directly proportional to the brevity of my report, so I skipped a good many of the things that could have been said. It was pointed out in our work session's discussions that countries which are embarking on industrialization now will have to be much more careful in considering energy as one of the principal factors of production and will have to avoid, if they are energy constrained or energy poor, going in the direction of industries that would complicate their energy problems. Dr. Moon can correct me if I misunderstood that Korea is now trying to shift from energy-intensive industries to others that are less energy intensive and more knowledge intensive.

As to the comment made by Amory Lovins [see chap. 18], it seems to me that the emphasis on some of the newer renewable energy sources may have the end result—unintended, I am sure—of being diversionary rather than productive. It seems to me that one really cannot put photovoltaic conversion today in the same category with coal; it is as if one would offer a 450 XL Mercedes to somebody who is dreaming of someday having a bicycle. During the Symposium, I talked with a gentleman from Great Britain, Mr. I. C. Price, whose company is actively engaged in developing photovoltaic water pumps. Now, I do not know—and I wish I did—what the cost factors of such pumps are, and also what their maintenance problems are. Ambassador Alejandoro Melchor, knowing that there is a pressing need to get away from reliance on imported oil and that coal is cheaper as a fossil fuel, brought into the discussions of our two joined groups a constructive proposal to make coal available on concessional terms. Obviously such a proposal, if carried out for a country, would have an immediate, direct impact on the balance of payments of that country. This does not mean that I am in any way hostile to the transfer of advanced technology, except that I have yet to see a photovoltaic cell in action somewhere in my very affluent Los Angeles neighborhood, and I therefore am not sure that this is a solution for today's problems in the Third World as a whole. If one considers very unusual situations, such as radio communication in remote places of Papua-New Guinea or television in the Sahara, then maybe the cost factor is relatively unimportant, as it is in space probes, but I think that unless the cost of photovoltaic conversion goes down markedly it is a bit premature to advocate its diffusion.

Coming to the question that was raised by Dr. Gibbons [see chap. 18]: as I spend a good deal of time in Southeast Asia and talk to people who are intensely interested in energy there, I think that the need for an increased flow of information on matters concerning energy is quite obvious. Distinguished academic scholars, engineers, businessmen, and government officials hear about developments to which they do not have easy access. It seems to me that if energy information centers were created in the Third World, this would be a very major contribution. I get questions on this very frequently, and while the us government has commercial attachés who double as petroleum attachés or energy attachés, they do not have at their disposal the great number of publications produced by the us Department of Energy—and, of course, the United States is not the only country that should share its knowledge. Under what auspices that sharing should be done is one of the issues that perhaps could be very fruitfully discussed at Symposium III, as one of the practical issues that are really important.

Finally, as for Mr. Pendse's point concerning the intriguing question of modifying development strategies—again, for the sake of brevity I did not go into this. I was tempted to quote the very thought-provoking comments made by José Goldemberg, who, in commenting on Mrs. Dunkerley's report said that three of the four options of her strategy were implemented quite successfully in Brazil, but the fourth one, which concerns modification of development strategies, is very difficult to achieve, since the planners are captives of well-entrenched ideas originating in the advanced industrial countries.

Closing Session

Introductory Remarks

Hans H. Landsberg
Senior Fellow
Resources for the Future

Ladies and gentlemen, I think we had a very lively session this morning, the more so since as Symposium participants, you had your first—and, I regret to say, your last—chance to break out of the more narrowly focused working groups and were finally able to range broadly over the many topics that made up the Symposium's agenda. I trust that you have by now worked off any frustration accumulated over the past three days' confinement. Thanks go to the session chairmen/integrators for having so rapidly produced summary reports of what went on in their particular conference rooms, and thanks also go to everybody here this morning for having contributed to a discusssion that was both spirited and civilized.

We now come to the final item on the agenda. A little more than a year ago I stood in this very same spot to do the very same thing I shall do at this point: introduce the next speaker, David Rose, who now divides his professional activities between the Massachusetts Institute of Technology and the East-West Center in Hawaii. In between he finds time to speak to a variety of audiences and, what is more important, to listen. As you will presently discover or rediscover, Dave is an excellent listener —that is, a listener with a highly developed digestive apparatus that almost instantaneously sorts information and puts it in the relevant storage departments. In this process he roams widely over continents and centuries, even millennia, and when he is through, he will have shown you (a) that you'd better look at most pieces of information from all sides and preserve a healthy skepticism toward "gee-whiz" attitudes, (b) that there are very few things that are totally new, and (c) that very often energy is only a proxy for more profound issues with which humans try to cope.

There has been plenty of food for him in this Symposium. Indeed, the conference has been characterized by one common denominator, which is overfeeding—some of it physical, at lunch and especially at dinner. To cope with this problem, there is a simple recipe which I learned from a friend of mine: physical exercise consisting of vigorously shaking one's head when offered food. To cope with intellectual overfeeding is harder. Closing one's ears or mind is not as easy for most of us, especially when the information is as interesting as it has been here over the past three days. That is where Dave comes in. I shall not try to describe him further. All the clever things that I could think of to introduce him I said a year ago, and these are, for better or for worse, preserved for posterity in last year's *Proceedings*. Since it is now in print, they are obviously true.

The only point I will add is that I consider him extremely courageous for trying the same stunt twice: to bring order out of seeming chaos. I use the word "seeming" deliberately, because in fact, the structure of our deliberations has been very neat—some might think even too neat. Someone once said, "All generalizations are wrong, including this one!" In setting down categories of countries, the Symposium structure surely has indulged in broad generalizations. Perhaps they were the right ones; perhaps not. The intriguing thing that emerges from the session chairmen/integrators' reports to which we have been listening is the degree to which there is similarity among problems in the energy field rather than diversity. Thus, there is hardly a country that does not put more efficient use of energy, or conservation, at the top of its list of objectives.

But here I go, exploiting the fact that for the moment I "own" this lectern by starting to do what our next speaker will be doing so much better. I shall now ask him to take over and will remind him that the conference organizers must have listened closely to him in the past and absorbed especially his call for a more just and participatory environment: one can hardly be more participatory than being both the opening and the closing speaker. Dave, tell us what we *really* have been talking about since we first sat down together Monday night.

CHAPTER TWENTY-FIVE

Closing Summary

David J. Rose
Professor of Nuclear Engineering
Massachusetts Institute of Technology

The weight of material in these *Proceedings* and respect for what others have already written in the many excellent papers and integrators' summaries disincline me to attempt to repeat what can be found pretty much self-contained in the various sections. Particularly, see the summaries of the seven working groups. Therefore, I will strain out what I think are some important common themes, commenting upon them and upon some additional topics that seem understressed. Where it seems suitable, I will arrange material as "conclusions" and "recommendations." But not everything fits such a mold.

Some of these topics are similar, even virtually identical, to those mentioned in my opening charge. That reflects in part my own intellectual fixations, but also in part two other things: first, some topics need repetition; second, the view at the finish sometimes differs from the view at the start—we see things from a new place, one hopes more clearly than before. The choice and order of presentation are mine alone; please do not blame the editors.

The mood, priorities, and findings (if any) of any symposium are determined to a large extent by who came and what their agenda was. Symposium II differs significantly from Symposium I in several aspects: an intent this time to concentrate on similarities and differences among the various regions of the Earth, plus extended discussions on specific major topics. The importance of all this was clear soon after (if not at) the conclusion of Symposium I, so this exercise at illuminating these major issues is, in that respect, a planned advance.

This brings me to the first of many topics.

Problem Identification, Problem Analysis, and Problem Resolution: Tension and Paradox

Those three words in sequence—identification, analysis, and resolution—could in an oversimplified way be used to describe the main purposes and activities of Symposium I in 1980, this Symposium II, and Symposium III to follow in May of 1982. Symposium I did what was planned for it. Symposium II went a long way toward its goal but more frequently went into dealing with tensions and paradoxes, many of which were not problems to be resolved but genuinely disparate conditions that must be accommodated and endured. "Resolution," then, becomes a much more complex activity than the usual connotation implies. This complexity widens the gap between conclusions that something needs attention and detailed recommendations of what to do about it. This is not to say that nothing can be done, but the response may be indirect and the focus of attention may shift from simple supposed fixes to dealing with underlying causes.

A striking example of this is the problem of nuclear arms and the threat of nuclear war. Conclusions about the problem abound, all agree on its seriousness, almost all desire relief, yet useful implementable recommendations are hard to find. Some of what we do here at this Symposium gets at the problems of international tensions and inequities and therefore indirectly helps to defuse nuclear arms, but the process is anything but straightforward.

Here is another example. The goals of low production costs and minimal environmental damage, both laudable, often conflict. Advances in this tricky area come not so much at the beginning, by simple formulas or simplistic exhortations, but through the formation and activities of national and international agencies devoted to the relevant issues. In 1970, about fifteen such national agencies existed, and in 1980 over a hundred. There was no single early recommendation, except to be concerned and express that concern by working hard on the issues. So it will sometimes be in the cases dealt with here, and in Symposium III; conclusions tend to come easier than recommendations.

Virtù

I use this uncommon word to capture a spirit, without precociously biasing the reader with overworked phrases.

The Italian statesman and political scientist Niccolò Machiavelli, who lived and wrote nearly five hundred years ago, has been unduly maligned; the adjective "machiavellian" these days sometimes connotes clever, even deceitful, manipulation. He was far better than that, and much of the art and skill of modern interpretive history developed from his work. In particular, he wrote about the importance—the essentiality—of a quality of public spirit which he called virtù, something whose presence permits great tasks to be accomplished. It was there in the days of the classic Greek and Roman republics and was not there in Rome a thousand years later, a thousand years ago, when the Romans mined their city for its marble, to make lime to patch their hovels.

The historians Montesquieu, Edward Gibbon, and Arnold Toynbee wrote of the same quality, and of social charity—*cāritās*—needed to address the problems similar to those we found in Symposium II and will find again in Symposium III. It is a quality without which good things don't happen, and with which much is possible.

Symposium II showed much of this quality and of its importance. Almost every working group and the plenary papers stressed the importance of international cooperation and the need to address the so-called North/South debate, not just in terms of geographical or political alignment but in rich/poor, urban/rural, and other terms.

There are conclusions implicit in these observations, and recommendations, too. These conclusions are often said, and although often ignored in practice, they are important and valid nevertheless. Besides pleas for cooperation there was a consensus that energy, food, the environment, resources, finances, international security, and other global issues are interconnected. As civilizations grow and expand, these issues expand also and leave less room on our finite earth for maneuver among them, less room for sloppy work, less room for sloppy thinking. The conclusion is that national responses, while necessary and important, are fundamentally insufficient; the strengthening of present international organizations and the creation of some new ones will be required, for which some suggestions appear below.

The More Just, Participatory, and Sustainable Society

Without apology, I present again this goal, endorsed explicitly in several of the papers and implicitly in most. Global security and tranquillity will be enhanced by direct, even-handed reduction in the weapons of mass destruction that we have in vast overabundance. Security is also enhanced by reducing the global insecurities, vast inequities, and misery that conspire to cause unrest and that contribute to international political instability.

We encounter paradoxes and tensions here. One is the competition between self-reliance and beneficial interdependence, both good things in the right context. The difficulty comes in taking either to extremes: self-reliance by those with plenty of resources can be used to defend autarky or selfishness, or to defend some international version of social Darwinism; interdependence can be used to defend lack of local initiative, or to defend inappropriate manipulation from outside. But on the whole, the national organizations that attend to the national aspects of energy (or food or finance, etc.) are much stronger than the international organizations that attend to the international aspects, although many international issues may dominate our future more than purely national ones.

From this follows a recommendation: that international groups be stimulated to address a number of particular international issues. We already have OPEC, the IAEA, and several other energy-related groups, and some at the Symposium called for groups to deal globally with bioenergy and coal. Whether such specific consortia are the best ones to inaugurate would need more study, but it seems to me that some others could work very well. One would relate to global or regional interchange programs of education and training—of which more later. Others could deal with topics that arise naturally in particular regions; one example is a study,* still in its formative stage, to illuminate the options for (and need for) electrification in the Asia-Pacific area, a rapidly developing region with considerable interdependence and commonality of interests.

At present, the United States spends less than one-third of one percent of its GNP on assistance to less industrialized countries. Larger contributions, particularly to programs of genuinely cooperative nature, would reward everyone handsomely, including the United States.

Time Perspectives

This topic was mentioned in my opening charge and needs only some specific comments here.

A conclusion. Virtually all of the groups at Symposium II had

*Initially suggested by the East-West Center in Honolulu and now extending to several participating countries.

their own views of appropriate time perspectives, and many of these perspectives were too short to accommodate important long-term issues. For example, the industrialized market-economy group tended to consider ten- or twenty-year time horizons, leading to consideration of how to develop coal, natural gas, and nuclear power. Other groups had longer time perspectives but tended generally to underemphasize long-term consequences (e.g., carbon dioxide buildup). People speak glibly of bridges to the future without considering the end on the other side.

A recommendation. Government and intergovernmental groups more consciously consider energy strategies adapted to a multiplicity of time perspectives and seek combinations of strategies that permit smoother transitions over time, matched to expected rates of social, economic, and technological change. Recasting some of the issues in terms of short- or long-term depletion of capital resources sometimes helps to put the issues into a framework to which economists can contribute. If replacements exist (with various associated money costs and other penalties), a more-or-less rational economic analysis can be made. But, if the depletable resource is not replaceable, things are much more difficult to deal with; present and future welfare and social equity all enter, as the economist T. Koopmans has explained very well in his 1975 Nobel Prize lecture. Such considerations will be essential if the global carbon dioxide problem turns out to be as serious as present indications suggest.

These ideas are cousins to the topic of uncertainty, exhumed from my opening remarks to this Symposium and found ubiquitously in its papers and working discussions. Uncertainty has constructive and destructive contexts; it is laudable in association with seeking better paths, and laudable as part of the humility that counteracts the certainty of arrogance; but wrongly used as vacillation, it is deadly. Developing large options and seeking new social consensus takes a long time; if signals change meanwhile in what seems to be a pattern of random noise because of successions of bad and conflicting assessments, then the hoped-for events die or are stillborn. A chicken can hatch its eggs in twenty-one days but must have a steady policy throughout; the job not only cannot be done any quicker but also cannot be done at all, even in forty-two days, with a policy of "one day on the egg, one day off."

Rational and Effective Utilization of Energy

A conclusion. All favor it; none oppose it; all agree that much can be done that is economically, socially, and environmentally attractive in this area, often called conservation. But a problem exists: it looks and sounds easy, but experience shows that it is not. It involves a multiplicity of actions and groups—it involves almost everything we do. It often involves initiatives and suggestions that have their origins elsewhere; hence either the ideas or the agents (or both) may be viewed ambiguously. The energy-using groups tend to be less well organized for the purpose of developing or implementing less energy-intensive options than is the supply side sector for providing energy. Saving energy produces financial returns some time later but often involves up-front money for groups not well organized or capable of raising it. The problem is ubiquitous: the words of this paragraph, up to here, were first written to apply to the problem of outreach from the Massachusetts State Energy Conservation Office to various disadvantaged ethnic groups in the city of Boston, but these words apply globally. The case of more energy-efficient cookstoves was mentioned several times at the Symposium.

This seeming simplicity (which arises because everyone is familiar with using energy) and actual complexity tend perversely to stimulate nice words and too little action. But plenty of opportunity exists, and all we have learned up to now about using energy and resources more effectively, less wastefully, more benignly, seems to me to be no more than a good start on a series of fine continuing enterprises.

The global energy bill is about 1 trillion dollars per year; gains in effectiveness of about 1 percent per year beyond those resulting from normal economic and technological causes seem possible without straining the social and technological fabric. That comes to a permanent gain of 10 billion dollars per year, each year.

A recommendation. Establish a global institute devoted to the relatively apolitical task of improving the effectiveness with which energy is used.

Less Industrialized Oil-Importing Countries, Rural Development, and Biomass

The three working groups on these three topics had much in common, as attested by their occasional joint sessions. The UN "Group of 77," now actually closer to 120, has in it most of the poor people of the world. Most of these poor people live not in urbia but in ruria, and their fuel source, meager as it is, is largely biomass. Desperate, understandable, and sympathizable pressures thus arise on a renewable but destructible resource. Bringing these pressing problems of about half the world's people to higher priority was one of the Symposium's chief accomplishments.

Let us focus in the next few paragraphs on the rural poor, and particularly on their use of biomass. Everyone concluded that biomass, properly used, makes a necessary and important contribution to the world's energy needs. Everyone agreed that it could be misused and overused. But beyond those too simple statements lay a host of different opinions about promises and perils. Smil, in the Energy for Rural Development group, presented dramatic evidence on the deleterious overuse of biomass in China, particularly (and, by induction, elsewhere)—a view supported at least in part by most participants in all other groups. But in view of the low efficiency with which biomass is generally used and the differing opportunities for growing it for fuel, some participants (Carioca and Arora, A. K. N. Reddy, Lumsdaine and Arafa, and others) suggested that much improvement is possible and overdue.

A few simple numbers help to show both the possibilities and the difficulties. Currently we use about 10 terawatts (TW) per year of commercial energy globally. The sun puts about 170,000 TW into the top of the atmosphere. Some is scattered by the atmosphere, reflected by clouds, and so forth, but nearly 90,000 TW reach the surface, and of this some 15,000 are absorbed on land surfaces (and to a large extent reradiated). There's plenty there for many uses, including one of my favorites for long-term adaption, solar photovoltaic devices. The numbers for biomass are much smaller but are still encouragingly large. According to several estimates, about 133 TW go into the biosphere, of which one-third is oceanic takeup, leaving about 90 for all the lands of the earth. Perhaps half of that goes into tropical and subtropical forests and crop lands (most of it into forests). Out of this flow, the world's people use something like 1 TW of biomass for fuel. This seems like a small fraction of what is available, but the fact that the principal users are distributed over only 5 percent of the world's land area stresses the resource regionally, up to and often beyond its limits.

This is one of the so-called problems of the commons—the world's common resources, mentioned by many at the Symposium. Overcropping, soil erosion, and other calamities lurk in the wings to punish at later times civilizations that have been imprudent. Plato, in Critias, said Greece lost its soil; the Greeks chopped down their trees for fuelwood and later put goats on the land, so Greece is made of much rock and little dirt. Biomass, properly used in moderation, is a goose that can lay golden eggs. Compare those terawatt numbers with the present flow of terawatts into (not out of) fossil fuel deposits: about 10^{-4} to 10^{-5} TW—terawatts which we use up a million times as fast as they were formed.

Almost everyone also agreed that food-growing land should not be converted to biomass farms (except for beneficial use of some, but not all, agricultural residues). This is a valid and important conclusion: the use of food grains to make gasohol in the United States has been a perversion and fortunately appears to be going out of fashion. The greatest pressure for using biomass as fuel arises mainly (but not exclusively) in countries where most of the good land is already being cultivated for food. This leaves the inherently lower yield lands for biomass. Schemes of intensive multiple coppicing and so forth will not apply there; if they had, the land could have been good for growing food. Woodlots for fuelwood, for windbreaks, for construction wood, and for many other good causes are fine, but the practical and sustainable limits for biomass here have never, in my opinion, been really well worked out. Carioca and Arora's table of possible regions is a step in that direction.

The tension between food and fuel has appeared not only in the use of biomass for energy. A decade ago, most food/fuel interactive discussions focused on methods of converting petroleum to food, via bacteria and other methods, just the inverse of many suggestions today. At this moment, the Soviet Union has under way experiments on converting an indigenous fuel—peat moss —to food.

Turning now to other issues of energy and less industrialized countries, inspect Figure 25-1, which shows a well-known trend. On the horizontal axis is Gross National Product per capita— GNP/capita—generally, but often oversimplifiedly taken as a measure of development. On the vertical axis is the fractional increase in energy that is used per unit of GNP—in the language of economists and scientists, the quantity $\delta E/\delta(\text{GNP})$. Data for the lowest GNP countries are most uncertain. The richest countries use proportionally less energy as their GNP increases, because they move to computerized, more highly optimized systems and less energy-intensive service industries. But as the least wealthy countries develop, their energy demands have appeared to increase more than proportionally. Some ascribe this to development routes that are robust but relatively simple and more energy intensive: steel making, for example. There is truth in this, but there is more also. The curve in the figure pertains to commercial energy; many countries that would fall at the left of the curve typically had been getting half their energy from noncommercial sources—biomass—and increasing commercial and industrial activity make demands for commercial fuels almost entirely. Thus, the commercial fuel demand would be expected to rise more than proportionally, as we see.

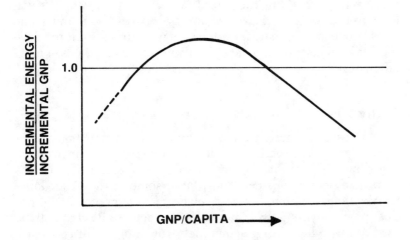

Figure 25-1. **Trend in the Ratio of Incremental Energy to Incremental GNP as a Function of GNP per Capita**

All these circumstances conspire to put the energy-deficit less industrialized countries into very difficult positions. Every nation in the world has trouble getting off oil, but none suffer more than the energy-deficit less industrialized nations, who now have about $400 billion per year total external debt and import about 1.5 billion barrels of oil per year.

In view of all these and many other implications, it is hard to make specific recommendations. Clearly, it would be to everyone's benefit if the countries currently living near the left margin of the figure were able to tunnel through the hill, so to speak, instead of having to climb over it. A role exists for the industrialized countries to collaborate in this endeavor: a global program, already mentioned, to improve efficiency, and other mutual education programs, to be mentioned later.

Other things can be done, too. Exploration for premium nonrenewable resources—oil and particularly gas—has been successful in a number of countries, and cooperative programs need expansion. That, and increased use of coal, are temporary measures, but they are sensible under present circumstances if accompanied with active programs to prepare for using less exhaustible sources.

Regional cooperation would save energy and money. Independent states do not surrender their independence by cooperative programs on the rational joint use of resources. Brazil and Paraguay collaborate on building the giant Itiapu dam on the Parana River. Opportunities exist elsewhere for hydro power—in India and Nepal, for example—and for other energy ventures: joint purchasing, regional distribution, new locally developed methods of more efficient use.

Electric Power—What to Do?

Almost everyone at the Symposium concluded that no matter what might be the total global demand for energy in the future, a larger fraction of it than at present would be electric. This comes from many basic considerations. First, some of the major long-term options are almost wholly electric: solar photovoltaic, nuclear power, hydro power, and so on. Second, while coal may be used to make liquid or gaseous fuels, much of it will go to producing electricity. Third, as use of energy becomes more precise, the ease of controlling electric power gives electricity a considerable advantage. Fourth, and contrary to some opinions, electricity when wisely used represents a fairly efficient use of its primary energy resource, compared with the generally low efficiency assignable to most fossil fuel use.

Everyone also concluded, easily enough, that electric power was expensive and capital intensive. Thus, the problems arise.

The question of centralized versus decentralized systems was discussed, with many opinions expressed. I think an increasing role is developing for decentralized ones, for several reasons: some of the new technologies adapt naturally to modest-size units—10 kw to 10 mw, let us say—via windmills, small hydro installations now purchasable "off the shelf," solar photovoltaic systems that will probably be available in one or two decades, and so forth. All this applies to both industrialized and less industrialized countries (see the papers and comments by Shupe, Alonso, and others). But now the questions of reliability, back-up and maintenance arise. At least locally, some interconnection is certainly desirable, even in the simplest installations. Regarding minimum practical size, systems larger than about 100 kw are much more liable to attract the attention of someone qualified to service them than are smaller ones.

Both the costs and the benefits tend to be distorted in debates about electrification. Quoted costs of $1,000 per household to connect rural dwellings to national grids would clearly preclude general adoption of electrification, but local grids arranged to provide community services would vastly improve the degree of intellectual connection of these people to the rest of the world.

Electric power for industrialized countries will pretty much take care of itself. The needs of the capital-poor less industrialized countries can probably be best met by selecting from a large range of available options, which fortunately start to become available, rather than by seizing on just a few. Thus, we can imagine small hydro plants for local well-watered areas (most of Asia, Africa, and South America being hydroelectrically much less developed than North America or Europe), coal (one hopes with fluidized-bed burners) in 50–100 mw-size devices, some wind power in particularly advantageous locations, increased use of natural gas in selected locations, and small nuclear reactors in states with well-developed infrastructures to deal with their complexity (of which more in the next major topic). Solar photovoltaic power looks promising to me personally, but when lower costs will make it generally attractive is not clear. The goals set by the us Department of Energy in the late 1970s will not be met. Ocean thermal energy conversion, power towers, and solar power satellites seem to me unattractive and in any event very expensive.

Nuclear Power

All agreed that nuclear power presents a particular set of problems and possibilities, that resolution of them is important, and that matters are much confused. It was also clear that nuclear fission power is here to stay, at least in a substantial part of the world.

But beyond that, and Professor Brooks's oral summary, quite a bit can be said. First, consider this question: if the present stagnation in some of the industrial countries persists, what might happen?

a. Will stagnation gradually spread to other countries with present aggressive programs—for example, Eastern Europe, France, or semiindustrialized countries? Or will such programs continue, largely insulated from trouble elsewhere?

Can nuclear power continue without, say, the United States and/or West Germany as principal partners?

b. If nuclear power continues under new circumstances, what will be the effect on the nonproliferation treaty (NPT) and its present adherents, and on the International Atomic Energy Agency (IAEA)? What changes in the NPT are negotiable to new situations? How would that affect safety, waste disposal, proliferation, and so on?

c. If the countries most interested in NPT drop out, then what happens to global stability? Do we get more nuclear bombs?

d. Consider the civilian and military nuclear programs. How are they separate; how are they joined? If civilian nuclear power were both necessary and sufficient for weapons, then our task would be easy—it would be worth giving up civilian nuclear power globally to insure getting rid of 10,000 megatons of bombs. But it isn't that simple. Civilian and military nuclear programs are separable—but will they be separated? Military programs can stand alone, and will do so if proliferating countries persist that way, but civilian nuclear power, undertaken to ease global energy problems and hence contribute to a better world, can be misused to promote great mischief. Working on energy is the task of working on paradoxes.

The opinion was voiced at the Symposium, and hotly disputed, that the NPT was dead as a consequence of the Israeli raid on the Iraqi reactor, the failure of the superpowers to work actively toward reducing their arsenals of nuclear arms, and other circumstances. I hope and believe that the NPT is not dead but requires strengthening.

Along that line, I think something useful can be done that would go far to resolving many difficulties—a nuclear fission power arrangement that satisfies the following conditions:

a. Natural uranium available, adequate for millennia at an acceptable price.

b. A reactor design in which accidents capable of releasing large amounts of radioactivity to the public were not just protected against but physically impossible.

c. Factory-built, quality-controlled construction.

d. No spent-fuel reprocessing; hence, no plutonium.

e. No breeder reactors.

f. Easier siting.

Can this be done?

Regarding point (a), research on uranium from sea water indicates that it could be produced now at $1,100/kg and perhaps in a few years at $450/kg. With a 1 percent burn-up (i.e., 10,000 megawatt-days per ton of natural uranium, or 40,000 MWd/ton from the enriched fuel with 0.1 percent U-235 in the tails, a reasonable assay for such expensive fuel), the higher price is less than the present price of coal on most international markets, on an energy basis. These are much higher numbers than conventionally considered but are quite within reason.

Regarding point (b), small reactors (200 megawatts-electric, for example) can be designed so that the consequences of a loss-of-coolant accident cannot cause a catastrophic meltdown, because the residual heat can be removed by natural circulation. Both water-cooled and gas-cooled designs have been put forward.

Regarding point (c), small reactors could be factory built, much more highly standardized than hitherto, and shipped on barges to coastal destinations, probably with much better quality control. They will still be expensive; estimates exceed $2,000 per kilowatt. But so are most other new energy options, and perhaps the economies of factory construction will counteract in part the penalty associated with small size.

Regarding points (d) and (e), uranium fuel availability and the projected rapidly rising costs of reprocessing would make fuel reprocessing not only unnecessary but unattractive. International control and inspection of spent fuel and its fate would be much easier.

Regarding point (f), small, safe reactors, with assured fuel supply and buyback through a reliable international agency, would permit more rational consideration of nuclear power in many parts of the world.

Incentives for any one country to develop this option or major parts of it in the face of present political, social, and economic uncertainties are, at best, modest. But it seems to me that the long-term gains in energy security, plus the much clearer separation of peaceful and warlike atoms, would be well worth the effort if some such scheme was found to be practicable and was adopted.

In a footnote to this section, it is worth noting that the prospects for practical controlled nuclear fusion power were not discussed at the Symposium. In my opinion, the prospects are much less certain than for, say, competitively priced solar photovoltaic power during the next half century, and therefore controlled fusion should not figure seriously in global energy planning during that period.

Pricing, Taxes, etc.

Almost all sectors—market economies, centrally planned economies, many developing countries—have pricing problems: the industrialized nations by public pressure, the centrally planned economies by public policy, the developing countries by subsidized cheap fuels. All these things are done for perceived good purposes, and all bring difficulties. The Soviet Union has remarked that price is a great educator.

High interest rates stifle development worldwide. That circumstance, plus the predictable but generally unplanned-for very high cost of modifying our global energy base from cheap fossil fuels to expensive ones or to even more expensive alternatives has thrown much of the world's energy sector into disarray. This disarray and the often ineffectual attempts to recover are, in my view, the natural consequences of too-short time perspectives, both in the past and continuing now. Another natural consequence of these factors is the almost worldwide difficulty of capital formation, an activity that invests in, and looks toward, the future. The new technological options are for at least the next thirty to forty years, but seldom do either the financial or the political outlooks extend that far.

Fuel subsidies in less industrialized countries, generally for the purpose of providing commercial fuels to the public at affordable cost, seem often to be too blunt an instrument of the intended purpose. Unless very particular care is taken, it stimulates profligate use, especially in sectors not originally intended to be subsidized at all. Thus arise more inequities.

Fuel subsidies, either direct or indirect, in industrialized countries also affect patterns of use, relatively harmlessly if plenty is available at relatively low cost anyway, but less harmlessly as the resource becomes more scarce, the demand has grown, and the price has risen too high for the support to continue.

The recommendation was made that all industrialized nations put a tax on oil: first, it stimulates conversion to other less critical fuels; second, it reduces the risk of supply disruption; third, it helps less industrialized countries to obtain an adequate share. But the industrialized nations need some relatively uniform agreement on this, if it is to be effective.

Oil-Exporting Less Industrialized Countries

Some asked whether OPEC's motivations were financial or political, as if they must be only one or the other. They are both. Oil, food, other things are used as policy instruments and have been for ages. Oil turns out to be, at least for a time, the paragon of such instruments.

Many OPEC and OPEC-like countries, experiencing a "gold-rush" phenomenon, try to avoid the fate of being mined out and then abandoned. Rather, they attempt to turn a depletable resource into the makings of a permanent sustainable society. Complex financial and industrial developments in OPEC countries, often based initially and logically on forays into the refining and petrochemical fields, are natural expressions of these trends and suggest that industrialized countries dependent on OPEC-like oil should take great care not to find themselves with a surplus of such facilities.

It was remarked that the recycling of petrodollars has not thrown the global economy into confusion, contrary to some earlier fears. Surely that could have been seen before: the amount is substantial, perhaps $200–$300 billion per year. But the same analysts who cried out failed to remark that even larger global military expenditures are being made, and those without any perceptible benefit to any ultimate consumers of the product. However, the burden of the petrodollars is unequally borne; the strain on the less affluent countries has been immense, as was already mentioned.

Better Information Transfer

Everyone concluded that we need new international institutions, stronger and better than before, to address these problems.

Consider education and the sharing of experience, for which global and regional needs were pointed out in virtually every session. What can we do without the intervention of smothering bureaucracies? Organizations such as national academies or universities in industrialized nations and their partners in less industrialized ones can start fraternal relationships at very modest cost compared with what is at stake, to assess and undertake tasks of mutual self-interest. Such associations could be made additionally effective by being given drawing rights, so to speak, on some R&D resources in the two countries.

Methods of information transfer, even of things that are completely in the public domain, could be beneficially extended, through overseas repository libraries, for example.

It has sometimes been claimed that international ownership of patent rights and such "legal" devices have impeded technological flow and maintained the technological dominance of the currently industrialized countries. Such statements at this Symposium were, even at their loudest, somewhat muted, but they have been seriously made elsewhere, and the topic is germane. To be sure, the maintenance of technological dominance by legalistic means is possible and has been practiced (although experience shows that the legal defense can often crumble or be circumvented). But the corollary conclusion that removing the legal barriers insures successful adoption elsewhere, or a shift of technological dominance from one group to another, is far from universally valid. In the mid 1950s the American Telephone and Telegraph Company, chiefly through the work of its research arm, Bell Telephone Laboratories, owned many quite basic patents related to solid-state devices (e.g., transistors) and other things useful to electric and communications technology. In an antitrust consent decree, it virtually gave away a large number of these. Some persons believed beforehand that the effect would be profound, but in fact not much happened. The commodity of value was not so much the patent rights, or even detailed drawings and instructions; it was scientific and technological experience, which would only flourish elsewhere through long-term involvement.

We learn from that and similar case studies that the international interaction must be two-way, not superficial, and long term.

A Collection of Recommendations

The individual work session summaries contain recommendations. Here are mine, listed below. Gathered together this way, they look almost simplistic, but at least they are not mutually inconsistent; they do involve a multitude of time perspectives, and they have some breadth. I will resist the temptation to categorize them as "national," "regional," "international," and so forth, because many apply at several levels of aggregation, just as things having to do with food and diet range from a single rice-farmer to the International Rice Institute to the UN Food and Agricultural Organization.

- An international group devoted to using energy more efficiently and effectively for presently recognized purposes, in cooperation with indigenous groups devoted to this task. It could be a formal treaty organization, or a less formal association between groups of countries. In fact, it could start with just one country offering to join with any other like-minded one; at this level, activity already exists. All countries could benefit, because all use energy less than optimally well.

- Increased international communication about energy—options, costs, effects, and so on—including education and training at both the professional and the technician levels, and dealing with a broad spectrum of energy-related problems ranging (for example) from a comparison of new options to local infrastructures.
- Better data collection, both nationally and internationally. Not just more undigestible data, please; we need better information on resources, patterns of use, and effects, interpreted so all can understand them.
- Expansion of present programs for discovering premium, non-renewable resources—oil, gas, coal—in oil-importing less industrialized countries, accompanied with comparable programs for long-term disconnection from such dependence.
- A uniform tax on gas and oil, especially in the industrialized countries. Also, all countries should adopt pricing policies that inhibit inappropriate or profligate use.
- Stronger reforestation programs, especially in populous, less industrialized countries.
- Minimal (preferably no) conversion of crop land from growing food to growing biomass for energy.
- Where substantial use of biomass makes sense, concentration preferentially not on sugar and starchy plants but rather on turning more plentiful cellulose into useful fuels.
- Establishment of regional cooperative programs for rational development of electric power.
- Acceleration by industrialized countries of their programs to shift away from fossil fuels use, in recognition of the very long time that the process will take.
- Increased support for and involvement by the IAEA in establishing more secure and more sustainable nuclear energy options.
- Involvement of UN agencies, such as the UN University, with educational programs involving undergraduate and graduate students in relevant programs, probably at selected participating universities. The UN Environmental Programme should also be strengthened, to study truly global effects particularly.

In closing, I reflect once more on how these problems need timely attention—attention not delayed by specious rationalization or sophistry until the eleventh hour and fifty-ninth minute, or even worse, until after the midnight bell. Machiavelli compared problems like we contemplate here, and especially misassessments of them, to tuberculosis—hard to diagnose but easy to cure at the beginning; easy to diagnose but very hard to cure at the end.

Acknowledgment

I thank Wayne Blasius of the Knoxville/Knox County Metropolitan Planning Commission for his assistance throughout the Symposium, in taking and collecting notes from work sessions that I could not attend, and in helping to select and arrange material for both my opening charge and this closing summary.

Closing Statement

Walter N. Lambert
Executive Vice President
The 1982 World's Fair

Let me first recognize a fact. This Symposium would not have happened, as far as the 1982 World's Fair is concerned, if it had not been for the foresight and wisdom of the board of directors of the Knoxville International Energy Exposition's management committee and of its president. I thus would like to take this opportunity to recognize those people who two years ago said that if we are going to have a world's fair in Knoxville and if we are going to have energy as its theme (a theme natural for a world's fair in Knoxville), then we must take that theme seriously—and they began to plan this Symposia Series. In particular, I would like to recognize and extend thanks to S. H. Roberts, President of the 1982 World's Fair.

From the very beginning, we who are involved in the Symposia Series have had a program committee working with us. Many of the members of that program committee are here today, and to them I also extend thanks. Much of the form, much of the shape, of this Series grew out of the work of that committee. It's an interesting committee, for it is made up of members scattered all over the world. It's a committee that has never met together in a single room except at the time of these Symposia. It is a committee that does work every single month—we send out assignments; they complete those assignments; and we take what they say seriously. And we are grateful to them.

In addition, the Series has a local organizing committee: a committee made up of people in the Oak Ridge community, the TVA community, and The University of Tennessee who meet with us to keep the Series' staff on track, most of the time. My thanks go to them as well.

And finally, the Series' staff members: I am deeply grateful to all of them, and to one in particular—Nelda Kersey, Assistant Vice President for Energy Programs of the 1982 World's Fair, without whom this Symposia Series could not have been accomplished.

And now, about this Series. When we started planning some time ago, we said, "It's very simple: you plan three meetings—in the first one you define the issues; in the second you analyze those issues; and in the third you resolve them." Simple. Straightforward. To the point. We said, "We'll consider the energy issues of the world in a nonpolitical context." And yet we have found, as we pursue this hopeful and somewhat naive course, that the energy issues of the world are technical, economic, political, human. And what we hope in this context is to provide a forum for all of those considerations. Strong differences of opinion have been expressed at this Symposium; we are pleased at that and hope this tenor will continue into the third Symposium.

Last year, a Symposium I participant told me that there are three possible outcomes for this Series; two of them are certain. First, we will certainly bring people together who have much to say to each other, and we will attract attention to global issues in the field of energy. And that's a good thing. Second, it is certain that we will indicate clearly that a world's fair can increase the general public's awareness of energy's importance in our lives. That's a good thing, too. The third outcome, if we are very, very lucky, is that we can do some real work to help reach a solution to the world's energy problem.

Today, at the end of the second of these three Symposia, we have moved far, in my opinion, toward achieving all three of the above outcomes. But the task is yet to be completed. Next year, at the third Symposium, which is to be held during the 1982 World's Fair, we'll bring together the continuation of our work. We will turn to a greater degree to the political leaders of the world and ask for their review of what we have been doing. And whether or not we succeed will depend on what happens then. We are deeply grateful for the suggestions we've received from you, the participants in this second Symposium, as we move toward the third.

In many cases, we have heard of the challenge of barriers —barriers to the export or import of goods or resources; barriers to the export or import of the human mind's energy. And so we are still working in an ambitious and perhaps somewhat naive mode in this undertaking. But we know that with the help of the participants in the first and second Symposia, we have reached one important conclusion: cooperation on an international basis is the only true answer to the energy problem. Furthermore, we know that the people who have arrived at this conclusion and will move ahead based upon it will make possible our success. The future of those human aspirations we hold most dear will be brighter because you have come together and contributed to the International Energy Symposia Series of the 1982 World's Fair.

Appendices

Symposium I Program

The 1982 World's Fair
INTERNATIONAL ENERGY SYMPOSIUM II
Increasing World Energy Production and Productivity

November 3–6, 1981

Tuesday, November 3, 1981

1:00 p.m. - 1:30 p.m.	**Concurrent Work Sessions** Introductions; discussions of ground rules, agenda, and time allocations for each task
1:30 p.m. - 3:00 p.m.	**Concurrent Work Sessions** Presentation of position paper; discussion of the nature of the problem
3:30 p.m. - 5:00 p.m.	**Concurrent Work Sessions** Initial evaluation of options
8:00 p.m. - 9:30 p.m.	**Opening Session** **Opening Charge** David J. Rose Professor of Nuclear Engineering Massachusetts Institute of Technology

Wednesday, November 4, 1981

8:30 a.m. - 10:00 a.m.   **Concurrent Work Sessions**
Discussion of institutional and
societal constraints

10:30 a.m. - 12:00 noon   **Plenary Session: Alternative
Energy Futures**

The Case for Electricity
The Honorable Umberto
Colombo
Chairman
Italian Atomic Energy Authority

**Appropriate Energy Strategies for
Industrializing Countries**
The Honorable Enrique V.
Iglesias
Secretary General
United Nations Conference on
New and Renewable Sources of
Energy

1:30 p.m. - 3:00 p.m.   **Concurrent Work Sessions**
Continued discussion of cons-
traints; discussion of information
and education considerations

3:30 p.m. - 5:00 p.m.   **Concurrent Work Sessions**
Presentation of case studies or
special papers; discussion

Thursday, November 5, 1981

9:00 a.m. - 10:00 a.m.   **Plenary Session: Development of
International Institutional
Arrangements for Increased
Energy Production and
Productivity**
D. W. Campbell and J. A. D.
Holbrook
International Energy Relations
Branch
Dept. of Energy, Mines and
Resources, Canada

10:30 a.m. - 12:00 noon   **Concurrent Work Sessions**
Final evaluation of options

Friday, November 6, 1981

8:30 a.m. - 10:00 a.m.   **Plenary Session**
Presentations by chairmen/integrators; comments by participants

Chairman: Hans H. Landsberg
Senior Fellow
Resources for the Future

10:30 a.m. - 12:00 noon   **Plenary Session**
Presentations by chairmen/integrators (continued); comments by participants

12:00 noon - 1:00 p.m.   **Closing Session**

Closing Summary
David J. Rose
Professor of Nuclear Engineering
Massachusetts Institute of Technology

Closing Statement
Walter N. Lambert
Executive Vice President
The 1982 World's Fair

Concurrent Work Session Chart

Work Session Topic	Chairman/Integrator	Position Paper Author	Case Study/Special Paper Author
The Role of Nuclear Power	**Harvey Brooks** Professor of Technology and Public Policy Harvard University	**Sigvard Eklund** Director General International Atomic Energy Agency	**H. G. MacPherson** Institute for Energy Analysis Oak Ridge Associated Universities
The Biomass Energy/ Nonenergy Conflict	**Philip Abelson** Editor *Science*	**J. O. B. Carioca** Coordinator, and **H. L. Arora** Vice-Coordinator Nucleus of Nonconventional Energy Sources, Federal University of Ceará, Brazil	**John W. Shupe** Director Hawaii Natural Energy Institute University of Hawaii at Manoa
Energy for Rural Development	**Ishrat H. Usmani** Inter-Regional Rural Energy Adviser Department of Technical Cooperation for Development United Nations	**Edward Lumsdaine** Director Energy, Environment, and Resources Center University of Tennessee. and **Salah Arafa** Professor of Physics and Solid-State Science American University at Cairo	**Vaclav Smil** Professor of Geography University of Manitoba

Industrialized Market-Economy Nations	Keichi Oshima Professor Emeritus University of Tokyo	Ulf Lantzke Executive Director, and Frederick W. Gorbet Director of the Office of Long-Term Cooperation International Energy Agency	David le B. Jones Deputy Secretary United Kingdom Department of Energy
Industrialized Nonmarket-Economy Nations	Ioan Ursu First Vice-President Romanian National Council for Science and Technology	László Kapolyi State Secretary Ministry of Industry Hungarian People's Republic	Naim Afgan Scientific Adviser Boris Kidrič Institute of Nuclear Science
Energy-Surplus Industrializing Nations	Mohammad Sadli Professor of Economics The University of Indonesia	Theodore H. Moran Director, Landegger Program in International Business Diplomacy Georgetown University School of Foreign Service	Hossein Askari President Askari, Jalal and Sheshunoff, International
Energy-Deficit Industrializing Nations	Guy J. Pauker Senior Staff Rand Corporation	Joy Dunkerley Senior Fellow Resources for the Future, and Emmanuel Mbi Economist World Bank	Hiwhoa Moon Coordinator of Economic Policies Office of the Prime Minister Republic of Korea

Symposium II Program
Members and Participants

PROGRAM MEMBERS

Philip H. Abelson has been Editor of *Science,* the weekly magazine of the American Association for the Advancement of Science, since 1962 and is currently President of the International Union of Geological Sciences. He also served as President of the Carnegie Institution from 1971 to 1978 and as President of the American Geophysical Union from 1972 to 1974. He has a background in physics, chemistry, geophysics, geochemistry, and medicine and has earned many distinctions for his contributions to these fields, including a 1945 award from the Navy for his work on uranium isotope separation. In 1972 he received the Kalinga Prize, the annual award made by UNESCO to honor outstanding contribution to the public understanding of science.

Naim Afgan is Scientific Adviser of the Boris Kidrič Institute of Nuclear Sciences at Vinca, Yugoslavia, and is Professor of Power Engineering at the University of Zagreb. He is also the Scientific Secretary of the International Centre for Heat and Mass Transfer in Belgrade. He has published extensively in energy-related areas and has served in consulting or advisory capacities to numerous government agencies and scientific organizations. He is currently a member of the Nuclear Energy Commission of the Yugoslav government.

Salah Arafa is a Professor of Physics and Solid-State Science at the American University in Cairo, Egypt. His professional interests are wide ranging and include rural energy systems. In pursuit of the latter, he has been involved for more than five years in an integrated field project on energy and rural development at the Egyptian village of Basaisa.

Harbans Lal Arora is a Professor of Physical and Analytical Chemistry and Vice-Coordinator of the Nucleus of Nonconventional Energy Sources at the Federal University of Ceará in Fortaleza, Brazil. He has been very active in researching and presenting material on emerging technologies to develop alternative energy sources, particularly biomass, and has served on various committees considering biomass energy production possibilities in Brazil. He is the author of numerous papers on various topics, including both solid-state physics and energy from biomass.

Hossein Askari is President of Askari, Jalal and Sheshunoff International. From 1978 to 1981 he was Adviser on International Economic and Financial Matters to the Minister of Finance and National Economy of Saudi Arabia and Adviser to the Executive Director of the International Monetary Fund; prior to that, he was Professor of Business at the University of Texas at Austin. He is the author of numerous books and articles, and his latest book entitled *Taxation and Tax Policies in the Middle East* is to be published in 1981–82 by Butterworth Publishers in England.

Harvey Brooks is at present the Benjamin Peirce Professor of Technology and Public Policy at Harvard University, where he has been a faculty member of the Kennedy School of Government for the past twenty years and has taught since 1950. His Harvard affiliations are extensive and include chairing the Program on Science, Technology, and Public Policy; serving on the Advisory Committee of the Center for Science and International Affairs; and being a member of the Faculty Council of the Harvard Institute for International Development. He also has numerous affiliations with other organizations and committees, such as the American Academy of Arts and Sciences, where he is Vice-Chairman of the Executive Committee; the National Academy of Science's Committee on Nuclear and Alternative Energy Systems, which he chaired from 1975 to 1980; and the Environmental Law Institute's Board of Trustees, the Institute on Man and Science's International Committee, and the Council on Foreign Relations, on all of which he serves as a member.

Donald W. Campbell has been Director General of the International Energy Relations Branch of Canada's Department of Energy, Mines and Resources since September of 1980. From 1977 to 1980, he was Director of the International Energy Policy Division of the Department of External Affairs, having previously served as the First Secretary of the Canadian High Commission in London (1971–74) and as Counsellor of the Canadian High Commission in Nairobi,

Kenya (1974–77). In the latter position, he concurrently was Canada's Deputy Permanent Representative to the United Nations Environment Programme. His career as a Canadian foreign service officer began when he joined its Department of External Affairs in 1964.

José Osvaldo Beserra Carioca is a Professor of Physical and Analytical Chemistry and Coordinator of the Nucleus of Nonconventional Energy Sources at the Federal University of Ceará in Fortaleza, Brazil. He is actively involved in alternative energy technology research and development and has contributed to several national and international committees and conferences on energy, including the recent United Nations Conference on New and Renewable Sources of Energy. He has presented numerous technical papers in Brazil, the United States, and the United Kingdom.

Umberto Colombo has since 1979 been President of the Italian Atomic Energy Authority. For the prior six years, he was associated with Montedison, first as Head of its Central Research and Corporate Strategies in Milan and then as Director General of its Research and Development Division. During this period, he also chaired the Committee for Scientific and Technological Policy at the Organisation for Economic Co-operation and Development (1972–75) and served with the European Industrial Research Management Association (1977–79). He has been published widely in the fields of geochemistry, industrial chemistry, and energy policy and contributed the Italian portions of the 1977 Workshop on Energy Strategies (WAES) study and the 1980 World Coal Study (WOCOL) publications.

Joy Dunkerley is a Senior Fellow of Resources for the Future in Washington, DC, and has devoted the past ten years to research and writing on energy conservation. From 1972 to 1974, she was a staff economist on The Ford Foundation Energy Policies Project, which produced the first comprehensive set of energy consumption scenarios for the United States, and she previously worked as a part-time consultant to the Foreign Policy Studies Section of the Brookings Institution in Washington, DC. As a member of Resources for the Future, she has completed three studies on the energy conservation lessons to be drawn from comparing US energy consumption patterns with those of other industrial countries. She is currently engaged in research on energy conservation possibilities in developing countries.

Sigvard Eklund has served as Director General of the International Atomic Energy Agency in Vienna, Austria, since 1961. Earlier in his career, he worked in Sweden at the Nobel Institute for Physics, the Research Institute for National Defense, and the Royal Institute of Technology. After 1950, he served AB Atomenergi as Director of Research, Director of Reactor Development, and Deputy to the Managing Director. He is a member of the Swedish Academy of Sciences and is the recipient of numerous honors, including the 1968 Atoms for Peace Award.

Frederick W. Gorbet has been Director of the Office of Long-Term Cooperation and Policy Analysis at the International Energy Agency since 1979. A Canadian citizen, he previously had served with the Canadian government from 1973 to 1976 in the Department of Energy, Mines and Resources as Assistant Senior Economic Advisor and then as Senior Advisor, Energy Policy Branch, following which he became Secretary to the Canadian Cabinet Committee on Economic Policy and Assistant Secretary to the Board of Economic Development Ministers. He has written extensively in the field of economics and energy.

J. Adam D. Holbrook is Chief of the North/South Energy Group in the International Energy Relations Branch of Canada's Department of Energy, Mines and Resources. Prior to his appointment in 1981 he was with the Program Branch of the Treasury Board of Canada with responsibility for the international development portfolio. Previous portfolios at the Treasury Board included science and technology and marine transportation. He joined the Treasury Board in 1975 from Telesat Canada, Canada's official satellite communications company, where he was the engineer in charge of the Satellite Control Centre.

Enrique V. Iglesias has since 1972 been the Executive Secretary of the Economic Commission for Latin America at the United Nations, where he was also appointed in early 1981 as Secretary-General of the UN Conference on New and Renewable Sources of Energy held in August of 1981 in Nairobi, Kenya. A Uruguayan economist, he has held posts in his country's government and in various inter-American organizations—during the mid 1960s, he was Technical Director of the National Planning Office in Uruguay and then President of the Central Bank of Uruguay, and he was the Uruguayan delegate to the Conference of the Latin American Free Trade Association, the Economic Commission for Latin America, and the Inter-American Committee on the Alliance for Progress. He has also held the Chair of Economic Development at the University of Montevideo, where he has been Director of its Institute of Economics, and since 1967 he has presided over the Governing

Council of the Latin American Institute for Economic and Social Planning.

David le B. Jones is currently responsible for divisions within the United Kingdom Department of Energy that deal with domestic energy policy coordination, international energy questions, energy conservation, and economics and statistics. He has served the British government in various capacities since 1947, including working on oil, gas, coal, and atomic energy issues during the 1950s and early 1960s, heading the branch of the Ministry of Power responsible for nationalization of the UK steel industry during the mid 1960s, and subsequently functioning in managerial and cabinet-level positions of the Department of Trade and Industry and the Department of Energy.

László Kapolyi is State Secretary of Ministry of Industry in the Hungarian People's Republic. He also serves as a member of the Hungarian Academy of Sciences and is a Titular Professor at the University of Budapest. He has written and been published in the field of systems and functional approaches to the development of mineral resources.

Hans H. Landsberg was formerly the Director of the Center for Energy Policy Research at Resources for the Future in Washington, DC, where he is currently a Senior Fellow. In this capacity he has chaired several important energy studies and has published numerous books and articles on energy. He has also served as a consultant to a number of government agencies and scientific organizations. Before joining Resources for the Future, he served as a consulting economist with Gass, Bell and Associates.

Ulf Lantzke has been located in Paris since 1974 as the Executive Director of the International Energy Agency (an autonomous agency set up within the framework of the Organisation for Economic Co-operation and Development following the 1973–74 oil crisis). Prior to that appointment, he had served since 1957 with the West German government in its Ministry for Economic Affairs and was appointed in 1968 as head of its Energy Department responsible for energy policy, iron and steel, and raw materials in the Ministry. He has a background in law and during the mid 1950s was employed in the legal service of the State of North Rhine-Westphalia.

Edward Lumsdaine is the current Director of the Energy, Environment, and Resources Center at The University of Tennessee, Knoxville, where he is also a Professor of Mechanical and Aerospace Engineering. He was previously the Director of the New Mexico Solar Energy Institute. As part of his career in aerospace engineering and alternative energy technologies, he has been a visiting professor to Cairo University in Egypt and Tatung Institute of Technology in Taipei, has consulted for a number of companies and national or international organizations, and has been published in such fields as solar energy, heat transfer, and fluid mechanics. For the past five years, he has also participated in an integrated energy systems project in the village of Basaisa, Egypt.

Herbert G. MacPherson currently works as a consultant to and part-time staff member of the Institute for Energy Analysis at Oak Ridge Associated Universities, where, prior to his retirement in 1975, he was Acting Director of the Institute. He has done extensive work and research in the field of nuclear reactors, first with the National Carbon Research Laboratories and then with Oak Ridge National Laboratory, where he was Deputy Director of the Laboratory from 1964 to 1970. He has also served as a Professor of Nuclear Engineering at The University of Tennessee.

Emmanuel Mbi is currently an Economist with the World Bank in Washington, DC, where he previously worked as a researcher and consultant for the World Bank's Urban Projects Department. He has also served as a researcher for Resources for the Future's Center for Energy Policy Research and as a procurement officer for the Department of Urban Development in Victoria, Cameroon. He has written extensively on issues concerning energy and urban development.

Hiwhoa Moon is currently the Coordinator of Economic Policies for the Office of the Prime Minister in the Republic of Korea. He has also served as Director of the Office of Research Planning and Coordination in the Korea Development Institute. He was a Country Economist with the World Bank in Washington, DC, from 1970 to 1977, where his concentration was on Europe, the Middle East, and North Africa.

Theodore H. Moran is Landegger Professor and Director of the Program in International Business Diplomacy at the Georgetown University School of Foreign Service. During 1977–78, he had responsibility for international energy and Persian Gulf security affairs on the Policy Planning Staff of the US Department of State. He notes that the arguments presented in his paper have benefited from discussions in 1980–81 in Mexico, Venezuela, the Persian Gulf, and Israel.

Keichi Oshima is a Professor Emeritus of the University of Tokyo, President of Industrial Research Institute in Japan, and Co-chairman of the International Energy Forum. From 1961 to 1981 he was a Professor in the University's Department of Nuclear Engineering, except for the period 1974–76 when he was on leave and served as Director for Science, Technology, and Industry at the Organisation for Economic Co-operation and Development. He is a member of numerous organizations and international committees, such as the Royal Swedish Academy of Engineering Science, and the UN Science and Technology Advisory Committee for Development.

Guy J. Pauker has been a senior staff member of the Rand Corporation in Santa Monica, California, since 1960. He is also currently the Executive Secretary of the Resource Systems Institute's Asia-Pacific Energy Studies Consultative Group at the East-West Center in Honolulu, Hawaii, and he is the editor of the recently published special issue of the journal *Energy* that dealt with national energy plans in the Asian Pacific region. Prior to 1960, he was a Professor of Political Science and Director of the Center for Southeast Asian Studies at the University of California at Berkeley.

David J. Rose has been a Professor of Nuclear Engineering at the Massachusetts Institute of Technology since 1958. He has also served as Director of the Office of Long Range Planning at Oak Ridge National Laboratory. For several years, he has been working with the World Council of Churches in an effort to deal with ethical issues of energy supply and demand, and he has recently become connected with the Resource Systems Institute of the East-West Center in Honolulu, Hawaii, as a Research Fellow.

Mohammad Sadli is a Professor of Economics at the University of Indonesia. He also has been Chairman of the Board of Indonesia's national oil company, Perta Mina, in addition to being Indonesia's Minister for Mines and Energy, and in 1974 he served as President of the Organization of Petroleum Exporting Countries. He is the author of numerous articles on economics.

John W. Shupe is a Professor of Civil Engineering and the Director of the Hawaii Natural Energy Institute at the University of Hawaii at Manoa, where he is also the Energy Research Coordinator. During 1977–78 he was on leave of absence from the university, serving as Scientific Adviser to the Assistant Secretary for Energy

Technology of the us Department of Energy. He has been Principal Investigator for several large-scale alternative energy research projects, has authored numerous technical reports and publications on energy-related topics, and has served in an advisory or consulting capacity to a number of national, state, and local boards or institutions concerned with environmental and energy issues.

Vaclav Smil immigrated from Czechoslovakia to Canada in 1969, where he is now a Professor of Geography at the University of Manitoba. His most recent research has dealt with developing countries' energy needs, including energy-related agricultural development and environmental issues in the People's Republic of China. He has been published extensively in these areas.

Ioan Ursu is currently First Vice-President of the National Council for Science and Technology in Bucharest, Romania. In addition to his academic work at the University of Bucharest, where he has been Head of the Physics Department since 1968, he has served terms of presiding over a number of national and international organizations, including the aforementioned Council, the State Committee for Nuclear Energy, the Romanian Committee for Physics, and the European Physical Society. He has also served in various capacities with the International Atomic Energy Agency and was Director of the Institute for Atomic Physics in Bucharest from 1968 to 1976. He has been published in the areas of nuclear materials and technologies, solid-state physics, and radiation/matter interaction.

Ishrat H. Usmani was appointed as Inter-Regional Advisor (Energy for Rural Areas) to the un Department of Technical Cooperation for Development in early 1980, prior to which he had been the Senior Energy Advisor to the un Environment Programme. He was Chairman of the Pakistan Atomic Energy Commission from 1960 to 1971, during which period he also served a one-year term in 1962–63 as Chairman of the International Atomic Energy Agency's Board of Governors and presided at the nuclear power sessions of the un "Atoms for Peace" Conferences held in 1964 and 1971. In addition, he was appointed as Secretary to Pakistan's Ministry of Education in 1968 and then as Secretary to its Ministry of Science and Technology in 1972. In his recent work and publications, he has promoted the concept of the electrification of rural villages in the developing countries of Asia, Africa, and Latin America through the integrated use of local resources such as solar, wind, and biogas energy at "Rural Energy Centers."

Such centers are being set up under his guidance in Sri Lanka, Senegal, and Pakistan.

PROGRAM PARTICIPANTS

Philip H. Abelson
Editor
Science
United States

Naim Afgan
Scientific Adviser
Boris Kidric Institute of Nuclear Sciences
Yugoslavia

James E. Akins
Former Ambassador to Saudi Arabia
United States

Marcelo Alonso
Executive Director
Florida Institute of Technology Research and Engineering
United States

Fernando Altmann Ortiz
Minister of Energy
Costa Rica

Salah Arafa
Professor of Physics and Solid-State Science
American University
Egypt

Harbans Lal Arora
Vice-Coordinator
Nucleus of Nonconventional Energy Sources
Federal University of Ceará
Brazil

Hossein Askari
President
Askari, Jalal and Sheshunoff International
United States

Professor Doctor Boettcher
International Bureau
Nuclear Research Facility, Julich
Federal Republic of Germany

Thinakorn Bhandugravi
Minister of Science, Technology, and Energy
Thailand

Harvey Brooks
Professor of Technology and Public Policy
Harvard University
United States

Donald W. Campbell
Director General
International Energy Relations Branch
Department of Energy, Mines and Resources
Canada

José Osvaldo Beserra Carioca
Coordinator
Nucleus of Nonconventional Energy Sources
Federal University of Ceará
Brazil

Umberto Colombo
Chairman
Italian Atomic Energy Authority

Joy Dunkerley
Senior Fellow
Resources for the Future
United States

Sigvard Eklund
Director General
International Atomic Energy Agency

M.S. Farrell
Deputy Director
National Energy Office
Department of National Development and Energy
Australia

Fereidun Fesharaki
Research Associate
East-West Resource Systems Institute
United States

S. David Freeman
Director
Tennessee Valley Authority
United States

John H. Gibbons
Director
Office of Technology Assessment
United States Congress

José Goldemberg
Director
Institute of Physics
The University of São Paulo
Brazil

Frederick W. Gorbet
Director of Long-Term Cooperation
International Energy Agency

Richard L. Grant
President
Boling Engineering and Construction Southeast, Inc.
United States

J. Adam D. Holbrook
Chief, North/South Energy Group
International Energy Relations Branch
Department of Energy, Mines and Resources
Canada

Enrique V. Iglesias
Secretary-General
United Nations Conference on
New and Renewable Sources of Energy

David le B. Jones
Deputy Secretary
Department of Energy
United Kingdom

Pierre Jonon
Director of Production Facilities
Electricité de France

László Kapolyi
State Secretary
Ministry of Industry
Hungarian People's Republic

Donald Klass
Vice President
Institute of Gas Technology
United States

Zsolt Kohalmi
First Secretary, Science Attaché
Embassy to the United States
Hungarian People's Republic

Shunsuke Kondo
Tokyo University
Japan

Hans H. Landsberg
Senior Fellow
Resources for the Future
United States

Ulf Lantzke
Executive Director
International Energy Agency

Robert L. Loftness
Director, Washington Office
Electric Power Research Institute
United States

Mans Lönnroth
Secretariat for Future Studies
Sweden

Amory B. Lovins
Friends of the Earth
United States

L. Hunter Lovins
Friends of the Earth
United States

Edward Lumsdaine
Director
Energy, Environment, and Resources Center
The University of Tennessee
United States

H. G. MacPherson
Consultant
Institute for Energy Analysis
Oak Ridge Associated Universities
United States

Margaret Maxey
Assistant Director
Energy Research Institute of South Carolina
United States

Emmanuel Mbi
Economist
World Bank

Alejandro D. Melchor
Board of Directors
Asian Development Bank
Philippines

Hiwhoa Moon
Coordinator of Economic Policies
Office of the Prime Minister
Republic of Korea

Theodore H. Moran
Director
Landegger Program in International Business Diplomacy
Georgetown University School of Foreign Service
United States

B. C. E. Nwosu
Assistant Director of Education (Science)
Federal Ministry of Education
Nigeria

Keichi Oshima
Professor Emeritus
University of Tokyo
Japan

Guy J. Pauker
Senior Staff
Rand Corporation
United States

D. R. Pendse
Economic Advisor
TATA Industries
India

Amulya Kumar N. Reddy
Professor of Chemistry
Indian Institute of Science

David J. Rose
Professor of Nuclear Engineering
Massachusetts Institute of Technology
United States

Sanga Sabhasri
Undersecretary of State
Ministry of Science, Technology, and Energy
Thailand

Mohammad Sadli
Professor of Economics
The University of Indonesia

Georgio Santuz
Undersecretary of State
Ministry of Public Works
Italy

Luis Sedgwick Baez
Executive Assistant to the Director General of Energy
Ministry of Energy and Mines
Venezuela

Allen C. Sheldon
Vice-President
Environment and Energy Resources
Aluminum Company of America
United States

John Shupe
Director
Hawaii Natural Energy Institute
University of Hawaii at Manoa
United States

Vaclav Smil
Professor of Geography
The University of Manitoba
Canada

Jean-Pierre Somdecoste
Electricité de France

Bogumil Staniszewski
Director
Institute for Thermal Engineering
Warsaw Technical University
Poland

Lewis E. Striegel
Deputy Administrator
Economic Regulatory Commission
United States Department of Energy

Ferenc Szidarovszky
Visiting Professor from Hungary
University of Arizona
United States

Gerald W. Thomas
President
New Mexico State University
United States

Ioan Ursu
First Vice-President
National Council for Science and Technology
Romania

Ishrat H. Usmani
Inter-Regional Rural Energy Adviser
Department of Technical Co-operation for Development
United Nations

Miguel S. Ussher
Director General
Planning Secretariat
Office of the President
Argentina

Dan Vamanu
National Council for Science and Technology
Romania

Alvin Weinberg
Director
Institute for Energy Analysis
Oak Ridge Associated Universities
United States

Macauley Whiting
Vice-President (retired)
Dow Chemical Corporation
United States

Symposium II Organizers, Contributors, and Staff

ORGANIZING COMMITTEE

The Organizing Committee is composed of representatives from major energy-related organizations in the Knoxville area. Primary responsibilities of the Organizing Committee are to advise Symposia personnel regarding overall operation in areas of finance, program management, facilities, services and hospitality, and communications.

Nobert J. Ackermann
Technology for Energy Corporation

Ben Adams
Adams Craft Herz Walker

William Bibb
United States Department of Energy

Tony Buhl
Technology for Energy Corporation

Harvey I. Cobert
Union Carbide Corporation—Nuclear Division

Pete Craven
Science Applications, Inc.

Kenneth E. DeBusk
The University of Tennessee

William E. Fulkerson
Oak Ridge National Laboratory

Ron Green
System Development Corporation

Robert F. Hemphill
Tennessee Valley Authority

Eugene Joyce
Attorney-at-Law

Walter N. Lambert
The 1982 World's Fair

Robert Landry
First Christian Church

Robert L. Little
The University of Tennessee

Lillian Mashburn
Goodstein, Hahn, Shorr & Associates

Wayne Range
United States Department of Energy

Don Riley
Clinch River Breeder Reactor

S. H. Roberts, Jr.
The 1982 World's Fair

Selma Shapiro
Oak Ridge Children's Museum

Ernest G. Silver
Oak Ridge National Laboratory

H. Brown Wright
Tennessee Valley Authority

PROGRAM COMMITTEE

The Program Committee is composed of internationally known
experts in energy-related areas. The primary responsibility of the
Committee is to screen participants and identify speakers for the
Symposia.

James E. Akins
Former Ambassador to Saudi Arabia

Kenneth E. Boulding
Chairman of the Board
American Association for the Advancement of Science
Professor Emeritus
University of Colorado

John S. Foster, Jr.
Vice President
Science & Technology
TRW, Inc.

John H. Gibbons
Director
Office of Technology Assessment
United States Congress

Denis Hayes
Executive Director
Solar Energy Research Institute

Hans H. Landsberg
Senior Fellow
Resources for the Future

Henry R. Linden
President
Gas Research Institute

Amory B. and L. Hunter Lovins
Friends of the Earth, Inc.

Guy J. Pauker
Senior Staff Member
Rand Corporation

David J. Rose
Professor
Massachusetts Institute of Technology

Mohammad Sadli
Professor
The University of Indonesia

John C. Sawhill
Director
McKinsey & Company, Inc.

Ishrat H. Usmani
Inter-Regional Energy Advisor
United Nations

Alvin M. Weinberg
Director
Institute for Energy Analysis

LOCAL CONTRIBUTORS

Beaty Chevrolet
Bechtel National, Inc.
Boeing Engineering and
    Construction, Inc.
Breeder Reactor Corporation
Burgin Dodge
Butcher Volkswagen
Delta Air Lines, Inc.
Hyatt Regency Hotel
Jim Cogdill Dodge
John Banks Buick
Oak Ridge Associated
    Universities

Rodgers Cadillac
Snider Motors
South Central Bell
System Development
    Corporation
Union Carbide Corporation -
    Nuclear Division
United American Bank
United States Post Office
West Knox Datsun

SYMPOSIUM II STAFF

Management Staff

Robert A. Bohm
Kim Bridges
Lillian A. Clinard
Mary R. English
Leigh R. Hendry
Richard D. Jacobs
M. Nan Lintz

Nelda T. Kersey
Walter N. Lambert
Edward Lumsdaine
Sheila W. McCullough
Katherine Murphy
Karen L. Stanley

Task Force Chairpersons

Tom Barnard
Robert Bohm
David Cash
Lillian Clinard
Ed Cureton
Leigh Hendry
Marian Kozar
Carroll Logan

Sheila McCullough
Katherine Murphy
Patsy Scruggs
Karen Stanley
Scott Tillery
John Underwood
Mary Lou Wardell
David Yarnell

Supporting Staff

The 1982 World's Fair Executive Staff

Jake F. Butcher, Chairman of the Board
S. H. Roberts., Jr., President
James E. Drinnon, Sr., Executive Vice President

The 1982 World's Fair Executive Vice Presidents

William R. Francisco
Edward S. Keen

Walter N. Lambert
George M. Siler

The 1982 World's Fair Vice Presidents

Jim Benedick
Peter H. Claussen
William C. Carroll
A. G. Forrester

Theotis Robinson, Jr.
Charles D. Smith
Charles K. Swan, III

The 1982 World's Fair Staff

Kelly Albro
Jane Bearfield
Mike Blachly
Jeannine Boyle
Carole Brailey
Traci Brakebill
Jon Brock
Alice Buckley
Jimmy Cannington
Cynthia Copeland
Paul Creighton
Cookie Crowson
Sue Duggin
Pamela Dunlap
Emmett Edwards
Jane Eppes
Stewart Evans
Cathy Farmer
Chuck Fleming
Lotta Gradin

Marc Grossman
Terry Haws
Cathy Higdon
Judy Holdredge
Pat Hutsell
Edith Johnson
Francis Jones
Marian Kozar
Ken Lett
Carroll Logan
Dora McCoury
Denise McKenzie
Clare Whelan-Maloney
Jean Miller
Gina Moore
Betty Neal
Rusty Nunley
Venice Peek
Reniee Peer
Sharon Phillips

Cherie Pratt

Barbara Ragland

Jack Rankin

Billie Richards

Sandy Richards

Gloria Richardson

Connie Robledo

William F. Roland

Bob Shelley

Bill Schmidt

Rich Sibley

Katie Smith

Pete Soukop

Ray Suarez

Lance P. Tacke

Jim Thorpe

Julia Walkei

Mary Nell Ward

Charlie Ware

Lee Ann Wells

Sharon Wells

Diane Wilde

The University of Tennessee
Energy, Environment, and Resources Center

William Clemons

Debra Durnin

Mary Ellen Edmondson

Joyce Finney

Nancy Gibson

Helen Hafford

Daniel Hoglund

Polly Koehler

Wallace Koehler

Bernie McGraw

Betty Moss

Carole Purkey

Robert L. Reid

Joe Stines

Carolyn Srite

Rica Swisher

Peggy Taylor

Celina Tenpenny

Joyce Troxler

Beverly Worman

Hyatt Regency Knoxville

Paul Sherbakoff
   General Manager

Pam Caldwell

Scott Caldwell

Willie Cannon

John Cardona

Mike Cowell

Arthur Davis

Barnes Goutermout

Sherri Harrell

George Hoch

Karl Holme

Bob Iantosca

Patti Lewis

Dan May

Richard May

Jo Marsh

Vaughn McCoy

C. J. McDaniel

Margaret Ogle

Ruth Hauk
Henrietta Pendergrass
Joyce Roth
Karen Stranathan

Sally Peach
Sanford Swann
Mary Lou Wardell
Susan White

US Department of Energy

William Bibb
John C. Bradburne, Jr.
Doris Brooks

Sidney Lanier
Lewis E. Striegel

Oak Ridge National Laboratory

Andre.v Loebl
Mitchell Olszewski

Tom Wilbanks

Union Carbide Corporation
Nuclear Division

Edward Aebischer
Lee Berry
Carol Grametbauer

Cindy R. Lundy
Fred Mynatt
Collin West

Institute for Energy Analysis

David Register

Project Management Corporation

Henry Piper

Tennessee Valley Authority

Jean Baldwin
Roger Bolinger
Reid Campbell
Ely Driver
Clara Dunn
L. B. Kennedy
Mary Knurr
Mary Longmire
James McBrearty

Linda Oxendine
Jean Pafford
Virgil Reynolds
Mary Troy
Barbara Turner
James Ward
Arthur B. Wardner
Brown Wright
Thomas Zarger

The University of Tennessee

Bob Allen
Robert Bledsoe
Houston Luttrell
Sheldon J. Reaven

Phil Ryan
William T. Snyder
David Van Horn

Knoxville/Knox County Metropolitan Planning Commission

Wayne Blasius

City of Knoxville

Randy Tyree, Mayor
Patricia G. Ball

Jim Humphrey
John Underwood

Miscellaneous Support

Beth Adams
Patti Anderson
Arnold Air Society
  Commando, Pledge Class
E. Ray Asbury
Ginny Baich
Betty Barber
Stephen Barbour
Ruth Barkley
Tom Barnard
Ruth Benn
Frieda Beretta
Helen Blackwell
Dyllis Blair
Lisa Bridges
Jim Burbank
Jan Carroll
John Casillo
David Cash
Lynn Clapp
Helen Collins
David Conklin

Nell Cook
Mary Craig
John Craig
Ralph Culvahouse
Neal Culver
George Deavours
Bob Drake
Gordon Dudley
Cissy Eckert
Mrs. Mack English
Jack Fagan
Tepe Fennelly
Betty Frierson
Jim Garland
Bobby Garret
Rick Griggs
David Hatfield
Russell M. Herndon
Hubert Hinote
James E. Hiteshew–
  Air Commando Cadets
Sheila Hoskins

Ronnie Huffaker
Mike Hutchison
Larry Hutsell
Shirey Irwin
Ruby Kaskuske
Mary Keaton
Kristopher Kendrick
Anne Lambert
Doyle Lewis
Randy Log
Betty Lowe
C. R. Loy
Grace Lu
Freddie Marsh
Sam Moon
Janie Morgan
Daniel Parrish
Mary Rankin
Robert Roach

Ralph Roark
Vera Roberts
Mildred Sammons
Sherry Sanders
Donna Schmidt
Karl Seger
Brenda Shell
Ron Shipe
Ed Shouse
Jeff Smola
Julian Spitzer
Jeanette Stevens
Dephine Stooksbury
Sally Tillery
Ken Upchurch
Carole Whitehead
Brown Wright
Lillian Zion